VII International Conference on Problems of Mathematical Physics and Mathematical Modelling (MPMM 2018)

Journal of Physics: Conference Series Volume 1205

Moscow, Russia
25 - 27 June 2018

ISBN: 978-1-5108-8721-3
ISSN: 1742-6588

Printed from e-media with permission by:

Curran Associates, Inc.
57 Morehouse Lane
Red Hook, NY 12571

proceedings.com

Some format issues inherent in the e-media version may also appear in this print version.

Institute of Physics
Dirac House, Temple Back
Bristol BS1 6BE UK

Phone: 44 1 17 929 7481
Fax: 44 1 17 920 0979

techtracking@iop.org

Additional copies of this publication are available from:

Curran Associates, Inc.
57 Morehouse Lane
Red Hook, NY 12571 USA
Phone: 845-758-0400
Fax: 845-758-2633
Email: curran@proceedings.com
Web: www.proceedings.com

TABLE OF CONTENTS

IOP Conf. Series: Journal of Physics: Conf. Series **1205** (2019) 011001 doi:10.1088/1742-6596/1205/1/011001

Preface

Mathematical models, mathematical methods and program codes for the prediction and analysis of experiments in recent years have become a necessary element not only in research projects, but also an integral part of many technological processes. The basis of many mathematical models are both partial differential equations and ordinary differential equations.

The aim of this conference is to bring together researchers with interests in: Analytical Theory of Differential Equations, Optimization Problems, Analytical and Solid Mechanics, Fluid Mechanics, Mathematical Modelling.

This annual conference was established at the National Nuclear Research University – Moscow Engineering Institute in the year 2012. The Conference was organized by three leading departments of the National Research Nuclear University MEPhI in the field of mathematical physics.

This year, above 120 participants from different countries including Belorussia, Czech Republic, France, Great Britain, Japan, Moldova, Russia, Slovakia, Spain, Sweden and the USA take part in the Conference with oral talks.

Valery V. Kozlov
Nikolay A. Kudryashov
Oleg V. Nagornov

IOP Conf. Series: Journal of Physics: Conf. Series **1205** (2019) 011002 doi:10.1088/1742-6596/1205/1/011002

Peer review statement

All papers published in this volume of *Journal of Physics: Conference Series* have been peer reviewed through processes administered by the proceedings Editors. Reviews were conducted by expert referees to the professional and scientific standards expected of a proceedings journal published by IOP Publishing.

IOP Conf. Series: Journal of Physics: Conf. Series **1205** (2019) 012001 doi:10.1088/1742-6596/1205/1/012001

The derivation of integration-fragmentation equations in the Becker–Döring case

S Adzhiev[1], I Melikhov[1] and V Vedenyapin[2]

[1] Lomonosov Moscow State University, Moscow, Russia
[2] Keldysh Institute of Applied Mathematics of Russian Academy of Sciences, Moscow, Russia; RUDN-University, Moscow, Russia

sergeyadzhiev@yandex.ru, vicveden@yachoo.com

Abstract. In the present paper the interconnection between the kinetic equations of evolution of particles distinguishing by masses (number of molecules forming them) or by other property was investigated in the Becker–Döring case. From the continuum integration-fragmentation equation we derived a new equation which we call the continuum Becker–Döring equation. From this equation we obtained the Becker–Döring system of equations and the continuum equation of the Fokker–Planck type (or of the Einstein–Kolmogorov type, or of the diffuse approximation). We clarified the form of the obtained equations basing on the physical sense of these conclusions. Due to unity of the kinetic approach the present work may be useful for specialists of various specialties, who studies the evolution of structures with differing properties.

1. Introduction

In the works of L. Boltzmann [1], J. Cl. Maxwell [2] and etc. appears the distribution function by velocities and coordinates. Now there is a multi-parameter description, which is due to the fact that it becomes available for experimental observation. The problem is to write the equations on the distribution function of particles (bodies) by these many parameters.

In the present paper we continue the line of works [3], [4] and others, devoted to mathematical modeling of emerging and growth of hierarchical structures, using the kinetic approach. In [3] a model of creation of nanostructures as a result of growth of nanoparticles in a supersaturated medium with their multistage modification was proposed. The multistage modification was doing in order to give nanoparticles the desired properties. The aggregation of nanoparticles in agglomerates and integration of aggregates with each other was taken into account and it was formulated the discrete and the continuum balance equations for the number of nanoparticles and aggregates, taking into account the possibility of their growth and fragmentation. The continuum equation has been given to the evolutionary equation of the Fokker–Planck type containing frequency functions. They describe the process of self-organization of aggregates by selection of more resistant forms. Moreover, each stage of modification is characterized by its own frequency functions. They subject to independent determination. Frequency functions for all types of impacts used for the creation of nanostructures can form the basis of a methodology of their optimal synthesis.

The Becker–Döring case [5] is the model when the only one molecule attaches or separates.

The formation of solids is well described by division into several stages – firstly primary particles emerge and grow, and then the stage of aggregation or the formation of agglomerates of the primary particles as a result of their collisions with each other comes, and then the formation of aggregates of them occurs, etc. Although all stages go at the same time, but firstly the process of formation of primary particles, then their aggregation, then aggregation of the secondary agglomerates dominates. Therefore, these processes are well modeled by generalizations of the Becker–Döring case that is considered by us.

2. The derivation of the Becker–Döring system of equations and equations of Fokker–Planck type from continuum integration-fragmentation equations

In this section we derive the equation of the Fokker–Planck type [3]–[4], which is called the Einstein–Kolmogorov equation [6] in probability theory, and the Becker–Döring system of equations [5] from the continuum integration-fragmentation equations.

We'll call a particle consisting of n molecules as n-mer. Let n-mers are defined by the distribution functions by the status parameters: spatial coordinates $\mathbf{x}$ and velocities (or momenta $\mathbf{p}$) of center of masses, body's mass nm (where m is the mass of one molecule or monomer), characteristics of body shape and etc.

Let consider the Becker–Döring system of equations:

$$\frac{dN_n}{dt}=\left(\alpha_{n-1}N_1N_{n-1}-\alpha_nN_1N_n\right)+\left(\beta_{n+1}N_{n+1}-\beta_nN_n\right) \text{ when } n\geq 2, \tag{1a}$$

where $N_n = N_n(t)$ is a number or concentration of n-mers, α_n is a frequency function (section) of integration of n-mer and monomer, and β_n is a frequency of fragmentation n-mer to monomer and $(n-1)$-mer. The system of equations (1a) is supplemented by the conservation law of the number of all molecules of substance, which forms crystals, in the system N_0:

$$\sum_{i=1}^{+\infty} iN_i(t)\equiv N_0 = const. \tag{1b}$$

Let take now instead n some property x, and let it changes not necessarily by one, but by some constant value c, then the system of equations (1) can be rewritten in the form

$$\frac{dN_x}{dt}=\left(\alpha_{x-c}N_cN_{x-c}-\alpha_xN_cN_x\right)+\left(\beta_{x+c}N_{x+c}-\beta_xN_x\right) \text{ при } x=2c,3c,\ldots, \tag{2a}$$

$$\sum_{i=1}^{+\infty} icN_{ic}(t)\equiv N_0 = const. \tag{2b}$$

Instead of the equation (2b) we can write the following equation:

$$\frac{dN_c}{dt}=-2\left(\alpha_cN_c^2-\beta_{2c}N_{2c}\right)-\sum_{i=2c}^{+\infty}\left(\alpha_{ic}N_cN_c-\beta_{(i+1)c}N_{(i+1)c}\right), \tag{3}$$

which is obtained by differentiation of (2b) with accordance to (2a).

Thus, we here begin to investigate more general equations than just the Becker–Döring system of equations: (2), and its generalizations are already applied for the simulation of the aggregation processes. In this very simple discrete model c is the number of molecules in the particle that attaches and fragmentize from the aggregate with the number of particles equal to x.

The histograms measuring by the experimenters, as a rule, are replaced by continuous functions. So it requires using a continuous description, and we need to write a continuum equation, an approximation of which is the equation of Fokker–Planck type. From discrete distributions and the system of equations (1) we can go to the continuum model by the introduction of continuous distribution function $\psi(x,t)$, where $x\in(0,+\infty)$, for which it is fulfilled accordance to the discrete distribution: for example, $\sum_{j=1}^{M} N_j = \int_0^M \psi(y,t)\,dy$ in the case of the system (1) and

$\sum_{j=1}^{Mc} N_{jc} = \int_0^M \psi(y,t)\,dy$ in the case of the system (2), where $M \in \mathbb{N}$. Here $\psi(t,x)dx$ is the number of particles having the property x from the interval dx. For the sake of simplicity of our consideration, we took the right border of the continuous distribution or the maximum value of the property x, described by a continuous distribution, equal to $+\infty$. Let now $\alpha(x)$ and $\beta(x)$ are continuous analogues of α_x and β_x. They are non-negative functions defined for $x > 0$.

Let us write the continuum integration-fragmentation equation:

$$\begin{aligned}\frac{\partial\psi(t,x)}{\partial t} &= \frac{1}{2}\int_0^x \left[A(x-y,y)\psi(x-y)\psi(y) - B(x-y,y)\psi(x)\right]dy - \\ &-\int_0^{+\infty}\left[A(x,y)\psi(x)\psi(y) - B(x,y)\psi(x+y)\right]dy.\end{aligned} \tag{4}$$

We'll search the function $A(x,y)$ in the form: $A(x,y)=\alpha(x)\cdot C(y)+\alpha(y)\cdot C(x)$, where $C(y)$ we'll find after substitution from the physical meaning of the Becker–Döring system of equations and equation of the Fokker–Planck type, that we obtain from following equation, which is the equivalent form of (4):

$$\begin{aligned}\frac{\partial\psi(t,x)}{\partial t} &= \int_0^{x/2} \left[A(x-y,y)\psi(x-y)\psi(y) - B(x-y,y)\psi(x)\right]dy - \\ &-\int_0^{+\infty}\left[A(x,y)\psi(x)\psi(y) - B(x,y)\psi(x+y)\right]dy,\end{aligned} \tag{5}$$

The terms with B in (5) are differed from terms with A, and therefore we need the substitution for B that corrects it, at least because in the equation of the Fokker–Planck type terms with α and β are the same with accuracy up to signs and multiplicative multipliers: for example, N_c, N_c^2 (see below).

So, let us consider the substitution

$$A(x,y)=\alpha(x)\cdot C(y)+\alpha(y)\cdot C(x),\ \operatorname{supp}(C(y))=[0,c], \tag{6a}$$

$$B(x,y)=\beta(x+y)\cdot(D(y)+D(x)),\ \operatorname{supp}(D(y))=[0,d]. \tag{6b}$$

Here in the condition of the equality $A(x,y)$ to zero if $y \notin [0,c]$ and the equality $B(x,y)$ to zero when $y \notin [0,d]$ we laid that only particles with non-negative property y, not exceeding c and d, respectively attach to and fragmentizes from the body with property x.

Let us make the substitution (6) in the equation (5), which is considered for $x \in (x_0,+\infty)$, where x_0 cannot be less than twice the value of the property attaching or fragmentizing particle, i.e. less than $\max(c,d)$. Then

$$\begin{aligned}\frac{\partial\psi(t,x)}{\partial t} &= \int_0^c \alpha(x-y)\psi(x-y)\psi(y)C(y)dy - \int_0^d \beta(x)\psi(x)D(y)dy - \\ &-\int_0^c \alpha(x)\psi(x)\psi(y)C(y)dy + \int_0^d \beta(x+y)\psi(x+y)D(y)dy =\end{aligned}$$

$$=\int_0^c\left[\alpha(x-y)\psi(x-y)-\alpha(x)\psi(x)\right]\psi(y)C(y)dy+$$

$$+\int_0^d\left[\beta(x+y)\psi(x+y)-\beta(x)\psi(x)\right]D(y)dy. \quad (7a)$$

We supplement the equation (7a) by the continuum conservation law of the number of all molecules in the system

$$\int_0^{+\infty} y\psi(y)dy \equiv N_0 = const, \quad (7b)$$

which is a consequence of equation (5).

The following simple lemma is valid.

Lemma 1. Let $c=d$, $C(c)=1$, $D(0)=0$, $D(c)=1$. Then the substitution of the sum of δ-functions: $\psi(x,t)=\sum_{i=1}^{+\infty}N_{ic}(t)\cdot\delta(x-ic)$, in (7) under the condition $\alpha(x)=\alpha_x$ and $\beta(x)=\beta_x$ for $x=c,2c,3c,\ldots$ gives (2).

The proof. If after substituting for each $x=2c,3c,4c,\ldots$ we integrate the two parts of equation (7a) on the interval that contains only one x, which is a multiple of c, we obtain (2) for $x=2c,3c,4c,\ldots$. The lemma is proved.

Definition. We'll name the equation (7) as the continuous the Becker–Döring equation.

So, here we have confined ourselves by the consideration of the unclosed equation (7), but we have obtained from it the Becker–Döring system of equations in Lemma 1, and we shall get a closed equation of the Fokker–Planck type.

According to (7)

$$\frac{\partial\psi(t,x)}{\partial t}=\int_0^c\left[\alpha(x-y)\left(\psi(x-y)-\psi(x)\right)+\left(\alpha(x-y)-\alpha(x)\right)\psi(x)\right]\psi(y)C(y)dy+$$

$$+\int_0^d\left[\beta(x+y)\left(\psi(x+y)-\psi(x)\right)+\left(\beta(x+y)-\beta(x)\right)\psi(x)\right]D(y)dy\approx$$

$$\approx\int_0^c\left[\left(\alpha(x)-\frac{d\alpha(x)}{dx}y\right)\left(-\frac{\partial\psi(x)}{\partial x}y+\frac{1}{2}\frac{\partial^2\psi(x)}{\partial x^2}y^2\right)\right]\psi(y)C(y)dy-$$

$$-\int_0^c\left[\left(-\frac{d\alpha(x)}{dx}y+\frac{1}{2}\frac{d^2\alpha(x)}{dx^2}y^2\right)\psi(x)\right]\psi(y)C(y)dy+$$

$$+\int_0^d\left[\left(\beta(x)+\frac{d\beta(x)}{dx}y\right)\left(\frac{\partial\psi(x)}{\partial x}y+\frac{1}{2}\frac{\partial^2\psi(x)}{\partial x^2}y^2\right)\right]D(y)dy+$$

$$+\int_0^d\left[\left(\frac{d\beta(x)}{dx}y+\frac{1}{2}\frac{d^2\beta(x)}{dx^2}y^2\right)\psi(x)\right]D(y)dy\approx$$

$$\approx-\frac{\partial\left(\alpha(x)\psi(x)\right)}{\partial x}\int_0^c y\psi(y)C(y)dy+\frac{1}{2}\frac{\partial^2\left(\alpha(x)\psi(x)\right)}{\partial x^2}\int_0^c y^2\psi(y)C(y)dy+$$

$$+\frac{\partial\left(\beta(x)\psi(x)\right)}{\partial x}\int_0^d yD(y)dy+\frac{1}{2}\frac{\partial^2\left(\beta(x)\psi(x)\right)}{\partial x^2}\int_0^d y^2D(y)dy.$$

We assume that

$$\int_0^c y\psi(y)C(y)dy \equiv \langle c\rangle N_{x\in(0,c]},\ \int_0^c y^2\psi(y)C(y)dy \equiv \langle c^2\rangle N_{x\in(0,c]}, \tag{8}$$

$$\int_0^d yD(y)dy \equiv \langle d\rangle \int_0^d D(y)dy,\ \int_0^d y^2D(y)dy \equiv \langle d^2\rangle \int_0^d D(y)dy, \tag{9}$$

where $N_{x\in(0,c]} \equiv \int_0^c \psi(y)C(y)dy$, if $y=n$ $\langle c\rangle$ and $\langle c^2\rangle$ are the average number and the average of the square of the number of molecules in an attaching particle, molecules in which is not more than c, $\langle d\rangle$ and $\langle d^2\rangle$ are the average number and the average of the square of the number of molecules in a fragmentizing particle, molecules in which is not more than d. Let $\int_0^d C(y)dy = \int_0^d D(y)dy = 1$. The equalities (8), (9) is valid, if $C(y)$ is the density of the conditional probability that an attaching to an aggregate particle has property equal to y, $D(y)$ is the density of the probability that an fragmentizing particle has property equal to y. The equation of the Fokker–Planck type takes the form:

$$\frac{\partial\psi}{\partial t} = -\frac{\partial\left(\left(\alpha\langle c\rangle N_{x\in(0,c]} - \beta\langle d\rangle\right)\psi\right)}{\partial x} + \frac{\partial^2\left(\left(\alpha\langle c^2\rangle N_{x\in(0,c]} + \beta\langle d^2\rangle\right)\psi\right)}{\partial x^2}. \tag{10}$$

So, from the equation (5) we obtained the equation of the Fokker–Planck type (10) by substitution (6), (8), (9). But this model is not closed, therefore we consider more simple: let us take $c=d$, $C(c)=1$, $D(x)=\delta(x-c)$, and instead ψ we substitute $N_c(t)\delta(x-c)+\psi(x,t)$, where $\mathrm{supp}(\psi(x))=[2c,+\infty)$, $N_{2c}(t)\equiv\psi(2c,t)$. Then we obtain the equation of the Fokker–Planck type on an unknown function $\psi=\psi(x,t)$, which has the form:

$$\frac{\partial\psi}{\partial t} = -\frac{\partial}{\partial x}J\left[N_c(t),\psi(x,t)\right],\ \text{при } x\ge 2c, \tag{11}$$

where $J\left[N_c(t),\psi(x,t)\right]$ is the stream of formation of clusters with the property x at time moment t: $J\left[N_c(t),\psi(x,t)\right] \equiv \left((\alpha N_c-\beta)\psi\right)c - \frac{\partial\left((\alpha N_c+\beta)\psi\right)}{\partial x}c^2$. The equation (11) is supplemented by the conservation law of the number of all molecules of substance, which forms crystals, in the system (N_0), which in accordance with (7b) can be rewritten in the form:

$$N_c(t) + \int_{2c}^{+\infty} x\psi(x,t)dx = N_0 = const, \tag{12}$$

From (11), (12) we obtain analogue of (3):

$$\frac{dN_c}{dt} = -2c\left(\alpha_c N_c^2(t) - \beta_c N_{2c}(t)\right) - \int_{2c}^{+\infty} J\left[N_c(t),\psi(y,t)\right]dy,$$

where $N_{2c}(t)\equiv\psi(2c,t)$, if the boundary conditions for (12) have the form:

$$J\left[N_c(t),\psi(x,t)\right]\Big|_{x=2c} \equiv \left(c(\alpha N_c-\beta)\psi - c^2\frac{\partial}{\partial x}\left((\alpha N_c+\beta)\psi\right)\right)\Big|_{x=2c} = \alpha_c N_c^2 - \beta_c N_{2c},$$

$$J\left[N_c(t),\psi(x,t)\right]\Big|_{x\to+\infty} \equiv \left(c(\alpha N_c-\beta)\psi - c^2\frac{\partial}{\partial x}\left((\alpha N_c+\beta)\psi\right)\right)\Bigg|_{n\to+\infty} = 0 .$$

Now, when we introduce the physical and mathematical sense of quantities $C(y)$ and $D(y)$, it is of interest the following proposition, which is worth to compare with Lemma 1.

Lemma 2. Let $c=d$, $C(x)=\delta(x-c)$, $D(x)=\delta(x-c)$. Then (7) gives the equation $\frac{\partial\psi(t,x)}{\partial t}=\left[\alpha(x-c)\psi(x-c)-\alpha(x)\psi(x)\right]\psi(c)+\left[\beta(x+c)\psi(x+c)-\beta(x)\psi(x)\right]$ for $x>c$, which is another form of (2a).

Considering a Taylor series up to second order for the increments of the functions in this equation, and assuming $\psi(c)=N_c$, we get the equation of the Fokker–Planck type (11). We have taken in (11) $x\geq 2c$ (instead of $x>c$) in order to correctly set the boundary conditions.

3. Conclusion

We wrote a new simple discrete model of the balance of the number of particles, which is some generalization of the Becker–Döring system of equations, and derived it and the continuum the Fokker–Planck equation, corresponding to it, from a continuum equation of integration-fragmentation. These models take into account the possibility of integration and fragmentation of bodies in the case when the agglomerate of particles attaches only aggregates of particles of the previous generation, and the integration of the bodies of one generation is not considered. This situation is described by some generalizations of the Becker–Döring case. Meanwhile we obtained the new continuum Becker–Döring equation (8), the approximation of which is the equation of the Fokker–Planck type. (8) is a unclosed model, but its development seems to be important, because the experimenters obtain histograms approximated by continuum functions, and the diffuse approximation of this equation, which we got clarified on the basis of the physical meaning of this derivation (on the basis of the physico-mathematical sense of functions $C(y)$ and $D(y)$) and which is often used [3]–[4].

Acknowledgments

This paper was financially supported by the Ministry of Education and Science of the Russian Federation on the program to improve the competitiveness of Peoples' Friendship University of Russia (RUDN-University) ''5-100'' among the world's leading research and educational centers in 2016–2020. It was also supported by the department of Mathematics of Russian Academy of Sciences, program 1.3.1 ''Problems of Mathematical Computational Physics''.

References

[1] Boltzmann L 1872 *Wien. Ber.* Weitere Studien über das Wärmegleichgewicht unter Gasmolekülen 66 pp 275–370; 1909 *Wissenschaftliche Abhandlungen* vol. 1 Barth, Leipzig pp 316–402.
[2] Maxwell J C 1868 *Phill. Mag.* On the Dynamical Theory of Gases 35 (235) pp 141–144.
[3] Melikhov I V 2006 *Physico-chemical evolution of solids* Moscow Binom (in Russian).
[4] Melikhov I V, Rudin V N, Kozlovskaya E D, Adzhiev S Z, Alekseeva O V 2016 *Theoretical Foundations of Chemical Engineering* Morphological memory of polymers andd their use in developing new materials technology 50 (3) pp 260–269.
[5] Becker R, Döring W 1935 *Ann. Phys.* Kinetische Behandlung der Keimbildung in ubersattingten Dampfer 24 pp 719–752.
[6] Kolmogorov A N 1938 *Uspekhi Mat. Nauk* On the analytic methods of probability theory 5 pp 5–41.

Symmetries of the equations of two-dimensional shallow water over a rough bottom

A V Aksenov and K P Druzhkov

Lomonosov Moscow State University, Moscow, Russia

E-mail: aksenov.av@gmail.com, Konstantin.Druzhkov@gmail.com

Abstract. The two-dimensional shallow water system over a rough bottom is considered. An overdetermined system of equations for finding the allowed symmetries is obtained. The consistency of this overdetermined system of equations is investigated. A general form of the solution of this system is obtained. The kernel of the symmetry operators is found. Cases where the kernel extensions of symmetry operators exist are presented. The corresponding classifying equations are given. The results of the group classification have indicated that the two-dimensional shallow water system above rough bottom cannot be linearized by point transformation in contrast to the one-dimensional shallow water system in cases of horizontal and inclined bottom profiles.

1. Introduction

In papers [1, 2], the one-dimensional shallow water system above rough bottom was considered and the group classification problem was solved.

In dimensionless variables, the two-dimensional shallow water system above rough bottom has the form [3]

$$\begin{aligned} &u_t + uu_x + vu_y + \eta_x = 0\,, \\ &v_t + uv_x + vv_y + \eta_y = 0\,, \\ &\eta_t + [u(\eta + h)]_x + [v(\eta + h)]_y = 0\,. \end{aligned} \tag{1}$$

Here $z = -h(x,y)$, $h(x,y) \geqslant 0$ is the bottom profile, $u = u(x,y,t)$, $v = v(x,y,t)$ are the velocities components averaged on the horizontal depth, and $\eta = \eta(x,y,t)$ is the deviation of the free surface.

In this paper, we solve the problem of group classification of the system (1) for all the bottom profiles.

2. Group classification

The task of the group classification is to find Lie symmetries admitted by the considered system of equations depending on the unknown function entering the system [4].

We seek symmetry operators of the system (1) as

$$X = \xi^1(x,y,t,u,v,\eta)\frac{\partial}{\partial x} + \xi^2(x,y,t,u,v,\eta)\frac{\partial}{\partial y} + \xi^3(x,y,t,u,v,\eta)\frac{\partial}{\partial t} + \\ + \eta^1(x,y,t,u,v,\eta)\frac{\partial}{\partial u} + \eta^2(x,y,t,u,v,\eta)\frac{\partial}{\partial v} + \eta^3(x,y,t,u,v,\eta)\frac{\partial}{\partial \eta}.$$

Applying the criterion of invariance [4], we obtain an overdetermined linear homogeneous system of determining equations, which is not given here because of its bulkiness. Consistency testing of this determining system leads to the following result

$$
\begin{aligned}
\xi^1 &= -\dot a x - 2By + 2k(t)\,, & \eta^1 &= (\dot a + 2C)u - 2Bv - \ddot a x + 2\dot k\,,\\
\xi^2 &= -\dot a y + 2Bx + 2l(t)\,, & \eta^2 &= (\dot a + 2C)v + 2Bu - \ddot a y + 2\dot l\,,\\
\xi^3 &= -2a(t) - 2Ct\,, & \eta^3 &= (2\dot a + 4C)\eta + \frac{x^2+y^2}{2}\dddot a - 2\ddot k x - 2\ddot l y + \dot f\,.
\end{aligned}
$$

For each $h(x,y)$ functions $a(t), k(t), l(t), f(t)$ and constants B, C are determined from the classifying equation

$$
\begin{aligned}
&\big(-\dot a x - 2By + 2k(t)\big)h_x + \big(-\dot a y + 2Bx + 2l(t)\big)h_y - (2\dot a + 4C)\cdot h(x,y) =\\
&= -\frac{x^2+y^2}{2}\cdot \dddot a + 2x\cdot \ddot k + 2y\cdot \ddot l - \dot f
\end{aligned}
$$

Next the results of the group classification will be shown.
1. Let $h(x,y)$ be arbitrary function and

$$X_1 = \frac{\partial}{\partial t}\,.$$

This operator appears for all bottom profiles.

The direct calculation shows that extensions of the Lie algebra of symmetries can be presented only in the following forms of the bottom
2. $h(x,y) = H(x) + a_0 y^2 + b_0 y$.
2.1. $H(x) = c_0 x^2 + d_0 x + e_0$. In this case unknown functions $a(t)$, $k(t)$, $l(t)$, $f(t)$ and constants B, C are found in the system of equations

$$
\begin{aligned}
&8a_0\dot a + 8a_0 C = \dddot a\,, \qquad (a_0 - c_0)B = 0\,,\\
&2b_0 B - 4d_0 C + 4c_0 k(t) - 3d_0\dot a = 2\ddot k\,,\\
&-2d_0 B - 4b_0 C - 3b_0\dot a + 4a_0 l(t) = 2\ddot l\,,\\
&-4e_0 C - 2e_0\dot a + 2d_0 k(t) + 2b_0 l(t) = -\dot f\,,
\end{aligned}
$$

that can be solved easily.
2.2. $H(x) = c_0 x^3 + d_0 x^2 + e_0 x + f_0$, $c_0 \neq 0$. In this case unknown functions $a(t)$, $k(t)$, $l(t)$, $f(t)$ and constants B, C are found in the system of equations

$$
\begin{aligned}
&8a_0\dot a + 8a_0 C = \dddot a\,, \qquad B = 0\,,\\
&-3\dot a b_0 + 4a_0 l(t) - 4b_0 C = 2\ddot l\,,\\
&4C + 5\dot a = 0\,, \qquad 8d_0(\dot a + C) - 12c_0 k(t) = \dddot a\,,\\
&e_0(4C + 3\dot a) - 4d_0 k(t) = -2\ddot k\,,\\
&f_0(4C + 2\dot a) - 2e_0 k(t) - 2b_0 l(t) = \dot f\,,
\end{aligned}
$$

that can be solved easily.

IOP Conf. Series: Journal of Physics: Conf. Series **1205** (2019) 012002 doi:10.1088/1742-6596/1205/1/012002

2.3. $H(x) = e_0 \ln(c_0 x - d_0) + f_0 x^2 + g_0 x + h_0$, $c_0 \neq 0$, $e_0 \neq 0$. In this case unknown functions $a(t)$, $k(t)$, $l(t)$, $f(t)$ and constants B, C are found in the system of equations

$$
\begin{aligned}
&8a_0\dot{a} + 8a_0 C = \dddot{a}\,, \qquad B = 0\,, \qquad \dot{a} + 2C = 0\,,\\
&-3\dot{a}b_0 + 4a_0 l(t) - 4b_0 C = 2\ddot{l}\,, \qquad \dddot{a} = 8f_0(\dot{a} + C)\,,\\
&-4c_0 g_0 C + 4d_0 f_0 C - 3c_0 g_0 \dot{a} + 4d_0 f_0 \dot{a} + 4c_0 f_0 k(t) = \frac{d_0 \dddot{a}}{2} + 2c_0 \ddot{k}\,,\\
&-4c_0 h_0 C + 4d_0 g_0 C - c_0 e_0 \dot{a} - 2c_0 h_0 \dot{a} + 3d_0 g_0 \dot{a} + 2c_0 g_0 k(t) -\\
&\qquad\qquad - 4d_0 f_0 k(t) + 2b_0 c_0 l(t) = -2d_0 \ddot{k} - c_0 \dot{f}\,,\\
&4d_0 h_0 C + 2d_0 h_0 \dot{a} + 2c_0 e_0 k(t) - 2d_0 g_0 k(t) - 2b_0 d_0 l(t) = d_0 \dot{f}\,,
\end{aligned}
$$

that can be solved easily.

2.4. $H(x) = e_0(c_0 x - d_0)\ln(c_0 x - d_0) + f_0 x^2 + g_0 x + h_0$, $c_0 \neq 0$, $e_0 \neq 0$. In this case unknown functions $a(t)$, $k(t)$, $l(t)$, $f(t)$ and constants B, C are found in the system of equations

$$
\begin{aligned}
&8a_0\dot{a} + 8a_0 C = \dddot{a}\,, \qquad B = 0\,,\\
&-3\dot{a}b_0 + 4a_0 l(t) - 4b_0 C = 2\ddot{l}\,,\\
&3\dot{a} + 4C = 0\,, \qquad c_0 k(t) + d_0(\dot{a} + 2C) = 0\,,\\
&8f_0\dot{a} + 8f_0 C = \dddot{a}\,,\\
&-3g_0\dot{a} - 4g_0 C + 4f_0 k(t) = 2\ddot{k}\,,\\
&2h_0(\dot{a} + 2C) - 2g_0 k(t) - 2b_0 l(t) = \dot{f}\,,
\end{aligned}
$$

that can be solved easily.

2.5. $H(x) = e_0(c_0 x - d_0)^2 \ln(c_0 x - d_0) + f_0 x^2 + g_0 x + h_0$, $c_0 \neq 0$, $e_0 \neq 0$. In this case unknown functions $a(t)$, $k(t)$, $l(t)$, $f(t)$ and constants B, C are found in the system of equations

$$
\begin{aligned}
&8a_0\dot{a} + 8a_0 C = \dddot{a}\,, \qquad B = 0\,,\\
&-3\dot{a}b_0 + 4a_0 l(t) - 4b_0 C = 2\ddot{l}\,,\\
&\dot{a} + C = 0\,, \qquad 2k(t)c_0 + d_0 C = 0\,,\\
&d_0(2c_0 k(t) + d_0 C) = 0\,,\\
&8f_0\dot{a} + 8f_0 C = \dddot{a}\,,\\
&-4g_0 C - 3g_0\dot{a} + 4f_0 k(t) = 2\ddot{k}\,,\\
&2h_0(\dot{a} + 2C) - 2g_0 k(t) - 2b_0 l(t) = \dot{f}\,,
\end{aligned}
$$

that can be solved easily.

2.6. $H(x) = e_0(c_0 x - d_0)^{i_0} + f_0 x^2 + g_0 x + h_0$, $c_0 \neq 0$, $e_0 \neq 0$, $i_0 \neq 0, 1, 2, 3$. In this case unknown functions $a(t)$, $k(t)$, $l(t)$, $f(t)$ and constants B, C are found in the system of equations

$$
\begin{aligned}
&8a_0\dot{a} + 8a_0 C = \dddot{a}\,, \qquad B = 0\,,\\
&-3\dot{a}b_0 + 4a_0 l(t) - 4b_0 C = 2\ddot{l}\,, \qquad \dddot{a} = 8f_0(\dot{a} + C)\,,\\
&-4c_0 g_0 C + 4d_0 f_0 C + 4c_0 f_0 k(t) - 3c_0 g_0 \dot{a} + 4d_0 f_0 \dot{a} - \frac{d_0 \dddot{a}}{2} - 2c_0 \ddot{k} = 0\,,\\
&-4c_0 h_0 C + 4d_0 g_0 C + 2c_0 g_0 k(t) - 4d_0 f_0 k(t) + 2b_0 c_0 l(t) - 2c_0 h_0 \dot{a} +\\
&\qquad\qquad + 3d_0 g_0 \dot{a} + 2d_0 \ddot{k} + c_0 \dot{f} = 0\,,\\
&\dot{a}(2 + i_0) + 4C = 0\,, \qquad c_0 i_0 k(t) + 2d_0 C + d_0 \dot{a} = 0\,,\\
&4d_0 h_0 C - 2d_0 g_0 k(t) - 2b_0 d_0 l(t) + 2d_0 h_0 \dot{a} = d_0 \dot{f}\,,
\end{aligned}
$$

IOP Conf. Series: Journal of Physics: Conf. Series **1205** (2019) 012002 doi:10.1088/1742-6596/1205/1/012002

that can be solved easily.
2.7. Let $H(x)$ be arbitrary function, excluding cases 2.1–2.6. In this case unknown functions $a(t)$, $k(t)$, $l(t)$, $f(t)$ and constants B, C are found in the system of equations

$$\begin{aligned}
&8a_0\dot{a} + 8a_0C = \dddot{a}\,, \qquad B = 0\,,\\
&2a_0l(t) = \ddot{l}\,, \qquad \dot{a} = 0\,,\\
&k(t) = 0\,, \qquad C = 0\,,\\
&2b_0l(t) = -\dot{f}\,,
\end{aligned}$$

that can be solved easily.
3. $h(x,y) = H(y - a_0x) + b_0x^2 + c_0xy + d_0x$, except case 2. In this case unknown functions $a(t)$, $k(t)$, $l(t)$, $f(t)$ and constants B, C are found in the system of equations

$$\begin{aligned}
&(1 + a_0^2)\dddot{a} - 8(b_0 + a_0c_0)(\dot{a} + C) = 4B(a_0^2c_0 - c_0 + 2a_0b_0)\,,\\
&2B(1 + a_0{}^2)H' - (4c_0\dot{a} + 4c_0C + 4B(b_0 + a_0c_0) - a_0\dddot{a})z =\\
&= -2(2b_0 + a_0c_0)k(t) - 2c_0l(t) + 2\ddot{k} + 2a_0\ddot{l} + d_0(2a_0B + 3\dot{a} + 4C)\,,\\
&\big((2a_0B - \dot{a})z - 2a_0k(t) + 2l(t)\big)H' - (2\dot{a} + 4C)H(z) =\\
&= \Big(-\frac{1}{2}\dddot{a} + 2c_0B\Big)z^2 + (2\ddot{l} - 2c_0k(t) + 2d_0B)z - 2d_0k(t) - \dot{f}\,,
\end{aligned}$$

where $z = y - a_0x$.
3.1. $H(z) = e_0z^2 + f_0z + g_0$. In this case unknown functions $a(t)$, $k(t)$, $l(t)$, $f(t)$ and constants B, C are found in the system of equations

$$\begin{aligned}
&(1 + a_0^2)\dddot{a} - 8(b_0 + a_0c_0)(\dot{a} + C) = 4B(a_0^2c_0 - c_0 + 2a_0b_0)\,,\\
&4B(1 + a_0{}^2)e_0 - (4c_0\dot{a} + 4c_0C + 4B(b_0 + a_0c_0) - a_0\dddot{a}) = 0\,,\\
&2B(1 + a_0{}^2)f_0 = -2(2b_0 + a_0c_0)k(t) - 2c_0l(t) + 2\ddot{k} + 2a_0\ddot{l} + d_0(2a_0B + 3\dot{a} + 4C)\,,\\
&2e_0(2a_0B - \dot{a}) - (2\dot{a} + 4C)e_0 = \Big(-\frac{1}{2}\dddot{a} + 2c_0B\Big)\,,\\
&(2a_0B - \dot{a})f_0 + 2e_0(-2a_0k(t) + 2l(t)) - f_0(2\dot{a} + 4C) = (2\ddot{l} - 2c_0k(t) + 2d_0B)\,,\\
&(-2a_0k(t) + 2l(t))f_0 - (2\dot{a} + 4C)g_0 = -2d_0k(t) - \dot{f}\,,
\end{aligned}$$

that can be solved easily.
3.2. $H(z) = e_0z^3 + f_0z^2 + g_0z + h_0$, $e_0 \neq 0$. In this case unknown functions $a(t)$, $k(t)$, $l(t)$, $f(t)$ and constants B, C are found in the system of equations

$$\begin{aligned}
&(1 + a_0^2)\dddot{a} - 8(b_0 + a_0c_0)(\dot{a} + C) = 4B(a_0^2c_0 - c_0 + 2a_0b_0)\,,\\
&B = 0\,, \qquad 4c_0\dot{a} + 4c_0C - a_0\dddot{a} = 0\,, \qquad 5\dot{a} + 4C = 0\,,\\
&-2(2b_0 + a_0c_0)k(t) - 2c_0l(t) + 2\ddot{k} + 2a_0\ddot{l} + d_0(3\dot{a} + 4C) = 0\,,\\
&-6a_0e_0k(t) - 4f_0C - 4f_0\dot{a} + 6e_0l(t) = -\frac{\dddot{a}}{2}\,,\\
&-4a_0f_0k(t) - 4g_0C - 3g_0\dot{a} + 4f_0l(t) = 2\ddot{l} - 2c_0k(t)\,,\\
&-2a_0g_0k(t) - 4h_0C - 2h_0\dot{a} + 2g_0l(t) = -2d_0k(t) - \dot{f}\,,
\end{aligned}$$

that can be solved easily.

IOP Conf. Series: Journal of Physics: Conf. Series **1205** (2019) 012002 doi:10.1088/1742-6596/1205/1/012002

3.3. $H(z) = g_0 \ln(e_0 z - f_0) + h_0 z^2 + i_0 z + j_0$, $e_0 \neq 0$, $g_0 \neq 0$. In this case unknown functions $a(t)$, $k(t)$, $l(t)$, $f(t)$ and constants B, C are found in the system of equations

$$
\begin{aligned}
&(1+a_0^2)\dddot{a} - 8(b_0 + a_0 c_0)(\dot{a} + C) = 0\,, \qquad \dot{a} + 2C = 0\,,\\
&B = 0\,, \qquad 4c_0\dot{a} + 4c_0 C - a_0\dddot{a} = 0\,, \qquad h_0 C = 0\,,\\
&-2(2b_0 + a_0 c_0)k(t) - 2c_0 l(t) + 2\ddot{k} + 2a_0\ddot{l} + d_0(3\dot{a} + 4C) = 0\,,\\
&-4a_0 e_0 h_0 k(t) - 4e_0 i_0 C + 4f_0 h_0 C - 3e_0 i_0 \dot{a} + 4f_0 h_0 \dot{a} + 4e_0 h_0 l(t) = \frac{f_0\dddot{a}}{2} + e_0(2\ddot{l} - 2c_0 k(t))\,,\\
&-2a_0 e_0 i_0 k(t) + 4a_0 f_0 h_0 k(t) - 4e_0 j_0 C + 4f_0 i_0 C - e_0 g_0 \dot{a} - 2e_0 j_0 \dot{a} + 3f_0 i_0 \dot{a} + 2e_0 i_0 l(t) =\\
&= 4f_0 h_0 l(t) - f_0(2\ddot{l} - 2c_0 k(t)) + e_0(-2d_0 k(t) - \dot{f}))\,,\\
&-2a_0 e_0 g_0 k(t) + 2a_0 f_0 i_0 k(t) + 4f_0 j_0 C + 2f_0 j_0 \dot{a} + 2e_0 g_0 l(t) - 2f_0 i_0 l(t) = -f_0(-2d_0 k(t) - \dot{f})\,,
\end{aligned}
$$

that can be solved easily.
3.4. $H(z) = g_0(e_0 z - f_0)\ln(e_0 z - f_0) + h_0 z^2 + i_0 z + j_0$, $e_0 \neq 0$, $g_0 \neq 0$. In this case unknown functions $a(t)$, $k(t)$, $l(t)$, $f(t)$ and constants B, C are found in the system of equations

$$
\begin{aligned}
&(1+a_0^2)\dddot{a} - 8(b_0 + a_0 c_0)(\dot{a} + C) = 0\,, \qquad 3\dot{a} + 4C = 0\,,\\
&B = 0\,, \qquad 4c_0\dot{a} + 4c_0 C - a_0\dddot{a} = 0\,, \qquad h_0\dot{a} = 0\,,\\
&-2(2b_0 + a_0 c_0)k(t) - 2c_0 l(t) + 2\ddot{k} + 2a_0\ddot{l} + d_0(3\dot{a} + 4C) = 0\,,\\
&-2a_0 e_0 g_0 k(t) + 4f_0 g_0 C + 2f_0 g_0 \dot{a} + 2e_0 g_0 l(t) = 0\,,\\
&-e_0 g_0 \dot{a} - 4a_0 h_0 k(t) - 4i_0 C - 3i_0 \dot{a} + 4h_0 l(t) = 2\ddot{l} - 2c_0 k(t)\,,\\
&-2a_0 e_0 g_0 k(t) - 2a_0 i_0 k(t) + 2e_0 g_0 l(t) - 4j_0 C - 2j_0 \dot{a} + 2i_0 l(t) = -2d_0 k(t) - \dot{f}\,,
\end{aligned}
$$

that can be solved easily.
3.5. $H(z) = g_0(e_0 z - f_0)^2 \ln(e_0 z - f_0) + h_0 z^2 + i_0 z + j_0$, $e_0 \neq 0$, $g_0 \neq 0$. In this case unknown functions $a(t)$, $k(t)$, $l(t)$, $f(t)$ and constants B, C are found in the system of equations

$$
\begin{aligned}
&(1+a_0^2)\dddot{a} - 8(b_0 + a_0 c_0)(\dot{a} + C) = 4B(a_0^2 c_0 - c_0 + 2a_0 b_0)\,,\\
&B = 0\,, \qquad 4c_0\dot{a} + 4c_0 C - a_0\dddot{a} = 0\,, \qquad \dot{a} + C = 0\,,\\
&-2(2b_0 + a_0 c_0)k(t) - 2c_0 l(t) + 2\ddot{k} + 2a_0\ddot{l} + d_0(3\dot{a} + 4C) = 0\,,\\
&-2a_0 e_0 k(t) + f_0 C + 2e_0 l(t) = 0\,, \qquad \dot{a} = 0\,,\\
&4a_0 e_0 f_0 g_0 k(t) - 4f_0^2 g_0 C - 2f_0^2 g_0 \dot{a} - 4e_0 f_0 g_0 l(t) = 0\,,\\
&-2a_0 e_0^2 g_0 k(t) + e_0 f_0 g_0 \dot{a} + 2e_0^2 g_0 l(t) - 4a_0 h_0 k(t) - 4i_0 C - 3i_0 \dot{a} + 4h_0 l(t) = 2\ddot{l} - 2c_0 k(t)\,,\\
&2a_0 e_0 f_0 g_0 k(t) - 2e_0 f_0 g_0 l(t) - 2a_0 i_0 k(t) - 4j_0 C - 2j_0 \dot{a} + 2i_0 l(t) = -2d_0 k(t) - \dot{f}\,,
\end{aligned}
$$

that can be solved easily.
3.6. $H(z) = g_0(e_0 z - f_0)^{k_0} + h_0 z^2 + i_0 z + j_0$, $e_0 \neq 0$, $g_0 \neq 0$, $k_0 \neq 0, 1, 2, 3$. In this case unknown

functions $a(t)$, $k(t)$, $l(t)$, $f(t)$ and constants B, C are found in the system of equations

$$\begin{aligned}
&(1+a_0^2)\dddot{a} - 8(b_0+a_0c_0)(\dot{a}+C) = 4B(a_0^2c_0 - c_0 + 2a_0b_0)\,,\\
&B=0\,, \qquad 4c_0\dot{a} + 4c_0C - a_0\dddot{a} = 0\,,\\
&-2(2b_0+a_0c_0)k(t) - 2c_0l(t) + 2\ddot{k} + 2a_0\ddot{l} + d_0(3\dot{a}+4C) = 0\,,\\
&\dot{a}(2+k_0) + 4C = 0\,, \qquad \dddot{a} = 8h_0(\dot{a}+C)\,,\\
&-2a_0e_0g_0k_0k(t) + 2e_0g_0k_0l(t) + 4f_0g_0C + 2f_0g_0\dot{a} = 0\,,\\
&-4a_0e_0h_0k(t) - 4e_0i_0C + 4f_0h_0C - 3e_0i_0\dot{a} + 4f_0h_0\dot{a} + 4e_0h_0l(t) = \frac{f_0\dddot{a}}{2} + e_0(2\ddot{l} - 2c_0k(t))\,,\\
&-2a_0e_0i_0k(t) + 4a_0f_0h_0k(t) - 4e_0j_0C + 4f_0i_0C - 2e_0j_0\dot{a} + 3f_0i_0\dot{a} + 2e_0i_0l(t) - 4f_0h_0l(t) =\\
&= -f_0(2\ddot{l} - 2c_0k(t)) + e_0(-2d_0k(t) - \dot{f})\,,\\
&2a_0f_0i_0k(t) + 4f_0j_0C + 2f_0j_0\dot{a} - 2f_0i_0l(t) = f_0(2d_0k(t) + \dot{f})\,,
\end{aligned}$$

that can be solved easily.

3.7. Let $H(z)$ be arbitrary function, excluding cases 3.1–3.6. In this case unknown functions $a(t)$, $k(t)$, $l(t)$, $f(t)$ and constants B, C are found in the system of equations

$$\begin{aligned}
&B=0\,, \qquad \dot{a}=0\,, \qquad C=0\,,\\
&-2(2b_0+a_0c_0)k(t) - 2c_0l(t) + 2\ddot{k} + 2a_0\ddot{l} + d_0(3\dot{a}+4C) = 0\,,\\
&l(t) = a_0k(t)\,, \qquad 2\ddot{l} - 2c_0k(t) = 0\,, \qquad -2d_0k(t) - \dot{f} = 0\,,
\end{aligned}$$

that can be solved easily.

4. $h(x,y) = \dfrac{g_0}{(x-a_0)^2}\cdot\dfrac{e^{h_0\arctan z}}{z^2+1} + c_0(x^2+y^2) + d_0x + e_0y + f_0$, $g_0 \neq 0$, $z = (y-b_0)/(x-a_0)$.
In this case unknown functions $a(t)$, $k(t)$, $l(t)$, $f(t)$ and constants B, C are found in the system of equations

$$\begin{aligned}
&\dddot{a} = 8c_0(\dot{a}+C)\,, \qquad Bh_0 = 2C\,,\\
&\ddot{l} - 2c_0l(t) = -2e_0C - d_0B - \frac{3}{2}e_0\dot{a}\,,\\
&\ddot{k} - 2c_0k(t) = e_0B - 2d_0C - \frac{3}{2}d_0\dot{a}\,,\\
&\dot{f} = 2f_0\dot{a} - 2d_0k(t) - 2e_0l(t) + 4f_0C\,,\\
&-2b_0h_0B - a_0h_0\dot{a} + 4a_0B - 2b_0\dot{a} + 2h_0k(t) + 4l(t) = 0\,,\\
&b_0h_0\dot{a} - 4b_0B - 4a_0C - 2a_0\dot{a} - 2h_0l(t) + 4k(t) = 0\,,
\end{aligned}$$

that can be solved easily.

5. Let $h(x,y)$ be a solution of the equation $(A_1x + A_2y + A_3)h_x + (A_1y - A_2x + A_4)h_y + A_5h = A_6(x^2+y^2) + A_7x + A_8y + A_9$ for some constants $A_1, ..., A_9$, among which there is a distinct from zero constant, excluding cases 1–4. In this case unknown functions $a(t)$, $k(t)$, $l(t)$, $f(t)$ and constants B, C are found in the system of equations

$$\begin{aligned}
\big(-\dot{a}x - 2By + 2k(t)\big)h_x + \big(-\dot{a}y + 2Bx + 2l(t)\big)h_y - (2\dot{a}+4C)\cdot h(x,y) =\\
= -\frac{x^2+y^2}{2}\cdot\dddot{a} + 2x\cdot\ddot{k} + 2y\cdot\ddot{l} - \dot{f}
\end{aligned}$$

Remark. Analysis of group classification results shows the impossibility of linearizing of the two-dimensional shallow water system above rough bottom with a point transformation.

IOP Conf. Series: Journal of Physics: Conf. Series **1205** (2019) 012002 doi:10.1088/1742-6596/1205/1/012002

3. Conclusion

Main results of the paper are the classification of bottom profiles where the kernel extensions of symmetry operators exist. Corresponding classifying equations for this profiles are derived. The distinctive property of a two-dimensional shallow-water system over a rough bottom is that it cannot be linearized by a point transformation, unlike a one-dimensional shallow-water system, when the bottom profiles are horizontal and inclined.

Acknowledgments

This work was supported by RFBR grants 18-01-00890.

References

[1] Aksenov A V and Druzhkov K.P. 2016 Zakony sokhranenija, simmetrii i tochnye reshenija sistemy uravnenij melkoj vody nad nerovnym dnom [Conservation laws, symmetries, and exact solutions of the shallow water system above rough bottom]. *Vestnik Natsionalnogo issledovateltskogo yadernogo universiteta MIFI* **5** 38-46

[2] Aksenov A .V and Druzhkov K P 2016 Conservation laws and symmetries of the shallow water system above rough bottom *J. Phys.: Conf. Ser.* 2016 **722** 1-7

[3] Stoker J J 1957 *Water Waves. The Mathematical Theory With Applications* (New York: Interscience Publishers)

[4] Ovsiannikov L V 1982 *Group Analysis of Differential Equations* (New York: Academic Press)

Self-similar solution for power-law liquid flow down an inclined plane

A V Aksenov, A D Chicherina and I S Chicherin
Lomonosov Moscow State University, Moscow, Russia
E-mail: aksenov.av@gmail.com, chicherina@imec.msu.ru, chicherin-ivan@rambler.ru

Abstract. Within the Stokes film approximation, unsteady plane-parallel spreading of a thin layer of a heavy non-Newtonian fluid along an inclined solid surface is studied. The forced spreading regime induced by mass supply is considered. An evolution equation for the film thickness is derived. The group classification problem is solved. Self-similar solution is constructed for the power time dependence on mass supply.

Keywords: non-Newtonian fluid, Stokes film approximation, evolution equation, symmetry operator, self-similar solution

1. Introduction

In different applications it is necessary to consider motion of non-Newtonian fluids. Such fluids often occur in nature and are used in industry. Mudflow, volcanic lava, oil, paint and lacquer, polymer melts and solutions, granulated material are the examples of non-Newtonian liquids. Nonlinear dependence of shear rate at each point of liquid volume on shear stress at the same point characterizes liquids with complex rheology properties. Fundamental characteristics of liquideous and solid-like objects, methods of experimental determination of rheology properties and detailed description of hydraulic flows are given in [1]. Determination of the principles of the non-Newtonian laminar flow is presented in [2]. The reference of rheological equations for different materials is stated in [3].

Slow spreading of a thin incompressible layer of a heavy non-Newtonian fluid along an inclined solid surface is studied at a given localized mass supply. The surface tension effect is neglected. A rheological equation is formulated in invariant terms [3]. An evolution equation for the film thickness is derived. The group classification problem is solved. Self-similar solution form is found in case when the law of mass supply in the film is a power function of time.

2. Rheological model of flow

A rheological equation is formulated in invariant terms [3] for two-dimension flow

$$\tau_{ij} = 2KI_2^{\frac{n-1}{2}} e_{ij},$$

where τ_{ij} are the components of the viscous stresses tensor, K and n are empirical constants, e_{ij} are the components of the deformation rate tensor and I_2 is the second invariant of the deformation rate tensor

$$I_2 = e^{ij}e_{ij} = u_x^2 + w_z^2 + \frac{1}{2}(w_x + u_z)^2.$$

3. The equations of motion and boundary conditions

The two-dimension slow source flow of a thin layer of a heavy non-Newtonian fluid is considered. The law of mass supply in the film is a power function of time. The flow is on an inclined solid surface. The angle of inclination of the layer equals α. The flow is described by equations of mass and momentum conservation. The system of equations is closed by rheological formula. We align coordinates so that the plane occupies $z = 0$, and x points downslope, origin of the coordinate system is related to the source. The flow is considered slow and laminar, all its characteristics depend on the spatial coordinates x and z and time t. The film thickness does not depend on coordinate z. The equation of motion in the projections on the coordinate axes has, respectively, the form

$$\begin{aligned} \rho\frac{du}{dt} &= \rho g\sin\alpha + \frac{\partial \tau_{xx}}{\partial x} + \frac{\partial \tau_{xz}}{\partial z}, \\ \rho\frac{dw}{dt} &= -\rho g\cos\alpha + \frac{\partial \tau_{zx}}{\partial x} + \frac{\partial \tau_{zz}}{\partial z}, \end{aligned} \tag{1}$$

where τ_{xx}, τ_{xz} and τ_{zz} are the components of the viscous stresses tensor. Let us write the conservation law

$$\frac{\partial u}{\partial x} + \frac{\partial w}{\partial z} = 0\,. \tag{2}$$

The boundary conditions at free surface $z = h$ are

$$p = \tau_{xz} = 0, \qquad \tau_{zz} = 0. \tag{3}$$

No-slip condition is suggested at the solid wall $z = 0$

$$u = 0, \qquad w = 0. \tag{4}$$

Let us write the equations of mass and momentum conservation in a dimensionless form. To remove dimensions we use L for the distance downslope from the source. Then the conservation law (2) gives

$$\begin{aligned} &(x, z) = L(x^*, \gamma z^*), \\ &(u, w) = U(u^*, \gamma w^*), \\ &t = \frac{L}{U}t^*, \qquad p = \rho g\sin\alpha\gamma L p^*, \\ &\tau_{ij} = \rho\nu\frac{U}{\gamma L}\tau^*_{ij}. \end{aligned}$$

Dimensionless values are marked with an asterisk. The thin boundary-layer approximation is invoked by neglecting terms of quadratic order in the cross-stream aspect ratio ($\gamma^2 \ll 1$), so the pressure field is hydrostatic. The characteristic flow speed U and an effective kinematic viscosity ν have the following form

$$U = \frac{\gamma^3 L^2 g\cos\alpha}{\nu}, \qquad \nu = \frac{K}{\rho}\left(\frac{U}{\gamma L}\right)^{n-1}.$$

Then the the asterisk decoration is lowered for convenience.

The dimensionless second invariant of the deformation rate tensor and the dimensionless components of the deformation rate tensor become

$$\begin{aligned} I_2 &= u_x^2 + w_z^2 + \frac{1}{2}\left(\gamma w_x + \frac{u_z}{\gamma}\right)^2, \\ e_{ij} &= \frac{1}{2}\begin{pmatrix} 2u_x & u_z + \gamma^2 w_x \\ u_z + \gamma^2 w_x & 2w_z \end{pmatrix}. \end{aligned} \tag{5}$$

IOP Conf. Series: Journal of Physics: Conf. Series **1205** (2019) 012003 doi:10.1088/1742-6596/1205/1/012003

When neglecting small terms in (5), we obtain

$$I_2 = \frac{1}{2}u_z^2 ,$$
$$e_{ij} = \frac{1}{2}\begin{pmatrix} 2u_x & u_z \\ u_z & 2w_z \end{pmatrix} .$$

Let us write equations of motion (1) in dimensionless form

$$\begin{aligned} \rho\frac{U^2}{L}\left(u_t + uu_x + wu_z\right) &= -\frac{\rho\nu U}{\gamma^2 L^2}p_x + \frac{\rho\nu U}{\gamma^2 L^2}\frac{tg\alpha}{\gamma} + \\ &+ \frac{\rho\nu U}{\gamma^2 L^2}\left(\gamma^2\tau_{xx,x} + \tau_{xz,z}\right), \\ \rho\frac{\gamma U^2}{L}\left(w_t + uw_x + ww_z\right) &= -\frac{\rho\nu U}{\gamma^3 L^2}p_z - \frac{\rho\nu U}{\gamma^3 L^2} + \\ &+ \frac{\rho\nu U}{\gamma L^2}\left(\tau_{xz,x} + \tau_{zz,z}\right). \end{aligned} \tag{6}$$

Let $S = tg\alpha/\gamma$ be a slope parameter that characterizes the angle of inclination of the underlying surface and the Reynolds number $Re = \delta^2\gamma^2\nu U/L$. Then the equations of motion (6) become

$$\begin{aligned} \gamma^2 Re\left(u_t + uu_x + wu_z\right) &= -p_x + S + \gamma^4\tau_{xx,x} + \tau_{xz,z} , \\ \gamma^4 Re\left(w_t + uw_x + ww_z\right) &= -p_z - 1 - \gamma^2\left(\tau_{xz,x} + \tau_{zz,z}\right) \end{aligned} \tag{7}$$

4. Equation for the film thikness

The flow is considered slow so that $\gamma^2 Re \ll 1$. We obtain the following system of equations by neglecting small terms in (7)

$$\begin{aligned} p_x &= S + \tau_{xz,z} , \\ p_z &= -1. \end{aligned} \tag{8}$$

We find the solution for the system of equations (8) taking into account boundary conditions (3) and (4) that are written in dimensionless form. Hence it follows that

$$\tau_{xz} = (S - h_x)(h - z).$$

On the other hand

$$\tau_{xz} = \frac{u_z^n}{B},$$

where $B = 2^{\frac{n-1}{2}}$. As a result we have a formula for the x–component of the velocity

$$u = \frac{n}{n+1}B^{\frac{1}{n}}(S - h_x)^{\frac{1}{n}}(h^{\frac{1}{n}+1} + (h - z)^{\frac{1}{n}+1}).$$

We use continuity equation (2) in dimensionless form to find $w = w(x, y, z)$. Thus

$$w|_{z=h} = \frac{B^{\frac{1}{n}}}{(n+1)(2n+1)}(S - h_x)^{\frac{1}{n}-1}h^{\frac{1}{n}+1}((n+1)hh_{xx} - (2n+1)(S - h_x)h_x). \tag{9}$$

The kinematic condition $w = h_t + uh_x$ is valid at $z = h = h(t, x)$

$$w|_{z=h} = h_t + \frac{nB^{\frac{1}{n}}}{(n+1)}(S - h_x)^{\frac{1}{n}}h_x h^{\frac{1}{n}+1}. \tag{10}$$

We set equal the right parts of expressions (9) and (10) and obtain the equation for the film thikness

$$(2n+1)h_t + B^{\frac{1}{n}}(S-h_x)^{\frac{1}{n}-1}h^{\frac{1}{n}+1}\left((2n+1)(S-h_x)h_x - hh_{xx}\right) = 0. \tag{11}$$

The statement of the problem for finding the surface shape $h(t,x)$ is completed by the formulation of an integral condition following from the mass supply law and the requirement of zero film thickness at the leading front x_f of the wetted area

$$\int_0^{x_f} h(x,t)dx = Q(t), \qquad h(t,x_f) = 0. \tag{12}$$

5. Group classification

Symmetry operators of equation (11) are found as

$$X = \xi^1\frac{\partial}{\partial x} + \xi^2\frac{\partial}{\partial t} + \eta\frac{\partial}{\partial h},$$

where $\xi^1 = \xi^1(t,x,h)$, $\xi^2 = \xi^2(t,x,h)$ and $\eta = \eta(t,x,h)$. Applying the criterion of invariance [4] we obtain the system of the determining equations. The analysis of the system solution shows that the kernel of the symmetry operators consists of the symmetry operator

$$X_1 = x\frac{\partial}{\partial x} - \frac{t}{n}\frac{\partial}{\partial t} + h\frac{\partial}{\partial h}. \tag{13}$$

6. Self-similar solution

The transformation group corresponding to symmetry operator (13) is the group of inhomogeneous dilations. The group invariants have the following form

$$I_1 = xt^n, \qquad I_2 = ht^n.$$

The solution of equation (11) is found in the form $h = t^{-n}F(\eta)$, where $\eta = Cxt^n$, $C = const$. As a result we have an ordinary differential equation for the determining of $F(\eta)$

$$\begin{aligned}(2n+1)n(\eta F' - F) + (2n+1)CB^{\frac{1}{n}}(S-CF')^{\frac{1}{n}}F'F^{\frac{1}{n}+1} - \\ - C^2B^{\frac{1}{n}}(S-CF')^{\frac{1}{n}-1}F''F^{\frac{1}{n}+2} = 0.\end{aligned} \tag{14}$$

Volume of the liquid bounded by the film surface shape according to condition (12) is calculated as

$$Q(t) = \frac{1}{C}t^{-2n}\int_0^{\eta_f} F(\eta)d\eta.$$

The value of η_f is a constant as $F(\eta_f) = 0$. Thus, the mass supply law in the film is a power function of time and it is written as

$$Q(t) = At^{-2n}, \qquad A = \frac{1}{C}\int_0^{\eta_f} F(\eta)d\eta = const > 0.$$

The value of the constant C is chosen so that the self-similar coordinate of the leading edge η_f is equal to unity. Next the constant C is defined by formula

$$C = A^{-1}\int_0^1 F(\eta)\,d\eta. \tag{15}$$

The solution of the equation (14) is defined by boundary condition $F(1) = 0$ and formula (15).

Asymptotic of the solution of equation (14) in the neighborhood of point $\eta = 1$ is found as

$$F(\eta) = F_0(1-\eta)^p + \cdots .$$

After the substitution of the sought solution asymptotic expansion in equation (14) we obtain

$$p = \frac{1}{n+2},$$
$$F_0^{1+\frac{2}{n}} C^{1+\frac{1}{n}} B^{\frac{1}{n}} = -(2n+1).$$

Further case of $0 < p < 1$ is considered. So

$$-1 < n < -\frac{1}{2}.$$

And as a result we have

$$F(\eta) \simeq \left(-\frac{2n+1}{C^{1+\frac{1}{n}} B^{\frac{1}{n}}}\right)^{\frac{n}{n+2}} (1-\eta)^{\frac{1}{n+2}}. \tag{16}$$

Equation (14) was solved numerically using the asymptotic (16) and condition (15).

Figure 1 shows the results of numerical solution of equation (14) at $A = 0.0085$, $n = -0.6$ and -0.7. It has the second order and it is solved numerically using Runge–Kutta method.

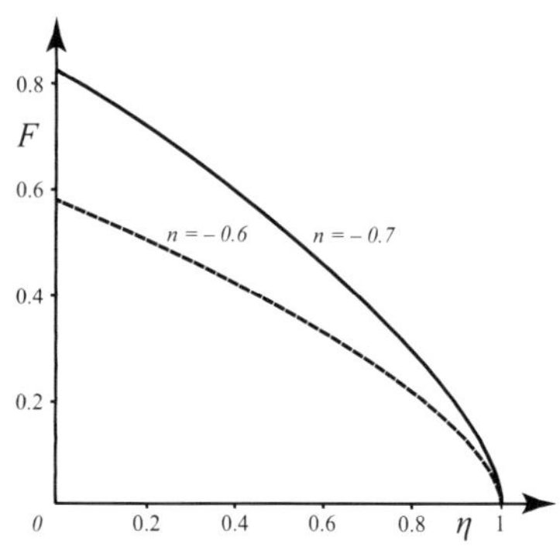

Figure 1. Function $F(\eta)$

7. Conclusion

The evolution equation for the film thickness is derived in the work. The group classification problem is solved. Self-similar solution form is found in case when the law of mass supply in the film is a power function. The results of numerical calculations are presented.

Acknowledgments

This work was supported by RFBR grant 18-01-00890.

References

[1] Wilkinson W L 1960 *Non-Newtonian Fluids. Fluid Mechanics, Mixing and Heat Transfer* (London: Pergamon Press)

[2] Shul'man Z P and Baikov V I 1979 *Rheodynamics and Heat and Mass Transfer in Film Flows* (Minsk: Science and Technology) [in Russian]

[3] Malkin A Ya and Isayev A I 2012 *Rheology. Concepts, Methods and Applications* (Toronto: ChemTec Publishing)

[4] Ovsiannikov L V 1982 *Group Analysis of Differential Equations* (New York: Academic Press)

Necessary conditions of point equivalence of second-order ODEs to the fifth Painlevé equation

Yu Yu Bagderina

Institute of Mathematics with Computer Center of Ufa Federal Research Center of Russian Academy of Sciences, 112 Chernyshevsky str., Ufa 450008, Russia

E-mail: `bagderinayu@yandex.ru`

Abstract. Equivalence problem for a projective type scalar second-order ODEs is considered with respect to invertible point changes of variables. In a recent paper the basis of invariants of this family of equations has been constructed. Here the invariants of such equations are used to find the necessary conditions of their equivalence to the fifth Painlevé equation.

1. Introduction

All second-order ordinary differential equations (ODEs) having no movable singularities other than poles, classified by Painlevé and Gambier, belong to the family

$$\frac{d^2y}{dx^2} = S(x,y)\Big(\frac{dy}{dx}\Big)^3 + 3R(x,y)\Big(\frac{dy}{dx}\Big)^2 + 3Q(x,y)\frac{dy}{dx} + P(x,y). \tag{1}$$

Once it is established that an ODE under study has this "Painlevé property", this necessitates mapping the equation to some standard form, in which its solution can be recognized. Hence the question arises of whether a given scalar second-order ODE is equivalent to anyone of the Painlevé equations PI–PVI. This equivalence problem has been solved for the first four Painlevé equations. For very particular cases of PV and PVI only some necessary equivalence conditions were found [1]. These equivalence conditions are stated in terms of the invariants of the family (1), because the equivalent equations have coincident sets of invariants. As usual, by invariants of a family of equations we mean the invariants of the group E of equivalence transformations of this family. For the family (1) the group E is generated by transformations

$$z = \xi(x,y), \qquad w = \eta(x,y), \qquad \det\frac{\partial(\xi,\eta)}{\partial(x,y)} \neq 0. \tag{2}$$

Its invariants applicable in the (degenerate) case of PV are described in the next Section.

The third and fifth Painlevé equations as well as their invariants have a similar structure. Both ODEs involve constant parameters α, β, γ, δ. By dilation of z, w two parameters of PIII

$$\frac{d^2w}{dz^2} = \frac{1}{w}\Big(\frac{dw}{dz}\Big)^2 - \frac{1}{z}\frac{dw}{dz} + \frac{\alpha w^2+\beta}{z} + \gamma w^3 + \frac{\delta}{w}$$

and one parameter of PV

$$\frac{d^2w}{dz^2} = \Big(\frac{1}{2w} + \frac{1}{w-1}\Big)\Big(\frac{dw}{dz}\Big)^2 - \frac{1}{z}\frac{dw}{dz} + \frac{(w-1)^2}{z^2}\Big(\alpha w + \frac{\beta}{w}\Big) + \gamma\frac{w}{z} + \delta\frac{w(w+1)}{w-1} \tag{3}$$

can be normalized. Hence, PV possesses three essential parameters while PIII has them a pair. As shown in [2, Section V], in order to obtain the criteria of the equivalence to PIII, it is sufficient to use seven invariants I_1, I_2, $\mathcal{D}_2 I_1$, J_1, J_2, J_3, J_4 (defined in the next Section). PV has one essential parameter more than PIII. So, we use the same set of the invariants and one more invariant J_5 for PV in Section 3. This suggests that the necessary conditions of equivalence to PV will be sufficient too, even if the proof is out of the question.

In [2] necessary conditions of equivalence to PIII, stated in terms of invariants, are imposed on an ODE $d^2y/dx^2 = f(x, y)$. The resulting equation coincides with the canonical form of PIII $d^2y/dx^2 = \alpha e^{x+y} + \beta e^{x-y} + \gamma e^{2(x+y)} + \delta e^{2(x-y)}$, which is to say that these conditions are furthermore sufficient. The same approach has been implemented in [1] for PIV. Here we do not apply it to (3), because in theory the necessary equivalence conditions can be obtained in terms of the invariants only, but they would be useless in view of their vast volume. In actual practice, we prefer to retain the dependence on w and an auxiliary variable H in Theorem 2 below. In Section 4 it is shown how to eliminate consecutively H and w from the equivalence conditions given by Theorem 2. This procedure leads either to transformation (2) relating ODEs (1) and (3) if they are equivalent, or to a contradiction which points to the fact that the equation (1) under study is not equivalent to PV.

2. Invariants of the fourth type equations (1)

Invariants of the family (1) are described completely in [3]. The six Painlevé equations fall into degenerate types of equations (1) with $J_0 = 0$. In particular, PII–PVI belong to the fourth type of equations (1). Their absolute invariants are expressed in terms of relative invariants

$$J_0 = \beta_2^2\gamma_{10} - \beta_1\beta_2(\gamma_{20} + \gamma_{11}) + \beta_1^2\gamma_{21}, \qquad j_0 = \frac{3}{\beta_1}\left(\frac{\beta_2}{\beta_1}\delta_{10} - \delta_{11}\right) + \frac{6\gamma_{10}}{\beta_1^2}\left(\gamma_{11} - \frac{\beta_2}{\beta_1}\gamma_{10}\right),$$
$$j_1 = \frac{5}{6}\left(2\beta_2\delta_{20} - \beta_1\delta_{30} - \frac{\beta_2^2}{\beta_1}\delta_{10}\right) + \left(\gamma_{20} + \gamma_{11} - 2\frac{\beta_2}{\beta_1}\gamma_{10}\right)\left(\gamma_{20} - \frac{2}{3}\gamma_{11} - \frac{\beta_2}{3\beta_1}\gamma_{10}\right),$$
$$j_2 = \frac{1}{\beta_1}\left(\delta_{20} - \frac{\beta_2}{\beta_1}\delta_{10}\right) + \frac{\gamma_{10}}{5\beta_1^2}\left(7\frac{\beta_2}{\beta_1}\gamma_{10} - 6\gamma_{20} - \gamma_{11}\right), \qquad j_3 = \frac{3}{5}\left(\frac{\delta_{10}}{\beta_1^3} - \frac{6\gamma_{10}^2}{5\beta_1^4}\right).$$

These latter are calculated using the values

$$\begin{aligned}
\alpha_0 &= Q_x - P_y + 2PR - 2Q^2, & \beta_1 &= \alpha_{1x} - \alpha_{0y} + R\alpha_0 - 2Q\alpha_1 + P\alpha_2,\\
\alpha_1 &= R_x - Q_y + PS - QR, & \beta_2 &= \alpha_{2x} - \alpha_{1y} + S\alpha_0 - 2R\alpha_1 + Q\alpha_2,\\
\alpha_2 &= S_x - R_y + 2QS - 2R^2, & \Gamma_0 &= 3\beta_2\gamma_{10} + \beta_1(\gamma_{20} - 4\gamma_{11}),\\
\gamma_{10} &= \beta_{1x} - Q\beta_1 + P\beta_2, & \delta_{10} &= \gamma_{10x} - 2Q\gamma_{10} + P(\gamma_{20} + \gamma_{11}) - 5\alpha_0\beta_1,\\
\gamma_{20} &= \beta_{1y} - R\beta_1 + Q\beta_2, & \delta_{20} &= \gamma_{20x} - R\gamma_{10} + P\gamma_{21} - 4\alpha_1\beta_1 - \alpha_0\beta_2,\\
\gamma_{11} &= \beta_{2x} - R\beta_1 + Q\beta_2, & \delta_{11} &= \gamma_{11x} - R\gamma_{10} + P\gamma_{21} - \alpha_1\beta_1 - 4\alpha_0\beta_2,\\
\gamma_{21} &= \beta_{2y} - S\beta_1 + R\beta_2, & \delta_{30} &= \gamma_{20y} - S\gamma_{10} + Q\gamma_{21} - 4\alpha_2\beta_1 - \alpha_1\beta_2.
\end{aligned}$$

The following statement describes explicitly equations (1) of the fourth type.

Theorem 1. *ODEs (1) of the fourth type satisfy the conditions*

$$\beta_1 \neq 0, \qquad J_0 = 0, \qquad j_0 = 0, \qquad j_1 \neq 0 \tag{4}$$

and have two algebraic basis invariants

$$I_1 = \frac{\Gamma_0}{\beta_1\sqrt{j_1}}, \qquad I_2 = \frac{5(2j_1j_3 + j_2^2)}{\sqrt{j_1}}. \tag{5}$$

Any invariant of equation (1) can be obtained from (5) by applying algebraic operations and the operators of invariant differentiation

$$\mathcal{D}_1 = \frac{1}{\sqrt{j_1}}(\beta_2\partial_x - \beta_1\partial_y), \qquad \mathcal{D}_2 = j_1^{1/4}\Big(\frac{5j_2}{2j_1}(\beta_2\partial_x - \beta_1\partial_y) - \frac{3}{\beta_1}\partial_x\Big).$$

Note that the hodograph transformation turns an equation (1) with $\beta_1 = 0$, $\beta_2 \neq 0$ to that with $\beta_1 \neq 0$. Criteria of the equivalence to the third Painlevé equation [2] can be stated in terms of the basis invariants (5) and invariant derivatives of I_1, namely, $\mathcal{D}_2 I_1$ and $\mathcal{D}_1^m I_1$. Instead of the latter invariants it is convenient to use the series of universal invariants introduced in [1]

$$J_1 = \frac{6}{5}\Big(\frac{1}{I_1^2} - 1\Big), \qquad J_{m+1} = (m + (m+1)J_1)J_m + \frac{1}{I_1}\mathcal{D}_1 J_m, \qquad m \in \mathbb{N}. \tag{6}$$

3. Equivalence to the fifth Painlevé equation

This section provides the necessary conditions of equivalence of an ODE of the second order to the fifth Painlevé equation. Here the case $\gamma = \delta = 0$ of PV reducible (via $z = \mathrm{e}^x$, $w = y$) to an autonomous ODE is not considered. The case of $\alpha = \beta = 0$, when PV reduces to PIII (see, e.g., [2]), is not studied separately.

Theorem 2. *If an ODE (1) of the fourth type is equivalent to the generic fifth Painlevé equation, then its invariants satisfy the conditions*

$$\det\frac{\partial(I_1, I_2)}{\partial(x, y)} \neq 0, \qquad I_{21} \equiv \mathcal{D}_2 I_1 \neq 0 \tag{7}$$

and there exists such solution

$$w = \eta(x, y), \qquad H = \zeta(x, y) \tag{8}$$

of the system

$$\begin{array}{l}
\mu_0 J_3 H^4 - 150(w+1)(w-1)^2 J_2 H^3 - 5(9w^2 - 10w + 9)(45w^2 - 94w + 45)J_1 H^2 \\
\qquad -15(w+1)(w-1)^2(375w^2 - 1006w + 375)H \\
\qquad +5625w^6 - 22590w^5 + 21495w^4 - 8548w^3 + 21495w^2 - 22590w + 5625 = 0, \\
7J_4 H^5 - 18(w+1)J_3 H^4 - 5(39w^2 - 50w + 39)J_2 H^3 \\
\quad -15(w+1)(53w^2 - 130w + 53)J_1 H^2 - 8(225w^4 - 390w^3 + 274w^2 - 390w + 225)H \\
\qquad +24(w+1)(75w^4 - 230w^3 + 358w^2 - 230w + 75) = 0, \\
5J_5 H^6 - 45(w+1)J_4 H^5 - 3(15w^2 - 106w + 15)J_3 H^4 + 900w(w+1)J_2 H^3 \\
\qquad +60w(87w^2 - 278w + 87)J_1 H^2 + 1440w(w+1)(9w^2 - 20w + 9)H \\
\qquad -32w(405w^4 - 630w^3 + 1106w^2 - 630w + 405) = 0.
\end{array} \tag{9}$$

that its substitution into relation

$$I_1 I_2 + \frac{3HM_1}{512w^3\mu_0} + \frac{1152 I_{21}^2 H^2 M_2}{5 I_1^9 M_0^2} = 0 \tag{10}$$

turns it into identity. And the substitution of (8) into

$$\begin{array}{l}
\dfrac{2^{24}3^5 5^2 I_{21}^2 H^3 \mu_0^3}{I_1^9 M_0^2}\alpha w^7(w-1)^2 + (6 - 5w)J_2 H^3 + 2(5w-3)(4-3w)J_1 H^2 \\
\qquad -(75w^3 - 176w^2 + 15w + 90)H + 75w^4 - 201w^3 + 67w^2 - 15w + 90 = 0,
\end{array}$$

$$\begin{array}{l}\frac{2^{24}3^5 5^2 I_{21}^2 H^3 \mu_0^3}{I_1^9 M_0^2}\beta w^2(w-1)^2+(5-6w)J_2H^3+2(3w-5)(4w-3)J_1H^2 \\ \qquad +(90w^3+15w^2-176w+75)H-90w^4+15w^3-67w^2+201w-75=0, \\ \frac{2^{24}3^5 5^4 I_{21}^2 H^3 \mu_0^2}{I_1^9 M_0^2}\frac{\gamma z w^7}{w-1}+7J_3H^4-25(w+1)J_2H^3-5(29w^2-50w+29)J_1H^2 \\ \qquad -40(w+1)(9w^2-28w+9)H+8(45w^4-110w^3+274w^2-110w+45)=0, \\ \frac{2^{28}3^5 5^4 I_{21}^2 H^3 \mu_0^2}{I_1^9 M_0^2}\frac{\delta z^2 w^7(w+1)}{(w-1)^2}-11J_3H^4+50(w+1)J_2H^3+5(37w^2-110w+37)J_1H^2 \\ \qquad +5(w+1)(81w^2-202w+81)H-405w^4+740w^3-1566w^2+740w-405=0\end{array} \tag{11}$$

leads to relations whence $z=\xi(x,y)$ and three constant parameters of PV can be found. This function $\xi(x,y)$ together with $\eta(x,y)$ in (8) define the change (2) relating ODEs (1) and (3).

In (9)–(11) we use the notation

$$\begin{array}{l}M_0=5(6J_2-7J_1^2)H^2\mu_1-2(5J_2+7J_1)H\mu_2+200w(5J_1+6)\mu_1+3(5J_1+6)H\mu_0\cdot \\ \quad\cdot[3000w(w+1)(w-1)^2-(105w^4-1220w^3-1706w^2-1220w+105)J_1H], \\ M_1=H\mu_1+2\mu_2+24w(J_1-5J_2)H^3\mu_0+(315w^4+60w^3-2798w^2+60w+315)H\mu_0 \\ \quad -1200w(w+1)(w-1)^2\mu_0, \\ M_2=(15J_1H^2\mu_1^2+12H\mu_1\mu_2-2\mu_2^2)I_1^2+36(45w^4-80w^3+326w^2-80w+45)H^2\mu_0\mu_1 \\ \quad +18(45w^4+20w^3+126w^2+20w+45)(45w^4-180w^3+526w^2-180w+45)H^2\mu_0^2 \\ \quad +216000w^2(w+1)(2(w-1)^2+(w+1)J_1H)H^3\mu_0^2, \qquad \mu_0=75w^2-158w+75, \\ \mu_1=90(w+1)(w-1)^2J_2H^3-40(9w^4+33w^3-92w^2+33w+9)J_1H^2-450(w+1)(w-1)^2\cdot \\ \quad\cdot(3w^2-2w+3)H-2025w^6+8910w^5-4935w^4-4412w^3-4935w^2+8910w-2025, \\ \mu_2=25(9w^4+168w^3-386w^2+168w+9)J_2H^3-300(w+1)(w-1)^2(3w^2-2w+3)J_1H^2 \\ \quad -4(3375w^6-5895w^5-21795w^4+48694w^3-21795w^2-5895w+3375)H \\ \quad +75(w+1)(w-1)^2(45w^4+540w^3-1874w^2+540w+45).\end{array}$$

Proof. Let find the relations on invariants of PV. The same relations hold for invariants of an ODE (1) equivalent to PV. Invariants (5), (6) and I_{21} of ODE (3) are given by

$$J_1=-\frac{2K_0K_2}{K_1^2},\quad J_2=\frac{2K_0^2K_3}{K_1^3},\quad J_3=-\frac{16K_0^3K_4}{K_1^4},\quad J_4=\frac{16K_0^4K_5}{K_1^5},\quad J_5=-\frac{32K_0^5K_6}{K_1^6}, \tag{12}$$

$$\begin{array}{l}\frac{I_{21}^2}{I_1^9}=-\frac{(w-1)^2\tilde{M}_0^2}{2^{15}3^4 5^2 w^2\mu_0^2K_1^7},\qquad \tilde{M}_0=5(3K_1K_3-7K_2^2)K_8+(7K_1K_2-5K_0K_3)K_7 \\ \qquad\qquad +100w(3K_1^2-5K_0K_2)K_8+3(3K_1^2-5K_0K_2)\mu_0\cdot \\ \quad\cdot[1500w(w+1)(w-1)^2K_1+(105w^4-1220w^3-1706w^2-1220w+105)K_2],\end{array} \tag{13}$$

$$\begin{array}{l}I_1I_2=\frac{K_1(w(w-1)^2\tilde{M}_2-4500\tilde{M}_1)}{3\cdot 2^{11}5^3w^3\mu_0K_0^3},\qquad \tilde{M}_2=\frac{6K_1K_8K_7-15K_2K_8^2-K_0K_7^2}{\mu_0(3K_1^2-5K_0K_2)} \\ +3(45w^4+20w^3+126w^2+20w+45)(45w^4-180w^3+526w^2-180w+45)\mu_0K_0 \\ \qquad +6(45w^4-80w^3+326w^2-80w+45)K_8+72000w^2(w+1)\mu_0\cdot \\ \cdot[(w-1)^2K_1-(w+1)K_2],\qquad \tilde{M}_1=2K_0K_7+K_1K_8-48w\mu_0(K_1K_2+5K_0K_3) \\ \quad +(315w^4+60w^3-2798w^2+60w+315)\mu_0K_0K_1-1200w(w+1)(w-1)^2\mu_0K_0^2,\end{array} \tag{14}$$

where we set

$$\begin{array}{l}K_0=A(3w-2)+B(3-2w)+\Gamma+\Delta(w+1), \\ K_1=-A(15w^2-15w+2)-B(2w^2-15w+15)+\Gamma(w+1)+\Delta(w^2+6w+1), \\ K_2=Aw(45w^2-60w+17)+B(17w^2-60w+45)+2\Gamma w+8\Delta w(w+1), \\ K_3=-Aw(315w^3-525w^2+231w-17)+B(17w^3-231w^2+525w-315) \\ \quad +2\Gamma w(w+1)+8\Delta w(w^2+6w+1),\end{array} \tag{15}$$

IOP Conf. Series: Journal of Physics: Conf. Series **1205** (2019) 012004 doi:10.1088/1742-6596/1205/1/012004

$$\begin{aligned}
K_4 &= Aw^2(315w^3-630w^2+378w-62)-B(62w^3-378w^2+630w-315)\\
&\quad+\Gamma w^2+16\Delta w^2(w+1),\\
K_5 &= -Aw^2(2835w^4-6615w^3+5040w^2-1320w+62)-B(62w^4-1320w^3+5040w^2\\
&\quad-6615w+2835)+\Gamma w^2(w+1)+16\Delta w^2(w^2+6w+1),\\
K_6 &= Aw^3(14175w^4-37800w^3+34965w^2-12720w+1382)+B(1382w^4-12720w^3\\
&\quad+34965w^2-37800w+14175)+2\Gamma w^3+128\Delta w^3(w+1),\\
K_7 &= 50(9w^4+168w^3-386w^2+168w+9)K_3+600(w+1)(w-1)^2(3w^2-2w+3)K_2\\
&\quad-4(3375w^6-5895w^5-21795w^4+48694w^3-21795w^2-5895w+3375)K_1\\
&\quad+75(w+1)(w-1)^2(45w^4+540w^3-1874w^2+540w+45)K_0,\\
K_8 &= 180(w+1)(w-1)^2K_3+80(9w^4+33w^3-92w^2+33w+9)K_2-450(w+1)(w-1)^2\cdot\\
&\quad\cdot(3w^2-2w+3)K_1-(2025w^6-8910w^5+4935w^4+4412w^3+4935w^2-8910w+2025)K_0,
\end{aligned}$$

$A=\alpha w^3(w-1)^3$, $B=\beta(w-1)^3$, $\Gamma=\gamma z w^3(w-1)$, $\Delta=4\delta z^2w^3$. For K_4, K_5, K_6 relations

$$\begin{aligned}
&16\mu_0K_4+300(w+1)(w-1)^2K_3-10(9w^2-10w+9)(45w^2-94w+45)K_2\\
&\quad+15(w+1)(w-1)^2(375w^2-1006w+375)K_1\\
&\quad-(5625w^6-22590w^5+21495w^4-8548w^3+21495w^2-22590w+5625)K_0=0,
\end{aligned} \tag{16}$$

$$\begin{aligned}
&56K_5+144(w+1)K_4-5(39w^2-50w+39)K_3-4(225w^4-390w^3+274w^2-390w+225)K_1\\
&\quad+15(w+1)(53w^2-130w+53)K_2+12(w+1)(75w^4-230w^3+358w^2-230w+75)K_0=0,
\end{aligned}$$

$$\begin{aligned}
&20K_6+90(w+1)K_5-6(15w^2-106w+15)K_4-225w(w+1)K_3+15w(87w^2-278w+87)K_2\\
&\quad-180w(w+1)(9w^2-20w+9)K_1+4w(405w^4-630w^3+1106w^2-630w+405)K_0=0
\end{aligned}$$

hold. Equalities (12) are solved for the values

$$K_2=-\frac{K_1^2J_1}{2K_0},\quad K_3=\frac{K_1^3J_2}{2K_0^2},\quad K_4=-\frac{K_1^4J_3}{16K_0^3},\quad K_5=\frac{K_1^5J_4}{16K_0^4},\quad K_6=-\frac{K_1^6J_5}{32K_0^5}. \tag{17}$$

With their substitution relations (16) take the form (9) while (13) becomes

$$\frac{I_{21}^2}{I_1^9}=-\frac{(w-1)^2M_0^2}{2^{19}3^45^2w^2H^2\mu_0^2K_1},\qquad H=\frac{K_1}{K_0}, \tag{18}$$

whence K_1 is expressed. Substituting K_1, (17) into (14) leads to (10). From (15) it follows

$$\begin{aligned}
&96Aw^2(w-1)\mu_0+2(5w-6)K_3+4(5w-3)(4-3w)K_2+(75w^3-176w^2+15w+90)K_1\\
&\quad-(75w^4-201w^3+67w^2-15w+90)K_0=0,\\
&96B(w-1)\mu_0+2(6w-5)K_3+4(3w-5)(4w-3)K_2-(90w^3+15w^2-176w+75)K_1\\
&\quad+(90w^4-15w^3+67w^2-201w+75)K_0=0,\\
&1200\Gamma w^2+56K_4+25(w+1)K_3-5(29w^2-50w+29)K_2+20(w+1)(9w^2-28w+9)K_1\\
&\quad-4(45w^4-110w^3+274w^2-110w+45)K_0=0,\\
&9600\Delta w^2(w+1)-176K_4-100(w+1)K_3+10(37w^2-110w+37)K_2\\
&\quad-5(w+1)(81w^2-202w+81)K_1+(405w^4-740w^3+1566w^2-740w+405)K_0=0.
\end{aligned}$$

Then on substituting (17) and K_1 expressed from (18), equalities (11) are obtained.

4. Application of theorem 2

When we have to examine a given ODE (1) for the equivalence to PV, first we should check the conditions (4), (7) for it. Then let note that combining relations (9) one can rewrite them in the form of the polynomials of the fourth degree in H

$$
\begin{aligned}
&P_4J_3H^4 - 150P_3J_2H^3 - 5P_2J_1H^2 - 15P_3P_1H + P_0 = 0,\\
&Q_4J_4H^4 - 30P_3Q_3J_3H^3 - 5Q_2J_2H^2 - 75P_3Q_1J_1H + Q_0 = 0,\\
&R_4J_5H^4 - 300P_3R_3J_4H^3 - 5R_2J_3H^2 - 105P_3R_1J_2H - R_0J_1 = 0,
\end{aligned} \quad (19)
$$

$$
\begin{aligned}
&P_0 = 5625w^6 - 22590w^5 + 21495w^4 - 8548w^3 + 21495w^2 - 22590w + 5625,\\
&P_1 = 375w^2 - 1006w + 375, \quad P_2 = (9w^2 - 10w + 9)(45w^2 - 94w + 45),\\
&P_3 = (w+1)(w-1)^2, \quad P_4 = 75w^2 - 158w + 75, \quad Q_0 = 2048w^3R_4,\\
&Q_1 = 1575w^6 - 8442w^5 - 24759w^4 + 73492w^3 - 24759w^2 - 8442w + 1575,\\
&Q_2 = 23625(w^8+1) - 142380(w^7+w) + 291060(w^6+w^2) - 350676(w^5+w^3) + 358790w^4,\\
&Q_3 = 1125w^4 - 3012w^3 + 2558w^2 - 3012w + 1125, \quad Q_4 = P_0,\\
&R_0 = 543375(w^8+1) - 3764250(w^7+w) + 8709120(w^6+w^2) - 9247110(w^5+w^3) + 7516706w^4,\\
&R_1 = 5175w^4 - 28950w^3 + 48254w^2 - 28950w + 5175,\\
&R_2 = 40365w^6 - 241380w^5 + 592083w^4 - 782072w^3 + 592083w^2 - 241380w + 40365,\\
&R_3 = 69w^2 - 151w + 69, \quad R_4 = 5175w^4 - 22650w^3 + 34966w^2 - 22650w + 5175.
\end{aligned}
$$

From (19) vanishing of three quadratic polynomials in H follows:

$$
5U_6H^2 + 15P_3U_5H + U_4 = 0, \quad 6P_3U_4H^2 + U_3H + 3P_3U_2 = 0, \quad U_2H^2 + 30P_3U_1H + 5U_0 = 0, \quad (20)
$$

$$
\begin{aligned}
&U_0 = -S_0R_2J_3 + 7S_1R_1J_2 + S_4R_0J_1, \quad S_0 = 5P_0Q_1J_1 - Q_0P_1, \quad S_1 = P_0Q_2J_2 - Q_0P_2J_1,\\
&U_1 = -10S_0R_3J_4 + 7S_2R_1J_2 + S_5R_0J_1, \quad S_2 = P_0Q_3J_3 - 5Q_0J_2, \quad S_3 = P_0Q_4J_4 - Q_0P_4J_3,\\
&U_2 = S_0R_4J_5 - 7S_3R_1J_2 - S_6R_0J_1, \quad S_4 = P_1Q_2J_2 - 5Q_1P_2J_1^2, \quad S_5 = P_1Q_3J_3 - 25Q_1J_1J_2,\\
&U_3 = S_1R_4J_5 - S_3R_2J_3 - S_8R_0J_1, \quad S_6 = P_1Q_4J_4 - 5Q_1P_4J_1J_3, \quad S_7 = P_2Q_3J_1J_3 - 5Q_2J_2^2,\\
&U_4 = S_2R_4J_5 - 10S_3R_3J_4 - S_9R_0J_1, \quad S_8 = P_2Q_4J_1J_4 - Q_2P_4J_2J_3, \quad S_9 = 5Q_4J_2J_4 - Q_3P_4J_3^2,\\
&U_5 = -S_5R_4J_5 + 10S_6R_3J_4 - 7S_9R_1J_2, \quad U_6 = -S_7R_4J_5 + 10S_8R_3J_4 - S_9R_2J_3.
\end{aligned}
$$

Values $U_0, \ldots, U_6$ satisfy the relations $U_0U_4 - U_1U_3 = U_2\tilde{U}_2$, $U_2U_6 - U_3U_5 = U_4\tilde{U}_4$ with $\tilde{U}_2 = 10S_1R_3J_4 - S_2R_2J_3 - S_7R_0J_1$, $\tilde{U}_4 = S_4R_4J_5 - S_6R_2J_3 + 7S_8R_1J_2$. Elimination of H from the equalities (20) leads to relations

$$
\begin{aligned}
&U_4[4050P_3^4U_2U_5^2 + 9P_3^2(4U_4^3 - 10U_3U_4U_5 - 20U_2U_4U_6 + 25U_2\tilde{U}_4U_6) + 5U_3^2U_6] = 0,\\
&U_2[16200P_3^4U_1^2U_4 + 9P_3^2(U_2^3 - 10U_1U_2U_3 - 20U_0U_2U_4 + 100U_0\tilde{U}_2U_4) + 5U_0U_3^2] = 0.
\end{aligned} \quad (21)
$$

Note that $J_1, \ldots, J_5$, $I_{21}^2I_1^{-9}$, I_1I_2 in (9)–(11), (19) are invariants of (1) depending on x, y.

Let both relations (21) have common factor, whence $w = \eta(x,y)$ can be found. With this substitution we try to factorize (20). If the left-hand sides of (20) have common factor, then one can find the expression for $H = \zeta(x,y)$. Otherwise system (19) (or, what is the same, system (9)) have no an appropriate solution (8), which means that ODE (1) is not equivalent to PV.

Substitution of (8) into relations (9), (10) should turn them into identities, while relations (11) should reduce to the system, whence one can find expression for $z = \xi(x,y)$ and constant parameters of equation (3). Nonconsistency of this system means that the considered ODE (1) is not equivalent to PV. If the system is consistent, then we have the change of variables (2) and we need only to verify that it relates the equation (3) and given ODE (1).

References

[1] Bagderina Yu Yu 2015 Equivalence of second-order ordinary differential equations to Painlevé equations *Theor. Math. Phys.* **182** 211–30

[2] Bagderina Yu Yu and Tarkhanov N N 2015 Solution of the equivalence problem for the third Painlevé equation *J. Math. Phys.* **56** 013507

[3] Bagderina Yu Yu 2013 Invariants of a family of scalar second-order ordinary differential equations *J. Phys. A: Math. Theor.* **46** 295201

IOP Conf. Series: Journal of Physics: Conf. Series **1205** (2019) 012005 doi:10.1088/1742-6596/1205/1/012005

On the application of systems of functions of special kind in mathematical physics

A N Barmenkov[1] **and N A Barmenkov**[1]

[1]National Research Nuclear University MEPhI (Moscow Engineering Physics Institute), 31 Kashirskoe shosse, 115409 Moscow, Russia

E-mail: anbarmenkov@mail.ru

Abstract. In the article considers the application of functional sequences of special kind for study of some boundary value problems of equations of mathematical physics. This sequences may be used in modelling of physical processes.

1. Introduction

Consider of functional sequences of special kind

$$\left\{Re[W(x)]^n;\ Im[W(x)]^n\right\}_{n=0}^{\infty}, \tag{1}$$

where $W(x)$ continuous complex-valued function of $x \in [a, b]$. In [1-2], the basic properties such sequences are proved. In [3] Minzoni A.A.proved completeness in $L_2[0; 2\pi]$ system of functions of special kind

$$\left\{e^{-np(x)}\cos(nt);\ e^{-np(x)}\sin(nt)\right\}_{n=0}^{\infty} \tag{2}$$

where $p(x)$ real function $p(x) \in C^{1,\alpha}$. The proof is based on the fact that this sequence is applied to the solution of the following boundary value problem of (see [4])

$$\triangle w(x, y) = 0, \tag{3}$$

$$w(x, p(x)) = f(x)$$

$$w(0, y) = w(2\pi, y)$$

$$w(x, y) - boundary,$$

as $y \to +\infty$. The solution of this problem is given in the form of a series

$$w(x, y) = \sum_{n=0}^{\infty}(A_n e^{-ny}\cos(nx) + B_n e^{-ny}\sin(nx)),$$

where the coefficients satisfy the relation

$$f(x) = w(x, p(x)) = \sum_{n=0}^{\infty} (A_n e^{-np(x)} \cos(nx) + B_n e^{-np(x)} \sin(nx)) \tag{4}$$

Moreover, the function $p(x)$ is subject to rather rigid conditions $p(x) \in C^{1,\alpha}[0; 2\pi]$. In fact, the sequence (2)(see Minzoni[3]) is a system of the form (1) for
$W(x) = e^{-p(x)+ix}$, i.e. $\{e^{-np(x)} \cos(nx); e^{-np(x)} \sin(nx)\}_{n=0}^{\infty}$, $x \in [0; 2\pi]$

2. Conditions of completeness of functional sequences in this kind in different function spaces

In fact, this result follows from the works of [1],[2], proved in a different way earlier than [3] and under more general conditions for the $p(x)$ function. The completeness of the system of functions (1) in different spaces is proved in [1], and the minimality in [2]. In [2], the minimality is proved and the sequence biorthogonally conjugate to (1) is given explicitly. The image of the segment $[0; 2\pi]$ when $W(x) = e^{-p(x)+ix}$, $x \in [0; 2\pi]$ displayed is the curve $\Gamma = W[0; 2\pi]$. In article [2] it is shown how the values of the interior angles at the corner points of the closed contour $\Gamma = W[0; 2\pi]$ are influenced by the "p" spaces $L_p[0; 2\pi]$, in which the sequence (2) is complete and minimal. In the simulation, the case when Γ is a smooth contour is most often used, because useful

Theorem 1. *Let $p(x)$ be a smooth real function, then*

1) if $p(0) = p(2\pi)$, then the sequence (2) is complete and minimal in $L_2[0; 2\pi]$

2) if $p(0) \neq p(2\pi)$, then the sequence (2) is densely complete in $L_2[0; 2\pi]$

Note that the sequence is densely complete in the corresponding space , which means that it remains complete in this space after removing any finite number of its elements.

If $p(0) \neq p(2\pi)$,then the contour Γ ($\Gamma = W[0; 2\pi]$) is open and the sequence (2) is densely complete in in any space $L_p[0; 2\pi]$, $p > 1$. If $p(0) = p(2\pi)$, then the contour Γ is closed, and the point 0 (origin) lies in the finite region bounded by this contour Γ, then this sequence (2) is complete and minimal in various functional spaces depending on the values of the angles at the angular points of the contour Γ (see [2]). At least for a smooth closed contour $\Gamma = W[0; 2\pi]$ the sequence (2) is complete and minimal in any space $L_p[0; 2\pi]$, $p > 1$.

Hence, the statement of Minzoni ([3], 1986) is a special case of the results ([1], 1977), ([2], 1981).

The following properties of the boundary value problem (3) of differential equations of mathematical physics are shown.

Boundary condition $p(0) \neq p(2\pi)$ of the problem (3) means the dense completeness of the sequence (2)in $L_2[0; 2\pi]$. Boundary condition $p(0) = p(2\pi)$ of the problem (3) means the completeness and minimality of the sequence (2)in $L_2[0; 2\pi]$.

Moreover, in this case, if equality (4) holds, then the article offers an effective way of calculating the coefficients A_n, B_n of this series. This method is analogous to the method for calculating the coefficients of a Fourier series in the classical trigonometric system. Namely, for the biorthogonal system to the sequence (2) the system $\{\varphi_n(t); \psi_n(t)\}$, $n = 0, 1, ...$

$\varphi_0(t) = \frac{1}{2\pi i} \frac{\chi\prime[W(t)]W\prime[(t)]}{\chi[W(t)]}$

$\varphi_n(t) = \frac{\chi\prime[W(t)]W\prime(t)}{\chi[W(t)]} \sum_1^n [C_k^{(n)} \chi^k(W(t)) - \overline{C}_k^{(n)} \chi^{-k}(W(t))]$,

IOP Conf. Series: Journal of Physics: Conf. Series **1205** (2019) 012005 doi:10.1088/1742-6596/1205/1/012005

$\psi_n(t) = -i\frac{\chi\prime[W(t)]W\prime(t)}{\chi[W(t)]}\sum_1^n[C_k^{(n)}\chi^k(W(t)) + \overline{C}_k^{(n)}\chi^{-k}(W(t))],$

where the function $\chi(w)$ conform maps the domain $Int\Gamma$ onto the unit disc $U = \{\zeta \in C\ |\zeta| < 1\}$, where $\chi(0) = 0$, $\chi\prime(0) > 0$ and $\chi\prime(W(t))$ is the angular boundary value of the derivative $\chi\prime(w))$ at the point $w = W(t)$.

The sequence of finite sets of $\{C_k^{(n)}\}_{k=1}^n$, $n = 0, 1, 2, ...$ complex constants and the form of the biorthogonal system are given in [2]. Then the coefficients A_n, B_n of the series (by the usual Fourier series formulas $A_n = \int_a^b f(t)\varphi_n(t)dt$, $B_n = \int_a^b f(t)\psi_n dt$) are calculated.

References

[1] Kazmin J A 1977 Closure of the linear shell of one system of functions *Siberian mathematical journal* **18** N4 R 799–805

[2] Barmenkov A N 1981 Minimality and orthogonality one system of functions *Siberian mathematical journal* **22** N1 8–26

[3] Minzoni A A 1986 On the completeness of the functions e-np(x) cos nx, e-np(x) sin nx for n≥0 and p(x) a 2π periodic function *Stadies in Applaid Mathematic* **75** N3 265–266

[4] Vequa I N 1959 Common analytical functions *M. Fizmatgiz*

IOP Conf. Series: Journal of Physics: Conf. Series **1205** (2019) 012006 doi:10.1088/1742-6596/1205/1/012006

Refinement of the Reactor Dynamics Mathematical Model

A V Baskakov and N P Volkov

National Research Nuclear University MEPhI, Kashirskoe shosse, 31,115409 Moscow, Russia

e-mail: npvolkov@mephi.ru

Abstract. A mathematical model of reactor dynamics described by a system of integro-differential equations consisting of an unsteady-state multivelocity kinetic transfer equation and a delayed neutrons balance equation is considered. To refine some parameters of this system a number of inverse problems of reactor dynamics with a new definition of integral form are solved. Statements of the linear inverse problem of recovery of a part of the function of inner neutron sources, and non-linear inverse problems of recovery of a part of the absorption coefficient are presented. The sought for parts of the said parameters are functions depending on the velocity and time variables. Theorems of existence and uniqueness of generalized solutions have been proved.

1. Introduction

Currently, construction of more and more nuclear generating stations calls for reliable, simple to control and safe nuclear power plants. This is an additional driver for the development of the fairly young and continuously evolving mathematical theory of nuclear reactors. This is evidenced by a number of works [1-14] dedicated to these topics. Researchers, together with engineers, aim to create such reactor models the operation of which has least dependence on the human factor. The issues on the agenda are the mathematical modeling of processes in nuclear reactors. To develop mathematical models researchers should realize that the nuclear reactor is to be considered a complex conglomerate of interacting processes; the contribution of each process in this or that case can hardly be neglected. However, our task is not to construct a holistic mathematical model describing the nuclear reactor as an object on the whole. The model under study is the reactor dynamics system given in [5]. Our task is to refine several parameters of this system using additional measurements of occurring processes. A refined model is required: 1) to have a better insight into how a specific plant works, what its structure and properties are; 2) to learn to control this object, or the processes running inside this object, and determine the best methods of control within specified objectives and criteria; 3) to predict the behavior of the reactor and evaluate the consequences of different methods and forms of action on this object.

2. Statement of linear inverse problem

Let us first examine a mathematical model of reactor dynamics without considering temperature feedback from [4], described by the system of integro-differential equations:

1) by the non-stationary anisotropic multi-speed kinetic transfer equation

$$\frac{\partial u}{\partial t}(x,v,t)+(v,\nabla_x)u(x,v,t)+\Sigma(x,v,t)u(x,v,t)=\sum_{k=1}^{N} z_k R_k(x,v,t)+\int_V J(x,v,v',t)u(x,v',t)dv'+$$
$$+F(x,v,t)\ ; \qquad (1)$$

2) by the delayed neutrons balance equation

$$\frac{\partial R_k}{\partial t}(x,v,t)=-z_k R_k(x,v,t)+\int_V J_k(x,v,v',t)u(x,v',t)dv' \quad \forall\ k=\overline{1,N}, \qquad (2)$$

where $(x,v,t)\in D=G\times V\times[0,T]$.

In this model the function $u(x,v,t)$ describes the density distribution of neutrons flying through point $x\in G$ with velocity $v\in V$ at the time moment $t\in[0,T]$. Functions $\Sigma(x,v,t)$, $J(x,v,v',t)$, $F(x,v,t)$ specify properties of the medium in which the mass transfer process is occurring. $\Sigma(x,v,t)$, at this, is the absorption coefficient, $J(x,v,v',t)$ is the dispersion index, $F(x,v,t)$ is the density of the radiation source. The kernels of the scattering integrals $J_k(x,v,v',t)$ specify densities of distribution of secondary neutrons, and the function $R_k(x,v,t)$ is the distribution density of the supports of the delayed neutrons of the k-th group. Strongly convex region G is the region of variation of space coordinates; the closed set V lying in the ball layer $\{0<v_0\le|v|\le v_1<\infty\}$ is the set of variation of velocities v of emitted particles.

Let us make a natural assumption that the process under consideration occurs in the absence of an external neutron source, i.e. the reactor walls to not transmit external radiation. Mathematically this is expressed by the following boundary condition:

$$u(x,v,t)=0, \qquad (x,v,t)\in\gamma_-\times[0,T], \qquad (3)$$

where $\gamma_-=\{(x,v)\in\partial G\times V:(v,n_x)<0\}$, and n_x is the outer normal to the boundary ∂G of region G at the point x.

In addition, let the initial conditions be specified:

$$u(x,v,0)=\varphi(x,v), \qquad (x,v)\in\overline{G}\times V, \qquad (4)$$

$$R_k(x,v,0)=R_{k0}(x,v), \quad \forall\ k=\overline{1,N}, \quad (x,v)\in\overline{G}\times V. \qquad (5)$$

Let us now turn to the statement of linear the inverse problem stating the over-specifying condition in the form:

$$\int_G \omega_{x^0}(x)u(x,v,t)\,dx=\chi(v,t), \quad (v,t)\in V\times[0,T]. \qquad (6)$$

From the physical viewpoint condition (6) means the internal measurement of the neutron population in a certain vicinity of the point $x^0\in G$. The finite function $\omega_{x^0}\in L_2(G)$, called the «instrument function» and specifying parameters of the instrument used to make the measurements at the point x^0 has a pronounced maximum. The support of this function, at this, $\mathbf{supp}\,\omega_{x^0}\subset G$.

Assume that it is necessary to refine the function of internal sources $F(x,v,t)$ in equation (1). On the basis of condition (6) look for this function in the following form:

$$F(x,v,t)=f(v,t)g_1(x,v,t)+h_1(x,v,t), \qquad (7)$$

where $f(v,t)$ is the sought for part of the source function, and $g_1(x,v,t)$, $h_1(x,v,t)$ are the a priori prescribed functions (correction function).

Inverse problem 1. To find the triple of functions $\{u(x,v,t), R_k(x,v,t), f(v,t)\}$, satisfying conditions (1)-(6), where $F(x,v,t)$ has form (7).

3. Solution of inverse problem 1

From the viewpoint of inversion theory inverse problem 1 is linear.

For further studies consider the following classes of functions examined in [6]:

1) $H_\infty(D)$ is the space of function u, lying in the class $L_\infty(D)$ together with its generalized derivatives u_t and $(v,\nabla_x)u$ on D and possessing traces $u_{|\Gamma_-}$ on $\Gamma_- = \gamma_- \times (0,T)$ from $L_\infty(\Gamma_-)$, i.e.

$$H_\infty(D) = \left\{ u \in L_\infty(D) : \frac{\partial u}{\partial t}, (v,\nabla_x)u \in L_\infty(D), u_{|\Gamma_-} \in L_\infty(\Gamma_-) \right\},$$

which is Banach relative to the norm

$$\|u\|_{H_\infty} = \left[\|u\|_{\infty,D} + \left\| \frac{\partial u}{\partial t} \right\|_{\infty,D} + \|(v,\nabla_x)u\|_{\infty,D} + \left\| u_{|\Gamma_-} \right\|_{\infty,\Gamma_-} \right] < \infty.$$

2) $L^t_\infty(D)$ is the space of functions R_k, belonging to the class $L_\infty(D)$ together with its generalized derivatives $\frac{\partial R_k}{\partial t}$ on D, i.e. $L^t_\infty(D) = \left\{ R_k \in L_\infty(D) : \frac{\partial R_k}{\partial t} \in L_\infty(D) \right\}$, which is Banach relative to the norm

$$\|R_k\|_{L^t_\infty} = \left[\|R_k\|_{\infty,D} + \left\| \frac{\partial R_k}{\partial t} \right\|_{\infty,D} \right] < \infty .$$

Here $\|\bullet\|_{\infty,D}$ is the norm in the space $L_\infty(D)$.

Definition 1. By *generalized solution of inverse problem 1* we mean the triple of functions $\{u(x,v,t), R_k(x,v,t), f(x,v)\}$, almost everywhere satisfying equations (1), (2) and conditions (3)-(6) and such, that $u \in H_\infty(D)$, $R_k \in L^t_\infty(D)$ and $f \in L_\infty(V \times [0,T])$.

Sufficient conditions of generalized solvability of the stated inverse problem 1 were obtained in the following theorem.

Theorem 1. Let $\Sigma \in L^t_\infty(D)$; $J \in L^t_\infty(D \times V)$; φ, $(v,\nabla_x)\varphi \in L_\infty(G \times V)$, $\varphi_{|\gamma_-} \in L_\infty(\gamma_-)$; $g_I, (v,\nabla_x)g_I \in L_\infty(D)$, $\left| \int_G \omega_{x^0}(x) g_I(x,v,t)\, dx \right| \geq g_{0I} > 0$; $h_I \in L^t_\infty(D)$; $J_k \in L^t_\infty(D \times V)$; $R_{k0} \in L_\infty(G \times V)$; $\forall\ k = \overline{1,\mathrm{N}}$; $\chi \in L^t_\infty(V \times [0,T])$; *and the matching conditions are fulfilled:*

$$\varphi(x,v) = 0 \text{ at } (x,v) \in \gamma_-, \text{ and } \int_G \omega_{x^0}(x)\varphi(x,v)\, dx = \chi(v,0), \text{ at almost all } v \in V. \tag{8}$$

Then inverse problem 1 has a unique generalized solution in the sense of definition 1.

Proof. To prove the theorem we used the method proposed in [8-10] for a kinetic transfer equation.

After integration by characteristics the system of equations (1)-(7) is reduced to the system of two integral equations:

$$u(y+v\xi,v,t) = \mathcal{A}(y,v,\xi,t) + \int_\eta^\xi \left[(Pu)(y+v\theta,v,\theta-\xi+t) + f(v,t) g_I(y+v\theta,v,\theta-\xi+t) + \right.$$
$$\left. + h_I(y+v\theta,v,\theta-\xi+t) d\theta \right], \tag{9}$$

$$f(v,t)=\left[\int_G\int_{\zeta_-}^{\zeta_+}\omega_{x^0}(y+v\xi)g_1(y+v\xi,v,t)\,d\xi dy\right]^{-1}\left\{\frac{\partial}{\partial t}\chi(v,t)+\int_G\int_{\zeta_-}^{\zeta_+}\omega_{x^0}(y+v\xi)\big[(v,\nabla)\varphi(y+v(\xi-t),v)-\right.$$

$$-(P\varphi+h_1)(y+v(\xi-t),v,0)-$$

$$-\int_{\xi-t}^{\xi}\left[\frac{\partial}{\partial t}(Pu+h_1)(y+v\theta,v,\theta-\xi+t)+\int_0^t f(v,\tau)(v,\nabla)g_1(y+v(\tau+\xi-t,),v,\tau)d\tau\right]-$$

$$\left.-\sum_{k=1}^N z_k\left[R_{k0}(y+v\xi,v)\exp\{-z_kt\}+z_k\int_{\zeta_-}^{\xi}R_{k0}(y+v\theta,v)\exp\{-z_k(\theta-\xi+t)\}d\theta-\right\}d\xi dy, \qquad (10)$$

where $y+v\xi=x$ is a characteristic straight line with a directing vector v, ξ is the parameter from segment $[\zeta_-,\zeta_+]$, at this $y+v\zeta_\pm\in\partial G$, and $(y+v\zeta_-,v)\in\gamma_-$;

$$\mathcal{A}(y,v,\xi,t)=\varphi(y+v(\xi-t),v)+\sum_{k=1}^N z_k\int_{\xi-t}^{\xi}R_{k0}(y+v\theta,v)\exp\{-z_k(\theta-\xi+t)\}d\theta, \qquad \eta=\xi-t$$

at $0\le t\le\xi-\zeta_-$, $\xi\in[\zeta_-,\zeta_+]$

and $\quad \mathcal{A}(y,v,\xi,t)=\sum_{k=1}^N z_k\int_{\zeta_-}^{\xi}R_{k0}(y+v\theta,v)\exp\{-z_k(\theta-\xi+t)\}d\theta$, $\quad\eta=\zeta_-$ at $\xi-\zeta_-<t\le T$,

$\xi\in[\zeta_-,\zeta_+]$;

$$(Pu)(y+v\xi,v,t)\equiv\sum_{k=1}^N z_k\int_0^t\exp\{-z_k(t-\tau)\}\int_V J_k(y+v\xi,v,v',t)u(y+v\xi,v',t)dv'd\tau+$$

$$+\int_V J(y+v\xi,v,v',t)u(y+v\xi,v',t)dv'-\Sigma(y+v\xi,v,t)u(y+v\xi,v,t).$$

It is easy to prove that the solution of inverse problem 1 will be the solution of the system of integral equations (9)-(10) and, conversely, the solution of the system of integral equations (9)-(10) is the solution of inverse problem 1, i.e. fulfills conditions (1)-(6) in the above said classes. Hence, inverse problem 1 has a unique generalized solution $u\in H_\infty(D)$, $R_k\in L_\infty^t(D)$ и $f\in L_\infty(V\times[0,T])$ when, and only when, the system of integral equations (9), (10) is unambiguously solvable in the specified class of functions. And this is possible if condition $\left|\int_G\omega_{x^0}(x)g_1(x,v,t)\,dx\right|\ge g_{01}>0$, where g_{01} as a certain constant is satisfied.

Further on, go over to investigating the solvability of the system of integral equations (9), (10). For this write it down in the operator form:

$$\{u,f\}=\mathcal{A}\{u,f\}=\{\mathcal{A}_1\{u,f\},\mathcal{A}_2\{u,f\}\},$$

where $\mathcal{A}_1$, $\mathcal{A}_2$ are the operators within the range of definition $H_\infty(D)\times L_\infty(V\times[0,T])$, acting on the pair of functions $\{u,f\}$ according to the rule defined by the right-hand sides of equations (9), (10) respectively. At this, the operator $\mathcal{A}$ continuously reflects the Banach space $H_\infty(D)\times L_\infty(V\times[0,T])$

into itself. As image $\mathcal{A}_1\{u,f\}$ has generalized derivatives belonging to the space $L_\infty(\boldsymbol{D})$, and its trace is an element of space $L_\infty(\Gamma_-)$. The image $\mathcal{A}_2\{u,f\}$ belongs to the space $L_\infty(\boldsymbol{V}\times[0,\boldsymbol{T}])$.

Investigating contractiveness of operator $\mathcal{A}$ we have that its second degree will be the contraction operator, providing $\left|\int_G \omega_{x^0}(x)g_1(x,v,t)\,dx\right| \ge g_{01} > 0$.

Therefore, by virtue of the Banach principle about the fixed point there exists a unique solution $\{u,f\}$ of the system of integral equations (9)- (10) from the space $H_\infty(\boldsymbol{D})\times L_\infty(\boldsymbol{V}\times[0,\boldsymbol{T}])$; which means that the inverse problem 1 also has a unique generalized solution $\{u,f\}$ at the found restrictions on the initial data of the problem, i.e. sufficient conditions of the unique generalized solution of inverse problem 1 have been met.

So, theorem 1 is completely proved.

Hence, on the basis of theorem 1 we can refine the density of internal sources, and, consequently, the entire mathematical model of reactor dynamics represented by equations (1), (2).

Remark 1. In theorem 1 conditions $\Sigma\in L^t_\infty(\boldsymbol{D})$; $J\in L^t_\infty(\boldsymbol{D}\times\boldsymbol{V})$, $h_1\in L^t_\infty(\boldsymbol{D})$ and $J_k\in L^t_\infty(\boldsymbol{D}\times\boldsymbol{V})$, $\forall\ k=\overline{1,N}$ can be replaced with conditions Σ, $(v,\nabla)\Sigma\in L_\infty(\boldsymbol{D})$, J, $(v,\nabla)J\in L_\infty(\boldsymbol{D}\times\boldsymbol{V})$, h_1, $(v,\nabla)h_1\in L_\infty(\boldsymbol{D})$, J_k, $(v,\nabla)J_k\in L_\infty(\boldsymbol{D}\times\boldsymbol{V})$ $\forall\ k=\overline{1,N}$, respectively. The results of theorem 1 will still stand, at this.

4. Nonlinear inverse problem

Now assume that we need to refine the absorption coefficient (the function of a complete macroscopic neutron cross-section) $\Sigma(x,v,t)$ in equation (1). Using conditions (6), look for this function in the following form:

$$\Sigma(x,v,t)=\sigma(v,t)g_2(x,v,t)+h_2(x,v,t), \tag{11}$$

where $\sigma(v,t)$ is the sought for part of function $\Sigma(x,v,t)$, and $g_2(x,v,t)$, $h_2(x,v,t)$ are the a priori prescribed functions (correction functions).

Inverse problem 2. To find the triple of functions $\{u(x,v,t),R_k(x,v,t),\sigma(v,t)\}$, satisfying conditions (1)-(6), where $\Sigma(x,v,t)$ has form (11).

From the view point of inverse theory inverse problem 2 is nonlinear.

Sufficient conditions of unambiguous solvability of inverse problem 2 have been obtained in the following theorem.

Theorem 2. *Let* $h_2\in L^t_\infty(\boldsymbol{D})$; $J\in L^t_\infty(\boldsymbol{D}\times\boldsymbol{V})$; $F\in L^t_\infty(\boldsymbol{D})$; φ, $(v,\nabla_x)\varphi\in L_\infty(\boldsymbol{G}\times\boldsymbol{V})$, $\varphi|_{\gamma_-}\in L_\infty(\gamma_-)$; $g_2\in L^t_\infty(\boldsymbol{D})$, $\left|\int_G \omega_{x^0}(x)g_2(x,v,t)\,dx\right|\ge g_{02}>0$; $J_k\in L^t_\infty(\boldsymbol{D}\times\boldsymbol{V})$; $R_{k0}\in L_\infty(\boldsymbol{G}\times\boldsymbol{V})$; $\forall\ k=\overline{1,N}$; $\chi\in L^t_\infty(\boldsymbol{V}\times[0,\boldsymbol{T}])$, $|\chi(v,t)|\ge\chi_0>0$; *and the matching conditions are fulfilled:*

$\varphi(x,v)=0$ *at* $(x,v)\in\gamma_-$, *and* $\int_G \omega_{x^0}(x)\varphi(x,v)\,dx=\chi(v,0)$, *at almost all* $v\in\boldsymbol{V}$.

Then inverse problem 2 has a unique generalized solution in the sense of definition 1 on a certain closed restricted set at a certain restriction on v_0 *and on* diam$\boldsymbol{G}$.

Sketch of the proof. Similar to the proof of theorem 1, reduce inverse problem 2, set by equations (1)-(6) under condition (11), to a system of integral equations. Providing $v_0^{-1}\text{diam}\boldsymbol{G} \le \boldsymbol{b}$, where $\boldsymbol{b}$ is a certain constant, the inverse problem 2 has a unique generalized equation when, and only when, the obtained system of integral equations of the second kind has a unique solution on the following set $\mathbf{U}\times\mathbf{\Omega} = \left\{\left\{u,\sigma\right\}: u \in H_\infty(\boldsymbol{D}),\ \sigma \in L_\infty(V\times[0,\boldsymbol{T}]),\ \|u\|_{H_\infty} \le \boldsymbol{k}_1,\ \|\sigma\|_{\infty,\boldsymbol{D}} \le \boldsymbol{k}_2\right\}$, where the constants $\boldsymbol{k}_1$, $\boldsymbol{k}_2$ depend on the norms of initial data of the problem. Examining the obtained system of integral equation further in the operator form proves that the second degree of the right hand sides of these systems will be a contraction operator on the set $\mathbf{U}\times\mathbf{\Omega}$.

This means that the system of integral equation has a unique solution belonging to the set $\mathbf{U}\times\mathbf{\Omega}$, and inverse problem 2 has a unique generalized solution. So, the inverse problem makes it possible to refine the absorption coefficient $\Sigma(\boldsymbol{x},\boldsymbol{v},t)$ in equation (1).

Remark 2. The results of theorem 2 will remain the same, if the conditions $J \in L^l_\infty(\boldsymbol{D}\times V)$, $F \in L^l_\infty(\boldsymbol{D})$, $h_2 \in L^l_\infty(\boldsymbol{D})$, $J_k \in L^l_\infty(\boldsymbol{D}\times V)$, $\forall\ \boldsymbol{k} = \overline{1,\mathrm{N}}$ are replaced by J, $(\boldsymbol{v},\nabla)J \in L_\infty(\boldsymbol{D}\times V)$, F, $(\boldsymbol{v},\nabla)F \in L_\infty(\boldsymbol{D})$, h_2, $(\boldsymbol{v},\nabla)h_2 \in L_\infty(\boldsymbol{D})$, J_k, $(\boldsymbol{v},\nabla)J_k \in L_\infty(\boldsymbol{D}\times V)$ $\forall\ \boldsymbol{k} = \overline{1,\mathrm{N}}$ respectively.

So, the work considers linear and nonlinear inverse problems for the system of reactor dynamics described by equations (1), (2). The proposed methods allow, at this, not only to synthesize the sought for parts of investigated parameters, but also to define phase conditions of the object under the study corresponding to them.

It should also be noted that as the proofs of the above given theorems inherently comprise the method of successive approximations this makes it possible to constructively create solutions of the studied inverse problem 1. The proposed methods make it possible to construct numerical algorithms of recovery of the said functions.

The work has been done within the framework of the Competitiveness Enhancement Program of the National Research Nuclear University MEPhI (Moscow Engineering and Physics Institute); № 02.a03.21.0005,27.08.2013.

References

[1] Marchuk G I 1961 *Methods for calculating nuclear reactors* (Moscow) 667 (In Russian)
[2] Albertoni S and Montagnini B 1966 On the spectrum of neutron transport equation in finite bodies *J. Math. Analys. Appl.* **13** 19
[3] Akcasu Z, Lellouche G S and Shotkin L M 1971 *Mathematical Methods in Nuclear Reactor Dynamics* (New York – London: Academic Press) 465
[4] Lewins J 1978 *Nuclear Reactor Kinetics and Control* (Pergamon Press) 264 Jefery Lewins
[5] Kryanev A V and Shikhov S B 1983 *Problems of nuclear reactor mathematical theory (Non-linear analysis)* (Moscow: Energoatomizdat) 280 (In Russian).
[6] Germogenova T A 1986 *Local properties of solutions of the transport equation* (Moscow: Nauka) 272 (In Russian)
[7] Kuznetsov Yu A and Morozov S F 1974 System of integro-differential equations for nuclear reactor kinetics *Differential equations* **10** 1491-1503 (In Russian)
[8] Prilepko A I and Volkov N P 1988 Inverse problems of finding parameters of a nonstationary kinetic transport equation from supplementary information on traces of the unknown function *Differential equations* **24** 136-146 (In Russian)

[9] Volkov N P 1992 On some inverse problems for time-dependent transport equation *Ill-posed problems in natural sciences* (Moscow: TVP-VSP) 431-438

[10] Volkov N P 2016 Solvability of certain inverse problems for the non-stationary kinetic transport equation *Computational mathematics and mathematical physics* 56 (Pleiades Publishing) 1598-1603

[11] Baskakov A V and Volkov N P 2013 On some conditions for the solvability of control problems for the non-stationary kinetic neutron transport equation *Electronic journal "MGOU Vestnik" (Series on physics and mathematics)* 1-10 (In Russian)

[12] Volkov N P 2017 Linear inverse problem of dynamics of the reactor *Journal of Physics: Conference Series* **788** 1-4

[13] Baskakov A V and Volkov N P 2017 On some control problems of the dynamics of a reactor *Journal of Physics: Conference Series* **937** N1 1-8

[14] Baskakov A V and Volkov N P 2018 On the control of transient processes in problems of reactor dynamics *Vestnik NIYaU Mephi* **7** N1 91-96

A new look at the double sine-Gordon kink-antikink scattering

E Belendryasova[1], V A Gani[1,2], A Moradi Marjaneh[3], D Saadatmand[4] and A Askari[5]

[1]National Research Nuclear University MEPhI (Moscow Engineering Physics Institute), 115409 Moscow, Russia
[2]National Research Center Kurchatov Institute, Institute for Theoretical and Experimental Physics, 117218 Moscow, Russia
[3]Young Researchers and Elite Club, Quchan Branch, Islamic Azad university, Quchan, Iran
[4]Department of Physics, University of Sistan and Baluchestan, Zahedan, Iran
[5]Department of Physics, Faculty of Science, University of Hormozgan, P.O.Box 3995, Bandar Abbas, Iran

E-mail: vagani@mephi.ru

Abstract.
We study the kink-antikink scattering within the double sine-Gordon model. In the numerical simulations we found a critical value $v_{\rm cr}$ of the initial velocity $v_{\rm in}$, which separates two different scenarios: at $v_{\rm in} < v_{\rm cr}$ the kinks capture each other and form a bound state, while at $v_{\rm in} > v_{\rm cr}$ the kinks pass through each other and escape to infinities. We obtain non-monotonous dependence of $v_{\rm cr}$ on the model parameter R. Besides that, at some initial velocities below $v_{\rm cr}$ we observe formation and interaction of the so-called oscillons (new phenomenon), as well as escape windows (well-known phenomenon).

1. Introduction

Field-theoretic models with polynomial and non-polynomial potentials are of great importance for various physical systems, e.g., in high energy physics, cosmology, condensed matter, etc., [1–4]. Significant progress has been made in studying topological solitons in models with one real scalar field [5–19], as well as in more complex models with two or more fields [20–34]. Both analytical and numerical methods are successfully applied to studying kink-antikink interactions. In particular, the collective coordinate method with one or more degrees of freedom enables to model kink-(anti)kink interactions [11, 35, 36]. Many interesting results has been obtained by numerical modeling of collision of kink and antikink, of kink with a defect [37], and of several kinks in one point [38–40].

In this work we study the (1 + 1)-dimensional double sine-Gordon (DSG) model [6, 41–43]. Although this model has been well-investigated, we have obtained new results. Firstly, we have found a non-monotonous dependence of the critical velocity $v_{\rm cr}$ on the model parameter R. Secondly, at some initial velocities from the range $v_{\rm in} < v_{\rm cr}$ in the final state we observed complex oscillating structures — oscillons. The latter phenomenon is rather new [16, 17] and was not reported for the DSG model previously.

2. Scattering of kinks

Consider the $(1+1)$-dimensional double sine-Gordon model. Within this model the dynamics of a real scalar field $\phi(x,t)$ is described by the Lagrangian

$$\mathcal{L} = \frac{1}{2}\left(\frac{\partial\phi}{\partial t}\right)^2 - \frac{1}{2}\left(\frac{\partial\phi}{\partial x}\right)^2 - V(\phi) \tag{1}$$

with the potential

$$V_{\rm R}(\phi) = \tanh^2 R\,(1-\cos\phi) + \frac{4}{\cosh^2 R}\left(1+\cos\frac{\phi}{2}\right), \qquad R > 0, \tag{2}$$

see figure 1a. From the Lagrangian (1) it is easy to obtain the equation of motion — partial differential equation of the second order:

$$\frac{\partial^2\phi}{\partial t^2} - \frac{\partial^2\phi}{\partial x^2} + \frac{dV}{d\phi} = 0. \tag{3}$$

In the static case $\phi = \phi(x)$, and we have the ordinary differential equation

$$\frac{d^2\phi}{dx^2} = \frac{dV}{d\phi} \quad \Leftrightarrow \quad \frac{d\phi}{dx} = \sqrt{2V}, \tag{4}$$

which yields static *kink* and *antikink* (see figure 1b):

$$\phi_{\rm k(\bar{k})}(x) = 4\pi n \pm 4\arctan\frac{\sinh x}{\cosh R}. \tag{5}$$

Note that the kink can be viewed as a superposition of two *subkinks* separated by distance R.

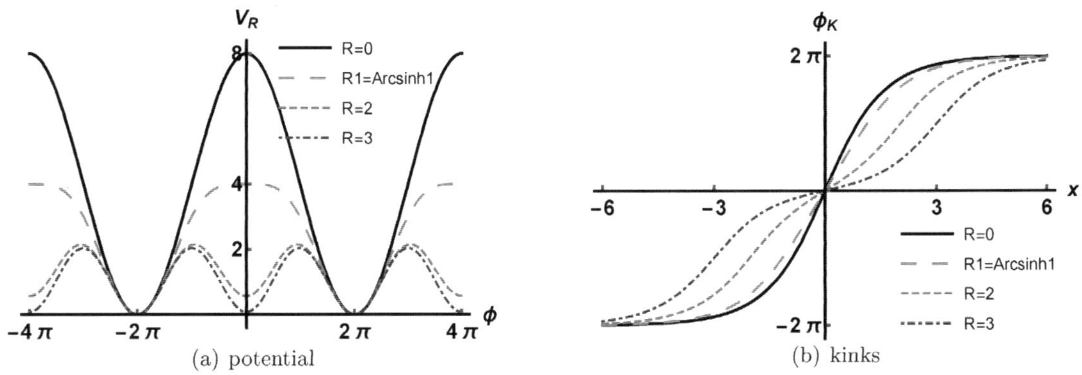

(a) potential (b) kinks

Figure 1. Potential (2) and kinks (5) from the sector $(-2\pi; 2\pi)$ for some values of R.

We performed numerical simulations of the kink-antikink scattering. To do this we solved the equation of motion (3) numerically using finite-difference method with the initial condition in the form of kink and antikink which were initially placed at $x = -\xi$ and $x = +\xi$, respectively, and moving towards each other with the velocities $v_{\rm in}$ in the laboratory frame of reference. (The moving kinks can be obtained from the static by the Lorentz transformations.) Thus the initial conditions for our simulations were taken from the following expression:

$$\phi(x,t) = \phi_{\rm k}\left(\frac{x+\xi-v_{\rm in}t}{\sqrt{1-v_{\rm in}^2}}\right) + \phi_{\rm \bar{k}}\left(\frac{x-\xi+v_{\rm in}t}{\sqrt{1-v_{\rm in}^2}}\right) - 2\pi. \tag{6}$$

In the numerical experiments we observed two different regimes of the collision: 1) at $v_{\rm in} < v_{\rm cr}$ the kink and antikink form *a bion* — a bound state, which then decays slowly, emitting small waves, see figure 2(a); 2) at $v_{\rm in} > v_{\rm cr}$ the kink and antikink pass through each other and escape to spatial infinities, see figure 2(b). The critical velocity $v_{\rm cr}$ depends on the parameter R, as it will be discussed below.

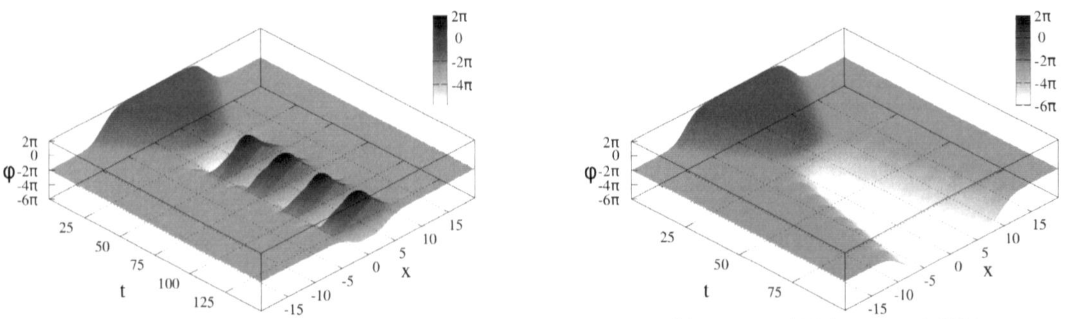

(a) formation of a bion, $v_{\rm in} = 0.2100$ (b) escape of kinks, $v_{\rm in} = 0.2500$

Figure 2. Space-time picture of the scattering for $R = 1.0$.

Besides that, at some initial velocities below the critical value we observed the so-called *escape windows* — intervals of the initial velocity within which the kinks collide several times and then escape to infinities instead of forming a bound state, see figure 3. This phenomenon can be explained by the resonant energy exchange between translational and vibrational modes of the kink (antikink), see, e.g., the review [44, 45].

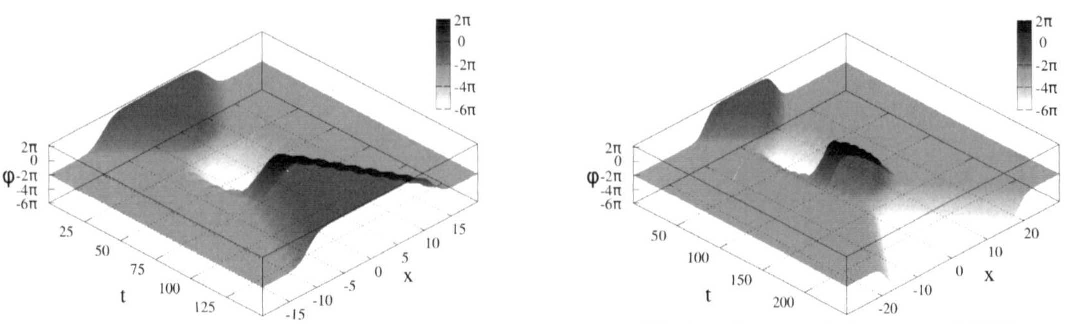

(a) two-bounce window, $v_{\rm in} = 0.2340$ (b) three-bounce window, $v_{\rm in} = 0.2380$

Figure 3. Escape of kinks after two and three collisions, $R = 1.0$.

We obtained the dependence of the critical velocity $v_{\rm cr}$ on the model parameter R, see figure 4. The curve $v_{\rm cr}(R)$ has several local maxima. (Note that this result has not been reported previously by other authors.) Such behavior could be a consequence of the complex structure of the DSG kink — the colliding kink and antikink can be viewed as composed objects constructed from subkinks. Pairwise interaction of subkinks can lead to non-monotonous dependence of $v_{\rm cr}$ on the model parameter, see, e.g., [36].

Another interesting and new phenomenon that we have obtained is formation and further interaction of *oscillons*, see figure 5. These oscillating structures can form bound states (figure 5(a)), or even escape to spatial infinities after several collisions (figure 5(b)).

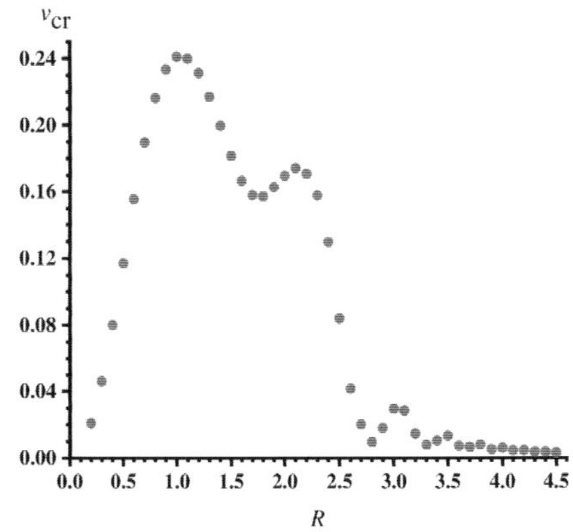

Figure 4. R-dependence of the critical velocity.

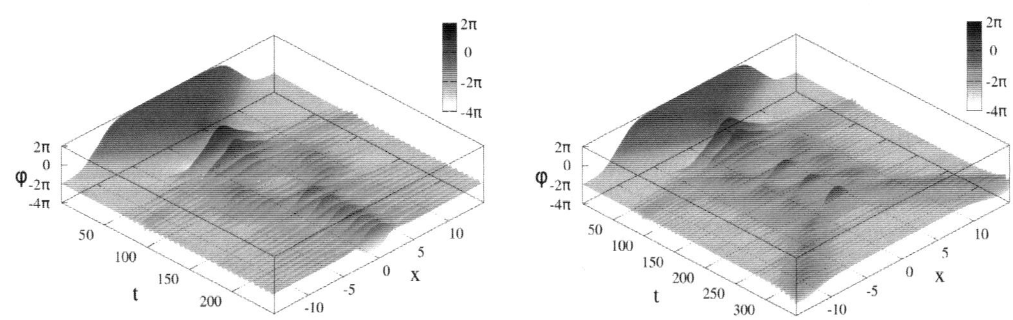

(a) $v_{\rm in} = 0.18472$ (b) $v_{\rm in} = 0.18470$

Figure 5. Oscillons in the final state.

3. Conclusion

We have studied the scattering of the kink and antikink of the double sine-Gordon model. Our main new results are the following.

- We have found that the critical value $v_{\rm cr}$ of the initial velocity depends on the model parameter R non-monotonously. This fact has not been reported for the DSG model by other authors. We can assume that such behavior of $v_{\rm cr}(R)$ could be a consequence of pairwise interaction of subkinks.
- At the initial velocities below the critical value, $v_{\rm in} < v_{\rm cr}$, we observed new phenomenon — formation of two oscillons in the final state. Their behavior seems to be rather complicated, they can form a bound state or, in some cases, can escape to spatial infinities after several collisions.

This research opens wide prospects for future work. For example, it would be interesting to study multikink collisions within the DSG model and to find maximal energy densities in such processes.

Acknowledgments
The research was supported by the MEPhI Academic Excellence Project (contract No. 02.a03.21.0005, 27.08.2013).

References

[1] Rajaraman R 1982 *Solitons and Instantons: An Introduction to Solitons and Instantons in Quantum Field Theory* (Amsterdam: North-Holland)
[2] Vilenkin A and Shellard E 2000 *Cosmic Strings and Other Topological Defects* (Cambridge U.K.: Cambridge University Press)
[3] Manton N and Sutcliffe P 2004 *Topological Solitons* (Cambridge U.K.: Cambridge University Press)
[4] Vachaspati T 2006 *Kinks and Domain Walls: An Introduction to Classical and Quantum Solitons* (Cambridge U.K.: Cambridge University Press)
[5] Lohe M A 1979 *Phys. Rev.* D **20** 3120–3130
[6] Gani V A and Kudryavtsev A E 1999 *Phys. Rev.* E **60** 3305–3309
[7] Gani V A, Kudryavtsev A E and Belova T I 1999 *Phys. Atom. Nucl.* **62** 895
[8] Gani V A, Kudryavtsev A E and Belova T I 1999 *Yad. Fiz.* **62** 956
[9] Belova T I, Gani V A and Kudyavtsev A E 2001 *Phys. Atom. Nucl.* **64** 140–142
[10] Belova T I, Gani V A and Kudyavtsev A E 2001 *Yad. Fiz.* **64** 143
[11] Gani V A, Kudryavtsev A E and Lizunova M A 2014 *Phys. Rev.* D **89** 125009
[12] Dorey P, Mersh K, Romanczukiewicz T and Shnir Y 2011 *Phys. Rev. Lett.* **107** 091602
[13] Khare A, Christov I C and Saxena A 2014 *Phys. Rev.* E **90** 023208
[14] Gani V A, Lensky V and Lizunova M A 2015 *J. High Energy Phys.* **2015** 147
[15] Gani V A, Lensky V, Lizunova M A and Mrozovskaya E V 2016 *J. Phys.: Conf. Ser.* **675** 012019
[16] Bazeia D, Belendryasova E and Gani V A 2017 *J. Phys.: Conf. Ser.* **934** 012032
[17] Bazeia D, Belendryasova E and Gani V A 2018 *Eur. Phys. J.* C **78** 340
[18] Belendryasova E and Gani V A 2017 *J. Phys.: Conf. Ser.* **934** 012059
[19] Belendryasova E and Gani V A 2019 *Commun. Nonlinear Sci. Numer. Simulat.* **67** 414–426
[20] Gani V A and Kudryavtsev A E 2001 *Phys. Atom. Nucl.* **64** 2043–2050
[21] Gani V A and Kudryavtsev A E 2001 *Yad. Fiz.* **64** 2130
[22] Lensky V A, Gani V A and Kudryavtsev A E 2001 *Sov. Phys. JETP* **93** 677–684
[23] Lensky V A, Gani V A and Kudryavtsev A E 2001 *Zh. Eksp. Teor. Fiz.* **120** 778
[24] Gani V A, Konyukhova N B, Kurochkin S V and Lensky V A 2004 *Comput. Math. Math. Phys.* **44** 1968
[25] Gani V A, Konyukhova N B, Kurochkin S V and Lensky V A 2004 *Zh. Vychisl. Mat. Mat. Fiz.* **44** 2069
[26] Gani V A, Ksenzov V G and Kudryavtsev A E 2010 *Phys. Atom. Nucl.* **73** 1889–1892
[27] Gani V A, Ksenzov V G and Kudryavtsev A E 2010 *Yad. Fiz.* **73** 1940
[28] Gani V A, Ksenzov V G and Kudryavtsev A E 2011 *Phys. Atom. Nucl.* **74** 771–777
[29] Gani V A, Ksenzov V G and Kudryavtsev A E 2011 *Yad. Fiz.* **74** 797
[30] Gani V A, Lizunova M A and Radomskiy R V 2016 *J. High Energy Phys.* **2016** 43
[31] Gani V A, Lizunova M A and Radomskiy R V 2016 *J. Phys.: Conf. Ser.* **675** 012020
[32] Radomskiy R V, Mrozovskaya E V, Gani V A and Christov I C 2017 *J. Phys.: Conf. Ser.* **798** 012087
[33] Gani V A, Kirillov A A and Rubin S G 2018 *JCAP* **2018** 042
[34] Gani V A, Kirillov A A and Rubin S G 2017 *J. Phys.: Conf. Ser.* **934** 012046
[35] Weigel H 2014 *J. Phys.: Conf. Ser.* **482** 012045
[36] Demirkaya A, Decker R, Kevrekidis P, Christov I and Saxena A 2017 *J. High Energy Phys.* **2017** 71
[37] Saadatmand D, Borisov D I, Kevrekidis P G, Zhou K and Dmitriev S V 2018 *Commun. Nonlinear Sci. Numer. Simulat.* **56** 62–76
[38] Moradi Marjaneh A, Gani V A, Saadatmand D, Dmitriev S V and Javidan K 2017 *J. High Energy Phys.* **2017** 28
[39] Moradi Marjaneh A, Askari A, Saadatmand D and Dmitriev S V 2017 *Eur. Phys. J* B **91** 22
[40] Saadatmand D, Dmitriev S and Kevrekidis P 2015 *Phys. Rev.* D **92** 056005
[41] Campbell D K, Peyrard M and Sodano P 1986 *Physica* D **19** 165–205
[42] Malomed B A 1989 *Phys. Lett.* A **136** 395–401
[43] Gani V A, Marjaneh A M, Askari A, Belendryasova E and Saadatmand D 2018 *Eur. Phys. J.* C **78** 345
[44] Belova T I and Kudryavtsev A E 1997 *Phys. Usp.* **40** 359–386
[45] Belova T I and Kudryavtsev A E 1997 *Usp. Fiz. Nauk* **167** 377–406

On inverse problem of determination of the coefficient in strongly degenerate parabolic equation

T I Bukharova[1], V L Kamynin[1] and A P Tonkikh[2]

[1]National Research Nuclear University MEPhI, Kashirskoe shosse, 31,115409 Moscow, Russia
[2]Bryansk State Academician I.G.Petrovsky University, Bezhitskaya Street, 14, 241036, Bryansk, Russia

E-mail: bukharova_t@mail.ru, vlkamynin2008@yandex.ru, a_tonkih@mail.ru

Abstract. We prove the existence and uniqueness theorems for inverse problem of determination of the lower coefficient in the Black-Scholes type equation with additional condition of integral observation. These results are based on the investigation of unique solvability of corresponding direct problem which is of independent interest. We give the example of the inverse problem for which the conditions of the theorems proved are fulfilled.

1. Introduction

In present work we study the questions of unique solvability of the inverse problem of determination of a pair of functions $\{u(t,x),\gamma(t)\}$ satisfying in the rectangle $Q\equiv[0,T]\times[0,l]$ the parabolic equation

$$u_t-a(t,x)u_{xx}+b(t,x)u_x+\gamma(t)u=f(t,x), \tag{1}$$

the initial and boundary conditions

$$u(0,x)=u_0(x),\ \ x\in[0,l];\ \ u(t,l)=0,\ \ t\in[0,T]; \tag{2}$$

and the additional condition of integral observation

$$\int_0^l u(t,x)\omega(x)\,\mathrm{d}x=\varphi(t),\ \ t\in[0,T]. \tag{3}$$

In our investigation we suppose that the leading coefficient $a(t,x)$ in the equation (1) is strongly degenerate for x=0, namely, we assume that $a(t,x)\sim a_0x^2,\ x\to 0,\ a_0=\text{const}>0.$

In what follows we use Lebesgue, Sobolev and Hölder spaces with corresponding norms in usual sense (see, for example, [1]). The derivatives are understood in generalized sense, the equalities and inequalities are satisfied almost everywhere.

We introduce notation

$$Q_\tau^\delta=[0,\tau]\times[\delta,l],\ 0\le\delta<l,\ 0\le\tau\le T;\ Q_\tau^0\equiv Q_\tau,\ Q_T^\delta\equiv Q^\delta,\ Q_T\equiv Q;\ \|\cdot\|_{L_q(0,l)}\equiv\|\cdot\|_q,\ q\in[0,\infty].$$

We assume that the functions occurring in the input data of the problem (1)–(3) are measurable and satisfy the following conditions:

$$x^2 a_1 \le a(t,x) \le a_2,\ (t,x)\in Q;\ a_1,a_2=\text{const}>0;\ a_x(t,x), \frac{a_x^2(t,x)}{a(t,x)}\in L_\infty(Q);$$
$$a(t,0)=0,\ t\in[0,T];\ \left\|a_x^2/a\right\|_{L_\infty(Q)}\le K_a; \tag{A}$$

$$\frac{b^2(t,x)}{a(t,x)}\in L_\infty(Q),\ \left\|b^2/a\right\|_{L_\infty(Q)}\le K_{b,a};\ \frac{f^2(t,x)}{a(t,x)}\in L_1(Q),\ \left\|f^2/a\right\|_{L_1(Q)}\le K_{f,a}(T); \tag{B}$$

$$u_0(x)\in \overset{0}{W}{}_2^1(0,l),\ \|u_0'\|_2\le M_1; \tag{C}$$

$$\omega(x)\in W_2^1(0,l),\ \omega(l)=0,\ \|\omega\|_\infty\le K_\omega,\ \left\|(a\omega)_x\right\|_{L_\infty(0,T;L_2(0,l))}\le K_{a,\omega}; \tag{D}$$

$$\varphi(t)\in W_\infty^1(0,T),\ \|\varphi\|_{L_\infty(0,T)}\le K_\varphi,\ \|\varphi'\|_{L_\infty(0,T)}\le K_\varphi^*; \tag{E}$$

$$\varphi(0)=\int_0^l u_0(x)\omega(x)\,\mathrm{d}x, \tag{F}$$

$$|\varphi(t)|\ge\varphi_0>0,\ t\in[0,T]. \tag{G}$$

Here a_1, a_2, K_a, K_ω, $K_{a,\omega}$, K_φ, φ_0, M_1 = const >0, $K_{b,a}$, $K_{f,a}(T)$, K_φ^* = const ≥ 0.

Definition 1. By generalized solution of the inverse problem (1)–(3) we mean a pair of functions $\{u(t,x),\gamma(t)\}$,

$$u(t,x)\in L_\infty\left(0,T;W_2^1(0,l)\right)\cap C^{0,\beta}(Q),\ \beta\in(0,1),\ u_t(t,x)\in L_2(Q),$$
$$au_{xx}^2\in L_1(Q),\ u_{xx}(t,x)\in L_2\left(Q^\delta\right)\ \forall\delta>0,\ \gamma(t)\in L_\infty(0,T),$$

satisfying equation (1) almost everywhere in Q and such that the function $u(t,x)$ satisfies the conditions (2), (3) in classical sense.

Remark 1. Conditions (A) and (B) are satisfied for the well-known Black-Scholes equation (see [2,3]). Hence, this equation can be considered in our investigation.

Remark 2. Equation (1) degenerates on the boundary of Q for x=0. From the well-known Fichera's theory it follows that whether or not boundary conditions should be given at the particular part of the boundary where the equation degenerates, depends on the sign of so-called Fichera function on that part of the boundary (see [4], [5, p. 20]). We impose such conditions on the input data that it is not necessary to specify the values of $u(t,x)$ for x=0 (these values cannot be arbitrary). But in our existence theorem we prove that the function $u(t,x)$ obtained in this theorem automatically satisfies the condition $u(t,0)$=0.

Inverse problems for degenerate parabolic equations of the type (1) are of great interest, for example, in financial mathematics (see Remark 1). In various settings but different from those considered in present paper they were investigated (including numerical calculations) in [6,7,8,9,10,11], etc.

Method of proving the existence and uniqueness theorems for inverse problem (1)–(3) in our paper is based on the study of the unique solvability of the direct problem (1)–(2) (when the function $\gamma(t)$ is assumed to be known) and on the estimates of its solution. We note that the investigation of the direct problem (1),(2) is of independent interest and the results obtained here for it are also new.

2. Unique solvability and estimates for the direct problem

Suppose that $\gamma(t)\in L_\infty(0,T)$ is a known function and consider the direct problem (1),(2) with this function $\gamma(t)$ in the equation (1). A generalized solution $u(t,x)$ of this problem will be understood in a sense of Definition 1.

Using the technique similar to [11] and the existence theorem for uniformly parabolic equations with two independent variables from [1] we prove the following existence and uniqueness theorems.

Theorem 1. Let the conditions (A)–(D) hold, $\gamma(t)\in L_\infty(0,T)$. Then the generalized solution of the direct problem (1),(2) is unique.

Theorem 2. Let the conditions (A)–(D) hold, $\|\gamma\|_{L_\infty(0,T)}\le K_\gamma$. Then there exists a generalized solution $u(t,x)$ of the direct problem (1),(2). For this solution we have $u(t,0)=0$, $t\in[0,T]$, and it satisfies the estimates

$$\sup_{0\le t\le T}\|u_x(t,\cdot)\|^2_{L_2(0,l)}+\|au^2_{xx}\|_{L_1(Q)}\le e^{\lambda^* T}\left[M_1^2+3K_{f,a}(T)\right], \tag{4}$$

$$\|u_{xx}\|^2_{L_2(Q^\delta)}\le\frac{e^{\lambda^* T}}{a_1\delta^2}\left[M_1^2+3K_{f,a}(T)\right]\quad\forall\delta\in(0,l), \tag{5}$$

$$\|u_t\|^2_{L_2(Q)}\le c_1\left[M_1^2+3K_{f,a}(T)\right], \tag{6}$$

$$|u(t_1,x_1)-u(t_2,x_2)|\le c_2\left(|x_1-x_2|^{1/2}+|t_1-t_2|^{1/6}\right),\ (t_1,x_1),(t_2,x_2)\in Q, \tag{7}$$

where $\lambda^*=3K_{b,a}+K_\gamma$, and c_1, $c_2=\text{const}>0$ depend only on $l,T,a_1,a_2,K_{b,a},K_{f,a}(T),M_1$ and K_γ.

3. Investigation of the inverse problem

Now we consider the inverse problem (1)–(3). Denote

$$F(t)=\int_0^l f(t,x)\omega(x)\,dx.$$

Theorem 3. Let the conditions (A)–(G) hold. Then there are no two different generalized solutions of the problem (1)–(3).

Sketch of the proof. Assume that there are two different solutions $\{u^{(1)}(t,x),\gamma^{(1)}(t)\}$ and $\{u^{(2)}(t,x),\gamma^{(2)}(t)\}$ of inverse problem (1)–(3). We put $v(t,x)=u^{(1)}(t,x)-u^{(2)}(t,x)$, $\eta(t)=\gamma^{(1)}(t)-\gamma^{(2)}(t)$.

Taking into account condition (3) and assumptions (A)–(G), after a series of calculations analogous to those carried out in the proof of the corresponding uniqueness theorem in [12] we obtain that the function $v(t,x)$ satisfies in Q the integro-differential equation

$$v_t-a(t,x)v_{xx}+b(t,x)v_x+\gamma^{(1)}(t)v=-\frac{1}{\varphi(t)}\int_0^l\left[(a\omega)_{yy}-(b\omega)_y\right]v\,dy\cdot u^{(2)}(t,x) \tag{8}$$

and the function $\eta(t)$ satisfies the relation

$$\eta(t)=\frac{1}{\varphi(t)}\int_0^l\left[(a\omega)_{xx}-(b\omega)_x\right]v(t,x)\,dx. \tag{9}$$

Using the well-known technique of energy estimates applied to the relation (8) we proof that $v(t,x)\equiv 0$ in Q. Then from (9) we obtain that $\eta(t)\equiv 0$ in $[0,T]$. Theorem 3 is proved.

In order to prove the existence of generalized solution of the inverse problem (1)–(3) we in addition to conditions (A)–(G) assume that

$$f(t,x)\in L_\infty(0,T;L_1(0;l)),\ (a\omega)_{xx},(b\omega)_x\in L_\infty(0,T;L_2(0;l)),$$

and

$$\|f\|_{L_\infty(0,T;L_1(0,l))} \le K_f, \tag{10}$$

$$\left\|(a\omega)_{xx}\right\|_{L_\infty(0,T;L_2(0,l))} \le K^*_{a,\omega}, \quad \left\|(b\omega)_x\right\|_{L_\infty(0,T;L_2(0,l))} \le K^*_{b,\omega}. \tag{11}$$

We derive the operator equation for the unknown function $\gamma(t) \in L_\infty(0,T)$.

Let pair of functions $\{u(t,x),\gamma(t)\}$ be a generalized solution of the inverse problem (1)–(3). We multiply equation (1) by $\omega(x)$ and integrate the result over the segment $[0,l]$. Taking into account conditions (2), (3) and also (A)–(G) we after integration by parts obtain the relation

$$\gamma(t) = \frac{1}{\varphi(t)}\left\{F(t) - \varphi'(t) + \int_0^l \left[(a\omega)_{xx} + (b\omega)_x\right] u \,\mathrm{dx}\right\}. \tag{12}$$

In view of this relation let us introduce the operator $\mathrm{A}: L_\infty(0,T) \to L_\infty(0,T)$ by the formula

$$\mathrm{A}(\gamma) = \frac{1}{\varphi(t)}\left\{F(t) - \varphi'(t) + \int_0^l \left[(a\omega)_{xx} + (b\omega)_x\right] u \,\mathrm{dx}\right\}, \tag{13}$$

where $\gamma(t)$ is an arbitrary function in $L_\infty(0,T)$ and $u \equiv u(t,x;\gamma)$ is a solution of the direct problem (1), (2) with given coefficient $\gamma(t)$ in the equation (1). Then the relation (12) can be written as

$$\gamma = \mathrm{A}(\gamma). \tag{14}$$

Remark 3. By virtue of Theorems 1 and 2 and assumptions (A)–(G) the operator A is well defined.

Lemma 1. Let conditions (A)–(G) hold. Then the operator equation (14) is equivalent to the inverse problem (1)–(3) in the following sense. If pair $\{u(t,x),\gamma(t)\}$ is a generalized solution of the inverse problem, then $\gamma(t)$ satisfies (14). Conversely, if $\gamma(t) \in L_\infty(0,T)$ is a solution of operator equation (14), and $u \equiv u(t,x;\gamma)$ is a solution of direct problem (1), (2) with this γ in the equation (1), then the pair $\{u(t,x;\gamma),\gamma(t)\}$ is a generalized solution of inverse problem (1)–(3).

The proof of this Lemma is standard (see, for example [11]).

Theorem 4. Let conditions (A)–(G), (10), (11) hold. Suppose that the value T_0 satisfies the inequality

$$\frac{1}{\varphi_0}\left\{K_f K_\omega + K^*_\varphi + \frac{l}{\pi} e^{1+3K_{b,a}T_0/2}\left(K^*_{a,\omega} + K^*_{b,\omega}\right)\left[M_1 + 3K_{f,a}(T_0)\right]^{1/2}\right\} \le \frac{2}{T_0}. \tag{15}$$

Then in the rectangle Q_{T_0} there exists a generalized solution $\{u(t,x),\gamma(t)\}$ of the inverse problem (1)–(3) wherein $u(t,0)=0$, $t \in [0,T]$. Moreover we have the estimate

$$\|\gamma\|_{L_\infty(0,T_0)} \le \frac{2}{T_0} \tag{16}$$

and the estimates (4)–(7) for $T=T_0$.

Sketch of the proof. We put $R_0 = 2/T_0$. Then we prove that the operator A is a completely continuous operator mapping the bounded convex closed set $B_{R_0} \equiv \left\{\gamma(t) \in L_\infty(0,T_0) : \|\gamma\|_{L_\infty(0,T_0)} \le R_0\right\}$ into itself. By the Schauder fixed point theorem (e.g. see [13, p. 193]) in this case the equation (14) has a solution $\gamma(t)$ from B_{R_0}, so the estimate (16) holds. Then the assertion of the theorem follows from Lemma 1.

Let us show that the conditions of our uniqueness and existence Theorems 3 and 4 are valid for the inverse problem for Black-Scholes equation.

Example. In the rectangle Q_{T_0} we consider the following inverse problem:

$$u_t - \sigma x^2 u_{xx} + \mu x u_x + \gamma(t) u = 0, \tag{17}$$

$$u(0,x) = u_0(x) \equiv x(l-x), \quad u(t,l) = 0, \tag{18}$$

$$\int_0^l u(t,x)(l-x)\,\mathrm{d}x = \frac{l^4}{12}. \tag{19}$$

Here σ, $\mu > 0$ are some given constants.

It is easy to check that the conditions (A)-(G), (10), (11) are fulfilled. By simple calculations we obtain that the condition (15) for the problem (17)–(19) can be written in the form

$$12\left(2\sigma + \frac{\mu}{\sqrt{3}}\right)\cdot\frac{1}{\sqrt{2}}\exp\left(1 + \frac{3\mu^2}{2\sigma}T_0\right) \le \frac{2}{T_0}. \tag{20}$$

Obviously the condition (20) is fulfilled for small T_0. In this case Theorems 3 and 4 are valid for the problem (17)–(19) and so the inverse problem (17)–(19) has a solution and this solution is unique.

The first and second authors were partially supported by the Program of competitiveness increase of the National Research Nuclear University MEPhI (Moscow Engineering Physics Institute); contract No.02.a03.21.0005, 27.08.2013.

References

[1] Kruzhkov S N 1979 Quasilinear parabolic equations and systems with two independent variables *Trudy Sem. im. Petrovskogo I G* **5** 217–272
[2] Black F and Scholes M 1973 The pricing of options and corporate liabilities. *J. Political Economy* **81** 637–659.
[3] Hull J 2005 Options, futures and other derivatives (NJ: Prentice Hall, upper Saddle River)
[4] Fichera G 1956 Sulle equazioni differenziali lineari ellitico-paraboliche del secondo ordine *Atti Accad. Nazionale dei Lincei. Mem. Cl. Sci. Fis. Mat. Natur. Ser. I(8)* **5** 1–30
[5] Oleĭnik O A and Radkevič E A 1973 Second Order Differential Equations with Nonnegative Characteristic Form (New York: AMS. Rhode Island and Plenum Press)
[6] Deng Z C and Yang L 2011 An inverse problem of identifying the coefficient of first-order in a degenerate parabolic equation *J. Computational Applied Math.* **235** 4404–4417
[7] Deng Z C and Yang L 2014 An inverse problem of identifying the radiative coefficient in a degenerate parabolic equation *Chinese Annals of Math. Ser. B* **35B** N3 355–382
[8] Bouchouev I and Isakov V 1999 Uniqueness, stability and numerical methods for the inverse problem that arises in financial markets *Inverse problems* **15** N3 95–116
[9] Jiang Lishang and Tao Yourshan 2001 Identifying the volatibility of underlying assets from option prices *Inverse problems* **17** N1 137–155
[10] Jiang Lishang, Chen Qihong, Wang Lijun and Zhang J E 2003 A new well-posed algorithm to recover implied local volatibility *Quantitative Finance* **3** N6 451–457
[11] Prilepko A I, Kamynin V L and Kostin A B 2018 Inverse source problem for parabolic equation with the condition of integral observation in time *Journal of Inverse and Ill-posed Problems* **26** N4 523–539
[12] Bukharova T I and Kamynin V L 2015 Inverse problem of determining the absorption coefficient in the multidimensional heat equation with unlimited minor coefficients *Computational Mathematics and Mathematical Physics* **55** N7 1183–1195
[13] Lyusternik L A and Sobolev V I 1982 *Kratkii kurs Functional'nogo Analiza* (Moskva: Vysh. Shkola)

Partly dissipative system with multizonal initial and boundary layers

V F Butuzov[1], **N N Nefedov**[1 4], **O E Omel'chenko**[2] **and L Recke**[3]

[1] Faculty of Physics, Lomonosov Moscow State University, 119991 Moscow, Russia
[2] Institute of Physics and Astronomy, University of Potsdam, Karl-Liebknecht-Strasse 24/25, 14476 Potsdam-Golm, Germany
[3] HU Berlin, Institut für Mathematik, Rudower Chaussee 25, 12489 Berlin-Adlershof, Germany

E-mail: butuzov@phys.msu.ru, nefedov@phys.msu.ru, oleh.omelchenko@gmail.com, recke@math.hu-berlin.de

Abstract. For a singularly perturbed parabolic - ODE system we construct the asymptotic expansion in the small parameter in the case, when the degenerate equation has a double root. Such systems, which are called partly dissipative reaction-diffusion systems, are used to model various natural processes, including the signal transmission along axons, solid combustion and the kinetics of some chemical reactions. It turns out that the algorithm of the construction of the boundary layer functions and the behavior of the solution in the boundary layers essentially differ from that ones in case of a simple root. The multizonal initial and boundary layers behaviour was stated.

1. Introduction and main results

We consider a coupled PDE-ODE system of the form

$$\varepsilon^2(\partial_t u - \partial_x^2 u) = F(u, v, x, t, \varepsilon), \qquad (x,t) \in D := (0,1) \times (0,T], \tag{1}$$

$$\varepsilon \partial_t v = f(u, v, x, t, \varepsilon), \qquad (x,t) \in D, \tag{2}$$

with homogeneous Neumann boundary conditions

$$\partial_x u(0,t,\varepsilon) = \partial_x u(1,t,\varepsilon) = 0, \quad t \in (0,T], \tag{3}$$

and initial conditions

$$u(x,0,\varepsilon) = u^0(x), \quad v(x,0,\varepsilon) = v^0(x), \quad x \in [0,1]. \tag{4}$$

Here, $\varepsilon > 0$ is a small parameter, F, f and u^0, v^0 are sufficiently smooth functions. The initial condition u^0 is supposed to be consistent with the boundary condition. Systems (1)–(2) are called *partly dissipative* reaction-diffusion systems, because the diffusion term appears only in the first of these equations. Such systems are used for mathematical modeling in chemical kinetics , theoretical biology and other applied sciences.

[4] Corresponding author

IOP Conf. Series: Journal of Physics: Conf. Series **1205** (2019) 012009 doi:10.1088/1742-6596/1205/1/012009

Our goal is to construct the asymptotics of boundary layer solutions to problem (1)–(4) in the case when the reduced equation

$$F(u,v,x,t,0)=0$$

obtained from equation (1) for $\varepsilon=0$, has a double root

$$u=\varphi(v,x,t).$$

This assumption can be reformulated in the following form.

(A1) There exist functions φ, h and F_1 such that

$$F(u,v,x,t,\varepsilon)=-h(v,x,t)(u-\varphi(v,x,t))^2+\varepsilon F_1(u,v,x,t,\varepsilon).$$

Some singularly perturbed problems with double roots of the reduced equation have been considered [1, 2, 3]. They showed that standard singular perturbation techniques developed for the case of a simple root of the reduced equation must be substantially modified in the case of a double root. The main difference is concerned with a more complicated three-zone structure of the initial and boundary layers.

In the present work it is shown that close to the initial time moment this solution behaves as a superposition of two initial layers with two different characteristic time scales

$$\sigma=t/\varepsilon \quad \text{and} \quad \tau=t/\varepsilon^2. \tag{5}$$

Respectively, two types of corner layers appear in the vicinity of the corner points $(x,t)=(0,0)$ and $(x,t)=(1,0)$.

We construct the asymptotics of boundary layer solution u, v to problem (1)–(4) in the form

$$\begin{aligned} u &= \overline{u}(x,t,\varepsilon)+Su(x,\sigma,\varepsilon)+\Pi u(x,\tau,\varepsilon)+Qu(\xi,t,\varepsilon)+Ru(\xi,\sigma,\varepsilon)\\ &+ Pu(\xi,\tau,\varepsilon)+\tilde{Q}u(\tilde{\xi},t,\varepsilon)+\tilde{R}u(\tilde{\xi},\sigma,\varepsilon)+\tilde{P}u(\tilde{\xi},\tau,\varepsilon), \end{aligned} \tag{6}$$

$$\begin{aligned} v &= \overline{v}(x,t,\varepsilon)+Sv(x,\sigma,\varepsilon)+\Pi v(x,\tau,\varepsilon)+Qv(\xi,t,\varepsilon)+Rv(\xi,\sigma,\varepsilon)\\ &+ Pv(\xi,\tau,\varepsilon)+\tilde{Q}v(\tilde{\xi},t,\varepsilon)+\tilde{R}v(\tilde{\xi},\sigma,\varepsilon)+\tilde{P}v(\tilde{\xi},\tau,\varepsilon), \end{aligned} \tag{7}$$

where the terms are the partial sum in some arbitrary orders of ε,

$$\xi=x/\varepsilon^{3/4}, \qquad \tilde{\xi}=(1-x)/\varepsilon^{3/4}, \tag{8}$$

and variables σ, τ are defined by formulas (5). Functions $\overline{u},\overline{v}$ are regular (or outer) asymptotics, whereas all other terms describe initial, boundary and corner layers. In particular, $Su, Sv, \Pi u, \Pi v$ are initial layers describing the behaviour of the solution close to the initial time moment $t=0$. Note that S- and Π-terms vary on different time scales corresponding to variables σ and τ, respectively. Moreover, these layers have qualitatively different decay estimates

$$|Su(x,\sigma,\varepsilon)|+|Sv(x,\sigma,\varepsilon)|\le c_0e^{-\kappa_0\sigma}$$
$$|\Pi u(x,\tau,\varepsilon)|+|\Pi v(x,\tau,\varepsilon)|\le c_0\Pi_{\kappa_0}(\tau,\varepsilon)$$

where

$$\Pi_\kappa(\tau,\varepsilon):=\frac{\sqrt{\varepsilon}\exp(-\sqrt{\varepsilon}\kappa\tau)}{1+\sqrt{\varepsilon}-\exp(-\sqrt{\varepsilon}\kappa\tau)}$$

IOP Conf. Series: Journal of Physics: Conf. Series **1205** (2019) 012009 doi:10.1088/1742-6596/1205/1/012009

is a special etalon function, and $c_0 > 0$ and $\kappa_0 > 0$ are some constants. Four other functions $Qu, Qv, \tilde{Q}u, \tilde{Q}v$ describe the behaviour of the solution close to the end points of the interval $0 \le x \le 1$. They will satisfy the estimate

$$|Qu(\xi, t, \varepsilon)| + |Qv(\xi, t, \varepsilon)| + |\tilde{Q}u(\xi, t, \varepsilon)| + |\tilde{Q}v(\xi, t, \varepsilon)| \le c_0 e^{-\kappa_0 \xi}.$$

Finally, functions

$$Ru, Rv, Pu, Pv, \tilde{R}u, \tilde{R}v, \tilde{P}u, \tilde{P}v$$

provide a description of the corner layers in the vicinity of the corner points $(x, t) = (0, 0)$ and $(x, t) = (1, 0)$. They satisfy the estimates

$$|Ru(\xi, \sigma, \varepsilon)| + |Rv(\xi, \sigma, \varepsilon)| + |\tilde{R}u(\xi, \sigma, \varepsilon)| + |\tilde{R}v(\xi, \sigma, \varepsilon)| \le c_0 e^{-\kappa_0(\xi+\sigma)}$$
$$|Pu(\xi, \tau, \varepsilon)| + |Pv(\xi, \tau, \varepsilon)| + |\tilde{P}u(\xi, \tau, \varepsilon)| + |\tilde{P}v(\xi, \tau, \varepsilon)| \le c_0 \Pi_{\kappa_0}(\tau, \varepsilon) e^{-\kappa_0 \xi}.$$

To construct a sequence of approximate solutions $U_n(x, t, \varepsilon)$, $V_n(x, t, \varepsilon)$ to problem (1)–(4) we need some other natural assumptions.

(A2) The equation

$$g(v, x, t) := f(\varphi(v, x, t), v, x, t, 0) = 0$$

has a solution $v = \overline{v}_0(x, t) \in C(\overline{D})$ *such that*

$$\partial_v g(\overline{v}_0(x, t), x, t) < 0 \quad \text{for all} \quad (x, t) \in \overline{D}.$$

(A3) The inequalities

$$h(\overline{v}_0(x, t), x, t) > 0 \quad \text{and} \quad F_1\left(\varphi(\overline{v}_0(x, t), x, t), \overline{v}_0(x, t), x, t, 0\right) > 0$$

are fulfilled for $(x, t) \in \overline{D}$.

(A4) Let $g(v, x, 0) \ne 0$ *for*

$$(v, x) \in \left\{(v, x) \,:\, (v - \overline{v}_0(x, 0))(v - v^0(x)) \le 0 \quad \text{and} \quad v \ne \overline{v}_0(x, 0), \;\; x \in [0, 1]\right\}.$$

(A5) Let

$$h(v, x, 0) > 0 \quad \text{and} \quad A(v, x) := F_1(\varphi(v, x, 0), v, x, 0, 0) - \partial_v \varphi(v, x, 0) g(v, x, 0) > 0$$

for $(v, x) \in \{(v, x) \,:\, (v - \overline{v}_0(x, 0))(v - v^0(x)) \le 0\}, \;\; x \in [0, 1]$.

(A6) Let

$$u^0(x) > \varphi\left(v^0(x), x, 0\right)$$

for $x \in [0, 1]$.

(A7) The inequalities

$$\partial_v \varphi(\overline{v}_0(x, t), x, t) > 0 \quad \text{and} \quad \partial_u f(\varphi(\overline{v}_0(x, t), x, t), \overline{v}_0(x, t), x, t, 0) > 0$$

are fulfilled for $(x, t) \in \overline{D}$.

Our main result is summarized in the following theorem.

Theorem 1. *Suppose that functions F, f, u^0 and v^0 are C^2-smooth. Moreover, suppose (A1)–(A7). Then, there exists $\varepsilon_0 > 0$ such that for all $\varepsilon \in (0, \varepsilon_0)$ there exists a solution $u(x,t,\varepsilon)$, $v(x,t,\varepsilon)$ to problem (1)–(4) and for each $(x,t) \in D$ we have*

$$\lim_{\varepsilon\to+0} u(x,t,\varepsilon) = \varphi(\overline{v}_0(x,t),x,t) \quad \textit{and} \quad \lim_{\varepsilon\to+0} v(x,t,\varepsilon) = \overline{v}_0(x,t).$$

Furthermore, if the functions F, f, u^0 and v^0 are C^{n+2}-smooth for some $n \geq 0$, then using the algorithm described in Section 2 one can construct approximate solution $U_n(x,t,\varepsilon)$, $V_n(x,t,\varepsilon)$ such that

$$\sup_{(x,t)\in\overline{D}} \Big(|u(x,t,\varepsilon) - U_n(x,t,\varepsilon)| + |v(x,t,\varepsilon) - V_n(x,t,\varepsilon)| \Big) = O\left(\varepsilon^{(n+1)/2}\right) \quad \textit{for} \quad \varepsilon \to +0.$$

2. Asymptotics of the solution to problem (1)–(4)

To construct the asymptotics we write the right-hand side of equation (1) in the following form

$$F = \overline{F} + SF + \Pi F + QF + RF + PF + \tilde{Q}F + \tilde{R}F + \tilde{P}F, \tag{9}$$

where

$$\begin{aligned} \overline{F} &= F(\overline{u}(x,t,\varepsilon), \overline{v}(x,t,\varepsilon), x, t, \varepsilon), \\ SF &= \Big[F(\overline{u}(x,t,\varepsilon) + Su(x,\sigma,\varepsilon), \overline{v}(x,t,\varepsilon) + Sv(x,\sigma,\varepsilon), x, t, \varepsilon) \\ &\quad -F(\overline{u}(x,t,\varepsilon), \overline{v}(x,t,\varepsilon), x, t, \varepsilon)\Big]_{t=\varepsilon\sigma}. \end{aligned} \tag{10}$$

$$\begin{aligned} \Pi F &= \Big[F(\overline{u}(x,t,\varepsilon) + Su(x,\sigma,\varepsilon) + \Pi u(x,\tau,\varepsilon), \overline{v} + Sv + \Pi v, x, t, \varepsilon) \\ &\quad -F(\overline{u}(x,t,\varepsilon) + Su(x,\sigma,\varepsilon), \overline{v} + Sv, x, t, \varepsilon)\Big]_{t=\varepsilon^{3/2}s,\ \sigma=\sqrt{\varepsilon}s}, \end{aligned} \tag{11}$$

$$QF = \Big[F(\overline{u}(x,t,\varepsilon) + Qu(\xi,t,\varepsilon), \overline{v} + Qv, x, t, \varepsilon) - F(\overline{u}(x,t,\varepsilon), \overline{v}, x, t, \varepsilon)\Big]_{x=\varepsilon^{3/4}\xi}, \tag{12}$$

$$\begin{aligned} RF &= \Big[F(\overline{u}(x,t,\varepsilon) + Su(x,\sigma,\varepsilon) + Qu(\xi,t,\varepsilon) + Ru(\xi,\sigma,\varepsilon), \overline{v} + Sv + Qv + \\ &\quad Rv, x, t, \varepsilon) - F(\overline{u}(x,t,\varepsilon) + Su(x,\sigma,\varepsilon), \overline{v} + Sv, x, t, \varepsilon) - F(\overline{u}(x,t,\varepsilon) \\ &\quad +Qu(\xi,t,\varepsilon), \overline{v} + Qv, x, t, \varepsilon) + F(\overline{u}(x,t,\varepsilon), \overline{v}, x, t, \varepsilon)\Big]_{x=\varepsilon^{3/4}\xi,\ t=\varepsilon\sigma}, \end{aligned} \tag{13}$$

$$\begin{aligned} PF &= \Big[F(\overline{u}(x,t,\varepsilon) + Su(x,\sigma,\varepsilon) + \Pi u(x,\tau,\varepsilon) + Qu(\xi,t,\varepsilon) + Ru(\xi,\sigma,\varepsilon) \\ &\quad +Pu(\xi,\sigma,\varepsilon), \overline{v} + Sv + \Pi v + Qv + Rv + Pv, x, t, \varepsilon) - F(\overline{u}(x,t,\varepsilon) \\ &\quad +Su(x,\sigma,\varepsilon) + Qu(\xi,t,\varepsilon) + Ru(\xi,\sigma,\varepsilon), \overline{v} + Sv + \Pi v + Rv, x, t, \varepsilon) \\ &\quad -F(\overline{u}(x,t,\varepsilon) + Su(x,\sigma,\varepsilon) + \Pi u(x,\tau,\varepsilon), \overline{v} + Sv + \Pi v, x, t, \varepsilon) \\ &\quad +F(\overline{u}(x,t,\varepsilon) + Su(x,\sigma,\varepsilon), \overline{v} + Sv, x, t, \varepsilon)\Big]_{x=\varepsilon^{3/4}\xi,\ t=\varepsilon^{3/2}s,\ \sigma=\sqrt{\varepsilon}s} \end{aligned} \tag{14}$$

IOP Conf. Series: Journal of Physics: Conf. Series **1205** (2019) 012009 doi:10.1088/1742-6596/1205/1/012009

(arguments of the v-terms are identical to the arguments of the corresponding u-terms, similar presentation we use for $\tilde{Q}F$, $\tilde{R}F$ and $\tilde{P}F$). .

Analogously, the right-hand side of equation (2) can be also decomposed by the similar way as follows

$$f = \overline{f} + Sf + \Pi f + Qf + Rf + Pf + \tilde{Q}f + \tilde{R}f + \tilde{P}f. \tag{15}$$

Then, we proceed by using boundary layer function method scheme as follows. Each term in formulas (6) and (7) we write as a formal series with respect to a fractional power of parameter ε (Recall that in the case of a simple root of the reduced equation one needs to work with a similar series containing integer powers of ε only!) Then we insert this into equations (1), (2) and replace the right-hand sides F and f by formulas (9) and (15). Collecting expressions with the same power of small parameter ε, we obtain equations for all terms in (6) and (7).

For example, for $\overline{u}$, $\overline{v}$ we obtain the system

$$\varepsilon^2(\partial_t \overline{u} - \partial_x^2 \overline{u}) = \overline{F} := -h(\overline{v}, x, t)(\overline{u} - \varphi(\overline{v}, x, t))^2 + \varepsilon F_1(\overline{u}, \overline{v}, x, t, \varepsilon), \quad (x,t) \in D, \tag{16}$$

$$\varepsilon \partial_t \overline{v} = \overline{f} := f(\overline{u}, \overline{v}, x, t, \varepsilon), \quad (x,t) \in D. \tag{17}$$

Equations for other terms can be written similarly.

Some of the terms in formulas (6) and (7) are determined from differential equations. In this case, one needs to equip the differential equations with initial and/or boundary conditions. They can be obtained if we insert expressions (6) and (7) into the boundary (3) and initial conditions (4) of the main problem. For example, first identity in (4) yields

$$\overline{u}(x,0,\varepsilon) + Su(x,0,\varepsilon) + \Pi u(x,0,\varepsilon) + Qu(\xi,0,\varepsilon) + Ru(\xi,0,\varepsilon)$$
$$+ Pu(\xi,0,\varepsilon) + \tilde{Q}u(\tilde{\xi},0,\varepsilon) + \tilde{R}u(\tilde{\xi},0,\varepsilon) + \tilde{P}u(\tilde{\xi},0,\varepsilon) = u^0(x).$$

Separating it by the above algorithm we obtain

$$\overline{u}(x,0,\varepsilon) + Su(x,0,\varepsilon) + \Pi u(x,0,\varepsilon) = u^0(x), \quad 0 \le x \le 1,$$
$$Qu(\xi,0,\varepsilon) + Ru(\xi,0,\varepsilon) + Pu(\xi,0,\varepsilon) = 0, \quad \xi \ge 0,$$
$$\tilde{Q}u(\tilde{\xi},0,\varepsilon) + \tilde{R}u(\tilde{\xi},0,\varepsilon) + \tilde{P}u(\tilde{\xi},0,\varepsilon) = 0, \quad \tilde{\xi} \ge 0.$$

Similarly, using the second identity in (4) and the first identity in (3) we get

$$\overline{v}(x,0,\varepsilon) + Sv(x,0,\varepsilon) + \Pi v(x,0,\varepsilon) = v^0(x), \quad 0 \le x \le 1, \tag{18}$$

$$Qv(\xi,0,\varepsilon) + Rv(\xi,0,\varepsilon) + Pv(\xi,0,\varepsilon) = 0, \quad \xi \ge 0, \tag{19}$$

$$\partial_x \overline{u}(0,t,\varepsilon) + \varepsilon^{-3/4} \partial_\xi Qu(0,t,\varepsilon) = 0, \quad 0 \le t \le T, \tag{20}$$

$$\partial_x Su(0,\sigma,\varepsilon) + \varepsilon^{-3/4} \partial_\xi Ru(0,\sigma,\varepsilon) = 0, \quad \sigma \ge 0, \tag{21}$$

$$\partial_x \Pi u(0,\tau,\varepsilon) + \varepsilon^{-3/4} \partial_\xi Pu(0,\tau,\varepsilon) = 0, \quad \tau \ge 0. \tag{22}$$

For some boundary layer functions we also use a standard condition implying their decay to 0 for the stretched variables, tending to infinity.

The regular part of the asymptotics $\overline{u}$ and $\overline{v}$ will be constructed in the form of a power series with respect to $\sqrt{\varepsilon}$

$$\overline{u}(x,t,\varepsilon) = \sum_{i=0}^{\infty} \varepsilon^{i/2} \overline{u}_i(x,t), \quad \overline{v}(x,t,\varepsilon) = \sum_{i=0}^{\infty} \varepsilon^{i/2} \overline{v}_i(x,t).$$

Inserting these formulas into equations (16)–(17) and expanding the left- and right-hand sides into series with respect to $\sqrt{\varepsilon}$, we obtain a recurrent sequence of equations for $\overline{u}_i(x,t)$ and $\overline{v}_i(x,t)$.

By the simple calculations we get, that $\overline{u}_0$ and $\overline{v}_0$ are determined due to assumption (A2), and higher order terms can be determined by means of assumption (A3).

Boundary and initial layers are also defined by proposed scheme.

For initial layers Su and Sv we obtain the following formal system

$$\varepsilon\partial_\sigma Su - \varepsilon^2\partial_x^2 Su = SF(x,\sigma,\varepsilon), \quad \partial_\sigma Sv = Sf(x,\sigma,\varepsilon), \quad (x,\sigma)\in(0,1)\times(0,\infty),$$

where function SF is defined by (10) and Sf has the analogous definition. Similarly, for initial layers Πu and Πv we write a system

$$\partial_\tau \Pi u - \varepsilon^2\partial_x^2 \Pi u = \Pi F(x,\tau,\varepsilon), \quad \varepsilon^{-1}\partial_\tau \Pi v = \Pi f(x,\tau,\varepsilon), \quad (x,\tau)\in(0,1)\times(0,\infty).$$

Here again the right-hand side ΠF is given by formula (11) and function Πf is defined by analogous expression.

We construct Su, Sv, Πu and Πv in the form

$$Su(x,\sigma,\varepsilon) = \sum_{i=0}^{\infty}\varepsilon^{i/2}S_i u(x,\sigma), \qquad Sv(x,\sigma,\varepsilon) = \sum_{i=0}^{\infty}\varepsilon^{i/2}S_i v(x,\sigma),$$

$$\Pi u(x,\tau,\varepsilon) = \sum_{i=0}^{\infty}\varepsilon^{i/2}\Pi_i u(x,\tau), \qquad \Pi v(x,\tau,\varepsilon) = \sum_{i=0}^{\infty}\varepsilon^{i/2}\Pi_i v(x,\tau).$$

For boundary layers Qu and Qv in the vicinity of the point $x=0$ we obtain a system of equations

$$\varepsilon^2\partial_t Qu - \sqrt{\varepsilon}\partial_\xi^2 Qu = QF(\xi,t,\varepsilon), \quad \varepsilon\partial_t Qv = Qf(\xi,t,\varepsilon), \quad (\xi,t)\in(0,\infty)\times(0,T), \qquad (23)$$

where stretched variable ξ is defined in (8), QF is a function given by formula (12) and Qf has analogous meaning. System (23) will be equipped with boundary condition (20) and standard decay condition.

We construct Qu and Qv in the form

$$Qu(\xi,t,\varepsilon) = \varepsilon^{3/4}\sum_{i=0}^{\infty}\varepsilon^{i/4}Q_i u(\xi,t), \quad Qv(\xi,t,\varepsilon) = \varepsilon^{3/4}\sum_{i=0}^{\infty}\varepsilon^{i/4}Q_i v(\xi,t).$$

To construct functions Ru and Rv we consider a system of equations

$$\varepsilon\partial_\sigma Ru - \sqrt{\varepsilon}\partial_\xi^2 Ru = RF(\xi,\sigma,\varepsilon), \quad \partial_\sigma Rv = Rf(\xi,\sigma,\varepsilon), \quad (\xi,\sigma)\in(0,\infty)^2, \qquad (24)$$

where the right-hand side RF is determined by formula (13) and Rf is defined by analogous expression. Similarly, we construct terms Pu and Pv considering system

$$\partial_\tau Pu - \sqrt{\varepsilon}\partial_\xi^2 Pu = PF(\xi,\tau,\varepsilon), \quad \varepsilon^{-1}\partial_\tau Pv = Pf(\xi,\tau,\varepsilon), \quad (\xi,\tau)\in(0,\infty)^2, \qquad (25)$$

where PF is defined by (14) and Pf has analogous definition.

Systems (24) and (25) will be equipped with the initial conditions (18) and (19), which make series Ru coupled with series Pu, and series Rv coupled with series Pv. Moreover, we also equip systems (24) and (25) with boundary conditions (21) and (22) as well as with the decay conditions at infinity.

We construct Ru, Rv, Pu and Pv in the form

$$Ru(\xi,\sigma,\varepsilon) = \varepsilon^{3/4}\sum_{i=0}^{\infty}\varepsilon^{i/4}R_iu(\xi,\sigma), \qquad Rv(\xi,\sigma,\varepsilon) = \varepsilon^{3/4}\sum_{i=0}^{\infty}\varepsilon^{i/4}R_iv(\xi,\sigma),$$

$$Pu(\xi,\tau,\varepsilon) = \varepsilon^{1/4}\sum_{i=0}^{\infty}\varepsilon^{i/4}P_iu(\xi,\tau), \qquad Pv(\xi,\tau,\varepsilon) = \varepsilon^{3/4}\sum_{i=0}^{\infty}\varepsilon^{i/4}P_iv(\xi,\tau).$$

Note, that the terms marked by "tilde" in formulas (6) and (7) can be determined by the analogous problems.

To prove our main Theorem we use the asymptotic method of differential inequalities [4, 5]. We modify functions U_n and V_n to construct the upper and lower solutions of problem (1)–(4). Using the known comparison theorem (see for e.q. [6]) we get the existence of the solution with the asymptotics constructed.

ACKNOWLEDGMENTS.
This work was supported by the RFBR (project N 18–01–00424) and by the program of cooperation of HU of Berlin and Lomonosov MSU.

References

[1] Butuzov V F 2013 *Mathematical Notes* **99** 60–70
[2] Butuzov V F 2016 *Mathematical Notes* **99** 210–21
[3] Butuzov V F 2016 *Sbornik: Mathematics* **207** 1100–26
[4] Nefedov N N 1995 *Differential Equations* **31** 668–771
[5] Nefedov N N 2013 *Lecture Notes in Computer Science* **8236** 62 - 72
[6] C.V. Pao, Nonlinear parabolic and elliptic equations, Plenum Press, New York 1992, xv+777 pp.

Mathematical models of transient processes for the distribution of resources in the system

E N Byhovets, S G Klimanov, A V Kryanev and D E Sliva
National Research Nuclear University "MEPhI", Russia

avkryanev@mephi.ru

Abstract. Problems for shares of resources to be distributed in conditions of constraints, including the process of dynamic change of shares are considered. The tasks of forming and dynamically changing the composition of portfolios are considered, in particular, in the presence of group restrictions. A mathematical model for predicting the dynamics of shares for a transition process from one equilibrium state of the system to another equilibrium state is proposed. The case of the presence of a noise chaotic component in such a model is considered. An example of the application of the proposed model of the dynamic process of resource allocation is given.

Keywords: resource allocation, constraints, group constraints, portfolio, effective portfolio, forecasting, equilibrium, transition process.

1. Introduction

In the formulation of many applied problems the unknown functions of time are the shares of the distributed resource over the objects of the system under consideration. Examples of such problems are the problems of restructuring the composition of effective portfolios [1-4] or the dynamics of the distribution of the priority coefficients of the quality indicators of a complex economic or technical system [5]. When solving this type of problem, it is often known, sometimes with some degree of uncertainty, the distribution of shares for the future fixed time. The article considers the problem of determining the transition process from the initial distribution of shares to their final distribution.

2. Schemes for the formation of an effective portfolio of shares of resource allocation in a system under constraints

Let, for example, resources be distributed between objects of the system. Then, based on preliminary estimates, and often on real opportunities that limit the amount of resources that can be distributed in the system under consideration, a priori restrictions are imposed on the volumes of distributed resources, and thus on the shares of the portfolio being formed. Such restrictions have the form of group restrictions:

$$a_j \leq x_{j1} + \cdots + x_{jm_j} \leq b_j,\ j = 1, \dots, m,\ m_1 + \cdots + m_m = n, \tag{1}$$

where m is the number of a priori groups, m_j is the number of objects in the j-th group, $a_j, b_j, 0 \leq a_j \leq b_j \leq 1$ is the lower and upper boundaries of the total fraction of the j-th group.
It is assumed that each object (each index $i = 1, \dots n$) belongs to one and only one of the groups (1). In order for the system of a priori constraints (1) to be consistent, it is necessary and sufficient that the conditions $\sum_{j=1}^{m} a_j \leq 1$, $\sum_{j=1}^{m} b_j \geq 1$, are satisfied. In particular, if $a_j = b_j, j = 1, \dots, m,$ then this means fixing the total share of resources included in the j-th group in the portfolio.
The system of a priori constraints (1) has often the form

$$\boldsymbol{a_j \leq x_j \leq b_j,\ j = 1, \dots, n,} \qquad (2)$$

that is, the constraints are superimposed separately on the share of resources for each object.
If there are a priori constraints to find effective portfolios, we have a two-criteria problem [6, 7]:

$$\begin{gathered} \sigma_p^2 = (W\boldsymbol{x},\ \boldsymbol{x}) - \min \\ m_p = (\boldsymbol{m}, \boldsymbol{x}) - \max \\ 0 \leq x_i \leq 1,\ i = 1, \dots, n,\ \textstyle\sum_{i=1}^{n} x_i = 1, \\ a_j \leq x_{j1} + \dots + x_{jm_j} \leq b_j, \quad j = 1, \dots, m, \\ m_p = (\boldsymbol{m}, \boldsymbol{x}) \in [m_p^*,\ m_{pmax}], \end{gathered} \qquad (3)$$

where m_p^*, σ_p^{*2} and $(m_{pmax}, \sigma_{pmax}^2)$ are determined by the solutions of the following problems: problem of forming a portfolio of minimum uncertainty of efficiency:

$$\begin{gathered} \sigma_p^2 = (W\boldsymbol{x},\ \boldsymbol{x}) - \min \\ 0 \leq x_i \leq 1,\ i = 1, \dots, n,\ \textstyle\sum_{i=1}^{n} x_i = 1, \\ a_j \leq x_{j1} + \dots + x_{jm_j} \leq b_j, \quad j = 1, \dots, m, \end{gathered} \qquad (4)$$

$m_p^* = (\boldsymbol{m},\ \boldsymbol{x}^*)$, $\sigma_p^{*2} = (W\ \boldsymbol{x}^*, \boldsymbol{x}^*)$, $\boldsymbol{x}^*$ is the solution of problem (4),
and the problem of forming a portfolio with the maximum average expected value of efficiency:

$$\begin{gathered} m_p = (\boldsymbol{m}, \boldsymbol{x}) - \max \\ 0 \leq x_i \leq 1,\ i = 1, \dots, n,\ \textstyle\sum_{i=1}^{n} x_i = 1, \\ a_j \leq x_{j1} + \dots + x_{jm_j} \leq b_j, \quad j = 1, \dots, m, \\ m_{pmax} = (\boldsymbol{m}, \boldsymbol{x}_{max}),\ \sigma_{pmax}^2 = (W\boldsymbol{x}_{max}, \boldsymbol{x}_{max}), \end{gathered} \qquad (5)$$

$\boldsymbol{x}_{max}$ is the solution of problem (5).
In problem (4) there is an effective portfolio of the least uncertainty satisfying a priori constraints (1), and in problem (5) - an effective portfolio with maximally expected efficiency, also satisfying a priori constraints (1). The problem of quadratic programming (4) is solved on the computer by one of the numerical methods well developed in computational mathematics.
The problem of linear programming (5) is effectively solved on a computer using a special algorithm that takes into account the specific nature of problem (5), (of course, problem (5) can be solved with the help of the well-known simplex method, but in this case it is inappropriate to apply the simplex method) [7].
As in the Markowitz problem [1], the set of points $(m_p,\ \sigma_p^2)$corresponding to the effective Pareto solutions of problem (3) on the plane $(m_p,\ \sigma_p^2)$ form an increasing convex downward curve.
The next modification of the Markowitz model is related to the situation when there is a strong correlation of capital investment efficiency in the investment objects under consideration. In this case, the covariance matrix W of the efficiency vector $\vec{R}$ is ill-conditioned or degenerated and, consequently, the Markowitz problem [1] and the investment portfolio optimization problem in the presence of a priori constraints (3) belong to the class of ill-posed problems which generates instability when solving them [8]. The economic reason for the instability of the problem of portfolio optimization is the fact that the risk of the portfolio may vary slightly with the allocation of resources between the objects whose efficiency is highly correlated. The instability of the problem of portfolio optimization from a computational point of view requires its regularization [8].

In addition to the two so far considered criteria - the expected value of efficiency and portfolio uncertainty, it is often desirable to optimize at least one more criterion. The most significant additional criterion is the volume of the transition process associated with the restructuring of the old portfolio composition into a new one. This additional criterion is especially important when the volume of transactions is large (restructuring of a "large" portfolio).
Mathematically, the criterion for the volume of the transition process $J_{dis}(\boldsymbol{x},\boldsymbol{x}_0)$ is described by the "distance" between the old and the new composition of the portfolio $\boldsymbol{x}_0 = (x_{10},\dots,x_{n0})^T$ and $\boldsymbol{x} = (x_1,\dots,x_n)^T$:

$$J_{dis}(\boldsymbol{x},\boldsymbol{x}_0) = \sum_{i=1}^{n} d_i \cdot (x_i \cdot S_p - x_{i0} \cdot S_0)^2 , \qquad (6)$$

where $S_0(S_p)$ is invested capital in the old (new) composition of the portfolio, d_i is positive numbers determined by the resource content of the objects.
It should be mentioned that in general $S_0 \neq S_p$, since at the date of portfolio restructuring a certain amount of resources can be output from (or into) it.
Taking into account the criterion of the volume of the transition process, the regularized problem of portfolio optimization has the form:

$$\sigma_p^2 = (W\boldsymbol{x},\boldsymbol{x}) + \alpha \cdot J_{dis}(\boldsymbol{x},\boldsymbol{x}_0) - \min,$$

$$0 \le x_i \le 1,\ i = 1,\dots,n,\ \sum_{i=1}^{n} x_i = 1, \qquad (7)$$

$$a_j \le x_{j1} + \dots + x_{jm_j} \le b_j, \qquad j = 1,\dots,m,$$

$$m_p = (\boldsymbol{m},\boldsymbol{x}) \in [m_p^*,\ m_{pmax}],$$

where $\alpha > 0$ is the regularization parameter which establishes the level of compromise between the portfolio risk and the criterion for the volume of the transition process.
Mathematically, problem (7) also belongs to the class of problems of quadratic programming and is solved on the computer with the help of a special algorithm, which allows one to choose the appropriate value of the regularization parameter $\alpha > 0$. When compiling an portfolio, which includes n objects, indexed $i = 1, \dots, n$, it is necessary to estimate the covariance matrix of a system of random variables

$$\boldsymbol{R}_{pre} = (R_{1pre},\dots,R_{npre})^T,$$

where R_{ipre} is the predicted effectiveness of the i-th investment object for the investor under consideration.
Let the investor use the prognostic algorithm (scheme) that can be applied to the already implemented dates in the past (perhaps not all) to obtain predicted values of efficiency. Thus, it is assumed that for the set of past dates the prediction algorithm allows to predict the values of the efficiencies $R_{ipre}(t_k), k = 1,\dots,N$ for a future interval of length T $(t_N = T)$.
Let R_{ipre} be the fulfilled values of the efficiencies. Then, assuming the stationarity of the covariance matrix W, for the vector of predictive efficiencies $\boldsymbol{R}_{pre}$ we obtain an estimation of its elements W_{ij}:

$$W_{ij} = \frac{1}{N} \cdot \sum_{k=1}^{N} (R_{ipre}(t_k) - R_{jpre}(t_k)), \qquad i,j = 1,\dots,n.$$

Let, for example, the predicted efficiencies of R_{ipre} for the current date belong to finite intervals:

$$R_{imin} < R_{ipre} < R_{imax}, \qquad i = 1,\dots n$$

and the random variables R_{ipre} $i=1,\dots,n$ do not pairwise correlate. Then, if the normal distribution law is postulated for each R_{ipre}, we have $W = diag(\sigma^2(R_{ipre}) = \frac{1}{N} \cdot \sum_{k=1}^{N} (R_{ipre}(t_k) - R_{jpre}(t_k))^2,\ i - 1,\dots,n.$
According to the introduced definition of predictive efficiencies and risks for the task of optimizing the portfolio, the estimation of the vector of expected mean values $M\boldsymbol{R}_{pre} = (MR_{1pre},\dots,MR_{npre})^T$ ^{T} is taken as the calculated values of efficiencies and the estimation of the covariance matrix of the vector $\boldsymbol{R}_{pre}$ is taken as the matrix W.

IOP Conf. Series: Journal of Physics: Conf. Series **1205** (2019) 012010 doi:10.1088/1742-6596/1205/1/012010

3. The scheme for calculating the dynamics of the transition process for the shares of resources allocation in the system

The predicted dynamics of the transition process for the shares of the distributed resource is described by the following composite system of equations:

$$\frac{dx_i(t)}{dt} = a_i - \alpha \cdot x_i, \ 0 < t < t^*, i = 1, \dots, n \tag{8}$$

with the initial condition

$$x_i(t) = x_{i0} \geq 0, i = 1, \dots, n, \ \sum_{i=1}^{n} x_{i0} = 1. \tag{9}$$

In system (8) the parameters $a_i, i = 1, \dots, n$ n are unknown and are to be determined.
Let's consider the following inverse problem for the system of differential equations (8).
It is required to find parameters $a_i, i = 1, \dots, n$ so that the normalization conditions $\sum_{i=1}^{n} x_i(t) = 1, \ 0 < t < t^*, x_i(t) \geq 0, i = 1, \dots, n,$ and the solution of the Cauchy problem (8) - (9) must satisfy the finite condition:

$$x_i(t^*) = x_i^* \geq 0, \ i = 1, \dots, n, \ \sum_{i=1}^{n} x_i^* = 1. \tag{10}$$

The solution of the inverse problem (8) - (10) has the form:

$$x_i(t) = x_i^* + (x_{i0} - x_i^*) \cdot \exp(-\alpha \cdot t), i = 1, \dots, n. \tag{11}$$

If the additive chaotic component $\varepsilon(t), M\varepsilon(t) = 0$ is present on the right-hand side of the original system (8), then the unbiased estimate for the solution of the inverse problems (8) - (10) will also be (11) and the estimate of the parameter α can be found from of the dynamics carrying out of the change $x_i(t), i = 1, \dots, n$ by the method of least squares. Below it is presented a numerical example of the dynamics of the shares of a distributed resource for a system of objects into which these resources are distributed. Table 1 presents the results of forecasting the shares of the distributed resource using the above scheme (the shares in Table 1 are given in percentages). Here the initial shares are defined by the equalities:
$x_{10} = 0.04, \ x_{20} = 0.21, x_{30} = 0.75, x_{40} = 0.00,$ and the shares of the final state by the equalities $x_1^* = 0.01, x_2^* = 0.03, x_3^* = 0.21, x_4^* = 0.75.$
Table 1 shows the fulfilled values of the shares of resource allocation at regular intervals in four objects (in parentheses are projected values of shares).

Table 1. The forecasted and the realized shares of the resource

4%	21%	75%	0%
3% (3%)	15% (15%)	61% (58%)	21% (24%)
3% (2.8%)	12% (12.2%)	50% (51%)	35% (34%)
2% (2%)	10% (10%)	43% (42%)	45% (46%)
2% (2%)	7% (7%)	35% (32%)	56% (59%)
2% (1.5%)	6% (6.5%)	29% (30%)	63% (62%)
1% (1.5%)	5% (5.5%)	25% (28%)	69% (65%)
2% (1%)	4% (4%)	24% (24%)	70% (71%)
1% (1%)	4% (4%)	22% (22%)	73% (73%)
1% (1%)	3% (3%)	21% (21%)	75% (75%)
1% (1%)	3% (3%)	21% (21%)	75% (75%)

4. Conclusion

The article presents mathematical models of transient processes for the distribution of resources in the system including the forecasting of the dynamic process of resource allocation in the transition from one equilibrium state to another equilibrium state. The mathematical model of predicting a dynamic process is based on the inverse problem for a system of ordinary differential equations taking into account the normalization and non-negativity of the shares for the entire considered period of time. An explicit solution of the inverse problem is obtained. It is specified that the obtained solution can be used as an estimation of the forecast in a dynamic mode with an unbiased chaotic component.

Acknowledgments
The work was supported by the Ministry of Education and Science of the Russian Federation (the basic part of the state assignment, project No. 1.9746.2017 / BC).

References

[1] Markowitz H M 1990 *Mean Variance Analysis in Portfolio Choice and Capital Markets* Basil: Blackwell
[2] Sharpe U F, Alexander G Dg and Baily D V 2001 *Investment* Moscow: Infra-M
[3] Rockafellar R T and Uryasev S 2000 Optimization of conditional value-at-risk J. Risk **2**(3)
[4] Lim Churlzu, Sherali Hanif D and Uryasev S 2010 Portfolio optimization by minimizing conditional value-at-risk via non differentiable optimization Computational Optimization and Applications **46**(3)
[5] Semenov S S, Voronov E M, Poltavskii AV and Kryanev A V 2015 *Methods of decision-making in problems assessing the quality and technical level of complex technical systems* Moscow: URSS (in Russian)
[6] Urlov F F and Shapkin E I 2007 *Selecting effective strategic decisions on the basis of multi-level and multi-criteria approach* N. Novgorod: Tutorial (in Russian)
[7] Nogin V D 2007 *Decision-making in many criteria* SPb: Publisher UTAS (in Russian)
[8] Kryanev A V and Fomenko M V 2004 Correctness of the formulation of the three-criteria problem of forming effective investment portfolios Bulletin of the Peoples' Friendship University series "Applied and Computer Mathematics" **3**(1)

Multidimensional singularly perturbed reaction-diffusion-advection problems with a balanced nonlinearity and their applications in the theory of nonlinear heat conductivity

M A Davydova and N N Nefedov

Department of Mathematics, Faculty of Physics, Lomonosov Moscow State University, 119991 Moscow, Russia

m.davydova@physics.msu.ru, nefedov@phys.msu.ru

Abstract. The stationary Lyapunov-stable solutions with internal transition layers (contrast structures) of multidimensional singularly perturbed reaction-diffusion-advection problems are investigated with the use of the modern methods of asymptotic analysis. On the basis of the modified boundary-function method, the asymptotic approximations of such solutions of an arbitrary order of accuracy in the case of a balanced nonlinearity are obtained. We propose a justified and effective algorithm that allows us to define and describe the localization region of the internal layer of contrast structure. We use this result to describe the thermal structures in the homogeneous nonlinear dissipative media.

1. Introduction

The present paper is aimed to the further development of the algorithms, set forth in [1-3], for the purpose of describing the stationary Lyapunov-stable solutions with internal transition layers (contrast structures) in the multidimensional singularly perturbed reaction-diffusion-advection problems. We define the corresponding solutions, as the solutions of stationary problems:

$$\varepsilon^2 \Delta u - f(\varepsilon \nabla u, u, x) = 0, \quad x = (x_1, ..., x_N) \in D \subset R^N, \tag{1}$$

$$u(x, \varepsilon) = g(x), \quad x \in S \tag{2}$$

where $\varepsilon > 0$ is a small parameter, the functions f, $g(x)$ and the boundary S are assumed to be sufficiently smooth, $\Delta = \sum_{k=1}^{N} \partial^2 / \partial x_k^2$ is the Laplace operator. The notation $\varepsilon \nabla u$ implies the dependence of the function f on the arguments $v_k := \varepsilon \partial u / \partial x_k$, $k = \overline{1, N}$. We assume that the function f satisfies the standard condition of the most quadratic growth with respect to ∇u.

One of the most significant applications of problem (1), (2) are the stationary problems of the nonlinear thermal conductivity. For example, the problem (1), (2) can describe the stationary stable states of dissipative media, characterized by the presence of spatial structures. In particular, the thermal structures, arising in a homogeneous dissipative medium with nonlinear characteristics under intense heating, are described by solutions with internal layers of a problem [4] of the type (1), (2):

$$\begin{aligned} &\varepsilon^2 \left(\Delta u - A(u)(\nabla u)^2 \right) = B(u, x), \quad x \in D \subset R^3, \\ &u(x, \varepsilon) = g(x), \quad x \in S, \end{aligned} \tag{3}$$

where $u(x)$ is a dimensionless temperature at the point x, $A(u) := -\partial \ln k(u) / \partial u$, $k(u) > \delta_0 > 0$ is a coefficient of the thermal conductivity, δ_0 is a some value. The function $B(u, x)$ describes the process of energy release, the function $g(x)$ determines the edge regime at the boundary S.

In the present paper, the problem (1), (2) is investigated in the framework of the schemes, described in [2] and [3] for the case of a balanced nonlinearity, and is illustrated by the application (3).

2. The basic conditions of existence and the formal asymptotics of the contrast structures

The defining condition in the problem (1), (2) is the condition on the nonlinearity f :

(C_1) *The degenerate equation* $f(0,u,x)=0$ *has isolated roots* $u=\varphi_i(x)$, $i=\overline{1,3}$, *satisfying the conditions* $\varphi_1(x)<\varphi_2(x)<\varphi_3(x)$, $f_u(0,\varphi_i(x),x)>0$, $i=1,3$, $f_u(0,\varphi_2(x),x)<0$ *for* $x\in D\cup S$.

In accordance with the algorithms of the papers [2] and [3], we define a set $\{\bar{\Omega}\}$ of sufficiently smooth closed surfaces in a domain D with the local coordinates (r,θ) in a $\delta-$neighborhood $\bar{\Omega}_\delta$ of each surface: $x\in\bar{\Omega}^\delta\mapsto(y,r)\in\bar{\Omega}\times[-\delta;\delta]$, where $y(\theta)\in\bar{\Omega}$, $\theta:=\{\theta_1,...,\theta_{N-1}\}\in\bar{\Theta}$ are curvilinear coordinates on the surface $\bar{\Omega}$, $r=\begin{cases} dist(x,y), & x\in D^{(+)}, \\ -dist(x,y), & x\in D^{(-)}, \end{cases}$ $D^{(+)}$ and $D^{(-)}$ are the exterior and interior subdomains of the domain D, separated by the surface $\bar{\Omega}$. In this coordinate system the representations $\partial u/\partial x_i=l^i(r,\theta)\partial u/\partial r+\sum_{j=1}^{N-1}q_j^i(r,\theta)\partial u/\partial\theta_j$, $i=\overline{1,N}$, are valid, where the functions $l^i(r,\theta)$, $q_j^i(r,\theta)$ are known. For convenience here and below, we retain the notations for the functions in variables (r,θ) as before.

By virtue of condition (C_1), the rest points $(\varphi_1(r,\theta),0)$ and $(\varphi_3(r,\theta),0)$, $(r,\theta)\in[-\delta;\delta]\times\bar{\Theta}$ should be classified as the saddles on the phase plane of associated system:

$$\frac{\partial\tilde{v}}{\partial\xi}=f(l^1(r,\theta)\tilde{v},...,l^N(r,\theta)\tilde{v},\tilde{u},r,\theta),\quad \frac{\partial\tilde{u}}{\partial\xi}=\tilde{v},\quad -\infty<\xi<+\infty, \tag{4}$$

where r and θ are treated as parameters. Let the following condition be satisfied.

(C_2) *For any* $(r,\theta)\in[-\delta;\delta]\times\bar{\Theta}$, *there exist the solutions* $\tilde{u}^{(\pm)}(\xi,r,\theta)$, $\tilde{v}^{(\pm)}(\xi,r,\theta)$ *of system (4) with the additional conditions:*

$$\tilde{u}^{(\mp)}(\mp\infty,r,\theta)=\varphi_i(r,\theta),\quad i=1,3,\quad \tilde{v}^{(\mp)}(\mp\infty,r,\theta)=0,\quad \tilde{u}^{(\mp)}(0,r,\theta)=\varphi_2(r,\theta). \tag{5}$$

For each surface $\bar{\Omega}\in\{\bar{\Omega}\}$ we define the function: $H(r,\theta):=\tilde{v}^{(+)}(0,r,\theta)-\tilde{v}^{(-)}(0,r,\theta)$, $(r,\theta)\in[-\delta;\delta]\times\bar{\Theta}$. The following requirement distinguishes the case of a balanced nonlinearity:

(C_3) *For any* $(r,\theta)\in[-\delta;\delta]\times\bar{\Theta}$, *the function* $H(r,\theta)$ *equals zero.*

The condition (C_3) can be interpreted as follows: on the phase plane of the system (4) there is a separatrix, connecting the saddles $(\varphi_1(r,\theta),0)$ and $(\varphi_3(r,\theta),0)$ for any surface $\bar{\Omega}\in\{\bar{\Omega}\}$.

We obtain the asymptotic approximation of a contrast structure by the C^1-matching of the boundary-layer asymptotics [5] on the transition surface $\Omega\in\{\bar{\Omega}\}$:

$$\begin{aligned} u^-(x,\varepsilon)&=\bar{u}^-(x,\varepsilon)+Qu^-(\xi,\lambda^*,\theta,\varepsilon),\\ u^+(x,\varepsilon)&=\bar{u}^+(x,\varepsilon)+\Pi u(\rho,\eta,\varepsilon)+Qu^+(\xi,\lambda^*,\theta,\varepsilon), \end{aligned} \tag{6}$$

where $\bar{u}^-(x,\varepsilon)=\varphi_1(x)+\varepsilon\bar{u}_1^-(x)+\ldots$, $\bar{u}^+(x,\varepsilon)=\varphi_3(x)+\varepsilon\bar{u}_1^+(x)+\ldots$ are regular series, $\Pi u(\rho,\eta,\varepsilon)=\Pi_0 u(\rho,\eta)+\varepsilon\Pi_1 u(\rho,\eta)+\ldots$ is a boundary series, describing the boundary layer in a neighborhood of the boundary S (see [5] for details), and $Qu^\pm(\xi,\lambda^*,\theta,\varepsilon)=Q_0u^\pm(\xi,\lambda^*,\theta)+\varepsilon Q_1u^\pm(\xi,\lambda^*,\theta)+\ldots$ are series, describing the boundary layers localized in a neighborhood of the transition surface, whose position is defined by the condition: $u(x,\varepsilon)=\varphi_2(x),\quad x\in\Omega$. We seek the equation of surface Ω in a local coordinate system, introduced in a neighborhood of some surface $\Omega_0\in\{\bar{\Omega}\}$ in a known way, in the form:

$$r=\lambda^*(\theta,\varepsilon)=\varepsilon(\lambda_1(\theta)+\varepsilon\lambda_2(\theta)+\ldots)\,. \tag{7}$$

The criterion for choosing of the surface Ω_0 will be formulated below.

Using the boundary-function method [1, 5], we obtain the problems for the terms of the series (6), where $\xi=\varepsilon^{-1}(r-\lambda^*(\theta,\varepsilon))$. In particular, taking into account the representation (7) and the condition (C_2), we have uniquely solvable nonlinear problems for sufficiently small ε:

$$\begin{gathered}\frac{\partial^2\tilde{u}^{(\pm)}}{\partial\xi^2}=f\left(l^1(\lambda^*,\theta)\frac{\partial\tilde{u}^{(\pm)}}{\partial\xi},\ldots,l^N(\lambda^*,\theta)\frac{\partial\tilde{u}^{(\pm)}}{\partial\xi},\tilde{u}^{(\pm)},\lambda^*,\theta\right),\\ \tilde{u}^{(\pm)}(0,\lambda^*,\theta)=\varphi_2(\lambda^*,\theta),\quad \tilde{u}^{(-)}(-\infty,\lambda^*,\theta)=\varphi_1(\lambda^*,\theta),\quad \tilde{u}^{(+)}(+\infty,\lambda^*,\theta)=\varphi_3(\lambda^*,\theta),\end{gathered} \tag{8}$$

where the notations are introduced: $\tilde{u}^{(-)}(\xi,\lambda^*,\theta):=\varphi_1(\lambda^*,\theta)+Q_0u^{(-)}(\xi,\lambda^*,\theta)$ for $-\infty<\xi\le 0$, $\tilde{u}^{(+)}(\xi,\lambda^*,\theta):=\varphi_3(\lambda^*,\theta)+Q_0u^{(+)}(\xi,\lambda^*,\theta)$ for $0\le\xi<+\infty$. For $n\ge 1$, we arrive to the linear problems:

$$\begin{gathered}\frac{\partial^2 Q_nu^{(\pm)}}{\partial\xi^2}=\sum_{k=1}^{N}l^k(\lambda^*,\theta)\tilde{f}_{v_k}(\xi,\lambda^*,\theta)\frac{\partial Q_nu^{(\pm)}}{\partial\xi}+\tilde{f}_u(\xi,\lambda^*,\theta)Q_nu^{(\pm)}+H_n^{(\pm)}(\xi,\lambda^*,\theta),\\ Q_nu^{(\pm)}(0,\lambda^*,\theta)=-\bar{u}_n^{(\pm)}(\lambda^*,\theta),\quad Q_nu^{(\pm)}(\pm\infty,\lambda^*,\theta)=0,\end{gathered} \tag{9}$$

whose solutions are written out in the explicit form by analogy with [5]. In the problems (9) we introduce the notations: $\tilde{f}(\xi,\lambda^*,\theta):=f\left(l^1(\lambda^*,\theta)\tilde{v}^{(\pm)},\ldots,l^N(\lambda^*,\theta)\tilde{v}^{(\pm)},\tilde{u}^{(\pm)},\lambda^*,\theta\right)$.

The coefficients of the expansion (7) are found from the condition of C^1– matching of the asymptotics (6) on the surface Ω:

$$\begin{gathered}\varepsilon\left.\left(\frac{\partial u^+}{\partial r}-\frac{\partial u^-}{\partial r}\right)\right|_{r=\lambda^*(\theta,\varepsilon)}=\left.\left(\tilde{v}^+(0,r,\theta)-\tilde{v}^-(0,r,\theta)\right)\right|_{r=\lambda^*(\theta,\varepsilon)}+\\ +\varepsilon\left.\left(\frac{\partial\varphi_3}{\partial r}(r,\theta)-\frac{\partial\varphi_1}{\partial r}(r,\theta)+\frac{\partial Q_1u^+}{\partial\xi}(0,r,\theta)-\frac{\partial Q_1u^-}{\partial\xi}(0,r,\theta)\right)\right|_{r=\lambda^*(\theta,\varepsilon)}+\ldots:=\\ H(\lambda^*(\theta,\varepsilon),\theta)+\varepsilon G_1(\lambda^*(\theta,\varepsilon),\theta)+\ldots=0.\end{gathered} \tag{10}$$

Taking into account the condition (C_3) and the representation (7), we have

$$\varepsilon G_1(0,\theta)+\varepsilon^2\left(M(\theta)\lambda_1(\theta)-\Phi_1(\theta)\right)+\ldots=0, \tag{11}$$

where the linear operator $M(\theta)$ will be defined below, $\Phi_1(\theta)$ is a known function.

The problem (8) for $\lambda^*=0$ corresponds to the problem (4), (5) for $r=0$. Therefore, the problem (8) is always solvable for $\lambda^*=0$, $\theta\in\overline{\Theta}$, and its solution $\tilde{u}(\xi,\theta)$ belongs to the functional class $C^2(-\infty,+\infty)$ and describes the internal layer of the contrast structure in the zeroth approximation.

The equation (11) is satisfied in the first approximation under the condition that determines the position of the transition surface in the zeroth approximation:

(C_5) *There exists a surface* $\Omega_0\in\{\overline{\Omega}\}$ *such that*

$$G_1(0,\theta):=\int_{-\infty}^{+\infty}\left(\sum_{k=1}^{N}\tilde{f}_{v_k}(\xi,\theta)\frac{\partial l^k}{\partial r}(0,\theta)\frac{\partial\tilde{u}}{\partial\xi}+\tilde{f}_r(\xi,\theta)\right)p(\xi,\theta)\tilde{v}(\xi,\theta)\xi d\xi+$$
$$+\int_{-\infty}^{+\infty}p(\xi,\theta)\tilde{v}(\xi,\theta)\sum_{k=1}^{N}\sum_{j=1}^{N-1}\tilde{f}_{v_k}(\xi,\theta)q_j^k(0,\theta)\frac{\partial\tilde{u}}{\partial\theta_j}d\xi-k(0,\theta)\int_{-\infty}^{+\infty}p(\xi,\theta)\tilde{v}^2(\xi,\theta)d\xi=0, \quad (12)$$

for $\theta\in\Theta_0$, *where* Θ_0 *is a region of change of the coordinate* θ *on the surface* Ω_0.

Here $k(0,\theta)$ is a sum of the principal curvatures of the surface Ω_0 at the point with the coordinate θ,

$$\tilde{f}(\xi,\theta):=f\left(l^1(0,\theta)\tilde{v},...,l^N(0,\theta)\tilde{v},\tilde{u},0,\theta\right),\ p(\xi,\theta)=\overline{p}(\xi,0,\theta):=\exp\left(-\int_0^{\xi}\sum_{k=1}^{N}\tilde{f}_{v_k}(\eta,r,\theta)l^k(r,\theta)d\eta\right)\Bigg|_{r=0}.$$

The periodic functions $\lambda_n(\theta)$, $n\geq1$, are determined by using the equation (11) from the linear differential problems for the equations:

$$M(\theta)\lambda_n(\theta)-\Phi_n(\theta)=0,\quad n\geq1,\quad \theta\in\Theta_0. \quad (13)$$

Here the functions $\Phi_n(\theta)$ are known, the linear elliptic operator $M(\theta)$ has the form:

$$M(\theta)=-\tilde{v}^{-1}(0,\theta)\left\{\int_{-\infty}^{+\infty}\tilde{v}^2(\eta,\theta)p(\eta,\theta)d\eta\sum_{j=1}^{N-1}m_j(0,\theta)\frac{\partial^2}{\partial\theta_j^2}+\right.$$
$$+\sum_{k=1}^{N}\sum_{j=1}^{N-1}q_j^k(0,\theta)\left[\frac{\partial\varphi_3}{\partial r}(0,\theta)\int_0^{+\infty}\tilde{f}_{v_k}(\eta,\theta)\tilde{v}(\eta,\theta)p(\eta,\theta)d\eta+\right.$$
$$\left.+\frac{\partial\varphi_1}{\partial r}(0,\theta)\int_{-\infty}^{0}\tilde{f}_{v_k}(\eta,\theta)\tilde{v}(\eta,\theta)p(\eta,\theta)d\eta\right]\frac{\partial}{\partial\theta_j}+$$
$$+\sum_{j=1}^{N-1}\left[2m_j(0,\theta)\int_{-\infty}^{+\infty}\partial\tilde{v}/\partial\theta_j\,\tilde{v}(\eta,\theta)p(\eta,\theta)d\eta+\overline{n}_j(0,\theta)\int_{-\infty}^{+\infty}\tilde{v}^2(\eta,\theta)p(\eta,\theta)d\eta\right]\frac{\partial}{\partial\theta_j}+$$
$$+\frac{\partial}{\partial r}\left[\int_{-\infty}^{+\infty}\tilde{v}(\eta,r,\theta)\overline{p}(\eta,r,\theta)\left(\sum_{k=1}^{N}\tilde{f}_{v_k}(\eta,\theta)\frac{\partial l^k}{\partial r}(r,\theta)\tilde{v}(\eta,r,\theta)+\right.\right.$$
$$\left.\left.+\tilde{f}_r(\eta,\theta)\right)\eta d\eta+\int_{-\infty}^{+\infty}\tilde{v}(\eta,r,\theta)\overline{p}(\eta,r,\theta)\sum_{k=1}^{N}\sum_{j=1}^{N-1}\tilde{f}_{v_k}(\eta,\theta)q_j^k(r,\theta)\frac{\partial\tilde{u}}{\partial\theta_j}(\eta,r,\theta)d\eta\right]\Bigg|_{r=0}+$$
$$\left.+\sum_{j=1}^{N-1}k_j^2(0,\theta)\int_{-\infty}^{+\infty}\tilde{v}^2(\eta,\theta)p(\eta,\theta)d\eta-k(0,\theta)\frac{\partial}{\partial r}\int_{-\infty}^{+\infty}\tilde{v}^2(\eta,r,\theta)\overline{p}(\eta,r,\theta)d\eta\Bigg|_{r=0}\right\},$$

where $\tilde{u}(\xi,r,\theta)$, $\tilde{v}(\xi,r,\theta)$ is a solution of the problem (4), (5), $m_j(0,\theta)$, $\bar{n}_j(0,\theta)$ are know function.

The condition that ensures the solvability of periodic problems for the equation (13) is formulated as follows (see [2] for details):

(C_6) *The operator* $M(\theta)$ *is positively invertible for* $\theta\in\Theta_0$ *on the set of positive periodic right-hand sides, or equivalently, the principal eigenvalue of the spectral problem for the equation* $M(\theta)\psi=\lambda\psi$, $\theta\in\Theta_0$ *with periodicity condition is positive.*

The existence of Lyapunov-stable solutions with the asymptotics (6), as the stationary solutions of the corresponding parabolic problem, can be proved on the basis of the asymptotic method of differential inequalities [6], by constructing the upper and lower solutions of problem (1), (2), using standard modifications of the partial sums of the n-th order of expansions (6) by analogy with [7, 8].

3. Example

We consider the problem (3) for $x\in D\subset R^2$, taking into account the results [8]. The dependence of the power of thermal sources from temperature is determined by the condition: the equation $B(u,x)=0$ has isolated roots $u=\varphi_i(x)$, $i=\overline{1,3}$, satisfying the conditions $\varphi_1(x)<\varphi_2(x)<\varphi_3(x)$, $B_u(\varphi_i(x),x)>0$, $i=1,3$, $B_u(\varphi_2(x),x)<0$ for $x\in D\cup S$. The solutions of problems (4), (5) can be represented in the explicit form:

$$\tilde{v}^{(\mp)}(\tilde{u}^{(\mp)},r,\theta)=\left(\int_{\varphi_i(r,\theta)}^{\tilde{u}^{(\mp)}} 2B(s,r,\theta)\exp\left(2\sum_{k=1}^{2}(l^k(r,\theta))^2\int_s^{\tilde{u}^{(\mp)}}A(\sigma)d\sigma\right)ds\right)^{1/2},\quad i=1,3. \tag{14}$$

Therefore, the condition

$$\int_{\varphi_1(r,\theta)}^{\varphi_3(r,\theta)} B(\xi,r,\theta)\exp\left(2\sum_{k=1}^{2}\left(l^k(r,\theta)\right)^2\int_{\xi}^{\varphi_2(r,\theta)}A(\eta)d\eta\right)d\xi=0,\quad (r,\theta)\in[-\delta;\delta]\times\bar{\Theta}$$

is equivalent to the condition (C_3) and marks out the case of a balanced nonlinearity for the problem (3). When this condition is satisfied, the equation (14) describe the separatrix, connecting the saddles $(\varphi_1(r,\theta),0)$ and $(\varphi_3(r,\theta),0)$ on the phase plane of associated system for $(r,\theta)\in[-\delta;\delta]\times\bar{\Theta}$.

Suppose, that the curve Ω_0 exists, and it can be described by an equation $R=\sigma(\theta)$, $0\le\theta<2\pi$ in some polar coordinate system (R,θ) with a pole inside D. Then, with respect to the function $\sigma(\theta)$, we obtain a nonlinear differential problem for the equation (12) under $N=2$ with the condition of periodicity:

$$\begin{aligned}&\int_{\varphi_1(0,\theta)}^{\varphi_3(0,\theta)} B_r(\tilde{u},0,\theta)exp\left(-2\int_{\varphi_2(0,\theta)}^{\tilde{u}}A(\lambda)d\lambda\right)\xi(\tilde{u})d\tilde{u}-\\&-\frac{\sigma^2+2\sigma'^2-\sigma''\sigma}{(\sigma'^2+\sigma^2)^{3/2}}\int_{\varphi_1(0,\theta)}^{\varphi_3(0,\theta)} exp\left(-2\int_{\varphi_2(0,\theta)}^{\tilde{u}}A(\lambda)d\lambda\right)\tilde{v}(\tilde{u},\theta)d\tilde{u}=0,\quad \sigma(\theta)=\sigma(\theta+2\pi),\end{aligned} \tag{15}$$

where the quadrature formula

$$\xi(\tilde{u})=\int_{\varphi_2(0,\theta)}^{\tilde{u}}\left(\int_{\varphi_i(0,\theta)}^{\lambda} 2B(s,0,\theta)\exp\left(2\int_s^{\lambda}A(\tau)d\tau\right)ds\right)^{-1/2}d\lambda,\quad i=1,\quad \xi\le0;\quad i=3,\quad \xi\ge0 \tag{16}$$

describes the internal layer of the contrast structure in the zero-th approximation, $\tilde{v}(\tilde{u},\theta)$ is the equation of the connecting separatrix, which follows from the equations (14) for $r=0$. Here the polar angle θ is chosen as one of the local coordinates, introduced in the neighborhood of the curve Ω_0. The equation (16) is obtained by using the equations (14) for $r=0$.

Suppose, that the boundary S is described by the equation: $R=\bar{\sigma}(\theta)$, $0\le\theta<2\pi$. Then, the solution of problem (3) in the neighborhood of the boundary S is described by the quadrature formula

$$\rho=-\int_{g(0,\theta)}^{\varphi_3(0,\theta)+\Pi_0 u}\left(\int_{\varphi_3(0,\theta)}^{\lambda}2B(s,0,\theta)\exp\left(2\int_s^{\lambda}A(\tau)d\tau\right)ds\right)^{-1/2}d\lambda,\quad 0\le\rho<+\infty$$

under the following condition on the choice of the edge regime for $g(0,\eta)>\varphi_3(0,\theta)$:

$$\int_{\varphi_3(0,\theta)}^{g(0,\theta)}B(s,0,\theta)\exp\left(2\int_s^{g(0,\theta)}A(\tau)d\tau\right)ds>0,\quad 0\le\theta<2\pi.$$

The results of the study of problem (3) can be used to create a numerical algorithm that uses asymptotic analysis to construct spatially inhomogeneous mashes when describing the internal layer of contrast structure [9], and also for the purpose of constructing the test examples.

The research was supported by the Russian Science Foundation (project no. 18-11-00042).

References

[1] Vasil'yeva A B and Butuzov V F 1990 *Asimptoticheskiye metody v teorii singulyarnykh vozmushcheniy* (Moskva: Vysshaya shkola)
[2] Nefedov N N and Sakamoto K 2003 *Hiroshima Mathematical Journal* vol 33 no 3pp 391- 432
[3] Davydova M A and Nefedov N N 2017 *Numerical Analysis and Its Applications Lecture Notes in Comput. Sci.* vol 10187 (Springer International Publishing AG) pp 277-285
[4] Davydova M A and Zakharova S A 2018 *Modeling and Analysis of Information Systems* vol 25 no 1 (Yaroslavl: Yaroslavl State University Press) pp 83-91
[5] Davydova M A 2015 *Mathematical Notes* (Consultants Bureau (United States)) vol 98 no 6 pp 909-919
[6] Nefedov N N *Numerical Analysis and Its Applications Applications Lecture Notes in Comput. Sci.* vol. 8236 (Springer International Publishing AG) pp 62–72
[7] Nefedov N N and Davydova M A 2012 *Differ. Equations* vol 48 (Maik Nauka/Interperiodica Publishing (Russia Federation)) no 5 pp 745-755
[8] Nefedov N N and Davydova M A 2013 *Differ. Equations* vol 49 (Maik Nauka/Interperiodica Publishing (Russia Federation)) no 6 pp 688-706
[9] Volkov V T and Nefedov N N 2013 *Numerical Analysis and Its Applications Lecture Notes in Comput. Sci.* vol 8236 (Springer International Publishing AG) pp 524-531

IOP Conf. Series: Journal of Physics: Conf. Series **1205** (2019) 012012 doi:10.1088/1742-6596/1205/1/012012

Cooling effect for evaporating of drops situated at high-conductivity substrate

S Z Dunin and O V Nagornov

National Research Nuclear University MEPhI (Moscow Engineering Physics Institute), 31 Kashirskoe shosse, 115409 Moscow, Russia

E-mail: nagornov@yandex.ru

Abstract. The temperature in evaporated liquid drop situated at high-conductivity substrate depends significantly on so called the evaporated number Γ. This number is responsible for the thermal properties of liquid, the vapor diffusion, and the concentration of the saturated vapor. We solve the problem on the evaporating drop at the solid substrate taking into account influence of the evaporated number on the vapor concentration and temperature in liquid and solid phases.

1. Introduction

Evaporation of drops is studied during more than hundred years. However, even spherically symmetric solution for sphere have many uncertainties [1]. The problem is significantly complicated for sessile drops [2] due to additional factors connected with the evaporation processes. Important effect of the drop evaporation results in the cooling effect at the surface. In majority of papers this effect is taken into account numerically with correction on experimental data.

Larson [2] calculated the temperature at the drop surface in approach of non-constant temperature. Simultaneous determination of the temperature and concentration at the surface drop was considered too [3]. In various papers the problems were solved in quadrature while the mass flux is extrapolated by equation $j(\theta_c) \approx F(\theta_c)(1 - (r/R)^{\lambda(\theta_c)-1})$ where the upper index depends on the edge angle θ_c [3, 4, 5]. The drop evaporation processes were studied by many authors [2, 3, 4, 5, 6, 7, 8]. Some results were obtained at small edge angle [3, 5]. The other approach is based on the numerical methods for total statement of the problem [6, 7, 8].

2. Mathematical model

In this paper we solve the problem analytically in the most wide statement so that the concentration, the temperature in liquid and solid phases are found out by quadrature without additional restrictions and extrapolations. The temperature depends on the heat conductivity k_L, and the evaporative number Γ. If the heat conductivity of substrate k_s is high comparing with ones for drop k_L, then the bottom part of the drop is isothermal. The convective terms and the gravitational effects can be neglected because the Pecle number $Pe << 1$, and $B_0 << 1$. The evaporation of the drop is assumed to be quasi-steady. Then we can use the Laplace equations $\triangle T_L = 0$, $\triangle c = 0$ for temperature T_L and concentration c together with the boundary equations

IOP Conf. Series: Journal of Physics: Conf. Series **1205** (2019) 012012 doi:10.1088/1742-6596/1205/1/012012

$T_L(\alpha, \theta = 0) = T_{tr}$, and $dc(\alpha, \theta = \pi)/d\theta = 0$. The solution of this problem in toroidal coordinates (α, θ) under condition $(1 - \Gamma)(T_L(\alpha, \theta_c) - T_{tr} << \Gamma \frac{\triangle c_{tr}}{c'_{tr}(T)})$ is derived analytically:

$$T_L(\alpha,\theta) - T_{tr} = \{2(\cosh\alpha + \cos\theta)\}^{\frac{1}{2}} \cdot$$
$$\cdot \int_0^\infty d\tau\, P_{-\frac{1}{2}+i\tau}(\cosh\alpha) \frac{\cosh\tau\theta_c}{\cosh\tau\pi} A_L \frac{\sinh\tau\theta}{\sinh\tau\theta_c},$$
$$T_L(\alpha, 0) = T_{tr}, \tag{1}$$
$$c(\alpha,\theta) - c_\infty = \{2(\cosh\alpha + \cos\theta)\}^{\frac{1}{2}} \cdot$$
$$\cdot \int_0^\infty d\tau\, P_{-\frac{1}{2}+i\tau}(\cosh\alpha) B \frac{\cosh\tau\theta_c}{\cosh\tau\pi} \frac{\cosh\tau(\pi - \theta)}{\cosh\tau(\pi + \theta_c)}.$$

Nuclei of integrals are found out from the following conditions: the Claiperon-Clausius equation and the energy ballance equation. We calculated the temperature inside the drop and the local mass flux at the drop surface. It is derived that maximum and minimum temperatures are at the drop edge and the drop top, respectively.

Let us consider cases of $\Gamma(\Gamma - 1 \ll 1)$ and temperature changes at the drop surface satisfied by equation $(1 - \Gamma)(T_L(\alpha, \theta_c) - T_{tr}) \ll \Gamma \frac{\triangle c_{tr}}{c'_{tr}(T)}$. For example, under temperature $295K$ and atmospheric pressure of $99.8kPa$, the evaporating number $\Gamma = 1.03$ for acetone in air. Then we can calculate nuclei B and A_L. So that temperature in liquid is

$$T_L(\alpha,\theta) - T_{tr} = -\frac{\Gamma \triangle c_{tr}}{c'_{tr}(T)} \{2(\cosh\alpha + \cos\theta)\}^{\frac{1}{2}} \cdot$$
$$\cdot \int_0^\infty d\tau\, P_{-\frac{1}{2}+i\tau}(\cosh\alpha) \frac{\tanh\tau\pi \sinh\tau\theta}{\cosh\tau(\pi - \theta_c) + \Gamma\sinh\tau(\pi - \theta_c)\tanh\tau\theta_c}. \tag{2}$$

Moreover, the rate of the drop volume due to evaporation is

$$-\frac{dV}{dt} = 4\pi DR\triangle c_{tr} \int_0^\infty d\tau\, P_{-\frac{1}{2}+i\tau}(\cosh\alpha) \cdot$$
$$\cdot \frac{1}{(1 + \Gamma\tanh\tau(\pi - \theta_c)\tanh\tau\theta_c)\cosh\tau\pi\cosh\tau\theta_c\cosh\tau(\pi - \theta_c)}. \tag{3}$$

Calculation of integrals (1)-(3) can be done in the complex plane τ, using the residual theory. The main input in the integral gives the poles near real axis τ. Functions $P_{-\frac{1}{2}+i\tau}(\cosh\alpha)$ can be expressed via hyper geometric function F and gamma function Γ [9]:

$$P_{-\frac{1}{2}+i\tau}(\cosh\alpha) = \frac{\Gamma(i\tau)}{\sqrt{\pi}\Gamma(0.5 + i\tau)} (2\cosh\alpha)^{i\tau} F(3/4 - i\tau, 1/4 - i\tau; 1 - i\tau; 1/\cosh^2\alpha) + CC$$

Substituting this function in integrals and using the resudial theory we can find temperature and concentration. The poles are in imaginary axis and can be determined by equation

$$\cos\lambda(\pi - \theta_c) - \Gamma\sin\lambda(\pi - \theta_c)\tan\lambda\theta_c = 0; \tau_d = \lambda.$$

The edge angle is connected with the evaporated number by the following equation

$$\theta_c = \frac{1}{\lambda} \arctan \frac{-(1-\Gamma)z - \sqrt{((1-\Gamma)z)^2 - 4\Gamma}}{2\Gamma}; z = \tan\lambda\pi \le 0.$$

We obtain the well-known result at $\Gamma = 0$ [4]:

$$\theta_c = -\frac{1}{\lambda} \arctan \frac{1}{\tan\lambda\pi}.$$

IOP Conf. Series: Journal of Physics: Conf. Series **1205** (2019) 012012 doi:10.1088/1742-6596/1205/1/012012

More higher terms can be determined at small Γ. Maximal and minimal temperatures are at the drop edge and top, respectively. The total heat flux along whole drop surface takes form:

$$Q_L(\theta_c) = -4\pi R \frac{k_L}{c'_{tr}(T)} \int_0^\infty d\tau \, \frac{\cosh\tau\theta_c}{\sinh\tau\pi\cosh\tau\pi} \frac{c'_{tr}(T)A_L}{\sinh\tau\theta_c}$$

$$= 4\pi R k_L \Gamma \frac{\triangle c_{tr}}{c'_{tr}(T)} \int_0^\infty d\tau \, \frac{1}{(1+\Gamma\tanh\tau(\pi-\theta_c)\tanh\tau\theta_c)\cosh\tau\pi\cosh\tau\theta_c\cosh\tau(\pi-\theta_c)}.$$

The rate of change of the edge angle is

$$-\dot{\theta}_c = \frac{\pi}{\rho R^2}(1+\cos\theta_c)\int_0^\infty d\tau \, \frac{1}{(1+\Gamma\tanh\tau(\pi-\theta_c)\tanh\tau\theta_c)\cosh\tau\pi\cosh\tau\theta_c\cosh\tau(\pi-\theta_c)}$$

The Marangoni forces can be defined through temperature distribution at the drop surface.

3. Conclusions

It is derived that the drop temperature and the evapoeration rate depend significantly on the evaporated number $\Gamma = c'_{T\infty} L D k_L^{-1}$. The temperature and concentration are obtained in explicit form. The mass flux depends on both the edge angle and the evaporated number Γ.

Acknowledgment. The authors acknowledge financial support of the Russian Foundation for basic research, grant # 18-29-10039.

References

[1] Farshid Chinia S, Amirfazlia A 2013 Understanding the evaporation of spherical drops in quiescent environment. Colloids and Surfaces A: Physicochem. Eng. Aspects 432, 82–88

[2] Larson R G 2014 Transport and deposition patterns in drying sessile droplets. AIChE J. 60(5), 1538

[3] David S, Sefiane K, Tadrist L 2007 Experimental investigation of the effect of thermal properties of the substrate in the wetting and evaporation of sessile drops. Colloids and Surfaces A: Physicochem. Eng. Aspects 298, 108-–114

[4] Deegan R D, Bakajin O, Dupont T F, Huber G, Nagel S R, Witten T A 2000 Contact line deposits in an evaporating drop. Phys. Rev. E 62, 756–765

[5] Dunn G J, Wilson S K, Duffy B R, David, S and Sefiane K 2009 The strong influence of substrate conductivity on droplet evaporation. J. Fluid Mech. 623, 329–351

[6] Saada M A, Chikh S, Tadrist L 2013 Evaporation of a sessile drop with pinned or receding contact line on a substrate with different thermophysical properties. Int. J. Heat Mass Transf. 58, 197-–208

[7] Chafea Bouchenna, Mebrouk Ait Saada, Salah Chikh, Lounes Tadrist 2017 Generalized formulation for evaporation rate and flow pattern prediction inside an evaporating pinned sessile drop Int. J. Heat Mass Transfer, vol.109, 482–500

[8] Kai Zhang, Liran Ma, Xuefeng Xu, Jianbin Luo, and Dan Guo 2014 Temperature distribution along the surface of evaporating droplets, Phys. Rev. E 89, 032404

[9] Lebedev N N 1965 Special functions and their applications [Silverman, R. A. (ed.)] Prentice-Hall, New Jersey

[10] Dunin S Z, Nagornov O V and Trifonenkov V P 2017 Model of spontaneous evaporating droplet on solid horizontal substrate. Journal of Physics: Conference Series 788, 012048

IOP Conf. Series: Journal of Physics: Conf. Series **1205** (2019) 012013 doi:10.1088/1742-6596/1205/1/012013

Multistability in dynamics of an encapsulated bubble contrast agent: coexistence of three attractors

I R Garashchuk[1], D I Sinelshchikov[2] and N A Kudryashov[3]

Department of Applied Maths, NRNU MEPhI, Kashirskoe shosse 31, Moscow, Russia

E-mail: [1] ivan.mail4work@yandex.ru, [2] disine@gmail.com, [3] nakudr@gmail.com

Abstract. In this work we discuss complex dynamics arising in a model describing behavior of an encapsulated bubble contrast agent oscillating close to an elastic wall. We demonstrate presence of three coexisting attractors in the system. We propose an efficient numerical procedure based on the continuation method that can be used to locate the area of coexistence of these attractors in the parameters space. We provide area of coexistence of three attractors obtained by means of the proposed procedure.

1. Introduction

Nowadays there is a rising interest towards physical models with multistability and hidden oscillations [1]. It is known that multistability is possible in models, describing microbubble contrast agent dynamics [2, 3], which are encapsulated micrometer size gas bubbles. Contrast agents are used in medical ultrasound imaging in order to improve echogenecity and more complicated applications related with targeted drug delivery and non-invasive therapy. Modern models of contrast agents are based on the Rayleigh–Plesset equation [4]. They are modified accordingly to take into account the influence of the shell, external ultrasound pressure field and some other effects like a blood vessel wall close to the oscillating bubble. Thus, the dynamical system being discussed has a form of a forced non–linear oscillator. Such system has no fixed points and according to classification in [1] all the attractors are hidden.

We believe that multistability is not beneficial for contrast agent applications, because it can lead to unpredictable behavior [2]. Locating the multistability areas can be useful, so they can be avoided in applications. Finding all the coexisitng attractors is a hard problem. Direct computation of the attractors' basins is a very CPU intense task. Locating a multistability area with the help of this approach would lead to such calculations for a lot of parameter values, which can be very time consuming. Thus, it would be interesting to have another method for localizing areas of multistability. A good candidate is the continuation method [1, 2, 5]. In earlier works on microbubble's dynamics only two coexisting attractors were presented [2, 3]. Here we focus on an area of multistability with three coexisting attractors.

2. The method applied

Here we consider the following model of an encapsulated gas bubble oscillating in the vicinity of an elastic wall [4]:

$$\left(1-\frac{1}{2}W-\frac{\dot{R}}{c}\right)R\ddot{R}+\frac{3}{2}\left(1-\frac{2}{3}W-\frac{\dot{R}}{3c}\right)\dot{R}^2=\frac{1}{\rho_1}\left[1+\frac{\dot{R}}{c}+\frac{R}{c}\frac{d}{dt}\right](P_r-P_{ac}\sin(\omega t)), \quad (1)$$

where

$$W=\frac{(\rho_1-\beta)R}{(\rho_1+\beta)d}+\frac{(\beta-\rho_3)R}{(\beta+\rho_3)(d+h)}-\frac{(\rho_1-\beta)(\beta-\rho_3)R}{(\rho_1+\beta)(\beta+\rho_3)h}, \quad (2)$$

$$P_r=\left(P_0+\frac{2\sigma}{R_0}\right)\left(\frac{R_0}{R}\right)^{3\gamma}-\frac{4\eta\dot{R}}{R}-\frac{2\sigma}{R}-P_0,$$

R is the bubble radius, $P_{\text{stat}}=100$ kPa is the static pressure, $P_v=2.33$ kPa is the vapor pressure, $P_0=P_{\text{stat}}-P_v$, P_{ac} is the magnitude of the pressure of the external field, $\sigma=0.0725$ N/m is the surface tension, $\rho_1=1000$ kg/m^3 is the density of the liquid inside the blood vessel, $\rho_2=1060$ kg/m^3 is the blood vessel wall density, $\rho_3=1000$ kg/m^3 is the density of the fluid surrounding the blood vessel, $\eta=0.001$ Ns/m^3 is the viscocity of the liquid, $c=1500$ m/s is the sound speed, $\gamma=4/3$ is the polytropic exponent, $\nu=0.5$ is the Poisson's ratio for an elastic wall, $\beta=\rho_2\nu/(1-\nu)$ is a characteristic of the wall, $h=1$ mm is the thickness of the wall, d denotes the distance between the wall and the center of the bubble and is close to R_0, shell parameters: $\chi=0.22$ N/m and $\kappa_S=2.5\cdot10^{-9}$ kg/s that correspond to the SonoVue contrast agent with equilibrium radius $R_0=1.72\mu$m [6]. For numerical calculations we use the following non–dimensional variables: $r=R/R_0$, $\tau=\omega_0 t$, where $\omega_0^2=3\kappa P_0/(\rho R_0^2)+2(3\kappa-1)\sigma/R_0+4\chi/R_0$ is the natural frequency of bubble oscillations [7]. We denote non–dimensional radial speed by u.

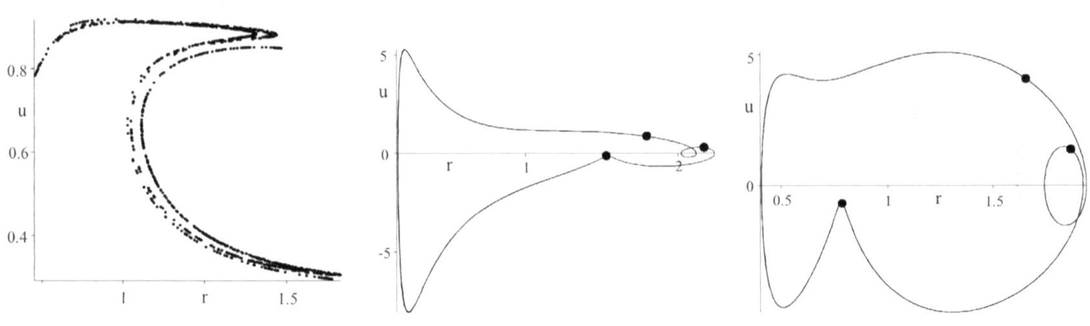

Figure 1. Three coexisting attractors at $P=2.35$ MPa, $\omega=5.567$ s^{-1}: chaotic (Poincare section) and two 3-periodic attrcators (phase portraits, bold dots represent points in the corresponding Poincare sections). We refer to them as follows: left – "first", middling - "second", right – "third".

Phase portraits of three coexisting attractors are given in Fig. 1. For convenience below we will refer to these attractors as "first", "second" and "third" in the ordering from left to right respectively as they are placed in Fig 1. Different attractors can be effectively distnguished by the value of the maximal Lyapunov exponent (below we will refer to it as λ_{max}), corresponding to each attractor. Basins of attraction corresponding to attractors in Fig. 1 are presented in Fig. 2. One can observe that these basins have quite complicated shapes. Consequently, random picking of several points for initial conditions could easily lead to finding not all of them.

Once we have found these attractors we can use the continuation tool to determine the initial conditions leading to each attractor for the other parameter values. To establish the borderline of coexistence of all three attractors, we can calculate the maximal of the Lyapunov exponents

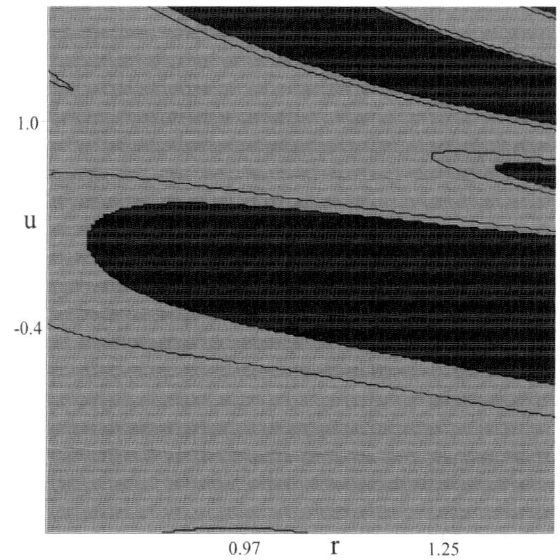

Figure 2. Bassins of attractors presented in Fig. 1: red – the first attractor (chaotic), black – the second attractor, gray - the third one.

of the attractors for consecutive values of (P, ω), while moving towards border in the parameter space. For the third attractor the bifurcation of disapperance is accompanied by the condition $\lambda_{max} = 0$. For the chaotic attractor when passing the point of disappearance the condition $\lambda_{max,n} \cdot \lambda_{max,n+1} < 0$ is met. Let us describe the numerical procedure of determining this borderline, which allows us to perform calculations only in a neighborhood of the borderline instead of scanning the entire parameters area and computing bassins of attractors.

Let us assume we have found a point (P_0^0, ω_0^0) which is close to the borderline of disappearance of one of the attractors. Such points can be obtained by using the very same procedure (which is described below) with larger step size by parameters and more rough restrictions on the precision of satisfying conditions of disappearance of the attractor. It allows us to get rough estimations of some points of the desired borderline; initially acquired attractors (see Fig. 1) can be the starting point to get these estimates:

(i) Start from the point (P_0^0, ω_0^0) taking initial conditions from the attractor, which is assumed to disappear when moving through the borderline (see illustration in Fig. 3)
(ii) Calculate λ_{max} on every step (P_i^0, ω_i^0) and move towards border with small steps while taking the initial conditions for each following step by the continuation method
(iii) Stop when the disappearance condition is met with precision high enough and save the final point (P_f^0, ω_f^0) as the point lying on the borderline
(iv) Choose (P_0^1, ω_0^1) close to (P_0^0, ω_0^0) along the desired borderline. Move from (P_0^0, ω_0^0) to (P_0^1, ω_0^1) using the continuation method in order to stay on the same attractor.
(v) Start the procedure from step one treating (P_0^1, ω_0^1) as the starting point

It allows us to move along the borderline to a desirable point (P^K, ω^K) and eventually outline the entire area. It is important that all the points (P_0^k, ω_0^k) are chosen inside the area of coexistence of all the attractors, and they do not fall out of this area on any step. This procedure allows us to avoid a lot of long computations of basins of attractors.

The area of coexistence of three attractors in parameters space (P, ω) obtained by this algorithm is shown in Fig. 3. Note that on the right borderline (blue) the first attractor

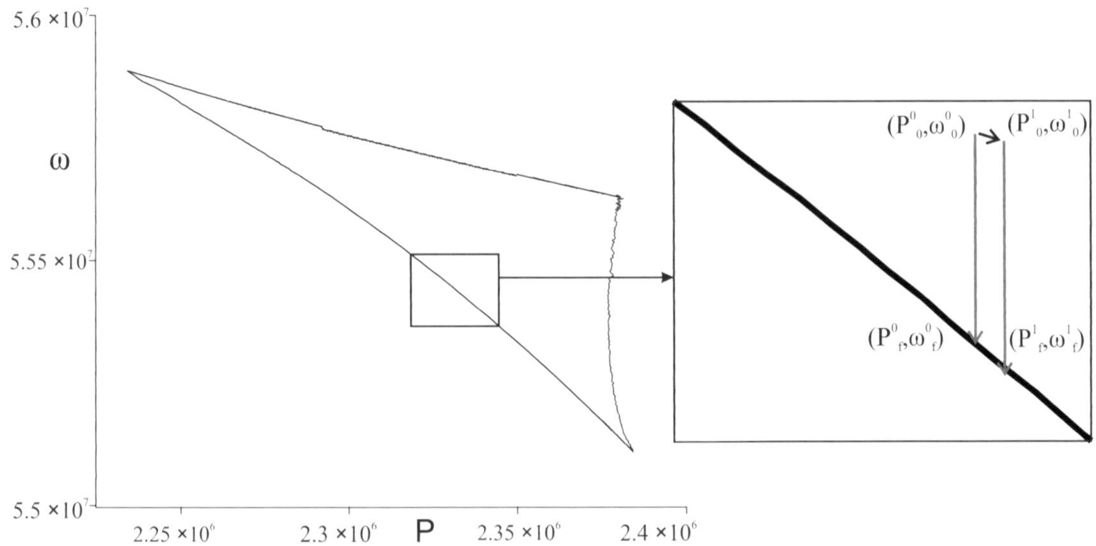

Figure 3. Area of coexistence of three attractors in parameters space (P, ω). Black borderline corresponds to disapperance of the third attractor, blue – to disappearance of the first one. The zoomed rectangle is used to illustrate the numerical procedure.

disappears, while on the others the disappearing attractor is the third one. Also, attractors presented in Fig. 1 do not stay the same in this entire area, but can deform quite significantly. For example, in the upper-left corner of this area the first attractor goes through reversed period-doubling cascade and becomes periodic (e.g. there exist values of (P, ω) for which this attractor is 4-periodic, while two others are still 3-periodic).

Although the region of parameters shown in Fig. 3 is small in comparison with the one investigated in [2], it shows that coexistence of more than two attractors is possible in this model. It is also possible that there are more small areas of coexistence of multiple attractors which have not been found yet. Note that the third attractor exists in a large region of parameters in higher pressures zone. In this work we are not interested in this area of parameters, because only two attractors coexist there (chaotic attractor disappears earlier) and we do not demonstrate it. The third attractor found here was missed in earlier work [2] due to its complicated area of existence (some areas of multistability with this attractor were mentioned as zones with only one attractor in [2]).

3. Conclusion

In this work we have found three coexisting hidden attractors in the model of a bubble contrast agent oscillating near an elastic wall. Phase portraits and basins of attraction are provided for certain value of parameters (see Fig.-s 1, 2). We suggest a numerical procedure that can help to establish the area of multistability with three coexisting attractors in (P, ω) parameters space. They key feature of the algorithm is that once we have localized all the coexisting attractors for one value of the parameters, we do not need to scan the initial conditions space for other parameters values anymore, but we can use the continuation approach instead. Thus we avoid very time consuming computations by means of the continuation method. The area of coexistence of the three attractors obtained by using this algorithm is shown in Fig. 3.

Acknowledgments
This work is supported by Russian Science Foundation, grant number 17-71-10241.

References

[1] Dudkowskia D, Jafari S, Kapitaniak T, Kuznetsov N V, Leonov G A and Prasad A 2016 *Physics Reports* **637**
[2] Garashchuk I R, Sinelshchikov D I and Kudryashov N A 2018 *Regular and Chaotic Dynamics* **23**
[3] Parlitz U, Engisch V, Scheffczyk C and Lauterborn W 1990 *The Journal of the Acoustical Society of America* **88**
[4] Doinikov A A, Aired L and Bouakaz A 2011 *Physics in Medicine and Biology* **56**
[5] Leonov G A, Kuznetsov N V and Mokaev T N 2015 *The European Physical Journal* Special Topics **224**
[6] Tu J, Guan J, Qiu Y and Matula T J 2009 *Journal of the Acoustical Society of America* **126**
[7] Garashchuk I R, Sinelshchikov D I and Kudryashov N A 2018 *EPJ Web of Conferences* **173**

Traveling waves and plasma acceleration in Quasi-Steady Plasma Accelerators (QSPAs) with longitudinal field

M B Gavrikov[1] **and A A Taiurskii**[1]
[1]Keldysh Institute of Applied Mathematics of the Russian Academy of Sciences, Miusskaya sq., 4, Moscow, 125047, Russia

E-mail: mbgavrikov@yandex.ru, tayurskiy2001@mail.ru

Abstract. We find and classify plane traveling waves in hot magnetized plasma in the case of zero transverse electric field (in the reference frame of background plasma motion). On the basis of these results, we evaluate the possibilities of using QSPAs with longitudinal magnetic field for plasma acceleration in plasma accelerator channels.

1. Introduction

A plane traveling wave is defined as a flow of a continuous medium whose parameters, at each time instant, depend on a single spatial coordinate x, while the profile of each parameter is moving as a whole along the axis x with a common constant phase velocity a. Such waves are solutions of the equations of motion of continuous media and have the form $f(\theta)$, where $\theta = x - at$ is the phase of the wave. In a plasma continuum subject to the equations of electromagnetic hydrodynamics (EMHD) [1], traveling waves are solutions of the following system of equations for the dimensionless functions $\boldsymbol{H}_\perp(\theta)$, $u(\theta) = U_\parallel(\theta) - a$ [2]:

$$\left(\frac{|u|}{2} - \beta^2\right)\boldsymbol{H}_\perp - \xi^2 |u| \frac{d}{d\theta}\left(|u| \frac{d\boldsymbol{H}_\perp}{d\theta}\right) + \Lambda\xi\beta|u|\left[\boldsymbol{k}, \frac{d\boldsymbol{H}_\perp}{d\theta}\right] + \boldsymbol{p} = 0, \tag{1}$$

$$|u| + K/|u|^\gamma + H_\perp^2 = 2, \tag{2}$$

where $\boldsymbol{k}$ is the unit vector along the axis x; all vectors have two components: along (||) and across ($\perp$) the vector $\boldsymbol{k}$; $\xi > 0$, $K \geq 0$, β are similarity numbers; $\boldsymbol{p} \perp \boldsymbol{k}$ is an arbitrary vector; $\Lambda = (\lambda_+/\lambda_-)^{1/2} - (\lambda_-/\lambda_+)^{1/2}$, $\lambda_\pm = m_\pm / e_\pm$, $m_\pm$, $e_\pm$ are masses and absolute values of the charges of plasma components. The parameter β is proportional to the longitudinal magnetic field; K is the plasma temperature; ξ^2 is inversely proportional to the number of plasma particles per unit length. The equations of electromagnetic hydrodynamics of plasma are a version of the "Extended-MHD" and "Inertial-MHD" equations [3]. In some special cases [4–6], system (1), (2) has exact solutions. In this paper, we obtain its solutions in the case of $\boldsymbol{p} = 0$, $K > 0$, assuming that the electrons and ions are ideal polytropic gases with a common adiabatic index $\gamma > 1$, and it is also assumed that $u > 0$.

Passing from the variable θ to θ/ξ in system (1), (2), we may assume that $\xi = 1$. However, it is more convenient to assume that $\xi > 0$ is arbitrary, in which case it becomes possible to pass to the limit in physically important relations. For instance, the MHD limit corresponds to $\xi \to 0$, and for

$\xi = 0$ system (1), (2) reduces to two algebraic equations with only a constant solution, except for the case of $\boldsymbol{p} = 0$, $\beta \neq 0$, when the solution is an Alfven wave [7]. Thus, in the MHD theory, there are no nontrivial traveling waves. In the Hall MHD limit, we have $\xi^2 = 0$, $\xi \neq 0$, and then system (1), (2) yields an easily integrable autonomous equation on the plane.

2. Integration of traveling wave equations

The integration of system (1), (2) is based on the following result.

Theorem 1. 1) Any solution of system (1), (2) satisfies the identity

$$\xi^2 u^2 \left(\dot{H}_y^2 + \dot{H}_z^2\right)/2 + u^2/8 + \beta^2\left(H_y^2 + H_z^2\right)/2 + \gamma K / \left(4(\gamma-1)|u|^{\gamma-1}\right) - H_y p_y - H_z p_z = \text{const}, \tag{3}$$

where $\boldsymbol{H}_\perp = (H_y, H_z)$, and the dot indicates differentiation in θ.

2) For $\boldsymbol{p} = 0$, any solution of system (1), (2) satisfies the identity

$$|u|\left(\dot{H}_y H_z - \dot{H}_z H_y\right) + \Lambda\beta\left(H_y^2 + H_z^2\right)/(2\xi) = \text{const}. \tag{4}$$

Any solution $w(\theta) = (H_y(\theta), H_z(\theta), u(\theta))$ of system (1), (2) lies on a spheroid S^2 obtained by revolving the curve $\Gamma : u + K/u^\gamma + H^2 = 2$ on the plane (H,u) about the axis u.

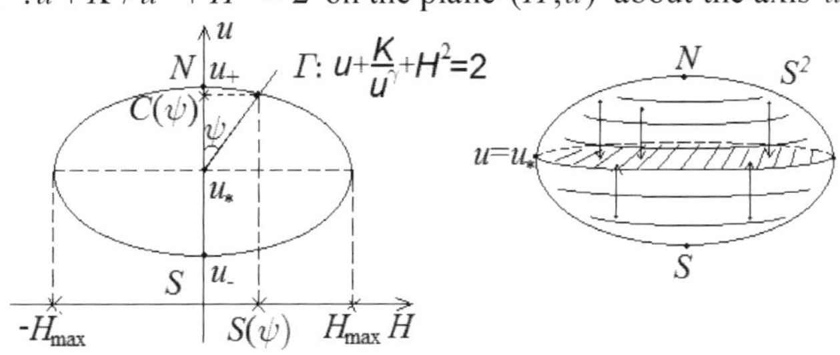

Figure 1. Spheroid S^2 obtained by rotating the curve Γ about the axis u; the projections of its hemispheres to the meridional plane.

Theorem 1 allows us to integrate system (1), (2), with $\boldsymbol{p} = 0$, in spherical coordinates on S^2:

$$H_y = S(\psi)\cos\varphi, \quad H_z = S(\psi)\sin\varphi, \quad u = C(\psi), \tag{5}$$

where ψ and φ are, respectively, the altitude and the longitude of a point on S^2, and $(S(\psi), C(\psi))$ are the coordinates of the point at which the curve Γ is crossed by the ray issuing from the point $(0, u_*)$ on the plane (H,u) and forming the angle ψ with the positive direction of the axis u. Here, $u_* = (\gamma K)^{1/(\gamma+1)}$ is the point of minimum of the function $u + K/u^\gamma$, $0 < u_- < u_* < u_+$ are the roots of the equation $u + K/u^\gamma = 2$, and $H_{\max} = \left(2 - u_* - K/u_*^\gamma\right)^{1/2} = \left(2 - (\gamma+1)(\gamma K)^{\frac{1}{\gamma+1}}/\gamma\right)^{1/2}$. The equatorial plane $u = u_*$ splits the spheroid S^2 into the northern and the southern hemispheres: $u \geq u_*$ and $u \leq u_*$, respectively, whose intersection coincides with the equator $u = u_*$. In spherical coordinates, the general solution of system (1), (2) with $\boldsymbol{p} = 0$ has the form

$$\pm\xi\int \frac{S'(\psi)C(\psi)d\psi}{\sqrt{2(E - \Phi(\psi))}} = \theta + \text{const}, \quad \frac{d\varphi}{d\theta} = \frac{I + (\Lambda\beta/2)S^2(\psi)}{\xi S^2(\psi)C(\psi)}, \tag{6}$$

where E, I are arbitrary constants, the prime indicates differentiation in ψ, and $\Phi(\psi)$ is the potential function

$$\Phi(\psi) = \frac{\beta^2}{2}\left(1 + \frac{\Lambda^2}{4}\right)S^2(\psi) + \frac{C^2(\psi)}{8} + \frac{\gamma}{4(\gamma-1)}\frac{K}{C(\psi)^{\gamma-1}} + \frac{I^2}{2S^2(\psi)}.$$

In order to obtain (6), we substitute (5) into the first integrals (3) and (4) and use the relations

$$\dot{H}_y = S'\dot{\psi}\cos\varphi - S\dot{\varphi}\sin\varphi, \quad \dot{H}_z = S'\dot{\psi}\sin\varphi + S\dot{\varphi}\cos\varphi,$$

where the prime indicates differentiation in ψ. Then, the integral (4) yields the second equation in (6). After the said substitution, the integral (3) with $\boldsymbol{p} = 0$ yields the identity

$$\xi^2 u^2\left[[S'\dot{\psi}]^2 + [S(\psi)\dot{\varphi}]^2\right]/2 + u^2/8 + \beta^2 S^2(\psi)/2 + \gamma K/(4(\gamma-1)u^{\gamma-1}) = \text{const}.$$

Taking $\dot{\varphi}$ from the second relation in (6) obtained above and inserting it into the above relation, we come to the identity $\xi^2[S'(\psi)\dot{\psi}]^2/2 + \Phi(\psi) \equiv E = \text{const}$, which brings us to the first equation in (6). The solution $\psi(\theta)$ lies in the potential well $\Phi(\psi) \le E$ and is obtained as a certain combination of the $\pm$-arcs defined by the integral in (6). Knowing $\psi(\theta)$, we can find $\varphi(\theta)$ from the second relation in (6). The said combination of the arcs depends on the sign of $S'(\psi)$ in the potential well. If the sign of $S'(\psi)$ alternates, then the derivative of the altitude $\psi(\theta)$ becomes infinite at some points and the traveling wave has singularities. Smooth (of class C^2) traveling waves can be divided into three classes: cold (the curve $w(\theta)$ lies on the northern hemisphere $u \ge u_*$); hot (the curve $w(\theta)$ lies in the southern hemisphere $u \le u_*$), and hybrid (the curve $w(\theta)$ lies on both hemispheres). Cold and hot waves can be studied by the methods of [6]. The hybrid waves are of the greatest interest for applications (see below). The structure of the potential function $\Phi(\psi)$ shows that there are four types of hybrid waves, as represented in figure 2.

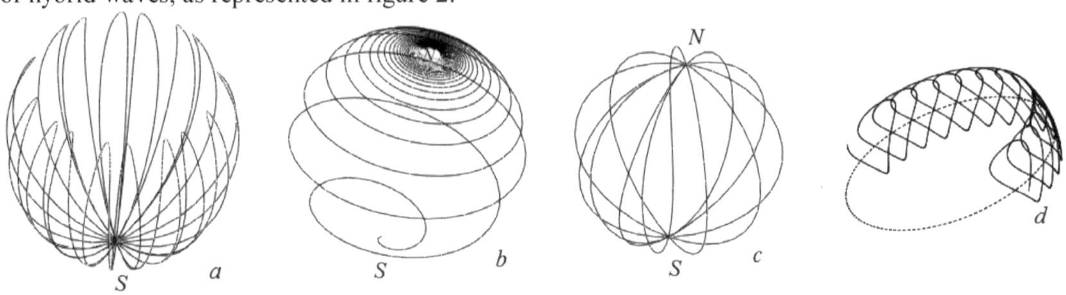

Figure 2. (a) multipetal rose; (b) spiral; (c) two-center rose; (d) multiloop spiral.

The waves in figure 2 take place for $E = \Phi(\pi/2)$. For the waves (*a*)–(*c*), we have $I = 0$, $u_* < 2\beta^2(1 + \Lambda^2/4)$, and the cases (*a*)–(*c*) correspond to the relations $\Phi(\pi/2) < \Phi(0)$, $\Phi(\pi/2) = \Phi(0)$, $\Phi(\pi/2) > \Phi(0)$, respectively. In the case (*d*), we have $I \ne 0$, $2\beta^2(1 + \Lambda^2/4) > u_* + 2I^2/H^4_{\max}$. It follows that hybrid traveling waves can exist only in a fairly strong longitudinal magnetic field. The wave (*a*) consists of a finite or countable number of petals issuing from the south pole and crossing the equator of the spheroid, while not reaching its north pole. The wave (*b*) is a spiral $w(\theta)$ on the spheroid issuing from the south pole and winding on the north pole, without reaching the latter (figure 2 shows a half of that spiral). The wave (*c*) consists of a finite or countable number of petals issuing from the south pole and crossing the north pole. In the case of countably many petals, the curve $w(\theta)$ forms a dense set on the spheroid. Finally, the wave (*d*) is realized as a multiple coil spiral whose directrix coincides with the equator.

3. Plasma acceleration in Quasi-Steady Plasma Accelerators (QSPAs)

Our results can be used to assess the possibilities of using QSPAs with longitudinal magnetic field for plasma acceleration. Such QSPAs are a subject of growing interest [8]. A QSPA is a plasma-filled space between two coaxial profiled electrodes. If the radius of the inner coaxial electrode is sufficiently large, then the QSPA can be approximately modeled by a plane channel with longitudinal magnetic field, in which the steady-state flows are described by plane traveling waves with zero phase speed $a=0$. For such flows, the transverse electric filed and the longitudinal magnetic field are constant. This plane channel model is a qualitatively correct reflection of the mechanisms of plasma acceleration caused by electrodynamic forces.

For definiteness, we assume that the plane walls of the channel are orthogonal to the axis z, while plasma acceleration occurs along the axis x. At the channel entrance, $x=0$, the following plasma parameters are given: ρ_0, p_0, H_0 (transverse magnetic field), U_0 (longitudinal speed), $\dot{H}_y(0)$, $\dot{H}_z(0)$ (current density). Then, we have $H_y(0)=H_0\cos\varphi_0$, $H_z(0)=H_0\sin\varphi_0$. It can be shown that if the characteristic scales of velocity and magnetic intensity are taken to be $U_0^*=CU_0$, $H_0^*=H_0M_A\sqrt{2C}$, where $2C=1+(\gamma M^2)^{-1}+(2M_A^2)^{-1}$, $M_A=U_0/\mathrm{v}_A$, $\mathrm{v}_A=H_0(4\pi\rho_0)^{-1/2}$ is the Alfven Mach number, $M=U_0(\rho_0/\gamma p_0)^{1/2}$ is the entering flow Mach number, then the dimensionless values of the velocity and the magnetic intensity in the steady-state plasma flow satisfy equations (1), (2) with $\beta=H_{\|}H_0^*/H_0$, $K=(\gamma M^2C^{\gamma+1})^{-1}$, $\xi=c(\omega_pL_0)^{-1}(C/2)^{-1/2}$, where $\omega_p=(4\pi\rho_0)^{1/2}(\lambda_+\lambda_-)^{-1/2}$ is the plasma frequency, L_0 is the characteristic length. It is not difficult to show that the constant K always satisfies the condition $(\gamma K)^{1/(\gamma+1)}<2\gamma(\gamma+1)^{-1}$. It is usually assumed that the entering plasma flow is subsonic, i.e., $M<1$, and therefore, $u(0)=C^{-1}<u_*=(\gamma K)^{1/(\gamma+1)}=M^{-2/(\gamma+1)}C^{-1}$, which implies that the point $(H_y(0),H_z(0),u(0))$ lies on the southern hemisphere of S^2. The possibilities of using QSPAs for plasma acceleration are largely determined by an answer to the following question. Is it possible to prescribe plasma parameters at the entrance, so as to obtain a supersonic plasma flow at the exit of the accelerator? In other words, can a QSPA function as a de Laval nozzle merely due to electrodynamic acceleration?

The mathematical problem reduces to finding the input parameters for which the integral curve of system (1), (2) on S^2 issuing from the point $(H_y(0),H_z(0),u(0))$ would cross the equator $u=u_*$ and enter the northern hemisphere with a possibly maximal acceleration $u_+/u(0)=Cu_+$. It is assumed that at the entrance, the transverse current density is known, and therefore, in dimensionless units, the values $\dot{H}_y(0)$, $\dot{H}_z(0)$ are given. Consider more closely the solution of the above problem for $\boldsymbol{p}=0$. Using the following the relations obtained in section 2:

$$\dot{H}_y(0)=S'(\psi_0)\dot{\psi}_0\cos\varphi_0-S(\psi_0)\dot{\varphi}(0)\sin\varphi_0,$$
$$\dot{H}_z(0)=S'(\psi_0)\dot{\psi}_0\sin\varphi_0+S(\psi_0)\dot{\varphi}(0)\cos\varphi_0,$$

where the latitude ψ_0 is uniquely determined by the values $u(0)$, $H_y(0)=(M_A\sqrt{2C})^{-1}\cos\varphi_0$, $H_z(0)=(M_A\sqrt{2C})^{-1}\sin\varphi_0$, we find that

$$S'(\psi_0)\dot{\psi}_0=\dot{H}_y(0)\cos\varphi_0+\dot{H}_z(0)\sin\varphi_0,\ S(\psi_0)\dot{\varphi}_0=\dot{H}_z(0)\cos\varphi_0-\dot{H}_y(0)\sin\varphi_0$$

and according to the above formulas we obtain the following values for the constants I and E:

$$I=\xi(M_A\sqrt{2C})^{-1}(\dot{H}_z(0)\cos\varphi_0-\dot{H}_y(0)\sin\varphi_0)-\Lambda\beta/(4CM_A^2),$$
$$E=\xi^2(\dot{H}_y(0)\cos\varphi_0+\dot{H}_z(0)\sin\varphi_0)^2/2+\Phi(\psi_0),$$

where $\Phi(\psi_0)=\beta^2(1+\Lambda^2/4)/(4CM_A^2)+1/(8C^2)+\gamma KC^{\gamma-1}/(4(\gamma-1))+I^2CM_A^2.$

The above considerations show that in order to ensure supersonic speed at the exit from the plane channel with a steady-state flow it is necessary and sufficient that the following relations are satisfied:

$$2\beta^2(1+\Lambda^2/4)>u_*+2I^2/H_{\max}^4,\quad E=\Phi(\pi/2), \tag{7}$$

where $H_{\max}^2=2-\dfrac{(\gamma+1)}{\gamma}u_*$, $\Phi(\pi/2)=\dfrac{\beta^2}{2}\left(1+\dfrac{\Lambda^2}{4}\right)H_{\max}^2+\dfrac{u_*^2}{8}+\dfrac{\gamma K}{4(\gamma-1)u_*^{\gamma-1}}+\dfrac{I^2}{2H_{\max}^2}.$

For simplicity, set $\dot{H}_y = \dot{H}_y(0)$, $\dot{H}_z = \dot{H}_z(0)$. Without the loss of generality, we can assume that $\varphi_0 = 0$. Then, the equation $E = \Phi(\pi/2)$ can be rewritten in the form

$$\frac{\xi^2}{2}\dot{H}_y^2 + \frac{1}{4CM_A^2}\left(\frac{1}{S(\psi_0)^2} - \frac{1}{H_{\max}^2}\right)\left(\xi\dot{H}_z - \frac{\Lambda\beta}{2M_A\sqrt{2C}}\right)^2 = \frac{\beta^2}{2}\left(1+\frac{\Lambda^2}{4}\right)\left(H_{\max}^2 - S(\psi_0)^2\right) - \left(\frac{1}{8C^2} + \frac{\gamma K C^{\gamma-1}}{4(\gamma-1)} - \frac{u_*^2}{8} - \frac{\gamma K}{4(\gamma-1)u_*^{\gamma-1}}\right). \tag{8}$$

Since $S(\psi_0)^2 < H_{\max}^2$ and the right-hand side of (8) is positive, we see that equation (8) specifies an ellipse on the plane $(\dot{H}_y, \dot{H}_z)$. The inequality (7) cuts out from the ellipse (8) either two open arcs or two separate points.

Thus, for the above problem to have a solution, the parameters $\dot{H}_y$, $\dot{H}_z$, φ_0, p_0, ρ_0, U_0, H_0 at the entrance to the channel, together with its length $\ell = L_0$, must satisfy conditions (7), which define (in the space of input parameters) a hypersurface whose intersection with the planes $(\dot{H}_y, \dot{H}_z)$ coincides with open arcs of the ellipse (8) or a pair of points. If the input parameters do not belong to this hypersurface, then the solution of system (1), (2) with the initial condition $(H_y(0), H_z(0), u(0))$ either remains subsonic $(u(x) < u_*)$ or reaches the speed of sound at a point of its gradient catastrophe, i.e., the velocity derivative $du(x)/dx$ becomes infinite and the x-derivatives of both magnetic field components $H_y(x)$, $H_z(x)$ have finite jumps. Therefore, since the said hypersurface has zero Lebesgue measure and is nowhere dense in the space of input parameters, there is only a narrow range of input parameters for which the flow becomes supersonic at the exit of the accelerator. Moreover, the physical length of the channel is also determined by the input parameters and the desired speed at its exit. For a channel of fixed length, the problem becomes more difficult and requires special consideration, with the range of input parameters ensuring supersonic speeds becoming even smaller.

An even more ambitious and interesting problem consists in determining the input parameters at the entrance ensuring a maximal possible supersonic speed, which is equal to u_+. Mathematically, this means that the curve on S^2 realizing the solution of system (1), (2) and issuing from the point $(H_y(0), H_z(0), u(0))$ should reach the north pole N. Our analysis shows that a necessary and sufficient condition for this can be stated in the form

$$2\beta^2(1+\Lambda^2/4) > u^*, \quad E = \Phi(\pi/2), \quad I = 0, \tag{9}$$

where $$u^* = \frac{1}{H_{\max}^2}\left(\frac{u_+^2}{2} + \frac{\gamma}{\gamma-1}\frac{K}{u_+^{\gamma-1}} - \frac{u_*^2}{2} - \frac{\gamma}{\gamma-1}\frac{K}{u_*^{\gamma-1}}\right).$$

Relations (9) specify an even narrower range of input parameters, since the dimension of the surface that should contain the desired input parameters is decreased by 1. Here, the channel length is not fixed.

4. Conclusion

Our analysis allows us to claim that for a channel with longitudinal field, the input parameters of a subsonic plasma flow at the entrance to the channel must lie within a fairly narrow range, in order to ensure its supersonic speed at the exit from the channel; to ensure a maximal possible supersonic speed at the exit, this range has to be further decreased, and it becomes even smaller, if the problem is considered for a channel of fixed length. It has been assumed here that $\boldsymbol{p} = 0$ in system (1), (2). The case $\boldsymbol{p} \neq 0$ requires special consideration, but the conclusions are expected to be more or less the same.

Acknowledgments
This work has been supported by the Russian Science Foundation (Project 16-11-10278).

5. Bibliography

[1] Gavrikov M B 1988 Linear waves in nonrelativistic magnetohydrodynamics *Preprint* 199 Keldysh Institute Preprints
[2] Gavrikov M B, Savel'jev V V and Tayurskiy A A 2010 Solitons in two-fluid` magnetohydrodynamics with nonzero electron interia *Izvestiya VUZ Applied Nonlinear Dynamics* **18** 4 pp 132–147
[3] Kimura K and Morrison P J 2014 On energy conservation in extended magnetohydrodynamics *Phys. Plasmas* **21** 082101
[4] Adlam J H Allen J E 1958 *Phil. Mag.* **3** p 448
[5] Lüst R 1959 *Fortschritter der Physik* **7** pp 503–558
[6] Gavrikov M B 1991 Aperiodic vibrations of cold plasma *Preprint* 33 Keldysh Institute Preprints
[7] Landau L D, Lifshitz E M and Pitaevskii L P 1984 *Electrodynamics of Continuous Media* vol 8 (2nd ed Butterworth-Heinemann)
[8] Brushlinskii K V, Kozlov A N, Konovalov V S 2015 Numerical models of steady-state and pulsating flows of self-ionizing gas in plasma accelerator channels *Computational Mathematics and Mathematical Physics* **55** 8 pp 1370–1380

IOP Conf. Series: Journal of Physics: Conf. Series **1205** (2019) 012015 doi:10.1088/1742-6596/1205/1/012015

On the Integrability of Ablowitz-Ladik models with local and nonlocal reductions

V S Gerdjikov

Department of Applied Mathematics, National Research Nuclear University MEPHI, 31 Kashirskoe Shosse, 115409 Moscow, Russian Federation

Institute for Advanced Physical Studies, New Bulgarian University, 21 Montevideo Street, Sofia 1618, Bulgaria

Institute of Nuclear Research and Nuclear Energy, Bulgarian Academy of Sciences

72, boul. Tsarigradsko chaussee, 1784 Sofia, Bulgaria

E-mail: vgerdjikov@gmail.com

Abstract. We briefly consider the Ablowitz-Ladik model (ALM) with special nonlocal reduction introduced in [4]. The nonlocal ALM also allows Lax representation. Next the expansions over the 'squared solutions' developed for the local case, are naturally extended to the nonlocal one. This allows to derive all fundamental properties of the nonlocal ALM including its Hamiltonian properties and action-angle variables.

1. Introduction

Besides the numerous integrable nonlinear evolution equations (NLEE) (see [1, 7, 17]) there also exist, a number of physically important difference nonlinear evolution equations (DEE), solvable by the inverse scattering transform (IST) [2, 3] known today as Ablowitz-Ladik model (ALM).

The ALM allows Lax pair of the form [2, 3]:

$$\psi(n+1,z) = L_n(z)\psi(n,z), \quad L_n(z) = E(z) + Q(n), \quad i\frac{\partial\psi(n)}{\partial t} + M(n,t)\psi(n,t) = 0, \tag{1.1}$$

$$M(n,t) = -f(z)\sigma_3 + \sigma_3 Q(n)Q(n-1) + \sigma_3 Q(n-1)E(z) - \sigma_3 E(z)Q(n). \tag{1.2}$$

Here $E(z) = \begin{pmatrix} z & 0 \\ 0 & z^{-1} \end{pmatrix}$, $Q(n) = \begin{pmatrix} 0 & q(n) \\ r(n) & 0 \end{pmatrix}$ and $f(z) = \frac{1}{2}(z - z^{-1})^2$ is the dispersion law of ALM. The compatibility condition is

$$i\frac{\partial L(n,z)}{\partial t} + L(n,z)M(n,z) - M(n+1,z)L(n,z) = 0. \tag{1.3}$$

It holds true identically with respect to z and for $r(n) = \epsilon q(n)^*$, $\epsilon = \pm 1$ provides the ALM:

$$i\frac{\partial q(n)}{\partial t} - q(n+1,t) + 2q(n,t) - q(n-1,t) + \epsilon|q(n,t)|^2(q(n-1,t) + q(n+1,t) = 0. \tag{1.4}$$

In [14, 15] the results on ALM were substantially extended. These include: i) the Hamiltonian formulation and the hierarchy of Hamiltonian structures of ALM; ii) the proof that AKNS idea of treating the ISM as generalized Fourier transform holds true also for ALM; iii) the proof of the complete integrability of ALM and derivation of its action-angle variables; iv) derivation of the classical and quantum R-matrices for ALM.

The present paper is a further development of our preprint [14] and the paper [15] proving that the results i)–iii) for ALM can be extended to the non-local version of ALM (nALM) recently introduced in [4]. In Section 2 we have collected the main results of the properties i)–iii) of ALM. Special attention here should be paid to the symplectic basis of squared solutions $\{\mathcal{P}(n,z), \mathcal{Q}(n,z)\}$ of the Lax operator (1.1) generalizing the one for the continuous case [13]. This basis allows one to treat the expansion coefficients of $\sigma_3\delta Q(n,z)$ over $\{\mathcal{P}(n,z), \mathcal{Q}(n,z)\}$ as the variations of the action-angle variables ALM. In the next Section 3 we briefly outline how the results i)–iii) can be extended to the nonlocal ALM. An important consequence of this is the explicit transformation of the symplectic form (3.12) into canonical form (3.13) thus allowing one to obtain the action-angle variables of nALM in terms of the scattering data. The last Section 4 contains discussion and conclusions.

2. Preliminaries

2.1. Lax representation of ALM

The class of integrable DEE is by now rather well studied [15, 5, 6, 23, 24]. The group of reductions of these DEE, unlike the reductions of NLEE [22] are not so well known. This is specially true for their nonlocal reductions [4], which recently started to attract attention. In this Section we remind the main results for ALM following [2, 3, 15].

Let us start with some known facts about the direct and inverse scattering problem for the system (1.1). In order to make the exposition simpler, we consider the case when the potential $Q(n) \in \mathcal{S}(\mathbb{Z}, \mathbb{C}^2)$, the space of complex-valued off-diagonal matrix sequences such that $\lim_{n\to\pm\infty} n^k Q(n) = 0$ for all $k = 0, 1, 2, \dots$. Imposing the condition $0 < \prod_{k=-\infty}^{\infty} |h(k)| < \infty$, $h(k) = 1 - q(k)r(k)$ one is able to prove the existence and the analyticity properties of the Jost solutions of (1.1), introduced by

$$\begin{aligned} &\lim_{n\to\infty} \psi(n,z)E^{-n}(z) = \mathbb{1}, && \lim_{n\to\infty} \phi(n,z)E^{-n}(z) = \mathbb{1}, \\ &\psi(n,z) = ||\psi^-, \psi^+||, && \phi(n,z) = ||\phi^+, \phi^-||, \end{aligned} \tag{2.1}$$

where ψ^+, ϕ^+, (ϕ^-, ψ^-) are analytic for $|z| > 1$ (or $|z| < 1$). The scattering matrix is introduced by

$$\phi(n,z) = \psi(n,z)T(z), \qquad T(z) = \begin{pmatrix} a^+ & -b^- \\ b^+ & a^- \end{pmatrix}, \qquad \det T(z) = v = \prod_{k=-\infty}^{\infty} h(k). \tag{2.2}$$

We will need also the symmetry properties of $T(z,t)$. It is easy to check, that for generic $Q(n)$ we have $L(n,z) = -\sigma_3 L(n,-z)\sigma_3$, which implies that the Jost solutions satisfy $\psi(n,z) = (-1)^n \sigma_3 \psi(n,-z)\sigma_3$. Thus the matrix elements of $T(z,t)$ satisfy:

$$a^{\pm}(z) = a^{\pm}(-z), \qquad b^{\pm}(z) = -b^{\pm}(-z). \tag{2.3}$$

We will request that $r(n) = \epsilon q^*(n)$, i.e. $Q(n) = \epsilon Q^\dagger(n)$; it is this last constraint that renders the compatibility condition (1.3) to the form (1.4). This constraint can be written down as

$$\sigma_\epsilon (L(n,z^*))^* \sigma_\epsilon^{-1} = L(n,\epsilon/z), \qquad \sigma_\epsilon = \begin{pmatrix} 0 & 1 \\ \epsilon & 0 \end{pmatrix}, \qquad \epsilon = \pm 1. \tag{2.4}$$

IOP Conf. Series: Journal of Physics: Conf. Series **1205** (2019) 012015 doi:10.1088/1742-6596/1205/1/012015

The consequences of the second symmetry on the scattering matrix and its matrix elements are:

$$\sigma_\epsilon T^*(z^*)\sigma_\epsilon^{-1} = T(\epsilon/z), \qquad a^+(z) = a^{-,*}(\epsilon/z^*), \qquad b^+(z) = \epsilon b^{-,*}(\epsilon/z^*). \tag{2.5}$$

An immediate consequence of the Lax representation (1.3) is the fact that the t-dependence of the scattering matrix $T(z,t)$ is given by:

$$i\frac{\partial T}{\partial t} + f(z)[\sigma_3, T(z,t)] = 0. \tag{2.6}$$

Thus we have shown that the nonlinear DEE (1.4) becomes *linear* evolution equation for the scattering matrix. In components eq. (2.6) becomes:

$$i\frac{\partial a^\pm(z)}{\partial t} = 0, \qquad i\frac{\partial b^\pm(z)}{\partial t} = \mp 2f(z)b^\pm(z,t). \tag{2.7}$$

An immediate consequence of eq. (2.7) is that ALM possesses an infinite set of integrals of motion. Indeed, both $a^+(z)$ and $a^-(z)$ can be viewed as the generating functionals of such integrals. In fact $a^+(z)$ and $a^-(z)$ are analytic functions of z for $|z| > 1$ and $|z| < 1$ respectively. Skipping the details (see, e.g. [14, 15]), we can express them via the dispersion relations:

$$\begin{aligned}
\ln a^+(z) &= \frac{1}{4\pi i}\oint_{S^1} \frac{d\zeta^2}{\zeta^2 - z^2}\ln(1+\epsilon|\rho^+(\zeta)|^2) + \sum_{j=1}^N \ln\frac{z^2 - z_{j,+}^2}{z^2 - z_{j,-}^2}, && |z|>1,\\
-\ln a^-(z) &= \frac{1}{4\pi i}\oint_{S^1} \frac{d\zeta^2\, z^2}{\zeta^2(\zeta^2 - z^2)}\ln(1+\epsilon|\rho^+(\zeta)|^2) + \sum_{j=1}^N \ln\frac{(z^2 - z_{j,+}^2)z_{j,-}^2}{(z^2 - z_{j,-}^2)z_{j,+}^2}, && |z|<1,
\end{aligned} \tag{2.8}$$

where

$$\rho^\pm(\zeta) = \frac{b^\pm(\zeta)}{a^\pm(\zeta)}, \qquad |\zeta| = 1, \qquad z_{j,-} = 1/z_{j,+}^*. \tag{2.9}$$

The points $\pm z_{j,+}$, $\pm z_{j,-}$, $j = 1,\dots N$ are the discrete eigenvalues of L.

2.2. The ISM is a generalized Fourier transform

The idea of AKNS [1] that the inverse scattering method is a generalized Fourier transform was extended to a large class of differential Lax operators [8]. It was extended also to the DEE generated by the discrete Lax operator (1.1) [14, 15]. To this end we introduce the fundamental analytic solutions by $\chi^\pm$ of (1.1); they are analytic for $|z| > 1$ and $|z| < 1$:

$$\begin{aligned}
\chi^+(n,z) &= ||\phi^+, \psi^+||, & \chi^-(n,z) &= ||\psi^-, \phi^-||,\\
\chi^+(n,z) &= \psi(n,z)T^-(z) = \phi(n,z)S^+(z), & \chi^-(n,z) &= \psi(n,z)T^+(z) = \phi(n,z)S^-(z),
\end{aligned} \tag{2.10}$$

where

$$S^+(z) = \begin{pmatrix} 1 & b^-/v \\ 0 & a^+/v \end{pmatrix}, \quad S^-(z) = \begin{pmatrix} a^+ & 0 \\ b^+ & 1 \end{pmatrix}, \quad T^+(z) = \begin{pmatrix} 1 & -b^- \\ 0 & a^- \end{pmatrix}, \quad T^-(z) = \begin{pmatrix} a^-/v & 0 \\ -b^+/v & 1 \end{pmatrix}.$$

Here $v = \prod_{k=-\infty}^{\infty} h(k)$ and obviously $S^-\hat{S}^+ = T^+\hat{T}^- = T(z)$; by $\hat{X}$ we have denoted the matrix inverse to X, i.e., $\hat{X} \equiv X^{-1}$.

The analytic properties of $a^\pm(z)$ allow one to reconstruct them from the set of reflection coefficients $\rho^\pm(\zeta) = b^\pm/a^\pm(\zeta)$ or $\tau^\pm(\zeta) = b^\mp/a^\pm(\zeta)$. Thus the minimal sets of scattering data which determine both the potential $Q(n)$ and the scattering matrix $T(z)$ consist of two parts:

the reflection coefficients on the continuous spectrum (the unit circle) $z \in S^1$ and additional data, characterizing the discrete spectrum of $L(n,z)$. More precisely:

$$\begin{aligned} \mathcal{T}_1 &= \{\rho^{\pm}(\zeta), \quad z \in S^1, \quad \pm z_{j,\pm}, \quad c_j^{\pm}, \quad j = 1,\dots,N\}, \\ \mathcal{T}_2 &= \{\tau^{\pm}(\zeta), \quad z \in S^1, \quad \pm z_{j,\pm}, \quad m_j^{\pm}, \quad j = 1,\dots,N\}, \end{aligned} \tag{2.11}$$

where $c_j^{\pm}$ and $m_j^{\pm}$ are the norming constants of the discrete eigenfunctions of $L(n,z)$; for more details see [14, 15].

Obviously, the scattering matrix $T(z)$ is recovered uniquely from either of the sets $\mathcal{T}_i$ using the dispersion relations (2.8). The recovery of the potential $Q(n)$ is based on the study of the mapping $\mathcal{F}_i\colon \mathcal{M} \leftrightarrows \mathcal{T}_i$, where $\mathcal{M}$ is the manifold of allowed potentials $Q(n)$. Such study is based on the Wronskian relations and is deeply related to the interpretation of the ISM to a generalized Fourier transform. These facts are detailed in [14, 15]. Here we will formulate the main results in gauge covariant formulation (see, [8, 18]). The mappings $\mathcal{F}_i$ are based on the expansions of the 'squared solutions' of $L(n,z)$ which can be introduced by:

$$\begin{gathered} \Psi^{\pm}(n,z) = va^{\pm}(z)P_0(\chi^{\pm}\sigma_{\mp}\hat{\chi}^{\pm}(n,z)), \qquad \Phi^{\pm}(n,z) = va^{\pm}(z)P_0(\chi^{\pm}\sigma_{\pm}\hat{\chi}^{\pm}(n,z)), \\ \sigma_+ = \begin{pmatrix} 0 & 1 \\ 0 & 0 \end{pmatrix}, \qquad \sigma_- = \begin{pmatrix} 0 & 0 \\ 1 & 0 \end{pmatrix}, \qquad \sigma_3 = \begin{pmatrix} 1 & 0 \\ 0 & -1 \end{pmatrix}, \end{gathered} \tag{2.12}$$

where $P_0(X)$ are the projectors onto the off-diagonal part of the 2×2 matrix X:

$$P_0(X) = \frac{1}{4}[\sigma_3, [\sigma_3, X]]. \tag{2.13}$$

Skipping the details we formulate the main results of [14, 15] in explicitly gauge covariant form.

The first important result there is the fact that the set of 'squared solutions' is a complete set of functions on $\mathcal{M}$. Therefore any elements of $\mathcal{M}$, including the potential $Q(n)$ and its variation $\sigma_3\delta Q(n)$ can be expanded over the squared solutions. The result is:

$$\begin{aligned} Q(n) &= \frac{i}{2\pi}\oint_{S^1} \frac{dz}{z}(\rho^+(z)\Psi^+(n,z) + \rho^-(z)\Psi^-(n,z)), \\ \sigma_3\delta Q(n) &= -\frac{i}{2\pi}\oint_{S^1} \frac{dz}{z}(\delta\rho^+(z)\Psi^+(n,z) - \delta\rho^-(z)\Psi^-(n,z)). \end{aligned} \tag{2.14}$$

where we have neglected the terms corresponding to the discrete spectrum and

$$\rho^{\pm}(z) = [\![Q(n), \Phi^{\pm}(n,z)h^{-1}(n)]\!], \qquad \delta\rho^{\pm}(z) = [\![\sigma_3\delta Q(n), \Phi^{\pm}(n,z)h^{-1}(n)]\!]. \tag{2.15}$$

By $[\![X(n), Y(n)]\!] = -[\![Y(n), X(n)]\!]$ above we have denoted is the skew-scalar product:

$$[\![X(n), Y(n)]\!] = \sum_{n=-\infty}^{\infty} \operatorname{tr}(X(n), [\sigma_3, Y(n)]). \tag{2.16}$$

Note that the sets of squared solutions are bi-orthogonal seta of functions with respect to the skew-scalar product [15]. We can also introduce the so-called symplectic basis [13, 15]:

$$\begin{aligned} \mathcal{P}(n,z) &= \frac{1}{2\pi}(\rho^+\Psi^+ + \rho^-\Psi^-)(n,z) = -\frac{1}{2\pi v}(\tau^+\Phi^+ + \tau^-\Phi^-)(n,z), \\ \mathcal{Q}(n,z) &= \frac{i}{2b^+b^-}(v\rho^+\Psi^+ - \tau^+\Phi^+)(n,z) = \frac{i}{2b^+b^-}(\tau^-\Phi^+ - v\rho^-\Psi^-)(n,z), \end{aligned} \tag{2.17}$$

which is orthogonal with respect to the skew-scalar product, i.e. for $z, z' \in S^1$ we have

$$\begin{aligned} &[[\mathcal{P}(n,z), \mathcal{P}(n,z')h^{-1}(n)]] = 0, \qquad [[\mathcal{Q}(n,z), \mathcal{Q}(n,z')h^{-1}(n)]] = 0, \\ &[[\mathcal{Q}(n,z), \mathcal{P}(n,z')h^{-1}(n)]] = -i\delta(\arg(z) - \arg(z')). \end{aligned} \tag{2.18}$$

The expansions over the symplectic basis take the form:

$$Q(n) = -i \oint_{S^1} \frac{dz}{z} \mathcal{P}(n,z), \qquad \sigma_3 \delta Q(n) = \oint_{S^1} \frac{dz}{z} \left(\delta\pi(z)\mathcal{Q}(n,z) - \delta\kappa(z)\mathcal{P}(n,z) \right), \tag{2.19}$$

where

$$\begin{aligned} \pi(z) &= -\frac{1}{2\pi} \ln(1 + \rho^+\rho^-(z)) = \frac{1}{2\pi}(\ln v - \ln(a^+(z)a^{+,*}(z^*)), \\ \kappa(z) &= -\frac{i}{2} \ln\left(\frac{b^+}{b^-}(z,t)\right) = \arg(b^+(z,t)), \qquad v = \prod_{k=-\infty}^{\infty} h(k), \end{aligned} \tag{2.20}$$

and $z \in S^1$. As we shall see below, $\pi(z)$ and $\kappa(z)$ provide the action-angle variables for ALM.

2.3. Hamiltonian Properties and Complete integrability of ALM

The simplest nontrivial Hamiltonian formulation of ALM is based on non-canonical Poisson brackets [14, 15] and the Hamiltonian:

$$\begin{aligned} &\{q(n), q^*(m)\} = (1 - \epsilon|q(n)|^2)\delta_{nm}, \\ &H_{\rm ALM} = \sum_{n=-\infty}^{\infty} \left(\epsilon(q(n)q^*(n+1) - q(n+1)q^*(n)) + \ln(1 - \epsilon|q(n)|^2) \right). \end{aligned} \tag{2.21}$$

The generating functional of integrals of motion is:

$$\ln a^+(z) = -\frac{1}{2\pi} \ln(1 + \epsilon|\rho^+(z)|^2) = \sum_{p=0}^{\infty} C_p z^{-2p}. \tag{2.22}$$

The first two integrals of motion are:

$$C_0 = \ln v = \sum_{n=-\infty}^{\infty} \ln(1 - \epsilon|q(n)|^2), \qquad C_1 = \sum_{n=-\infty}^{\infty} (q(n)q^*(n+1) - q(n+1)q^*(n)). \tag{2.23}$$

The Poisson brackets (2.21) are dual to the symplectic form [15]:

$$\Omega^{(0)} = \sum_{n=-\infty}^{\infty} \frac{\delta q(n) \wedge \delta q^*(n)}{1 - \epsilon|q(n)|^2} = [[\sigma_3 \delta Q(n) \mathop{\wedge}_{,} \sigma_3 \delta Q(n)]]. \tag{2.24}$$

Using the expansion of $\sigma_3\delta Q(n)$ (2.19) over the 'symplectic basis one finds [14, 15]:

$$\Omega^{(0)} = 2i \int_{S^1} \frac{dz}{z} \delta\pi(z) \wedge \delta\kappa(z). \tag{2.25}$$

Thus we conclude that the action-angle variables of the ALM can be expressed in terms of the scattering data as follows:

$$\pi(z) = -\frac{1}{2\pi} \ln(1 + \epsilon|\rho^+(z)|^2), \qquad \kappa(z) = -\frac{i}{2} \ln\left(\frac{b^+(z)}{b^-(z)}\right) = \arg(b^+(z)), \qquad z \in \mathbb{S}^1. \tag{2.26}$$

In addition we conclude also that

$$\{C_n, C_m\} = 0. \tag{2.27}$$

3. Nonlocal Difference Nonlinear Equations

Here, on the example of the ALM we will describe the effect of the nonlocal reductions on the Lax representations of the DEE mentioned above. One can check that if $r(n) = \epsilon q^*(-n)$, i.e.

$$Q(n) = \epsilon Q^\dagger(-n), \qquad h(n) = \det L(n) = 1 - \epsilon q(n) q^*(-n) = h^*(-n). \tag{3.1}$$

then the Lax operator satisfies

$$\sigma_\epsilon (L(n, z^*))^* \sigma_\epsilon^{-1} = L(-n, 1/z) = h(-n) L^{-1}(-n, -z). \tag{3.2}$$

Therefore for the scattering matrix elements we get (see also [4]):

$$\begin{aligned} &a^+(z) = e^{2i\theta} a^{+,*}(z^*), && a^-(z) = e^{2i\theta} a^{-,*}(z^*), && b^+(z) = \pm e^{2i\theta} b^{-,*}(z^*), \\ &\rho^+(z) = \tau^{+,*}(z^*), && \rho^-(z) = -\tau^{-,*}(z^*), && z \in S^1, \end{aligned} \tag{3.3}$$

where $\theta = \arg(v)$, $\rho^\pm(z) = \frac{b^\pm}{a^\pm}(z)$ and $\tau^\pm(z) = \frac{b^\mp}{a^\pm}(z)$. The nonlocal reduction (3.1) does not influence the analyticity properties of the Jost solutions and the minimal sets of scattering data (2.2). One can again introduce the FAS as in (2.10) and analyze the Wronskian relations.

Remark 1 *Of course, one should keep in mind that all these quantities: the scattering matrix $T(z)$, FAS $\chi^\pm(n, z)$, reflection coefficients $\rho^\pm(z)$, $\tau^\pm(z)$ etc. are substantially different from the their 'local' analogs in the previous Section, because of the nonlocal reduction (3.1).*

Thus most of the constructions and results for the 'local' case can be extended to the nonlocal one, keeping in mind Remark 1 above. The first important one is the expansions over the 'squared solutions' of $L(n, z)$ which again are introduced by:

$$\Psi^\pm(n, z) = v a^\pm(z) P_0(\chi^\pm \sigma_\mp \hat{\chi}^\pm(n, z)), \qquad \Phi^\pm(n, z) = v a^\pm(z) P_0(\chi^\pm \sigma_\pm \hat{\chi}^\pm(n, z)), \tag{3.4}$$

and over the symplectic basis, which is formally provided by (2.17). The projector $P_0(X)$ is the same as in (2.13). The 'squared solutions' (3.4), along with the symplectic basis, preserve their main property, i.e. they are complete set of functions on $\mathcal{M}_{\rm nl}$, which of course has different complex structure as compare to $\mathcal{M}$. We can derive the expansions

$$\begin{aligned} Q(n) &= \frac{i}{2\pi} \oint_{S^1} \frac{dz}{z} (\rho^+(z) \Psi^+(n, z) + \rho^-(z) \Psi^-(n, z)), \\ \sigma_3 \delta Q(n) &= -\frac{i}{2\pi} \oint_{S^1} \frac{dz}{z} (\delta\rho^+(z) \Psi^+(n, z) - \delta\rho^-(z) \Psi^-(n, z)). \end{aligned} \tag{3.5}$$

where

$$\rho^\pm(z) = [\![Q(n), \Phi^\pm(n, z) h^{-1}(n)]\!], \qquad \delta\rho^\pm(z) = [\![\sigma_3 \delta Q(n), \Phi^\pm(n, z) h^{-1}(n)]\!], \tag{3.6}$$

where $[\![X(n), Y(n)]\!]$ is the skew-scalar product in (2.16). Similar expansions over the dual set of squared solutions $\Phi^\pm(n, z)$ with coefficients $\tau^\pm(z)$ hold true.

Like in the local case, one can interpret the mapping of $Q(n)$ onto the set of reflection coefficients $\rho^\pm(z)$ or the dual set $\tau^\pm(z)$ as generalized Fourier transform. Of course the expansions (3.5) are derived assuming that the Lax operator $L(n, z)$ has no discrete eigenvalues.

One can derive also the expansions over the symplectic basis with the results:

$$Q(n) = -i \oint_{S^1} \frac{dz}{z} \mathcal{P}(n, z), \qquad \sigma_3 \delta Q(n) = \oint_{S^1} \frac{dz}{z} \left(\delta\pi(z) \mathcal{Q}(n, z) - \delta\kappa(z) \mathcal{P}(n, z) \right), \tag{3.7}$$

where

$$\begin{aligned} \pi(z) &= -\frac{1}{2\pi} \ln(1 + \rho^+ \rho^-(z)) = -\frac{1}{2\pi} \ln(1 + \tau^+ \tau^-(z)) = \frac{1}{2\pi} \ln \left(\frac{v}{a^+(z) a^-(z)} \right), \\ \kappa(z) &= -\frac{i}{2} \ln \left(\frac{b^+}{b^-}(z) \right) = \arg(b^+(z)), \qquad z \in S^1. \end{aligned} \tag{3.8}$$

IOP Conf. Series: Journal of Physics: Conf. Series **1205** (2019) 012015 doi:10.1088/1742-6596/1205/1/012015

3.1. Hamiltonian Properties and Complete integrability of the nonlocal ALM

The simplest nontrivial Hamiltonian formulation of the nonlocal ALM (nALM) is based on non-canonical and nonlocal Poisson brackets and the Hamiltonian:

$$\begin{aligned} &\{q(n), q^*(m)\} = (1 - \epsilon q(n) q^*(-n))\delta_{nm}, \\ &H_{\rm nALM} = \sum_{n=-\infty}^{\infty} \left(\epsilon(q(n)q^*(-n+1) - q(-n+1)q^*(n) + \ln(1 - \epsilon q(n)q^*(-n))\right). \end{aligned} \tag{3.9}$$

The generating functional of integrals of motion is, as in the local case:

$$\ln a^+(z) = -\frac{1}{2\pi}\ln(1 + \epsilon|\rho^+(z)|^2) = \sum_{p=0}^{\infty} \tilde{C}_p z^{-2p}. \tag{3.10}$$

The first two integrals of motion are:

$$\tilde{C}_0 = \sum_{n=-\infty}^{\infty} \ln(1 - \epsilon q(n)q^*(-n)), \qquad \tilde{C}_1 = \sum_{n=-\infty}^{\infty} (q(n)q^*(-n+1) - q(n+1)q^*(-n)). \tag{3.11}$$

The Poisson brackets (3.9) are dual to the symplectic form:

$$\tilde{\Omega}^{(0)} = \sum_{n=-\infty}^{\infty} \frac{\delta q(n) \wedge \delta q^*(-n)}{1 - \epsilon q(n) q^*(-n)} = [\![\sigma_3 \delta Q(n) \underset{'}{\wedge} \sigma_3 \delta Q(n)]\!]. \tag{3.12}$$

Using the expansion of $\sigma_3\delta Q(n)$ over the symplectic basis (3.7) one finds :

$$\Omega^{(0)} = 2i \int_{S^1} \frac{dz}{z} \delta\pi(z) \wedge \delta\kappa(z). \tag{3.13}$$

Thus we conclude that the action-angle variables of the ALM can be expressed in terms of the scattering data as follows:

$$\pi(z) = \frac{1}{2\pi}\ln\left(\frac{v}{a^+(z)a^-(z)}\right), \qquad \kappa(z) = -\frac{i}{2}\ln\left(\frac{b^+(z)}{b^-(z)}\right) = \arg(b^+(z)), \qquad z \in \mathbb{S}^1. \tag{3.14}$$

Formally the expressions for the action-angle variables are similar to the ones for the local case. Of course one should keep in mind the Remark 1. Again we conclude also that

$$\{\tilde{C}_n, \tilde{C}_m\} = 0. \tag{3.15}$$

Discussion and conclusions

We formulated ideas that will be developed and elaborated in next publications. They can be extended in several directions.

First, like the ALM model, the nALM possesses a hierarchy of Hamiltonian structures which is generated by the recursion operators $\Lambda_\pm$ which have the squared solutions as eigenfunctions, see [14, 15]. Combined with the expansions of $Q(n)$ and $\sigma_3\delta Q(n)$ over the squared solutions one is able to derive all fundamental properties of nALM in terms of the recursion operators $\Lambda_\pm$. It seems that nALM, like ALM, possesses also classical and quantum R matrices.

Another challenge is related to the multicomponent models. Indeed, ALM can be viewed as integrable discrete approximation to the nonlinear Schrödinger (NLS) equation [16]. Recently it was demonstrated, see that several classes of multicomponent NLS and N-wave equations such as [11, 10], also allow integrable nonlocal reductions [12, 9, 20, 21, 19]. Constructing integrable discrete multicomponent analogs of ALM is a difficult task, though some results in this direction are already known [23, 24]. Constructing nonlocal multicomponent analogs of ALM however is a bigger challenge.

Acknowledgements
I am grateful to prof. G. Grahovski for useful discussions and careful reading of the manuscript.

References

[1] Ablowitz M.J., Kaup D.J., Newell A.C. and Segur H. (1974) *The inverse scattering transform – Fourier analysis for nonlinear problems*, Studies in Appl. Math. **53**, 249–315.
[2] Ablowitz M. J. and Ladik J. F. (1975) *Nonlinear differential-difference equations*, J. Math. Phys. **16** 598–603.
[3] Ablowitz M. J. and Ladik J. F. (1976) *Nonlinear differential-difference equations and Fourier Analysis*, J. Math. Phys. **17** 1011-1018.
[4] Ablowitz M. J. and Musslimani, Z. (2014) *Integrable discrete $\mathcal{PT}$ symmetric model*, Phys. Rev. E **90** 032912.
[5] Ablowitz, M. J., Prinari B. and Trubach, D. A. (2004) *Discrete and Continuous Nonlinear Schrödinger Systems*, LMS Lecture Note Series **302**, Cambridge University Press, Cambridge.
[6] Doktorov E. V., Matsuka N. P. and V. M. Rothos, (2004) *Dynamics of the Ablowitz-Ladik soliton train*, Phys. Rev. E **69**, 056607 (7 pp.).
[7] Faddeev L. D. and Takhtadjan L. A. (1987) *Hamiltonian Method in the Theory of Solitons*, Springer Verlag, Berlin.
[8] Gerdjikov V. S. (1986) *Generalised Fourier transforms for the soliton equations. Gauge covariant formulation*, Inverse Problems **2**, 51–74.
[9] Gerdjikov V. S. (2017) On nonlocal models of Kulish-Sklyanin type and generalized Fourier transforms. Eds.: K. Georgiev, M. Todorov, I. Georgiev. Advanced Computing in Industrial Mathematics, Studies in Computational Intelligence **681**, pp 37-52. **arXiv:1710.03705v1 [nlin.SI]**;
Gerdjikov V. S. (2017) Kulish-Sklyanin type models: integrability and reductions. Theoretical and Mathematical Physics **192** (2): 1097–1114.
[10] Gerdjikov V. S. and G. G. Grahovski G. G. (2010) *Multi-component NLS models on symmetric spaces: Spectral properties versus representations theory*, SIGMA **6**, paper 044 (29 pages).
[11] Gerdjikov V. S., Grahovski G. G. and Kostov N. A. (2001) *Reductions of N-wave interactions related to low–rank simple Lie algebras. I: $\mathbb{Z}_2$- reductions*, J. Phys. A: Math & Gen. **34** 9425–9461; (E-print: nlin.SI/0006001).
[12] Gerdjikov V. S., Grahovski G. G., Ivanov R. I. and Kostov N. A. (2001) *N-wave interactions related to simple Lie algebras. $\mathbb{Z}_2$- reductions and soliton solutions*, Inverse Problems **17** 999–1015; (E-print: nlin.SI/0009034).
[13] Gerdjikov V. S. and Khristov E. Kh. (1980) *On the evolution equations solvable with the inverse scattering problem. II. Hamiltonian structures and Backlund transformations.* Bulgarian J. Phys. **7,** No.2, 119–133.
[14] Gerdjikov V. S., Ivanov M. I. and Kulish P. P. (1980) *Complete integrability of the difference evolution equations.* JINR preprint E2–80–882, Dubna, USSR.
[15] Gerdjikov V. S., Ivanov M. I. and Kulish P. P. (1984) *Expansions over the "squared" solutions and difference evolution equations.* J. Math. Phys. **25,** n. 1, 25–34.
[16] Gerdjikov V. S. and Saxena A. (2017) *Complete Integrability of Nonlocal Nonlinear Schrödinger equation*, J. Math. Phys. **58**, 013502 (E-print: arXiv:1510.0480).
[17] Gerdjikov V. S., Vilasi G. and Yanovski A. B. (2008) *Integrable Hamiltonian Hierarchies. Spectral and Geometric Methods*, Lecture Notes in Physics **748**, Springer Verlag, Berlin, Heidelberg, New York.
[18] Gerdjikov V. S. and Yanovsky A. B. (1984) *Gauge covariant formulation of the generating operator. I. The Zakharov–Shabat system.* Phys. Lett. A, **103A**, n. 5, 232–236.
[19] Grahovski G. G., Mohammed A. J. and Susanto H. (2018) *Nonlocal reductions of the Ablowitz-Ladik equation*, Theor. Math. Phys. **196**, 1260–1277 (E-print: *arXiv:1711.08419*).
[20] Gürses M. (2017) Nonlocal Fordy–Kulish equations on symmetric spaces Physics Letters **A 381** (21), 1791-1794.
[21] Gürses M. and Pekcan A. (2018) Nonlocal nonlinear Schrödinger equations and their soliton solutions Journal of Mathematical Physics **59** (5), 051501.
[22] Mikhailov A. V. (1981) *The reduction problem and the inverse scattering problem*, Physica D **3** 73–117.
[23] Tsuchida T. (2018) On a new integrable generalization of the Toda lattice and a discrete Yajima–Oikawa system. **ArXiv:1808.03261v2 [nlin.SI]**.
[24] Vakhnenko O. O. (2018) Integrable nonlinear Schrödinger system on a lattice with three structural elements in the unit cell J. Math. Phys. **59**, 053504.

A method of two-dimensional filtering of modulated matrix data sequences

V G Getmanov[1], I I Astapov[2], N S Barbashina[2], A D Gvishiani[1], A N Dmitrieva[2], M N Dobrovolsky[1], D V Peregoudov[1], R V Sidorov[1], A A Soloviev[1], V E Chinkin[1], V V Shutenko[2] and I I Yashin[2]

[1]Geophysical Center of the Russian Academy of Sciences (GC RAS), Molodezhnaya St., 3, Moscow, 119296, Russia
[2]National Research Nuclear University MEPhI ((Moscow Engineering Physics Institute)), Kashirskoe shosse 31, Moscow, 115409, Russia

r.sidorov@gcras.ru

Abstract. The method of two-dimensional filtering of modulated matrix data sequences is proposed. The structure of a two-dimensional filter based on operations of element-by-element matrix multiplications is developed. An application of the two-dimensional filtering of the URAGAN muon hodoscope matrix data sequence for the elimination of muon flux diurnal modulations is discussed.

1. Introduction and posing of the problem

A method for two-dimensional filtering of sequences of modulated matrix data, consisting of sums of initial and modulated components, is proposed. The need for such filtering – the elimination of component modulations in the matrix data sequences – arises in many problems of experimental physics and various applications associated with the requirements for the separation of time-varying spatially-temporal oscillatory processes: for example, for example, in the analysis of periodic secular and seasonal processes of ice formation in the polar regions; in the studies related to the influence of cyclic solar activity on the dynamics of slow climatic changes of given regions of the Earth's surface; in sea disturbance studies using satellite data etc. Here an application associated with filtering of diurnal modulations in muon fluxes (MF) in the URAGAN [1] muon hodoscope (MH) matrix data is discussed.

MF reaching the MH aperture-type detector are subject to temporal and spatial modulations, which can be divided into periodic: solar 11- and 22- year and 27-day modulations [2], diurnal modulations due to the Earth rotation [3] and aperiodic ones – due to possible Forbush decreases [4] and atmospheric phenomena influences [5], including temperature and pressure influences [6]. In some cases, modulations can distort the useful information about the processes in the heliosphere contained in the MF.

In the MH, two-dimensional MF intensity distribution functions are measured based on the count of registered muons for a set of discrete solid angles for a given time. The measured values of the distribution function are placed into the MH matrices and denoted as $Y(i, j, Tn)$, where $i = \overline{1, N_1}$,

$j = \overline{1, N_2}$, T is a time step, $n = \overline{0, N_f - 1}$. The indices i and j correspond to the azimuth and zenith angles with given discreteness and ranges.

Due to the rotation of the Earth, the contents of the MH matrices are subject to diurnal modulations with a frequency $\Omega_0 = 2\pi / T_0$, T_0 is a day duration. As an example, hourly MH matrices recorded in June-July 2015 were analyzed, T_0 = 3600 sec, N_f =1451. For the data in the MH matrices, corrections were made for temperature and pressure [7]. On figure 1, index 1 is the graph of the component of the function $Y(i_0, j_0, Tn)$ for i_0=30, j_0=45, $\overline{n}_1 \le n \le \overline{n}_2$, $\overline{n}_1$=200, $\overline{n}_2$=1451; this function of measurements contains noise and modulation diurnal components. Filtering, eleminating the noise components, was performed usind a one-dimensional digital low pass Batterworth filter with an order N_B =8 and a cutoff frequency $\omega_c = 1.2\Omega_0$; the software module from this filter was obtained from [9]. On figure 1, index 2 is the result of noise component filtering, and the remaining modulation diurnal components can be seen.

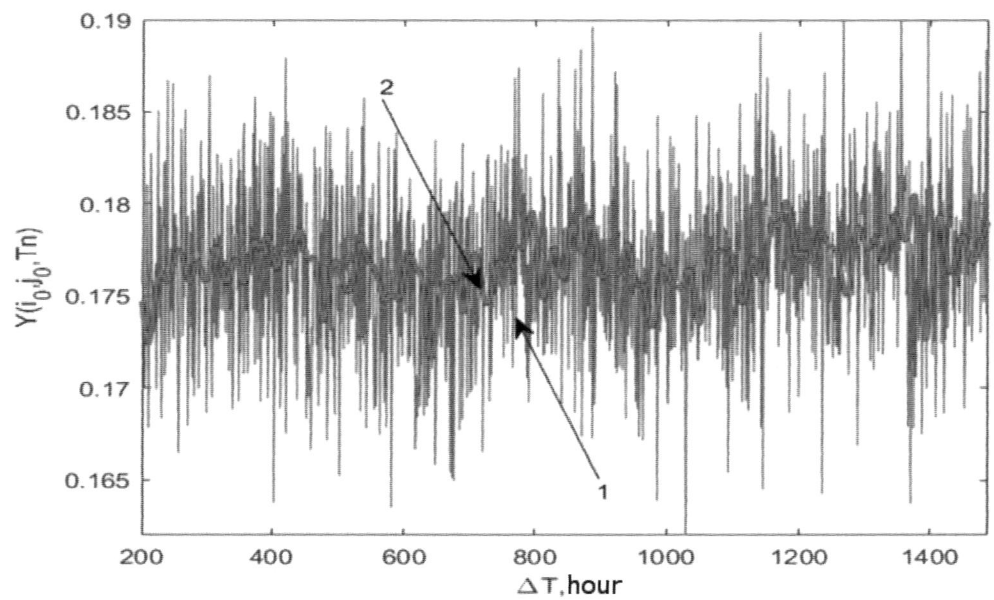

Figure 1. $Y(i_0, j_0, Tn)$ component function plot (1), the result of Butterworth filtering (2).

To clarify the spectral composition of a set of signals in the MH matrices $Y(i, j, Tn)$, where $i = \overline{1, N_1}$, $j = \overline{1, N_2}$, the total MF intensity function $S = S(Tn)$, muon·sec^{-1}, was introduced. We calculate its discrete Fourier transform (DFT) $C(k)$ for a given N

$$S(Tn) = \frac{1}{N_1 N_2} \sum_{i=1}^{N_1} \sum_{j=1}^{N_2} Y(i, j, Tn), \; C(k) = (1/N) \sum_{n=0}^{N-1} S(Tn) \exp(-2\pi nk / N), \; k = \overline{0, N-1}. \quad (1)$$

The physical meaning of a function $S(Tn)$ is obvious. The logarithmic DFT spectrum for the introduced function, denoted as $LC(\Delta Fk) = 20\log_{10}(C^*(k)C(k))$, allows to estimate the spectral component amplitudes for $S(Tn)$. On figure 2, the LC plot (dB) is displayed against the frequency ΔFk Hz, where $\Delta F = 1/NT$ is the DFT resolution, N=1024, ΔF = 2.7126 Hz, k=0, 1, …, 260. Two spectral peaks can be seen, indicating the 1-day and 0.5-day periods, with the frequencies of $1.1574{\cdot}10^{-5}$ Hz and $2.3118{\cdot}10^{-5}$ Hz, respectively. The presence of at least two spectral components allows us to conclude that daily modulations in $S(Tn)$ are not sinusoidal, which is quite consistent with their physics.

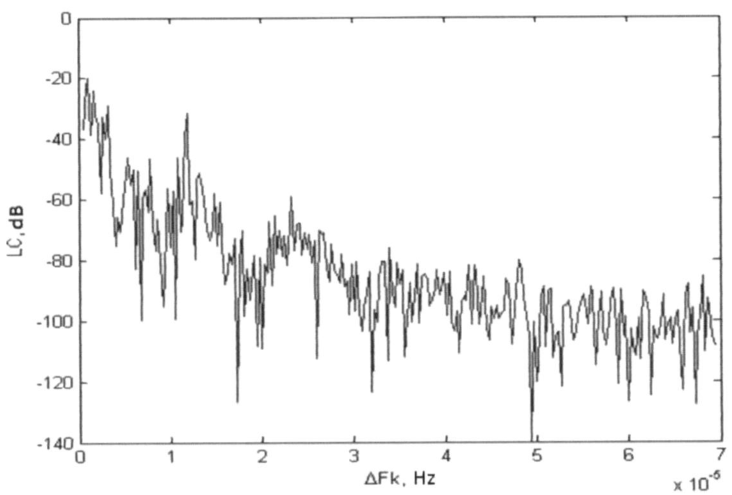

Figure 2. DFT spectrum of the function of total intensity of muon fluxes

In this paper, the task is to develop a method for two-dimensional filtering of sequences of modulated matrix data and the implementation of an application of filtering the MF daily modulations in MH matrix sequences – the formation of corresponding matrix sequences in which these variations are absent.

2. Method of two-stage two-dimensional low-pass filtering

To eliminate modulations, we use one-dimensional digital low-pass filters. To the input functions $\boldsymbol{Y(i, j, Tn)}$ we apply the standard procedures for digital one-dimensional low-pass filtering with weights $b_r(i, j), r = 1,...,r_0$, $a_s(i, j), s = 0,...,s_0$, the parameters r_0, s_0 define the orders of the filters. The filter cutoff frequencies $\omega_c(i, j)$ must satisfy the condition $\omega_c(i, j) < \Omega_0$. We obtain the weight values from [9]; let us formulate the difference equation, in which $\boldsymbol{Y_F(i, j, Tn)}$ is a digital filter output

$$\boldsymbol{Y_F(i, j, Tn)} = -\sum_{r=1}^{r_0} \boldsymbol{b_r(i, j) Y_F(i, j, T(n-r))} + \sum_{s=0}^{s_0} \boldsymbol{a_s(i, j) Y(i, j, T(n-s))}. \quad (2)$$

The use of digital one-dimensional low-frequency filters (2) for the task posed involves two problems:

1) the need to provide small time costs for performing one-dimensional filtering operations by (2);

2) the need to eliminate the arising phase shifts that occur in $Y_F(i, j, Tn)$ due to the recurrence relations (2). We realize the filtering of the sequence of matrices from MH based on the two-step method.

At the first stage we form the matrices $\boldsymbol{B_1,...,B_{r_0}}$, $A_0,...,A_k$ consisting of the weight elements $b_r(i, j)$, $a_s(i, j)$ for (2). With their help we form the structure of a two-dimensional filter on the basis of the difference equation in the matrix form:

$$Y_{1F}(Tn) = -\sum_{r=1}^{r_0} B_r \circ Y_{1\phi}(T(n-r)) + \sum_{s=0}^{s_0} A_s \circ Y(T(n-s)) \qquad (3)$$

where " $\circ$ " denotes the operation of element-wise multiplication of matrices. Using [9], a superfast matrix element-wise multiplication is realized. The matrix sequence $Y_{1F}(Tn)$ is output for the filter (3).

At the second stage we perform the elimination of phase shifts. Based on (1), we introduce the total intensity function $S_{1F}(Tn)$ for $Y_{1F}(i, j, Tn)$. Let us construct a functional $F(S, S_F, n_d)$ and find an optimal phase shift n_d°:

$$S_{1F}(Tn) = \frac{1}{N_1 N_2}\sum_{i=1}^{N_1}\sum_{j=1}^{N_2} Y_{1F}(i, j, Tn),\; F(S, S_{1F}, n_d) = \sum_{n=n_1}^{n_2}(S(Tn) - S_{1F}(T(n-n_d)))^2,$$

$$n_d^\circ = \arg\{\min_{1\le n_d \le n_{d0}} F(S, S_{1F}, n_d)\} \qquad (4)$$

The result of the second stage of filtering is defined as the sequence of matrices in which the phase shift correction n_d° is performed: $Y_F(i, j, Tn) = Y_{1F}(i, j, T(n - n_d^\circ))$.

3. Filtering of daily modulations in muon fluxes for observation of the hydroscope of URAGAN hodoscope

The examination of the sequence of MH matrices obtained during June-July 2015 was continued.

1. The MH matrices $Y_{1F}(Tn)$ were calculated using the structure (3) based on Butterworth filters of order $N_B = 8$ and $\omega_c = 1.2\Omega_0$; the $S_{1F}(Tn)$ function with phase shifts was formed. On figure 3, curve 1 depicts the total intensity function $S = S(Tn)$ for MH against the time Tn, sec, for the interval $\bar{n}_1 \le n \le \bar{n}_2, \bar{n}_1 = 200,\ \bar{n}_2 = 1451$; diurnal modulations with an amplitude of ≈ 0.0015 are seen; the mean intensity value is ≈ 0.2; curve 2 is the plot for the filtered function $S_{1F}(Tn)$ with a phase shift. On curve 2, the diurnal modulations are absent. With the application of optimization $F(S, S_F, n_d)$, the phase shift estimate $n_d^\circ = 39$ was obtained, and the shift correction was realized.

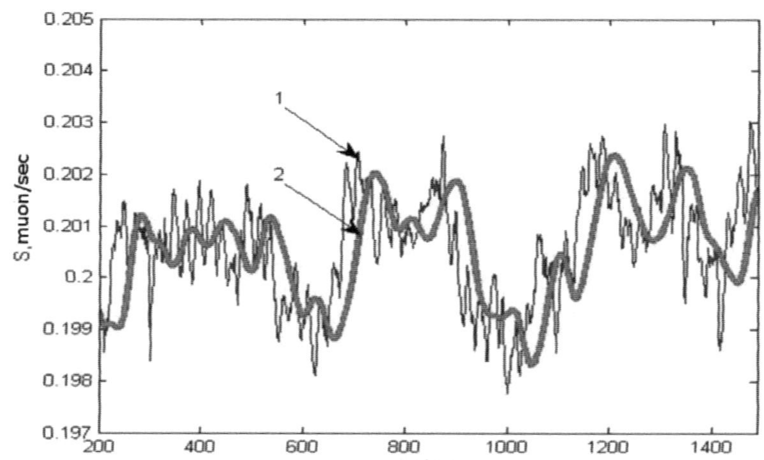

Figure 3. The result of the work of a two-dimensional matrix filter for eliminating diurnal variations of muon fluxes

2. The computational experiments were performed, which allowed to determine that: time costs of the proposed method of two-dimensional filtering, on average, an order of magnitude less than the time costs that are realized for filtering based on one-dimensional filters: phase shift estimates are ≈1–2 grad, approximately, and the model sinusoid modulation amplitude decrease ≈10–20 times.

The method of two-dimensional filtering of sequences of modulated matrix data developed in the article is more effective in comparison with traditional approaches [7, 8] in terms of time costs and possibilities for eliminating phase shifts.

4. Conclusions

1. The proposed method of two-dimensional filtering of sequences of modulated matrix data proved to be workable.

2. The filtering of the diurnal modulations of muon fluxes for the matrix observations of the URA-GAN MH was successfully implemented.

3. On the basis of computational experiments, it was established that: the time costs of the proposed two-dimensional filtering method are, on average, one order of magnitude less than the time costs that are realized for filtering based on one-dimensional filters; the errors of phase shift estimates are ≈1–2 grad; the amplitudes of the model sinusoidal modulations decrease ≈10–20 times.

4. The proposed two-dimensional filtering method can be applied to many problems of experimental physics and other applications related to the elimination in the sequences of matrix data of modulation components.

5. References

[1] Yashin I I et al 2015 *J. Phys.: Conf. Ser.* **632** 012086.
[2] Okazaki Y et al 2008 *Astrophys. J.* **681** 1.
[3] Poirier J, Catanach T 2011 *Proc. 32-nd Int. Cosmic Ray Conf. (Beijing)* (Under the Auspices of the International Union of Pure and Applied Physics (IUPAP))vol 11 p 173.
[4] Braun I, Engler J, Horandel J R and Milke J 2009 *Adv. Space. Res.* **43**(4) 480–488.
[5] Karapetyan G G 2014 *Phys. Rev.* D**89** 093005.
[6] Dmitrieva A N et al 2011 *Astropart. Phys.* **34**(6) 401-411.
[7] Yashin I I et al 2015 *Adv. Space. Res.* **56**(12) 2693–2705.
[8] Dmitrieva A et al. 2015 *J. Phys.: Conf. Ser* **632** (2015) 012054.
[9] Garello R (ed) 2013 *Two-dimensional Signal Analysis* (London: Wiley-ISTE).
[10] Dyakonov V P, Abramenkova I V 2002 *Matlab. Image and Signal processing* (in Russian, Saint Petersburg: Piter)
[11] *Image Processing Toolbox*. http://matlab.exponenta.ru/imageprocess (last accessed: 10.09.2018).

Statistically significant performance testing of Julia scientific programming language

M N Gevorkyan[1], A V Demidova[1], A V Korolkova[1] and D S Kulyabov[1,2]

[1]Department of Applied Probability and Informatics,
Peoples' Friendship University of Russia (RUDN University),
6 Miklukho-Maklaya str., Moscow, 117198, Russia
[2]Laboratory of Information Technologies, Joint Institute for Nuclear Research,
6 Joliot-Curie, Dubna, Moscow region, 141980, Russia

E-mail: `gevorkyan-mn@rudn.ru`, `demidova-av@rudn.ru`, `korolkova-av@rudn.ru`, `kulyabov-ds@rudn.ru`

Abstract. In this article, we compare the support for SIMD instructions for Julia and Fortran. The comparison is carried out according to the methodology described in work of T. Kalibera, R. E. Jones. The first part of the article gives a brief description of this technique. We emphasize on the practical implementation using Python, NumPy and SciPy. The second part of the article briefly discusses the syntactic capabilities of Fortran and Julia to work with SIMD processor extensions. Specific code snippets are given. Next, the performance of Julia and Fortran is compared for arithmetic operations on arrays of small length. The results are presented in tabular and graphical form.

1. Introduction

In this paper, we benchmark Fortran and Julia [1, 2, 3, 4, 5] SIMD instructions implementation. We carry out benchmarking guided by the methodology described in papers [6, 7]. This technique allows to obtain the statistically significant estimator of average time and to determine the optimal number of measurements.

In the first part of our paper, we provide the brief mathematical introduction in technique [6, 7]. We mainly focus on how to implement this technique using Python [8, 9] with Numpy [10], SciPy [11] and Matplotlib [12] libraries. Our source code is open and available at `https://bitbucket.org/mngev/fortran-vs-julia`.

In the second part of the article, we describe Fortran (GNU Fortran) and Julia languages SIMD vectorization capabilities and give some code snippets. In the last part of the paper we introduce benchmark results and give some summary.

2. Statistically significant estimator of program execution time

A naive approach to measuring the execution time of a program (or subroutine or function) is to repeatedly call the program and find the average time of its execution. Sometimes the first few runs of the program are not taken into account, as they are needed for warm-up. The number of execution is usually chosen heuristically. If we want to obtain statistically significant results, this approach cannot be considered satisfactory. In our work, we use the [6, 7] methodology.

These papers are freely available on the Internet, so we will focus on the implementation of the methods proposed in these works using the Python language and its libraries [13, 14]

2.1. Experiment's levels

Before starting an experiment, one should determine the maximum number of *experiment levels*. In our case, we will measure the execution time of the function. The first level is r_1 function calls inside the program, the second level is r_2 executions of program and the third is r_3 completions of the program source code. In general case it is possible to introduce any number of experiment levels.

At the first stage, a preliminary experiment with a heuristic choice of the number of levels $n+1$ and the number of tests r_i at each i-th level is carried out. Each test gives a value of the measurement time $X_{j_{n+1}j_n \ldots j_2 j_1}$, where $j_i = 1, \ldots, r_i$. For example, if we measure the time spent on the third call of a function when the program is run for the second time using the fourth compiled executable file, it will be X_{423}. To store the results of measurements we used a multidimensional NumPy array. Every element of this array may be considered independent random number.

At each level, a one-dimensional sample of the obtained time measurements is considered.

Level $i = 1$. We fix the indexes of previous higher levels and get a sample of the following type:

$$\mathbf{X}_{j_{n+1}j_n \cdots j_3 j_2 \bullet} = (X_{j_{n+1}j_n \cdots j_2 1}, X_{j_{n+1}j_n \cdots j_2 2}, \ldots, X_{j_{n+1}j_n \cdots j_2 r_1}).$$

The total number of such samples will be $r_{n+1} \cdot r_n \cdot \ldots \cdot r_3 \cdot r_2$ for all possible combinations of higher indexes. When using NumPy arrays to obtain, for example, $\mathbf{x}_{42\bullet}$, it is sufficient to take a slice by the last dimension: `x[3, 1,:]` (the indexing starts from zero).

Level $i = 2$. To study the results of the second level of execution, consider a sample of $\mathbf{x}_{j_{n+1}j_n \cdots j_3 j_2 \bullet}$ and find its average value. As a result, for a fixed set of indexes $j_{n+1}j_n \ldots j_3 j_2$ we get a scalar value $\bar{X}_{j_{n+1}j_n \cdots j_3 j_2 \bullet}$. By changing the index j_2 we get a sample for the second level of the experiment.

$$\bar{X}_{j_{n+1}j_n \cdots j_3 j_2 \bullet} = \mathrm{M}[\mathbf{X}_{j_{n+1}j_n \cdots j_3 j_2 \bullet}] = \frac{1}{r_1}\sum_{j_1=1}^{r_1} X_{j_{n+1}j_n \cdots j_2 j_1},$$

$$\mathbf{X}_{j_{n+1}j_n \cdots j_3 \bullet\bullet} = (\bar{X}_{j_{n+1}j_n \cdots j_3 1 \bullet}, \bar{X}_{j_{n+1}j_n \cdots j_3 2 \bullet}, \ldots, \bar{X}_{j_{n+1}j_n \cdots j_3 r_2 \bullet})$$

The total number of such samples will be $r_{n+1} \cdot r_n \cdot \ldots \cdot r_3$ for all possible combinations of indexes greater than 2. When you are working with NumPy arrays, the calculation of $\bar{X}_{j_{n+1}j_n \cdots j_3 j_2 \bullet}$ is made with functions `mean` by applying it to the last dimension: `np.mean (X, axis=-1)`. The result is an array with the number of dimensions reduced by one. For the following levels, similar actions have to be taken.

Level $i = n$.

$$\bar{X}_{j_{n+1}j_n \underbrace{\bullet \ldots \bullet}_{n-1}} = \mathrm{M}[\mathbf{X}_{j_{n+1}j_n \underbrace{\bullet \ldots \bullet}_{n-1}}] = \frac{1}{r_{n-1}}\sum_{j_{n-1}=1}^{r_{n-1}} \bar{X}_{j_{n+1}j_n j_{n-1} \underbrace{\bullet \ldots \bullet}_{n-2}}$$

$$\mathbf{X}_{j_{n+1} \underbrace{\bullet \ldots \bullet}_{n}} = (\bar{X}_{j_{n+1}1 \underbrace{\bullet \ldots \bullet}_{n-1}}, \bar{X}_{j_{n+1}2 \underbrace{\bullet \ldots \bullet}_{n-1}}, \ldots, \bar{X}_{j_{n+1}r_n \underbrace{\bullet \ldots \bullet}_{n-1}})$$

Level $n+1$

$$\bar{X}_{j_{n+1} \underbrace{\bullet \ldots \bullet}_{n}} = \mathrm{M}[\mathbf{X}_{j_{n+1} \underbrace{\bullet \ldots \bullet}_{n}}] = \frac{1}{r_n}\sum_{j_n=1}^{r_n} X_{j_{n+1}j_n \underbrace{\bullet \ldots \bullet}_{n-1}}$$

$$\mathbf{X}_{\underbrace{\bullet \ldots \bullet}_{n+1}} = (\bar{X}_{1 \underbrace{\bullet \ldots \bullet}_{n}}, \bar{X}_{2 \underbrace{\bullet \ldots \bullet}_{n}}, \ldots, \bar{X}_{r_{n+1} \underbrace{\bullet \ldots \bullet}_{n}})$$

IOP Conf. Series: Journal of Physics: Conf. Series **1205** (2019) 012017 doi:10.1088/1742-6596/1205/1/012017

Finally, to get the mean time for all tests, we should find the mean for the entire data set (the grand mean).

$$\bar{X} = \mathrm{M}[\bar{X}\underbrace{\bullet\ldots\bullet}_{n+1}] = \frac{1}{r_{n+1}}\sum_{j_{n+1}=1}^{r_{n+1}} X_{j_{n+1}\underbrace{\bullet\ldots\bullet}_{n}}.$$

2.2. Visual analysis

After the measurements are completed, one should proceed to a visual evaluation of the data, which will give an opportunity to estimate the number of pre-runs required for the system warm-up. We denote the number of pre-runs for each level by $c_1, c_2, \ldots, c_{r_i}$. For each level i one should draw three plots.

- The first plot is the run-sequence plot. The horizontal axis represents the test number from 1 to r_i, and the vertical axis — $X_1, \ldots, X_{r_i}$. Additionally, one may build a similar plot with mixed values $X_1, \ldots, X_{r_i}$. The analysis of the resulting image will help to determine how many pre-runs the system requires to warm-up.
- The second plot is the lag-plot. One axis represents the values $X_1, \ldots, X_{r_i-l}$ and on the second axis — $X_l, \ldots, X_{r_i}$, where the integer l specifies the lag. If the numbers $X_1, \ldots, X_{r_i}$ are independent random values, the points on the chart will be placed chaotically, without visible structure.
- The third plot is the auto-correlation function plot (ACF-plot). It allows to estimate the correlation between a sample of $\mathbf{X} = X_1, \ldots, X_{r_i-l}$ and its version with lag $\mathbf{X}_l = X_l, \ldots, X_{r_i}$. The value of the autocorrelation is determined by the formula

$$\mathrm{cor}(\mathbf{X}, \mathbf{X}_l) = \frac{\sum\limits_{k=1}^{r_i-l}(X_k - \bar{X})(X_{l+k-1} - \bar{X}_l)}{\sqrt{\sum\limits_{k=1}^{r_i-l}(X_k - \bar{Y})^2 \sum\limits_{k=l}^{r_i}(X_k - \bar{X}_l)^2}}$$

$$\bar{X} = \frac{1}{r_i - l}\sum_{k=1}^{r_i-l} X_k, \quad \bar{X}_l = \frac{1}{r_i - l}\sum_{k=l}^{r_i} X_k.$$

  To draw the plot along the horizontal axis, the lag values l are plotted, and along the vertical axis, the values $\mathrm{cor}(\mathbf{X}, \mathbf{X}_l)$ for each corresponding l are plotted.

The last two plots are called correlograms and allow us to determine whether the available time measurements can be considered independently distributed random variables.

To build the first plot, you may use the `scatter` function of the Matplotlib library to visualize points cloud. To mix the elements in the array, the `np.random.shuffle` function is used.

By using the same `scatter` function in conjunction with array slices we can build the second plot. The values `X[l:]` are plotted on one axis and the values `X[:-l]` are plotted on the other axis, where `l` specifies the lag. It is also useful to draw the bisector of the first quarter of the coordinate axes.

To plot the autocorrelation graph, the use of the `acorr` function will give an incorrect result, since it uses a different algorithm for calculating autocorrelation in signal processing. So the calculation of autocorrelation must be implemented independently. The result of its work can be displayed on the chart using the `vlines` function.

2.3. Biased and unbiased estimators

S_i^2 is biased estimator for variance at level i. It is calculated by following formulas

$$S_1^2 = \frac{1}{\prod\limits_{k=2}^{n+1} r_k} \frac{1}{r_1 - 1} \sum_{j_{n+1}=1}^{r_{n+1}} \cdots \sum_{j_1=1}^{r_1} (X_{j_{n+1}j_n \ldots j_2 j_1} - \bar{X}_{j_{n-1}j_n \ldots j_2 \bullet})^2;$$

$$S_i^2 = \frac{1}{\prod\limits_{k=i+1}^{n+1} r_k} \frac{1}{r_i - 1} \sum_{j_{n+1}=1}^{r_{n+1}} \cdots \sum_{j_i=1}^{r_i} (\bar{X}_{j_{n+1}j_n \ldots j_i \underbrace{\bullet \ldots \bullet}_{i-1}} - \bar{X}_{j_{n+1}j_n \ldots j_{i+1} \underbrace{\bullet \ldots \bullet}_{i}})^2, \; i = 2, \ldots, n;$$

$$S_{n+1}^2 = \frac{1}{r_{n+1} - 1} \sum_{j_{n+1}=1}^{r_{n+1}} (X_{j_{n+1} \underbrace{\bullet \ldots \bullet}_{n}} - \bar{X}_{\underbrace{\bullet \ldots \bullet}_{n+1}})^2;$$

Despite the cumbersome mathematical formulas, the computation of s_i^2 in terms of NumPy arrays is a trivial task. For example, let us measure time with three levels of experiment. As a result of measurements we obtain a three-dimensional array `X`. To calculate S_i^2, one need to use `mean` function for 2 and 3 dimensions, and then `var` function to find the unbiased sample variance

```
res = np.mean(X, axis=(-2, -1))
res = np.var(res, ddof=1, axis=-1)
```

where the parameter `axis=(-2, -1)` indicates that the function should use only last tow dimensions of the array, and the parameter `ddof=1` indicates that we need an unbiased sample variance.

When we have S_i^2 we can use iterative process to calculate the unbiased estimators T_i^2:

$$T_1^2 = S_1^2, \; T_i^2 = S_i^2 - \frac{S_{i+1}^2}{r_{i-1}}, \; i = 2, \ldots, n+1.$$

Estimators T_i^2 allow to calculate the optimal number of repetitions for each level of the experiment

$$n_1 = \sqrt{c_1 \frac{T_1^2}{T_2^2}}, \; n_i = \sqrt{\frac{c_i}{c_{i-1}} \frac{T_i^2}{T_{i+1}^2}},$$

where c_i is the number of repetitions, needed for the level i warm-up.

3. SIMD vector operations in Fortran and Julia

3.1. SIMD extensions

SIMD extensions (single instruction stream / multiple data stream) can be found in most modern processors. SIMD extensions are a set of special registers and instructions that can be applied to these registers in parallel mode. The most common modern SIMD extensions are AVX (Advanced Vector Extensions) of different revisions. For the measurements we used Intel Haswell processor with AVX2 support, which gives 16 registers with a total volume of 256 bits. Modern AMD Zen processors also support AVX2 instructions.

SIMD vectorization is often the easiest way to improve the performance of numerical methods for solving systems of differential equations [15]

3.2. Syntax for SIMD operations in Fortran and Julia
Let's consider several ways how to add one array to another in Fortran and Julia languages. Let's begin from Fortran and set three dynamic arrays and then fill them with random numbers.

```
integer, parameter :: array_size = 8
real, allocatable, dimension(:) :: V, U, Res
integer :: i
allocate(A(array_size), B(array_size), C(array_size))
call random_number(V); call random_number(U)
```

In the first example, let's add two arrays elementwise in a loop and write the result to the third array

```
do i = 1,array_size,1
  Res(i) = V(i) + U(i)
end do
```

One may also use the `concurrent` statement added to the Fortran 2008 [16, 17, 18] standard to explicitly point out the loop iterations independence:

```
do concurrent(i = 1:array_size,1)
  Res(i) = V(i) + U(i)
end do
```

In the second example, we perform the same summation, but without cycles, using so-called vector notation

```
Res = V + U
```

We intentionally use dynamic arrays in order to provide a more fair comparison of languages, as in the Julia standard arrays can change size during execution.

There are three options for similar actions in Julia. Two of them are comparable with Fortran:

```
const array_size = 8
V = Vector{Float32}(array_size)
U = Vector{Float32}(array_size)
Res = Vector{Float32}(array_size)
rand(V); rand(U); rand(Res)
for i in 1:array_size
  Res[i] = V[i] + U[i]  # iterative summation
end
Res = V + U  # same without loop
```

The third option allows using `@simd` and `@inbounds` macros to explicitly specify the usage of SIMD instructions.

```
@simd for i in 1:20
  @inbounds Res[i] = V[i] + U[i]
end
```

3.3. Benchmarking
Benchmarking is carried out for GNU Fortran [19] version 7.3.0. To get a log file about the vectorization process, we use the `-fopt-info-vec-all=file.log` command line option during the compilation. We implement above examples as separate functions and use the `cpu_time` intrinsic subroutine to measure time.

Tests are performed for arrays with 32-bits floating point numbers. The length of the array is calculated from the total volume of AVX registers (256 bits). The division by 32 bits gives exact 8. The time required for 10^4 function calls is measured. We perform 3 experiment levels, where $r_1 = 100$, $r_2 = 20$ and $r_3 = 20$. The total amount is $100 \cdot 20 \cdot 20$ measurements.

We also perform separate measurements for each of the three possible optimization levels: `O1`, `O2` and `-O3`.

	Overall mean		**Confident interval sizes**		**Optimum**	
	Iter.	Vect.	Iter.	Vect.	Iter.	Vect.
O1	0.000436	0.000246	0.000009	0.000011	34	27
O2	0.000405	0.000374	0.000010	0.000009	41	29
O3	0.000400	0.000379	0.000008	0.000009	33	36

Table 1. Results for Fortran. Confident interval was calculated for $\alpha = 0.01$

The benchmark results are organized in three tables 1. The first column of each table contains measurements for iterative summation and the second column contains data for vectorized summation. The last table contains the optimal number of repetitions for each experiment level which are calculated as described in first part of this paper. For overall mean we also plot bar charts 1.

The obtained measurement data show that when using the first level of optimization, the compiler automatically applies SIMD instructions to those functions where non-index summation was used. At a higher level of optimization, the compiler automatically detects loop independence and uses SIMD instructions even for iterative summation.

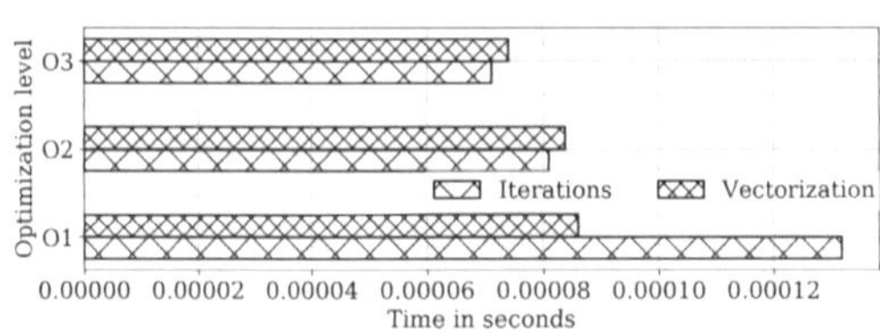

Figure 1. Overall mean from Table 1

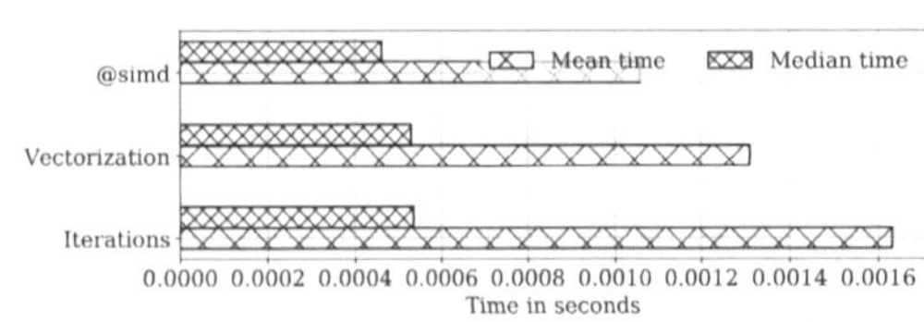

Figure 2. Overallmean from Table 2

Julia language is implemented as JIT-compiler, so there is no explicit compilation stage. So we left only 20 program executions and 100 functions calls. Also, there are no levels of optimization. The results of measurements are presented in the table 2, and the plot 2.

	Iterations	**Vectorization**	`@simd`
Overall mean	0.00163201	0.00131059	0.00105772
Conf. interval	0.0011087924109	0.00087504387	0.00071035994
Optimal	11	11	11

Table 2. Results for Julia. Confident interval was calculated for $\alpha = 0.01$

The plot shows that the average time of Julia program is almost two orders of magnitude grater than Fortran. However, the median time is less than the average Fortran program execution time by an order of magnitude. If we consider the run-sequence plot, it becomes clear that a some values of measurements differ greatly from the average time. It is these anomalous measurements that lead to an increase in the average time.

As mentioned in the official documentation of the Julia language, the use of non-index operations in the Julia does not provide any significant increase in productivity (see median time). However, the use of the `@simd` macro allows to get slightly better loop performance.

4. Conclusion

In this article, we have compared the implementation of SIMD extensions in the GNU Fortran standard compiler and language Julia JIT compiler. Julia language in showed worse results, but there are a few caveats to be made.

First of all, the Julia language is in the process of intensive development and has not yet reached even the first version. Therefore, the results of measurements in any case can not be considered as final.

Also Julia language implements a lot of high level features from interpreting languages. Due to this, the development of software on Julia is potentially faster.

Acknowledgment

The publication has been prepared with the support of the "RUDN University Program 5-100" and funded by Russian Foundation for Basic Research (RFBR) according to the research project No 16-07-00556.

[1] Bezanson J, Edelman A, Karpinski S and Shah V B 2017 *SIAM Review* **59** 65–98
[2] Bezanson J, Karpinski S, Shah V B and Edelman A 2012 (*Preprint* 1209.5145)
[3] Kulyabov D, Gevorkyan M, Korolkova A and Sevastianov L 2017 *CEUR Workshop Proceedings* **1995** 93–99
[4] Balbaert I, Sengupta A and Sherrington M 2016 *Julia: High Performance Programming* ISBN 9781787125704
[5] Kwon C 2016 *Julia Programming for Operations Research: A Primer on Computing* ISBN 9781533328793
[6] Kalibera T and Jones R E 2012 Quantifying Performance Changes with Effect Size Confidence Intervals Technical Report 4–12 University of Kent URL http://www.cs.kent.ac.uk/pubs/2012/3233
[7] 2013 *Rigorous Benchmarking in Reasonable Time* (New York: ACM)
[8] Rossum G 1995 Python reference manual Tech. rep. Amsterdam, The Netherlands, The Netherlands
[9] 2018 Python home site URL https://www.python.org/
[10] 2018 Numpy home site URL http://www.numpy.org//
[11] Jones E, Oliphant T, Peterson P *et al.* 2001– SciPy: Open source scientific tools for Python URL http://www.scipy.org/
[12] 2018 Matplotlib home site URL https://matplotlib.org/
[13] Unpingco J 2016 *Python for Probability, Statistics, and Machine Learning* (Springer International Publishing) ISBN 9783319307176
[14] Haslwanter T 2016 *An Introduction to Statistics with Python: With Applications in the Life Sciences* Statistics and Computing (Springer International Publishing) ISBN 9783319283166
[15] Gevorkyan M N, Demidova A V, Korolkova A V and Kulyabov D S 2018 *Distributed Computer and Communication Networks* (*Communications in Computer and Information Science* vol 919) ed Vishnevskiy V M and Kozyrev D V (Cham: Springer International Publishing) pp 532–546 ISBN 978-3-319-99447-5
[16] Brainerd W S 2015 *Guide to Fortran 2008 Programming* (London Heidelberg New York Dordrecht: Springer) ISBN 9781447167594
[17] Hanson R J and Hopkins T 2013 *Numerical computing with modern Fortran* (Philadelphia: SIAM) ISBN 9781611973112
[18] Chapman S J 2018 *Fortran for Scientists and Engineers* (New York: McGraw-Hill Education) ISBN 9780073385891
[19] 2018 Gnu fortran URL https://gcc.gnu.org/fortran/

IOP Conf. Series: Journal of Physics: Conf. Series **1205** (2019) 012018 doi:10.1088/1742-6596/1205/1/012018

Self-excited relaxation oscillations in networks of modified FitzHugh–Nagumo oscillators

S D Glyzin[1], **A Yu Kolesov**[1] **and N Kh Rozov**[2]
[1] P.G. Demidov Yaroslavl State University, Sovetskaya str., 14, Yaroslavl, 150000, Russia
[2] Lomonosov Moscow State University, Leninskiye Gory, Main Building, Moscow, 119991, Russia

E-mail: glyzin@uniyar.ac.ru; kolesov@uniyar.ac.ru; fpo.mgu@mail.ru

Abstract. A modification of the well-known FitzHugh–Nagumo model with a fast and a slow variable is proposed. The existence and stability of a nonclassical relaxation cycle in this system are studied. The slow component of the cycle is asymptotically close to a discontinuous function, while the fast component is a δ-like function. Self exciting oscillations in a chain of diffusively coupled neurons as well as in one-dimensional ring of unidirectionally coupled neurons are studied. The existence of an arbitrarily large number of traveling waves for this chain has been shown. In order to illustrate the increase in the number of stable traveling waves, numerical methods are involved.

1. Introduction

For an analytical study of self-oscillations in an isolated neuron, various simplified models are used which take into account only external manifestations of neural activity (changes in membrane potential and generation of spikes), but do not reflect intracellular reasons for these phenomena. Two-dimensional singularly perturbed systems are used as simplified neuron models. In the system

$$\varepsilon\dot{u} = f(u,v), \quad \dot{v} = g(u,v), \tag{1}$$

$u = u(t)$ is membrane potential, $v = v(t)$ is current, and $\varepsilon > 0$ is a small positive parameter. In view of the relaxation character of electrical activity in the cell, we shall impose on the right-hand sides $f, g \in C^\infty$ the standard restrictions [1] ensuring existence of a stable relaxation cycle.

Some models of the form (1) can be deduced from the Hodgkin–Huxley model [2] under several additional assumptions, and others can be expressed independently. One of the simplest mathematical models for the functioning of an individual neuron is the well-known FitzHugh–Nagumo (FHN) model (see [3, 4]), whose underlying system has the form

$$\varepsilon\dot{u} = v + u - u^3/3 + c, \quad \dot{v} = a - u - bv, \tag{2}$$

where $0 < \varepsilon \ll 1$, $a, b = \text{const} > 0$, and $c = \text{const} \in \mathbb{R}$. In [3] the system (2) was used as a phenomenological model reproducing the most important properties of the Hodgkin–Huxley model. It was then shown in [4] that the same system of equations describes electrical oscillations in an oscillator with a tunnel diode, provided that current-voltage characteristic of the latter can be approximated by a cubic parabola. Finally, it should be noted that, for appropriately chosen parameters, the system (2) admits a stable relaxation cycle.

The class of models (1) also contains the well-known Morris–Lecar system [5]. Yet another example of a system (1) is given by the two-dimensional version of the Hindmarsh–Rose system [6], which has the form

$$\varepsilon\dot{u} = v - au^3 + bu^2 + I, \quad \dot{v} = c - du^2 - v. \tag{3}$$

The system (3) is usually taken with $\varepsilon = 1$, $a = 1$, $b = 3$, $c = 1$, $d = 5$, $I = 2.7$. But if we fix these values a, b, c, d, and I and decrease ε, then the stable cycle in this system becomes a relaxation cycle.

The two-dimensional models (2), (3) have at least two significant drawbacks. The first is that the classical relaxation oscillations arising in these models do not precisely describe the oscillations of the membrane potential of an actual neuron. In principle, there exists a remedy for this. For instance, the papers [7] and [8] proposed a modification

$$\varepsilon\dot{u} = v - g(u), \qquad \dot{v} = a - u - v, \tag{4}$$

($0 < \varepsilon \ll 1$, $a = \text{const} > 0$) of the model (2) in which the component $u = u(t)$ exhibits δ-shaped oscillations, which better correspond to what we have in reality.

Consider the properties of the function $g(u) \in C^\infty(\mathbb{R})$ from (4). The nonlinear characteristic $i = f(u)$ must satisfy the following constraints.

Condition 1. There exists $u = u_* > 0$ such that

$$\begin{gathered} g(0) = 0, \qquad g'(u) > 0 \text{ for } u \in (-\infty, u_*), \qquad g'(u) < 0 \text{ for } u \in (u_*, +\infty), \\ g'(u_*) = 0, \qquad g''(u_*) < 0, \qquad a - u_* - g(u_*) > 0. \end{gathered} \tag{5}$$

Condition 2. For $u \to +\infty$, we have the asymptotic representation

$$g(u) = \alpha_0 + \sum_{k=1}^{\infty} \frac{\alpha_k}{u^k}, \qquad \alpha_0 > 0, \tag{6}$$

which remains valid after being differentiated with respect to u any number of times.

As an example of such a function, we can use

$$g(u) = c_1 u \exp(-u) + c_2(1 - \exp(-u)), \qquad c_1, c_2 = \text{const} > 0. \tag{7}$$

Indeed, Condition 2 is satisfied; moreover, the coefficient α_0 in (6) is equal to c_2, while the other α_k, $k \geq 1$, vanish. Furthermore, u_* in (5) is given by $u_* = 1 + c_2/c_1$, while the requirement $a - u_* - g(u_*) > 0$ is equivalent to the condition

$$\Big[a - u - c_1 u \exp(-u) - c_2(1 - \exp(-u))\Big]\Big|_{u=1+c_2/c_1} > 0 \tag{8}$$

on the parameters a, c_1, and c_2.

Let us analyze the existence and stability of a nonclassical (or impulsive type) relaxation cycle in system (4). According to the terminology used in [9], this is cycle $(u, v) = (u_*(t, \varepsilon), v_*(t, \varepsilon))$ of period $T_*(\varepsilon)$ in system (4) such that, as $\varepsilon \to 0$, the component $v_*(t, \varepsilon)$ converges pointwise to a discontinuous function, $T_*(\varepsilon)$ tends to a finite limit $T_* > 0$, and the component $u_*(t, \varepsilon)$ varies in a δ-like manner with time.

Before formulating the corresponding rigorous result, we introduce some notation. Define

$$v_* = g(u_*),\ \ v_{**} = 2\alpha_0 - v_*,\ \ u_{**} = \varphi(v)|_{v=v_{**}},\ \ x^*_{int} = v_* - v_{**},\ \ x^*_{max} = v_* - \alpha_0, \tag{9}$$

where $u = \varphi(v)$, with $v \in (-\infty, u_*]$ being the unique root of the equation $g(u) = v$ on the interval $(-\infty, u_*]$. Note that due to properties (5) and (6), we have $x^*_{int} > 0$, $x^*_{max} > 0$ and $u_{**} < u_*$, and the quantity

$$T_* = \int\limits_{u_{**}}^{u_*} \frac{g'(u)du}{a - u - g(u)}. \tag{10}$$

is positive.

In addition to constants (9) and (10), we will need functions $u_*(t)$ and $v_*(t)$, where $u_*(t) = \varphi(v_*(t))$ and $v_*(t)$ is determined by solving the Cauchy problem

$$\dot{v} = a - \varphi(v) - v, \qquad v|_{t=0} = v_{**}. \tag{11}$$

It is easy to see that, when continued from the interval $0 \le t \le T_*$ to the entire axis t according to the T_*-periodicity law, these functions are discontinuous at the points $t = kT_*$, $k \in \mathbb{Z}$. The following statement holds [7].

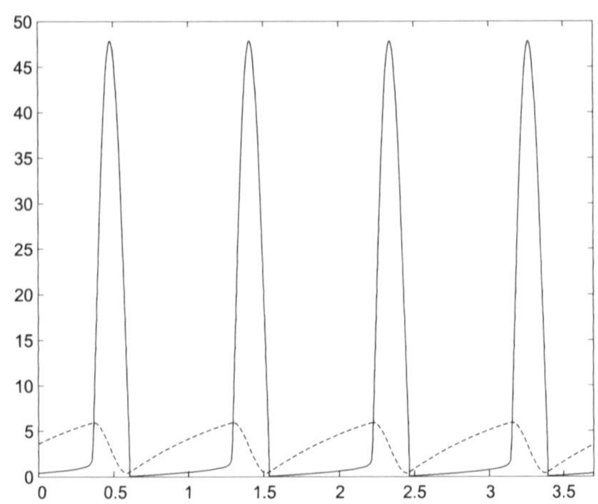

Figure 1. Relaxation oscillations of the system (4)

Theorem 1. *Let Conditions 1 and 2 hold. Then there is a sufficiently small* $\varepsilon_0 > 0$ *such that, for all* $0 < \varepsilon \le \varepsilon_0$*, system* (4) *has an exponentially orbitally stable relaxation cycle*

$$\Gamma_*(\varepsilon) = \{(u, v) :\ \ u = u_*(t, \varepsilon),\ \ v = v_*(t, \varepsilon),\ \ 0 \le t \le T_*(\varepsilon)\} \tag{12}$$

of period $T_*(\varepsilon)$*, where* $u_*(0, \varepsilon) \equiv u_* + 1$*. Moreover,*

$$\lim_{\varepsilon \to 0} T_*(\varepsilon) = T_*, \tag{13}$$

$$\lim_{\varepsilon \to 0} \int\limits_0^{t_*(\varepsilon)} u_*(t, \varepsilon)dt = x^*_{int}, \qquad \lim_{\varepsilon \to 0} \max_t (\sqrt{\varepsilon}\, u_*(t, \varepsilon)) = x^*_{max}, \tag{14}$$

$$\lim_{\varepsilon\to 0} \max_{\delta_1\le t\le T_*(\varepsilon)-\delta_2} \Big(|u_*(t,\varepsilon)-u_*(t)|+|v_*(t,\varepsilon)-v_*(t)|\Big)=0. \tag{15}$$

Here, $t_*(\varepsilon)=O(\sqrt{\varepsilon})$ is the first positive root of the equation $u_*(t,\varepsilon)=u_*+1$ and $\delta_1,\delta_2\in(0,T_*/2)$ are arbitrary fixed constants.

A visual representation of the properties of the impulsive-type relaxation cycle is given by Fig. 1, which shows its components as functions of time. These plots were obtained with the help of the numerical integration of system (4) in the case of (7) and (8) with $\varepsilon=0.005$, $a=12$, $c_1=10$, $c_2=3$ (the solid and broken curves depict $u_*(t+c,\varepsilon)$ and $v_*(t+c,\varepsilon)$, respectively, where $c\in\mathbb{R}$ is a phase shift).

Of great interest is the problem of interaction of oscillators of the form (4). The following paragraphs examine two such problems. In the first of them, a chain of diffusively coupled oscillators of the form (4) is considered. In the second, a ring of unidirectionally coupled oscillators is studied.

2. Self-exciting oscillations in a chain of diffusively coupled neurons

Consider a one-dimensional chain of diffusively coupled neurons with Neumann boundary conditions. Assume that every individual neuron is described by system (4). As a result, we obtain the system

$$\varepsilon\dot u_j=v_j-g(u_j)+d(u_{j+1}-2u_j+u_{j-1}),\qquad \dot v_j=a-u_j-v_j,\qquad j=1,\dots,m, \tag{16}$$

where $u_{m+1}=u_m$, $u_0=u_1$, and $d=\text{const}>0$.

The system (16) admits a homogeneous or synchronous cycle

$$u_1\equiv\ldots\equiv u_m=u_*(t,\varepsilon),\qquad v_1\equiv\ldots\equiv v_m=v_*(t,\varepsilon), \tag{17}$$

where $(u_*(t,\varepsilon),v_*(t,\varepsilon))$ is the periodic solution of system (4) given by Theorem 1. Thus, the question arises as to whether this cycle is stable within the framework of the diffusion model.

To solve this problem, we linearize system (16) on the cycle (17) to obtain the system

$$\begin{gathered}\varepsilon\dot h_{1,j}=h_{2,j}-g'(u_*(t,\varepsilon))h_{1,j}+d(h_{1,j+1}-2h_{1,j}+h_{1,j-1}),\qquad \dot h_{2,j}=-h_{1,j}-h_{2,j},\\ j=1,\dots,m,\qquad h_{1,m+1}=h_{1,m},\qquad h_{1,0}=h_{1,1}.\end{gathered} \tag{18}$$

Applying the Fourier method over the eigenvectors of the difference Laplacian to (18), more precisely, setting

$$h_{r,j}=\sum_{k=0}^{m-1} g_{r,k}(t)\cos\left(\frac{\pi k}{2m}(2j-1)\right),\qquad j=1,\dots,m,\qquad r=1,2. \tag{19}$$

we see that the pairs $(g_{1,k}(t),g_{2,k}(t))$, $k=0,1,\dots,m-1$, in (19) satisfy the system

$$\varepsilon\dot g_1=g_2-(g'(u_*(t,\varepsilon))+z)g_1,\qquad \dot g_2=-g_1-g_2 \tag{20}$$

at $z=z_k$, where

$$z_k=4d\sin^2\left(\frac{\pi k}{2m}\right),\qquad k=0,1,\dots,m-1. \tag{21}$$

For convenience, the parameter z in (20) is assumed to vary continuously on a certain interval $z_1\le z\le z_2$, where $z_2>z_1>0$. Let $\mu_1(\varepsilon,z)$, and $\mu_2(\varepsilon,z)$ denote the multipliers of this system.

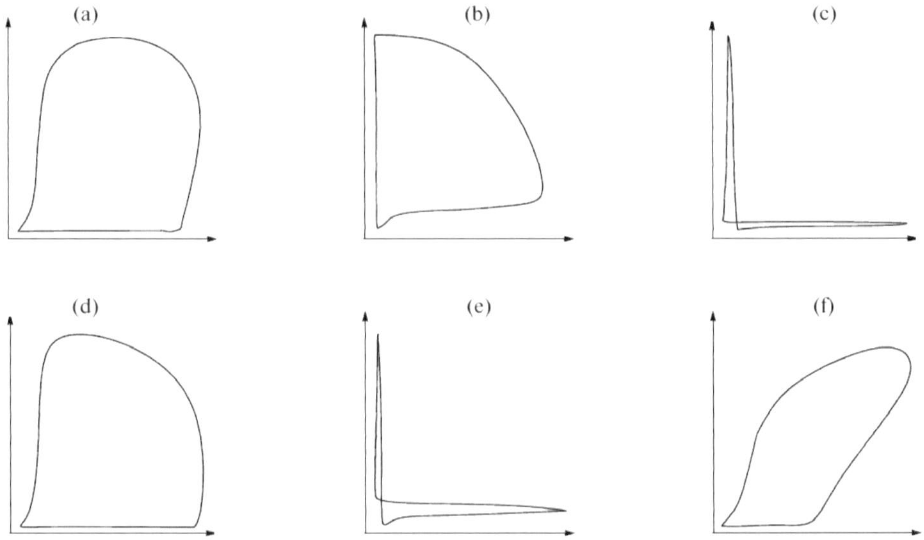

Figure 2. Coexisting stable inhomogeneous cycles of the system (16)

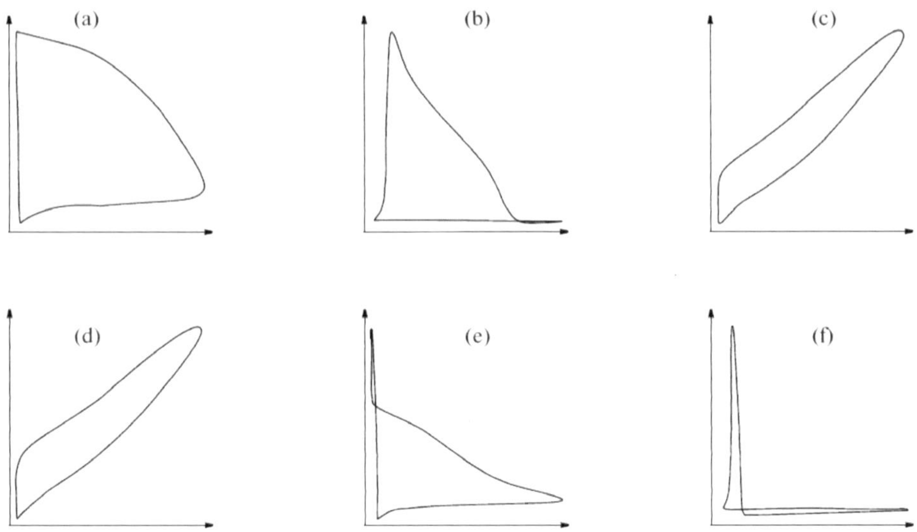

Figure 3. Coexisting stable inhomogeneous cycles of the system (16)

Lemma 1. *The limiting equalities*

$$\lim_{\varepsilon\to 0}\mu_1(\varepsilon,z)=\exp\left(-\int\limits_0^{T_*}\left(1+\frac{1}{g'(u_*(t))+z}\right)dt\right)<1,\qquad \lim_{\varepsilon\to 0}\mu_2(\varepsilon,z)=0, \tag{22}$$

hold uniformly in $z\in[z_1,z_2]$. *Here,* T_* *is the constant given by* (10) *and* $u_*(t)$ *is the function from* (15).

This lemma allows us to prove the statement about the stability of a homogeneous cycle. The stability spectrum of homogeneous cycle (17) of system (16) consists of the multipliers of system (20) at the discrete values of z given by (21). Among these multipliers, there is only one that equals to one, while the others, due to (22), are less than one in absolute value. Thus, the following result is true.

Theorem 2. *For any fixed* $d > 0$ *and all sufficiently small* $\varepsilon > 0$, *the homogeneous cycle* (17) *of system* (16) *is exponentially orbitally stable.*

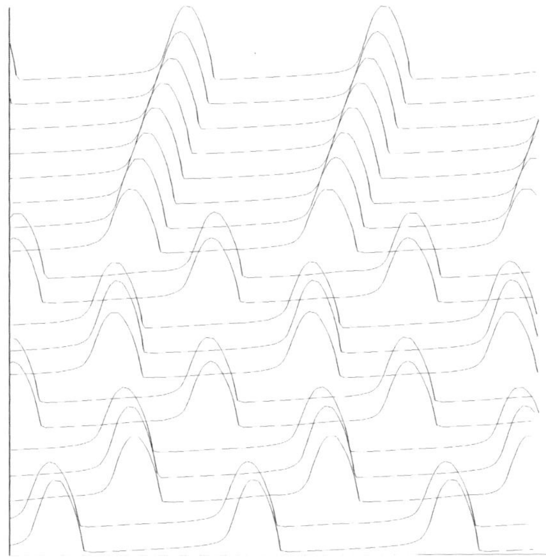

Figure 4. Spatial patterns in the evolution of the spikes for $u = u_j(t)$.

Interestingly, with a suitable choice of parameters in system (16), along with the stable homogeneous cycle, there are other attractors. For example, a numerical analysis of system (16) with conditions (7) and (8) for $m = 20$, $a = 5$, $c_1 = 3$, $c_2 = 2$, $\varepsilon = 0.005$, and $d = 0.01$ has revealed that it has 24 coexisting stable inhomogeneous cycles. Figures 2 and 3 show the projections of some of these cycles onto the plane (u_1, u_{18}) (the other cycles are obtained from these ones by making the substitutions $u_{m+1-j} \to u_j$, $v_{m+1-j} \to v_j$, $j = 1, \dots, m$).

The evolution of the spikes in $u = u_j(t)$ with increasing j, i.e., the propagation of excitation waves along the neural chain deserves a special investigation. For the cycle presented in Fig. 3c, the corresponding spatial pattern is shown in Fig. 4. More specifically, this figure depicts the graphs of the curves$\{(t, u) : u = u_j(t)\}$, $j = 1, \dots, m$ in parallel planes. The relative positions of these graphs suggest that the excitation waves are divided into groups, so-called packets of pulses. The other stable inhomogeneous cycles found have a similar spatial structure.

3. Self-exciting oscillations in a continual ring of unidirectionally coupled neurons

Let us now consider a discrete ring chain of unidirectionally coupled neurons assuming that each separate neuron is described by system (4). As a result we have system

$$\varepsilon \dot{u}_j = v_j - g(u_j) + \frac{\mu m}{2\pi}(u_{j+1} - u_j), \qquad \dot{v}_j = a - u_j - v_j, \qquad j = 1, \dots, m, \tag{23}$$

where $u_{m+1} = u_1$, $\mu > 0$ is the coupling parameter. Then, given $m \gg 1$, we approximate magnitude $2\pi j/m$ by continuous index $x \in [0, 2\pi] \,(\bmod 2\pi)$ and replace component $m(u_{j+1} -$

$u_j)/2\pi$ in (23) with derivative $\partial u/\partial x$. As a result, we get the boundary value problem

$$\varepsilon \frac{\partial u}{\partial t} = v - g(u) + \mu \frac{\partial u}{\partial x}, \quad \frac{\partial v}{\partial t} = a - u - v, \\ u(t, x + 2\pi) \equiv u(t, x), \quad v(t, x + 2\pi) \equiv v(t, x), \tag{24}$$

which represents a mathematical model of continual ring of unidirectionally coupled neurons.

We will consider special periodic solutions of problem (24), where the so-called traveling waves are denoted as

$$u = u_n(\xi, \varepsilon, \mu), \qquad v = v_n(\xi, \varepsilon, \mu), \qquad \xi = \omega_n(\varepsilon, \mu)t - nx. \tag{25}$$

Here $\omega_n(\varepsilon, \mu) > 0$, $n \in \mathbb{N}$ and 2π-periodic by ξ functions $u_n(\xi, \varepsilon, \mu)$, and $v_n(\xi, \varepsilon, \mu)$ satisfy system

$$(\varepsilon\omega_n(\varepsilon, \mu) + n\mu) \frac{du}{d\xi} = v - g(u), \qquad \omega_n(\varepsilon, \mu) \frac{dv}{d\xi} = a - u - v. \tag{26}$$

Established theorem 1 allows one to deal with the issue of the existence of traveling waves (25), (26). For this purpose, let us consider the auxiliary system

$$\widetilde{\varepsilon} \frac{du}{d\xi} = v - g(u), \qquad \omega \frac{dv}{d\xi} = a - u - v, \tag{27}$$

assuming that $0 < \widetilde{\varepsilon} \ll 1$ and parameter $\omega > 0$ is of order of one and changes within a finite segment. Applying the mentioned theorem to system (27), we make sure that, for all small enough values of $\widetilde{\varepsilon} > 0$, it permits the periodic solution

$$u = u(\xi, \widetilde{\varepsilon}, \omega), \qquad v = v(\xi, \widetilde{\varepsilon}, \omega), \qquad u(0, \widetilde{\varepsilon}, \omega) \equiv u_* + 1 \tag{28}$$

of period $T(\widetilde{\varepsilon}, \omega)$. At the same time,

$$T(\widetilde{\varepsilon}, \omega) = \omega T_* + O(\widetilde{\varepsilon}^{1/4}), \qquad \widetilde{\varepsilon} \to 0, \tag{29}$$

where T_* is the constant (10). In turn, asymptotic representation (29) implies that equation

$$T(\widetilde{\varepsilon}, \omega)|_{\widetilde{\varepsilon} = \varepsilon\omega + n\mu} = 2\pi \tag{30}$$

for defining frequency ω has at least one solution

$$\omega = \omega_n(\varepsilon, \mu), \qquad \omega_n(\varepsilon, \mu) = 2\pi/T_* + O((\varepsilon + \mu)^{1/4}), \qquad \varepsilon, \mu \to 0.$$

It is necessary to add that the triplet of functions $\omega_n(\varepsilon, \mu)$, $u_n(\xi, \varepsilon, \mu)$, and $v_n(\xi, \varepsilon, \mu)$, where

$$u_n(\xi, \varepsilon, \mu) = u(\xi, \widetilde{\varepsilon}, \omega)|_{\widetilde{\varepsilon} = \varepsilon\omega_n(\varepsilon,\mu) + n\mu,\, \omega = \omega_n(\varepsilon,\mu)}, \; v_n(\xi, \varepsilon, \mu) = v(\xi, \widetilde{\varepsilon}, \omega)|_{\widetilde{\varepsilon} = \varepsilon\omega_n(\varepsilon,\mu) + n\mu,\, \omega = \omega_n(\varepsilon,\mu)},$$

is the sought value; i.e., it turns equations in (26) into correct equalities. The analysis performed above leads to the following statement.

Theorem 3. *For any natural N small enough values $\varepsilon_N > 0$ and $\mu_N > 0$ can be found such that, for all $0 < \varepsilon \le \varepsilon_N$, $0 < \mu \le \mu_N$, boundary value problem* (24) *permits traveling waves* (25), (26) *with numbers $n = 1, \dots, N$.*

It is worth noting that, in the case

$$\mu = \varepsilon d, \qquad d = \text{const} > 0, \tag{31}$$

boundary problem (24) may also have traveling waves of the form

$$u = u_n(\xi, \varepsilon), \qquad v = v_n(\xi, \varepsilon), \qquad \xi = \omega_n(\varepsilon)t + nx. \tag{32}$$

Here, as earlier $\omega_n(\varepsilon) > 0$, $n \in \mathbb{N}$, and 2π-periodic by ξ functions $u_n(\xi, \varepsilon)$, and $v_n(\xi, \varepsilon)$ are defined from the system similar to (26)

$$\varepsilon(\omega_n(\varepsilon) - nd)\,\frac{du}{d\xi} = v - g(u), \qquad \omega_n(\varepsilon)\,\frac{dv}{d\xi} = a - u - v. \tag{33}$$

Functions $\omega_n(\varepsilon)$, $u_n(\xi, \varepsilon)$, and $v_n(\xi, \varepsilon)$ from (32) are defined by the same scheme as described above. Namely, let us consider auxiliary system similar to (27)

$$\varepsilon(\omega - nd)\,\frac{du}{d\xi} = v - g(u), \qquad \omega\,\frac{dv}{d\xi} = a - u - v, \tag{34}$$

where parameters ω, and $d > 0$ are of the order of one and inequality $nd < \omega$ holds. Theorem 1 guarantees the existence of a periodic solution of system (34) similar to (28)

$$u = u_n(\xi, \varepsilon, \omega), \qquad v = v_n(\xi, \varepsilon, \omega), \qquad u_n(0, \varepsilon, \omega) \equiv u_* + 1 \tag{35}$$

of period

$$T_n(\varepsilon, \omega) = \omega T_* + O(\varepsilon^{1/4}), \qquad \varepsilon \to 0. \tag{36}$$

for any small enough value of $\varepsilon > 0$. Then (36) implies that the equation $T_n(\varepsilon, \omega) = 2\pi$, which is similar to (30), has at least one solution as follows:

$$\omega = \omega_n(\varepsilon) = 2\pi/T_* + O(\varepsilon^{1/4}), \qquad \varepsilon \to 0. \tag{37}$$

As for functions $u_n(\xi, \varepsilon)$, and $v_n(\xi, \varepsilon)$ in (32), (33), they are obtained from (35) after the substitution of equation (37). In this way, the following statement is established.

Theorem 4. *Assume that for some natural N inequality $Nd < 2\pi/T_*$ is fulfilled. Then under condition (31) and for all small enough $\varepsilon > 0$ boundary value problem (24) permits traveling waves (32), (33) with numbers $n = 1, \ldots, N$.*

The problem of stability of traveling waves in metrics of the phase space $(u, v) \in W_2^1 \times W_2^1$ (W_2^1 – Sobolev space of 2π-periodic functions) is reduced to studying the spectrum of some boundary value problems. In the case of traveling waves (25), (26), the procedure for deriving corresponding spectral problem is the following. Let us proceed to traveling spatial variable $\xi = \omega_n(\varepsilon, \mu)t - nx$ in problem (24), then linearize it at the equilibrium state $u = u_n(\xi, \varepsilon, \mu)$, $v = v_n(\xi, \varepsilon, \mu)$. Then, let us substitute equalities $u = h_1(\xi)\exp(\lambda t)$, $v = h_2(\xi)\exp(\lambda t)$, $\lambda \in \mathbb{C}$ into the obtained linear system. As a result, in order to define $h_1(\xi)$, $h_2(\xi)$, and λ we have

$$\begin{gathered} \varepsilon\lambda h_1 + (\varepsilon\omega_n(\varepsilon, \mu) + n\mu)\,\frac{dh_1}{d\xi} = h_2 - g'(u_n(\xi, \varepsilon, \mu))h_1, \\ \lambda h_2 + \omega_n(\varepsilon, \mu)\,\frac{dh_2}{d\xi} = -h_1 - h_2, \qquad h_j(\xi + 2\pi n) \equiv h_j(\xi), \qquad j = 1, 2. \end{gathered} \tag{38}$$

A study of location of Eigenvalues λ in problem (38), and a similar boundary value problem for traveling waves (32), (33), represents a separate and still unsolved problem. Therefore, we will confine with results of numeric analysis of boundary value problem (24), which show the principal possibility of the existence of stable traveling waves.

When performing the corresponding numeric experiment, in contrast to (23), we approximate the first derivative by x with symmetric difference operator, i.e., we assume that

$$u_j(t) = u(t,x)|_{x=2\pi j/m}, \qquad u_j(t) = u(t,x)|_{x=2\pi j/m},$$
$$\frac{\partial u}{\partial x}(t,x)|_{x=2\pi j/m} \approx \frac{m}{4\pi}\,(u_{j+1}(t) - u_{j-1}(t)), \quad j = 1,\ldots,m.$$

Finally, we obtain the following system for variables $u_j(t)$, and $v_j(t)$ after the replacement $\mu/2\pi \to \mu$:

$$\varepsilon \dot{u}_j = v_j - g(u_j) + \frac{m}{2}\,\mu\,(u_{j+1} - u_{j-1}), \qquad \dot{v}_j = a - u_j - v_j, \qquad j = 1,\ldots,m, \tag{39}$$

where $u_0 = u_m$, and $u_{m+1} = u_1$.

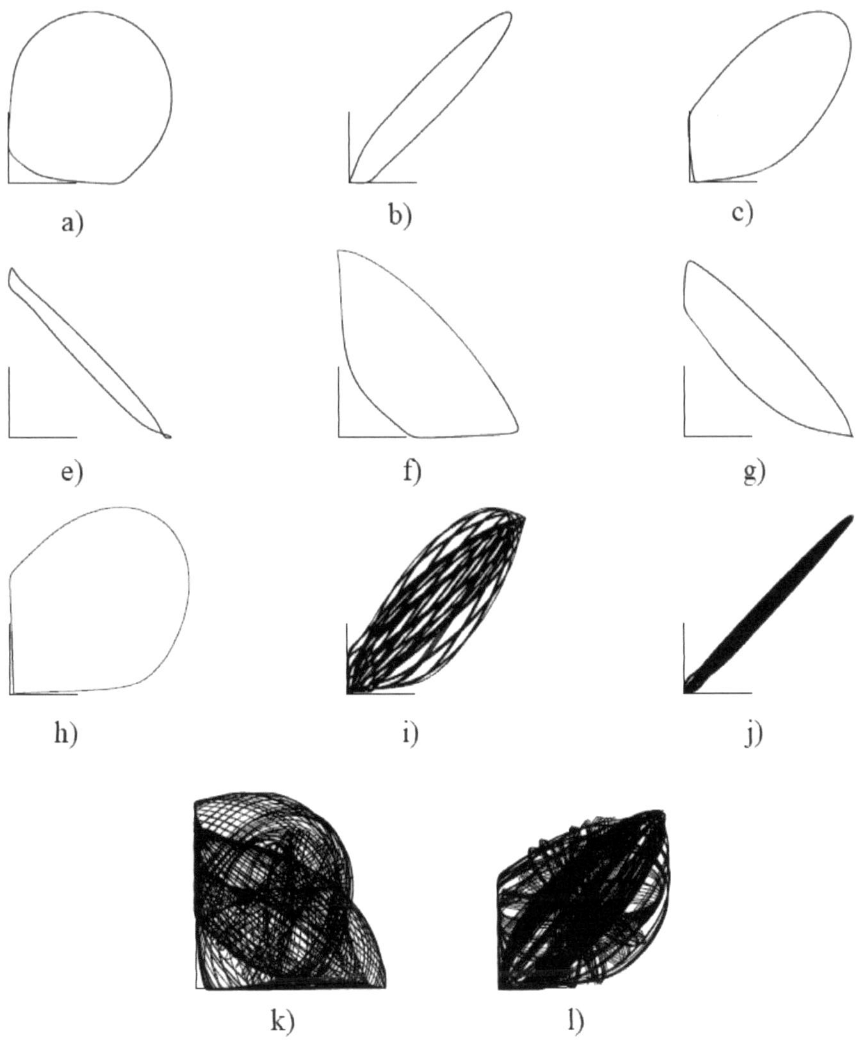

Figure 5. Traveling-wave type solutions of the system (39)

Numerical analysis of system (39) was performed by the fourth order Runge–Kutta method with a constant step of $h = 10^{-4}$ under conditions (6), (7) and, at values of parameters $m = 21$, $a = 15$, $c_1 = 3$, $c_2 = 1$, $\varepsilon = \mu = 0.01$, (software package Tracer 3.70 developed by D. S. Glyzin was applied). It has been established that, at the given set of parameters, it has seven stable periodic solutions of traveling wave type and four stable two-dimensional invariant tori. Figures 5 show projections of the given attractors to plane (u_1, u_{18}) (the first seven images correspond to cycles and the remaining four correspond to invariant tori).

4. Conclusions

To conclude, we note that the new mathematical model proposed for an individual neuron has rich contents. Systems with nonclassical relaxation oscillations give us oscillations of membrane potential of δ-like shape. This feature corresponds to the real biological objects. Moreover, the diffusion chain (16) and one-dimensional ring (24) of unidirectionally coupled neurons corresponding to system (4) exhibits nontrivial dynamics, namely, the buffer phenomenon. In this context, we recall that the buffer phenomenon (i.e., the coexistence of any prescribed finite number of attractors) is typical for neuron systems, as suggested by the results of [10] – [15].

Acknowledgments

This work is supported by RFBR according to the research project no. 18-29-10055.

References

[1] Mishchenko E F and Rozov N Kh 1980 Differential equations with small parameters and relaxation oscillations, Math. Concepts Methods Sci. Engrg., vol. 13 *Plenum Press, New York*
[2] Hodgkin A L, Huxley A F 1952 A quantitative description of membrane current and its application to conduction and excitation in nerve *Journal of Physiology* **117** 500–544
[3] FitzHugh R. 1961 Impulses and physiological states in theoretical models of nerve membrane *Biophys. J.* **1** 445 – 466.
[4] Nagumo J, Arimoto S and Yoshizawa S 1962 An active pulse transmission line simulating nerve axon *Proc IRE* **50** 2061 – 2070.
[5] Morris C and Lecar H 1981 Voltage oscillation in barnacle giant muscle fiber *Biophys. J.* **35** 193–213
[6] Hindmarsh J L and Rose R M 1982 A model of the nerve impulse using two first-order differential equations *Nature* **296** 162–164
[7] Glyzin S D, Kolesov A Yu and Rozov N Kh 2014 On a Modification of the FitzHugh–Nagumo Neuron Model *Comput. Math. Math. Phys.* **54** 443–461
[8] Glyzin S D, Kolesov A Yu and Rozov N Kh 2016 Nonclassical Relaxation Oscillations in Neurodynamics *Automatic Control and Computer Sciences* **50** 571–585
[9] Mishchenko E F, Kolesov Yu S, Kolesov A Yu and Rozov N Kh 1995 Periodic Motions and Bifurcation Processes in Singularly Perturbed Systems (Fizmatlit, Moscow, 1995) [in Russian].
[10] Glyzin S D, Kolesov A Yu and Rozov N Kh 2012 Buffer phenomenon in neurodynamics *Doklady Mathematics* **85** 297–300.
[11] Glyzin S D, Kolesov A Yu and Rozov N Kh 2012 Relaxation self-oscillations in neuron systems: III *Differ. Equations* **48** 159–175.
[12] Glyzin S D, Kolesov A Yu and Rozov N Kh 2012 Discrete autowaves in neural systems *Comput. Math. Math. Phys.* **52** 702–719.
[13] Glyzin S D, Kolesov A Yu and Rozov N Kh 2013 Relaxation self-oscillations in Hopfield networks with delay *Izvestiya: Mathematics* **77** 271–312.
[14] Glyzin S D, Kolesov A Yu and Rozov N Kh 2014 The buffer phenomenon in ring-like chains of unidirectionally connected generators *Izvestiya: Mathematics* **78** 708–743
[15] Glyzin S D, Kolesov A Yu and Rozov N Kh 2015 Self-excited relaxation oscillations in networks of impulse neurons *Russian Math. Surveys* **70** 383–452.

IOP Conf. Series: Journal of Physics: Conf. Series **1205** (2019) 012019 doi:10.1088/1742-6596/1205/1/012019

On the possibility of constructing relativistic quantum mechanics on the basis of the definition of the functions of differential operators

A V Golovin[1] **and V M Lagodinskiy**[2]

[1] Physics Faculty, St. Petersburg State University, 198504 St. Petersburg, Russia
[2] Petersburg State University of Aerospace Instrumentation, 190000 St. Petersburg, Russia

E-mail: lagodinskiy@mail.ru

Abstract. A new definition of the function of a differential operator that leads to local operators of infinite order is proposed. It allows one to obtain an expression for the square root of the differential operator (the Hamiltonian of a free spinless particle) and to determine the relativistic Schrödinger equation as a close analog of the non-relativistic Schrödinger equation. It is shown that this equation does not lead to difficulties of the Klein-Gordon equation. Boundary conditions that lead to self-adjoint boundary problems similar to Sturm-Liouville problems, periodic boundary value problems, and singular boundary value problems are determined. Some problems of relativistic quantum mechanics are solved.

It is known that the Klein-Gordon equation (KGE) leads to significant difficulties, for example, to solutions corresponding to the states of a free particle with negative energy, complex energy values of a pionic atom with a nuclear charge $Z > 68$, the paradox of Klein and others. Two-particle KGE is impossible. It is easy to see that the difference between the KGE and the non-relativistic Schrödinger equation (NRSE) is related to the fact that the NRSE is obtained by a formal substitution

$$\varepsilon \to i\frac{\partial}{\partial t}, \qquad \mathbf{p} \to -i\nabla$$

(we use a system of units in which the speed of light c and Planck's constant $\hbar$ are equal to unity) in the expression of energy through a pulse, and KGE — the same substitution in the expression bf of the square of the energy via momentum. As a result, KGE is an equation of hyperbolic type, while NRSE is a parabolic type. It can be assumed that the KGE describes a physical field, which, as is well known, does not hold for NRSE, KGE is the basis of quantum field theory, rather than relativistic quantum mechanics (RQM). Apparently, the basis of RQM should be an equation similar to NRSE, that is, it should be determined by standard replacement not in the expression for the square of the energy, but in the expression for the energy:

$$\varepsilon = \sqrt{m^2 + \mathbf{p}^2}. \tag{1}$$

However, this leads to a somewhat unusual equation [1,2]:

$$\left(i\frac{\partial}{\partial t} - \sqrt{m^2 - \nabla^2}\right)\Psi(t, \mathbf{r}) = 0. \tag{2}$$

In this equation, there appears an operator having the form of a square root of the differential operator. In an attempt to avoid the need to define this operator mathematically correctly, it is usually [1,2] the equation (2) is declared inadmissible, referring its asymmetry with respect to time and spatial coordinates and non-locality. This gives grounds for abandoning the equation (2), replacing it with the KGE and obtaining all of the above difficulties.

Let's consider, whether the reasons of refusal of the equation (2) are valid. The asymmetry (2) is related to the real difference between time and space, corresponding to the difference in signs under the squares of the time differentials and spatial coordinates in the expression for the square of the interval: $ds^2 = dt^2 - dx^2 - dy^2 - dz^2$. The formula (1) is also not symmetric with respect to the time and space components of the energy-momentum 4-vector. In fact, for the invariance of an equation with respect to certain transformations, it suffices that the set of solutions of this equation is invariant under these transformations [3]. And the statement about the nonlocality of the operator is due to the fact that usually such an operator is defined as pseudo-differential [4], that is, using Fourier or Fourier-Bessel transforms. This definition is used in some works [5,6], which leads to very cumbersome calculations with not always correct results and does not allow to recognize this approach as correct and expedient. It is necessary to clarify the concept of a local operator: it is a linear operator that any function that vanishes on some region of a linear metric space in which it is defined associates a function defined on this region and equal to zero on it [7]. Any self-adjoint differential operator is non-local, regardless of its order, since its definition includes boundary conditions. Obviously, the use of the Fourier transform corresponds to periodic boundary conditions. Therefore, if we define a non-relativistic Hamiltonian as a pseudo-differential operator, its application to other boundary problems would be impossible. In fact, the non-relativistic Schrödinger equation (NRSE) includes a Hamiltonian defined as the local operator $H = -\nabla^2/(2m)$ on functions $\psi(\mathbf{r})$ defined and twice differentiable on some domain $X \subset \mathbb{R}^3$, and the boundary conditions are introduced additionally so that they correspond to a given physical problem and that the boundary value problem is self-adjoint. A self-adjoint operator (with its spectrum) appears as a result of solving a self-adjoint boundary problem, they do not operate in the process of solving it. The operator in our equation must be local, so that it can be applied to solving various physical, that is, self-adjoint boundary problems. Obviously, linear differential operators of finite order defined on functions corresponding to the number of times differentiate are local. But we need to define a local operator of infinite order, and it must have the form and properties of the square root of the second-order operator, that is, it must correspond to a particular function of a numerical variable. The definition of the function of a differential operator that leads to a local linear differential operator of infinite order was proposed by one of the co-authors of this paper [8].

Let the analytic function $f(z)$ (symbol) be holomorphic in some neighborhood of zero, has branch points, each of which is connected by a cut with an infinitely distant point, and the extensions of the cuts beyond the branch points intersect at the origin, $x \in Q \subset \mathbb{R}^k$ D is a linear local differential operator (ordinary or partial derivatives), and $u(x)$ is a function such that all its images are defined on the domain Q $(D^n u)(x)$, $n = 0, 1, 2, \ldots$, and the set of limit points of the sequence

$$\{\gamma_n(x_0)\} = \left\{ |(D^n u)(x_0)|^{1/n} \exp\left(i\frac{\varphi_n}{n}\right) \right\},$$

where $\varphi_n = \arg[(D^n u)(x_0)]$, is bounded and does not intersect the set of cut points of $f(z)$ for all $x_0 \in Q$. Then the function is defined on Q by an analytic continuation of the function

$$\big(f(\boldsymbol{\alpha} D)u\big)(x) = \sum_{n=0}^{\infty} \frac{f^{(n)}(0)}{n!} \boldsymbol{\alpha}^n (D^n u)(x), \qquad \forall x \in Q$$

by the real parameter $\boldsymbol{\alpha}$ from $\boldsymbol{\alpha} = 0$ to $\boldsymbol{\alpha} = 1$:

$$\big(f(D)u\big)(x) = \sum_{i=0}^{\infty} \frac{(1-\boldsymbol{\alpha}_p)^i}{i!} \sum_{k=0}^{\infty} \frac{(\boldsymbol{\alpha}_p - \boldsymbol{\alpha}_{p-1})^k}{k!} \cdots \sum_{l=0}^{\infty} \frac{\boldsymbol{\alpha}_1^l f^{i+k+\ldots+l}(0)}{l!} (D^{i+k+\ldots+l}u)(x). \tag{3}$$

We set $f(z) = \sqrt{m^2 + z^2}$, where $m > 0$ (rest mass of the particle). We choose a holomorphic branch in the following way: $f(0) = m$, cut the cuts along the imaginary axis from $z = i$ to $z = i\infty$ and from $z = -i$ to $z = -i\infty$. $D = -i\nabla$. The corresponding equation has the form (2). Let $u(\mathbf{r}) = \exp(i\mathbf{pr})$, $\forall \mathbf{r} \in Q \subset \mathbb{R}^3$, ($Q$ —open set), $\mathbf{p} \in \mathbb{C}^3$. Then, if $\mathbf{p} \in \mathbb{R}^3$,

$$H(-i\nabla)u(\mathbf{r}) = f(-i\nabla)u(\mathbf{r}) = \sqrt{m^2 - \nabla^2}\exp(i\mathbf{pr}) = \sqrt{m^2 + \mathbf{p}^2}\exp(i\mathbf{pr}), \qquad \forall \mathbf{r} \in Q.$$

Note that if $f(-i\nabla)$ was defined as a pseudo-differential operator, then $Q = \mathbb{R}^3$ would be necessary for such a result. The absence of this necessity shows the localness of our definition. If $u(\mathbf{r}) = \exp(\boldsymbol{\lambda}\mathbf{r})$, $\forall \mathbf{r} \in Q \subset \mathbb{R}^3$, then if $|\boldsymbol{\lambda}| < m$,

$$H(-i\nabla)u(\mathbf{r}) = \sqrt{m^2 - \nabla^2}\exp(\boldsymbol{\lambda}\mathbf{r}) = \sqrt{m^2 - \boldsymbol{\lambda}^2}\exp(\boldsymbol{\lambda}\mathbf{r}), \qquad \forall \mathbf{r} \in Q,$$

and if $|\boldsymbol{\lambda}| \geqslant m$, then the function $u(\mathbf{r}) = \exp(\boldsymbol{\lambda}\mathbf{r})$ is not included in the domain of the operator $H(-i\nabla)$.

It is interesting that in contrast to operators of finite order the operator $H(-i\nabla)$ has an inverse, since the function $f^{-l}(z)$ also decomposes into a Taylor series and its holomorphic branch can be defined with those same cuts. For example, for $p \in \mathbb{R}$

$$H^{-1}(-i\nabla)u(\mathbf{r}) = (m^2 - \nabla^2)^{-1/2}\exp(\mathbf{pr}) = (m^2 + \mathbf{p}^2)^{-1/2}\exp(i\mathbf{pr}), \qquad \forall \mathbf{r} \in Q,$$

It was shown in [8] that the equation (2) is relativistically invariant, that is, the set of its solutions is invariant under Lorentz transformations. It should be called the relativistic Schrödinger equation (RSE). From this follows the continuity equation:

$$\frac{\partial}{\partial t}\rho(t,\mathbf{r}) = -\nabla \cdot \mathbf{j}(t,\mathbf{r}),$$

where

$$\boldsymbol{\rho}(t,\mathbf{r}) = |\boldsymbol{\Psi}(t,\mathbf{r})|^2 + m^2|H^{-1}(-i\nabla)\boldsymbol{\Psi}(t,\mathbf{r})|^2 + |\mathbf{V}(-i\nabla)\boldsymbol{\Psi}(t,\mathbf{r})|^2,$$
$$\mathbf{j}(t,\mathbf{r}) = \boldsymbol{\Psi}^*(t,\mathbf{r})\mathbf{V}(-i\nabla)\boldsymbol{\Psi}(t,\mathbf{r}) + \boldsymbol{\Psi}(t,\mathbf{r})\mathbf{V}^*(-i\nabla)\boldsymbol{\Psi}^*(t,\mathbf{r}).$$

These expressions include the velocity operator

$$\mathbf{V}(-i\nabla) = -i\nabla(m^2 - \nabla^2)^{-1/2} = i\left(\sqrt{m^2 - \nabla^2}\,\mathbf{r} - \mathbf{r}\sqrt{m^2 - \nabla^2}\right)$$

in complete agreement with the principles of quantum mechanics and the mechanics of the theory of relativity. Let us consider in more detail the one-dimensional stationary RSE:

$$\left(\varepsilon - \sqrt{m^2 - \nabla_x^2}\right)\psi(x) = 0, \qquad \forall x \in Q \subset \mathbb{R}. \tag{4}$$

where $\nabla_x = d/dx$, A, $B \in \mathbb{C}$. It has solutions of the kind

$$\psi(x) = \exp\left(i\sqrt{\varepsilon^2 - m^2}\,x\right), \qquad \forall x \in Q.$$

If $\varepsilon \in (0, m)$, then p — is purely real, if $\varepsilon \in (0, m)$, then p — is purely imaginary, if $\varepsilon \leqslant 0$, then such a value p does not exist, therefore, the problem of negative energies of a free particle do not exist. Thus, a free stationary RSE, a linear differential equation of infinite order, has either two linearly independent solutions, or none. It is not difficult to verify that if $\psi_1(x)$ and $\psi_2(x)$ are solutions of the equation (2) on $Q \subset \mathbb{R}$, then Q is constant function

$$W(x) = \psi_1(x)V\psi_2(x) - \psi_2(x)V\psi_1(x),$$

where

$$V = -i(m^2 - \nabla_x^2)^{-1/2}\nabla_x.$$

Apparently, the function $W(x)$ is the relativistic analogue of the non-relativistic Wronskian. Its form makes it possible to find the boundary conditions that lead to self-adjoint boundary problems for the equation (4). If $Q = (a, b)$, then the corresponding Hilbert space is defined by the scalar product:

$$(\psi_1, \psi_2)_a^b = \int_a^b [\psi_1^*(x\psi_2(x) + m^2(H^{-1}\psi_1^*)(x)(H^{-1}\psi_2)(x) + (V^*\psi_1^*)(x)(V\psi_2)(x)]\, dx. \qquad (5)$$

It is easy to show that a regular self-adjoint problem can be given by boundary conditions that take the function $W(x)$ to zero at the end points:

$$\psi(a)\cos\boldsymbol{\alpha} + i(V\psi)(a)\sin\boldsymbol{\alpha} = 0, \qquad \psi(b)\cos\boldsymbol{\beta} + i(V\psi)(b)\sin\boldsymbol{\beta} = 0,$$

where $\boldsymbol{\alpha}$, $\boldsymbol{\beta} \in \mathbb{R}$ (similar to the Sturm-Liouville problem [10]) or periodic boundary conditions that ensure the equality $W(a) = W(b)$. These problems were solved in [9]. The spectrum of these problems is purely point-wise, bounded from below, each value of the spectrum is a single-valued function, the eigenfunctions form a complete system of functions that is orthogonal in the sense of the scalar product (5). A singular boundary value on $Q = [0, \infty)$ is given by conditions of the form

$$\psi(0)\cos\boldsymbol{\alpha} + i(V\psi)(0)\sin\boldsymbol{\alpha} = 0,$$

where $\boldsymbol{\alpha} \in \mathbb{R}$, and the condition that the solution tends to zero as $x \to \infty$.

This problem has both a continuous and a point spectrum, the latter is either empty or consists of a single value. All values of the spectrum are single-valued, all solutions of the continuous spectrum are orthogonal to the solution of the point spectrum in the sense (5) with $a = 0$, $b = \infty$ and orthogonal to each other in the sense that

$$\int_0^\infty (\psi_1, \psi_2)_0^\infty dx = \boldsymbol{\delta}(\varepsilon_1 - \varepsilon_2),$$

$\psi_i(x)$ — solution corresponding to the value ε_i of the continuous spectrum. The boundary problem on $Q = \mathbb{R}$ is given, just as for NSE, without boundary conditions, only by the condition of boundedness. The spectrum is also bounded from below, purely continuous, each its value is twofold.

The proposed formalism makes it possible to obtain solutions to problems, the application of which to KGE is either impossible or leads to paradoxical results. As an example, we give a solution of the problem of the S-states of the point spectrum of the negativ $\boldsymbol{\pi}$-meson in Coulomb field of atomic nucleus with charge Ze. For a radial wave function, we have a stationary equation:

$$\left(\varepsilon + Ze^2r^{-1} - \sqrt{m^2 - \nabla_x^2}\right)\boldsymbol{\chi}(r) = 0, \qquad \forall r > 0.$$

As shown in work [9], the spectrum of S-states has the form

$$\varepsilon_n = m(n^2 + Z^2\boldsymbol{\alpha}^2)^{-1/2}.$$

The spectrum is real, pointwise, bounded below, for small Z, it is approximately described by the non-relativistic Balmer formula.

Consider the two-particle one-dimensional RSE [11]:

$$\left(i\frac{\partial}{\partial t} - \sqrt{m_1^2 - \frac{\partial^2}{\partial x_1^2}} - \sqrt{m_2^2 - \frac{\partial^2}{\partial x_2^2}}\right)\boldsymbol{\Psi}(t, x_1, x_2), \qquad \forall t,\, x_1,\, x_2 \in \mathbb{R}.$$

Note that the time variable is one. The problem of reflection of a particle from an ideal mirror of finite mass is given by the condition that the solution vanishes for all t if $x_1 \geqslant x_2$ and the continuity condition of the solution. As shown in work [10], From this equation and boundary condition follow the laws of conservation of energy and momentum of this physical system and the value of the transmitted pulse

$$k = 2\varepsilon(p_1\varepsilon_2 - p_2\varepsilon_1)[\varepsilon^2 - (p_1 + p_2)^2]^{-1}.$$

We have obtained the same result as in classical relativistic mechanics. In this case, the laws of conservation of energy and momentum are satisfied.

In [9-12], solutions of some physical problems were obtained, the application of which KGE leads to paradoxical results, or is impossible. All solutions are exact, without divergences.

So, apparently, the use of the local definition of the function of the differential operator proposed in [9] allows us to formulate the basis for a new version of relativistic quantum mechanics free from the difficulties of the generally accepted theory based on KGE. We can assume that RSE is the basis of mathematically correct and physically plausible relativistic quantum mechanics of spinless particles [13].

References

[1] Dirac P A M 1979 *The principles of quantum mechanics* 408
[2] Bjorken D and Drell D 1976 *Relativistic Quantum theory* **2** *Relativistic Quantum Field*
[3] Olver P 1989 Application of Lie groups in the theory of differential equations Moscow Mir 683
[4] Dubinsky Yu A 1982 UMN 37:5(227) 97-137; Russian Math. Surveys, 37:5 109–153
[5] Tzara C A study of relativistic Coulomb problem in momentum space. Phys. Lett. A111 343-348.
[6] Lucha W, Rupprecht H and Schoberl F F 1992 Spinless Salpeter equation as a simple matrix eigenvalue problem. Phys. Rev. D. **45** 1233
[7] Lagodinski V M 2011 Functions of operators and differential equations of infinite order. Definitions and properties. Lambert Academic Publishing. Saarbrucken, Deutschland, 111
[8] Golovin A V and Lagodinski V M 2009 Vestnik SPbU ser.4 Physics, Chemistry iss. 2 143-149
[9] Golovin A V and Lagodinski V M 2012 Vestnik SPbU ser.4 Physics, Chemistry iss. 4 3-13
[10] Golovin A V and Lagodinski V M 2011 Vestnik SPbU ser.4 Physics, Chemistry iss. 2 3-14
[11] Golovin A V and Lagodinski V M 2012 Vestnik SPbU ser.4 Physics, Chemistry V.1(59) iss. 2 156-165
[12] Golovin A V and Lagodinski V M 2016 Vestnik SPbU ser.4 Physics, Chemistry V.4(62) iss. 3 249-263
[13] Golovin A V and Lagodinski V M 2018. *Functions of differential operators and relativistic Shrödinger equation.* 7th International conference "Problems of Mathematical Physics and Mathematical Modelling". Book of abstracts (Moscow, NRNU MEPhI. 25-27 June 2018)

The accuracy of the theories based on statistical physics for the thermodynamic modeling of state parameters of dense pure gases (fluids)

S A Gubin[1,2] and S B Victorov[3]

[1]National Research Nuclear University "MEPhI", Moscow, Russia
[2]State Centre «Interphysica», Moscow, Russia
[3]OpenSearchServer, Paris, France

gubin_sa@mail.ru

Abstract. A comparison of the accuracy of calculations of the parameters of the state of dense supercritical gases (fluids) using the equations of state (EOS), obtained using the three best one fluid theories HMSA/C, MRCSR and KLRR-T, based on the given potentials of intermolecular interactions and the laws of statistical mechanics is carried out. The best agreement of the calculated state parameters with Monte Carlo data was obtained for the EOS based on the HMSA/C and KLRR-T theories.

1. Introduction

A reliable model of the wide-range equation of state (EOS) (or equation of thermodynamic potential) is needed to calculate the thermodynamic properties of the gas-liquid phase (included supercritical fluid) in the range from normal conditions to pressures of 100 GPa and temperatures to15000 K. Special EOS models should be used for thermodynamic calculations in conditions where the gas is super-dense. Such EOS models allow us to calculate problems with chemical reaction in physics and chemistry of shock and detonation waves, astrophysical problems about the inner layers of giant planets and many others.

Most EOS models do not have a rigorous theoretical justification, i.e. they are empirical or semi-empirical. The functional form and constants of the used parameters of the EOS model are selected from the condition of the best agreement of the calculated values with the available experimental data. The use of empirical and semi-empirical EOS models does not allow to predict correctly of the state parameters outside the experimental range of such EOS. In recent decades interest has increased significantly to theoretically substantiated EOS models based on the predefined intermolecular interactions potential $\varphi(r)$ (r – distance between molecules). Calculation of thermodynamic parameters of the chemical system is carried out by such EOS models which are based only on the interaction potentials of molecules. No other information on the properties of the substances is required. Theoretical models of EOS provide reliable prediction of thermodynamic properties of dense gases, demonstrating agreement with the results of computer simulation by the Monte Carlo (MC) and molecular dynamics (MD) methods.

Direct computer simulation using MC method allows us to calculate accurately the thermodynamic properties of a system of molecules interacting with each other with a given potential. Therefore, MC modeling is a computer experiment to confirm the reliability of the theoretical models of EOS obtained on the basis of theories with a given intermolecular potential. However, calculations using the MC or MD methods require considerable computer time. At the same time, the rate of calculation of thermodynamic properties of matter on the basis of exact theories for thermodynamic potentials or EOS based on statistical mechanics is thousands of times higher than the rate of calculations by MC and MD methods, while maintaining high accuracy of calculations. Therefore, it is advisable to develop special theories based on statistical mechanics to obtain theoretical models of EOS based on the given interaction potentials of molecules for the modeling of thermodynamic properties of substances.

2. Problem statement

Only one component fluid systems are considered in the work. The extension of the theory for multicomponent mixtures (different molecules) is carried out by applying the van Der Waals effective potential model [1 - 4]. The ensemble of identical molecules with a spherical symmetric potential of intermolecular interaction Exp-6 [2, 3, 5] is considered at fixed temperature T, specific volume V and number of particles N. Only pair interactions are considered.

A reliable model of the full thermodynamic potential of one fluid can be obtained in the presence of a theory that allows to reasonably calculate the excess energy of the Helmholtz $\Delta F^{ex}(T,V,N,\varphi(r))$ system using the Exp-6 potential. The thermodynamic properties of the system are determined by the differentiation of the Helmholtz energy of the ensemble of molecules at given values T, V, N. In this case the Helmholtz energy $F(T,V,N,\varphi(r))$ (also any other thermodynamic characteristic obtained by the Helmholtz energy differentiation) of the system is the sum of the Helmholtz energy of an ideal gas and the excess value of the Helmholtz energy due to intermolecular interaction. Thus, the main task of the considered theories is to calculate the exact excess energy of the Helmholtz $\Delta F^{ex}(T,V,N,\varphi(r))$.

Today there are three theories providing the highest accuracy of calculation of thermodynamic properties of one fluid, whose molecules interact with a given potential in a wide range of pressures and temperatures including high densities. There are variational perturbation theory MCRSR [5, 6], the theory of integral equations HMSA/C [2, 7] and the KLRR perturbation theory [3, 8]. These theories are developed on the basis of statistical mechanics and have deep physical justifications. Calculations based on these theories are consistent with the results of the MC simulation. It makes possible to calculate reliably the thermodynamic properties of a fluid in a wide range from normal conditions to extremely high pressure (hundreds of GPa) and temperature (up to 15000 K) thousands of times faster than MC calculations and MD modeling. Powerful computer codes CHEQ [5], CHEETAH [2], CARTE [8], TDS [9] are developed on the basis of these theories. The variational perturbation theory MCRSR [6] was applied in computer codes CHEQ [5] and CARTE [8]. The KLRR-T perturbation theory [3] was used to create our modified perturbation theory THEOSTAR [4] for the TDS code [9], HMSA/C [2] for CHEETAH.

Due to the fact that different modeling assumptions are used in the construction of such theories, the calculation based on different theories with the same potential can lead to different results for the thermodynamic parameters of the substance. Hence the question arises which of the theories gives the most acceptable model of EOS and the real definition of the thermodynamic characteristics of the system. The theory providing a better agreement with the results of the computer experiment for the same intermolecular potential is correct.

3. Objective

The aim of this work is to compare the accuracy of the calculations for the three best physically one fluid theories based on statistical mechanics (variational perturbation theory MCRSR [6], the theory of integral equations HMSA/C [2] and modified perturbation theory KLRR-T [7, 11]) to determine the most accurate theory. The excess value of the Helmholtz energy of the studied system ΔF^{ex} is different for each theory.

The basic idea of any perturbation theory (variational theory too) is to represent the potential $\varphi(r)$ of a system as a sum of the basic potential $\varphi_0(r)$ and a small perturbation $\varphi_1(r)$: $\varphi(r) = \varphi_0(r) + \varphi_1(r)$.

The excess energy of Helmholtz of the investigated one-component system ΔF^{ex} is represented by the terms of the zero and first order expansion into infinite series for variational theory and perturbation theory

$$\Delta F^{ex} = \Delta F_0 + 2\pi\rho_N N\int_0^\infty \varphi_1(r) r^2 g_0(r) dr \,. \tag{1}$$

Here ΔF_0 – the excessive Helmholtz energy of some basic system, $\varphi_1(r)$ – perturbating potential, $g_0(r)$ – radial distribution function of molecules in basic system, $\rho_N = N/V$ – the number density. Only terms of the zero and first order of series expansion for excess Helmholtz energy are used for the investigated one fluid system. It is shown that the higher order terms can be neglected in expansion (1) because of their mutual compensation (the series members have alternating signs).

Under the assumption that the excess value of the Helmholtz energy of the basis system coincides with the hard sphere fluid, the expression for the excessive ΔF^{ex} in the variational perturbation theories is represented as:

$$\Delta F^{ex} \le \Delta F_{HS} + 2\pi\rho_N N\int_0^\infty \varphi_1(r) r^2 g_{PY}(r,d) dr$$

Here g_0=$g_{PY}(r,d)$ is the radial distribution of hard sphere one fluid. It is supposed that the excessive Helmholtz energy of the reference system coincides with the excessive Helmholtz energy of a certain effective hard sphere fluid ΔF_0=ΔF_{HS}. Next the variational parameter (for example, the diameter of hard spheres d) is selected. The value of the variational parameter is determined in such way that ΔF^{ex} reaches a minimum at fixed T, V and N . The found minimum value ΔF^{ex} corresponds to the equilibrium state of the system (for data T, V and N). The need for multiple numerical integration and differentiation leads to significant computer time costs in the application of MCRSR, variational theories, but, of course, these costs are not comparable with the time of MC or MD calculation.

The problem that necessarily arises when using a hard sphere fluid in variational and perturbation theories is the choice of the hard spheres diameter on which all the properties of the basis system are depend. Different ways of solving this problem are proposed in various perturbation theories. The value of ΔF^{ex} is calculated after determining the diameter of hard spheres. All necessary thermodynamic parameters of the system can be expressed in terms of the corresponding first and second derivatives of the excessive Helmholtz energy ΔF^{ex}.

Direct calculation ΔF^{ex} for all values of T and V based on known properties of the basis system (ΔF_0 = ΔF_{HS} и g_0=$g_{PY}(r,d)$) is carried out using a reasonable decomposition of the given potential $\varphi(r)$ on basic and perturbing potentials in the theories of perturbations (in contrast to variational theory).

The problem of break point choosing of the thermodynamic potential is successfully solved in [4], arises in perturbation theories, in addition to the problem of choosing the diameter of hard spheres. The Laplace transform is used to find $g_{PY}(r,d)$) in deriving perturbation theories, as well as in variational theories.

The theoretical model of EOS one fluid with potential Exp-6 (KLRR-T) implemented in the form of computational subroutines is based on the version of the theory of perturbation THEOSTAR proposed by us [4]. A high-precision approximation in the form of a polynomial of the 7th order (error less than 5×10^{-7}) is proposed by us in this model for the calculation of the energy Helmholtz term. This term is responsible for the attraction forces of molecules. The contribution of this term, which was neglected in [10], is taken into account for calculating the total Helmholtz energy. High-speed algorithms and routines that significantly reduce the calculation time on the computer are developed.

The one-component theory HMSA/C based on the application of the Ornstein Zernike equation in conjunction with the closing equation HMSA allows to calculate accurately the radial distribution functions of molecules according to the given parameters of intermolecular potentials, pressure, excess internal energy. These data are sufficient to find the excess energy of Helmogolz using general thermodynamic relations (Maxwell's differential equation and the definition of thermodynamic potentials).

4. Discussion

The accuracy of the theories was tested by comparing the results with those obtained by the MC simulation. The statistics of comparisons of the calculations accuracy on the best theories HMSA/C, MRCSR and the improved version of the theory KLRR-T [7, 11] for a one fluid with the potential Exp-6 are shown in table 1. The values of relative deviations of compressibility factor (dimensionless pressure) Z and dimensionless excessive internal energy U^* computed by the presented theories from the MC simulations [2] are given in table 1. Dimensionless values of pressure Z and excess internal energy U^* were calculated using the following relations:

$$Z \equiv \frac{PV}{Nk_B T} = 1 + \rho^* \left(\frac{\partial F^*}{\partial \rho^*} \right)_{T^*}$$

$$U^* \equiv \frac{\Delta U}{Nk_B T} = \beta^* \left(\frac{\partial F^*}{\partial \beta^*} \right)_{\rho^*} = -T^* \left(\frac{\partial F^*}{\partial T^*} \right)_{\rho^*} \qquad (2)$$

Dimensionless parameters in expansion (2) are marked by the index *: ρ^* – density, k_B – Boltzmann constant, $\beta^* = \frac{1}{k_B T}$ here. The same characteristic of the exp-6 potentials for three different values of parameter determining the repulsive stiffness α were used in these calculations on the basis of MC and compared theories, table 1. The comparison was performed for MC points taken from [2]. These points cover the entire wide range of temperatures, pressures and densities. The calculations on the best theories give close deviations from the MC data (the comparison of the results in the table 1).

Table 1. The statistics of the relative deviations of calculated compressibility factor Z and excess internal energy U^* of a one component supercritical fluid from the MC results

	Over 55 MC points		Over 19 MC points for α = 11.5		Over 18 MC points for α = 13.5		Over 18 MC points for α = 15.5	
Method		δU*, %	δZ, %	δU*, %	δZ, %	δU*, %	δZ, %	δU*, %
			Relative average deviation, %					
KLRR-T	0.60	0.87	0.55	0.60	0.50	0.75	0.75	1.26
MCRSR	2.07	2.46	2.92	1.87	1.78	2.29	1.52	3.24
HMSA/C	0.72	1.89	0.46	1.19	0.72	1.69	0.98	2.85
			Relative root mean square deviation, %					
KLRR-T	0.90	1.44	0.62	0.76	0.56	0.99	1.32	2.16
MCRSR	2.33	3.29	3.10	2.14	1.86	2.85	1.79	4.45
HMSA/C	0.84	3.64	0.54	1.32	0.80	2.28	1.10	5.78
			Relative maximum deviation, %					
KLRR-T	5.30	7.00	1.48	1.72	1.05	2.96	5.30	7.00
MCRSR	5.06	11.56	4.11	3.87	2.40	4.82	5.06	11.56
HMSA/C	2.43	23.42	1.08	2.20	1.63	7.62	2.43	23.42

Note that the variational theory and HMSA/C theory use functions that correct excess Helmholtz energy so that the excess pressure and internal energy are in good agreement with the results of Monte Carlo computer experiments [2]. Whereas there are no correcting functions in the improved version KLRR-T [7, 11]. Consequently the theory MCRSR is less accurate compared to an improved version of the KLRR-T [7, 11] and HMSA/C [2] theory even with the use of the correction functions (table 1).

Calculations based on our improved version of the KLRR theory [7, 11] are more precise in comparison with estimations based on the theory HMSA/C.

We presented an EOS model that allows for reliably computing of the thermodynamic properties for systems over a wide range of temperatures and densities covering both the high-pressure area up to hundreds gigapascals and the region of moderate pressures. High predictive abilities of the presented EOS model allow one to apply it to solving a variety of practical and theoretical tasks in geophysics and astrophysics and in the chemistry and high energy physics.

Acknowledgments
This work is supported by the Russian Science Foundation under grant 16-19-00188.

References

[1] Ree F H 1983 *J. Chem. Phys.* **78** 409
[2] Fried L E, Howard W M 1998 *J. Chem. Phys* **109** 7338
[3] Kang H S, Lee C S, Ree T, Ree F H 1985 *J. Chem. Phys.* **82** 414
[4] Victorov S B, El-Rabii H, Gubin S A, Maklashova I V, Bogdanova Y A 2010 *J. of Energetic Materials* **28** 35
[5] Ree F H 1984 *J. Chem. Phys.* **81** 1251
[6] Ross M J 1979 *Chem. Phys.* **71** 1567
[7] Anikeev A A, Viktorov S B, Gubin S A 2014 *J. Phys. Chem. B* **8** 56
[8] Dubois V, Desbiens N, Auroux E 2010 *Chem. Phys. Let.* **494** 306
[9] Victorov S B, Gubin S A, Maklashova I V, Revyakin I I 2001 *Proc. of the 32nd Int. Ann. Conf. of ICT.* 69/1
[10] Byers-Brown W, Horton T V 1988 *Mol. Phys.* **63** 125
[11] Victorov S B, Gubin SA 2006 *Proc. of the 13th Int. Detonation Symp* 1118

Axial closed-form texture component approximating the canonical normal distribution

T M Ivanova

Department of High Mathematics, National Research Nuclear University, 31 Kashirskoe Shosse, Moscow 115409, Russia

e-mail: ivatatiana@gmail.com

Abstract. Easily calculated texture components are necessary for solving problems of quantitative texture analysis (QTA). In the present work we obtain closed-form approximation for the axial component of the canonical normal distribution on SO(3) with central and non-central scattering.

1. Introduction

Modeling and interpreting the texture of polycrystalline materials by the component fit method is based on the use of standard functions [1]. The latter describe the shape, the position in the orientation space and the scattering of texture components. The texture components in the form of the canonical normal distribution (CND) on the rotation group SO(3) [2] are widely used in quantitative texture analysis
[3] – [6].

Let the quaternion $R[\vec{n},\omega] = \cos\frac{\omega}{2} + \vec{n}\sin\frac{\omega}{2}$ determines the rotation g about the axis $\vec{n}$ by the angle ω. The peak component of the canonical normal distribution with central scattering described by the parameter ε is given by the series [2] – [3]:

$$f^{peak}(R[\vec{n},\omega],\varepsilon) = \sum_{l=0}^{\infty}(2l+1)\exp\left(-l(l+1)\varepsilon^2\right)\frac{\sin\left((2l+1)\frac{\omega}{2}\right)}{\sin\frac{\omega}{2}}. \quad (1)$$

The corresponding axial component with the axis $\vec{n}_A$ can be represented by the series of Legendre polynomials

$$f^{axial}(\vec{n}_A, g, \varepsilon) = \sum_{l=0}^{\infty}(2l+1)\exp\left(-l(l+1)\varepsilon^2\right)P_l\left((g\vec{n}_A,\vec{n}_A)\right), \quad (2)$$

Application of texture components (1) – (2) in the solution of the QTA problems is associated with large computational costs, as well as the necessity of truncating the series depending on the sharpness of the texture [4] – [6].

A computationally convenient approximation for the distribution (1) was obtained in [7]:

$$f_{approx}^{peak}(R[\vec{n},\omega],\varepsilon^2)=\frac{\sqrt{\pi}}{\varepsilon^3}\cos^{-4}\left(\frac{\omega}{2}\right)\exp\left\{-\frac{1}{\varepsilon^2}\tan^2\left(\frac{\omega}{2}\right)\right\}. \quad (3)$$

In the case of a non-central character of the scattering, the canonical normal distribution on SO (3) is represented by the triple series of generalized spherical harmonics [2]:

$$f^{peak}(g,\vec{\varepsilon})=\sum_{l=0}^{\infty}\sum_{m=-l}^{l}\sum_{n=-l}^{l}C_l^{mn}(\vec{\varepsilon})T_l^{mn}(g).$$

The coefficients of the series can be found only numerically, but its approximation has a rather simple form [7]:

$$f_{approx}^{peak}(g,\vec{\varepsilon})=$$

$$=\frac{\sqrt{\pi}}{\varepsilon_1\varepsilon_2\varepsilon_3\cos^4\left(\frac{\omega}{2}\right)}\exp\left\{-\frac{\tan^2\left(\frac{\beta}{2}\right)}{\cos^2\left(\frac{\alpha+\gamma}{2}\right)}\left[\frac{\sin^2\left(\frac{\alpha-\gamma}{2}\right)}{\varepsilon_1^2}+\frac{\cos^2\left(\frac{\alpha-\gamma}{2}\right)}{\varepsilon_2^2}\right]-\frac{\tan^2\left(\frac{\alpha+\gamma}{2}\right)}{\varepsilon_3^2}\right\}, \quad (4)$$

where $\vec{\varepsilon}=\{\varepsilon_1,\varepsilon_2,\varepsilon_3\}$ is a vector of scattering parameters, $\{\alpha,\beta,\gamma\}$ are the Euler angles of the rotation g, associated with the quaternion $R[\vec{n},\omega]$ by the relation

$$\cos\frac{\omega}{2}+\vec{n}\sin\frac{\omega}{2}=\cos\frac{\beta}{2}\cos\frac{\alpha+\gamma}{2}+\sin\frac{\beta}{2}\left(\vec{i}\sin\frac{\alpha-\gamma}{2}+\vec{j}\cos\frac{\alpha-\gamma}{2}\right)+\vec{k}\cos\frac{\beta}{2}\cos\frac{\alpha+\gamma}{2}.$$

However, the texture contains not only the peak components generally. Usually, the model includes axial components as well. And quite often the texture is described by them only. This paper is devoted to obtaining a convenient approximation of the axial component for the canonical normal distribution.

2. The relationship between the peak and axial components

The axial component is invariant with respect to rotations about a fixed axis. It is obtained by averaging the peak component over all possible rotations around the axis. If these rotations are represented by quaternion $R_A=\cos\frac{\omega_A}{2}+\vec{n}_A\sin\frac{\omega_A}{2}$, then

$$f^{axial}(\vec{n}_A,R[\vec{n},\omega],\varepsilon)=\frac{1}{2\pi}\int_0^{2\pi}f^{peak}(R_A^{-1}R,\varepsilon)\,d\omega_A. \quad (5)$$

As a result of averaging, $\vec{n}_A$ becomes the axis of the texture.

3. Approximating function for the axial component of the central normal distribution

The approximation of the axial component of the canonical normal distribution (2) is obtained by applying the averaging to the distribution (3). In this case, as an argument of the peak component, we take the angle $\tilde{\omega}$, corresponding to the product of the rotations $g_A^{-1}g$:

$$\cos\frac{\tilde{\omega}}{2}=\cos\frac{\omega_A}{2}\cos\frac{\omega}{2}+(\vec{n}_A,\vec{n})\sin\frac{\omega_A}{2}\cos\frac{\omega}{2}. \quad (6)$$

Substituting (6) into (3) and performing the integration in (5), we obtain

$$f_{approx}^{axial}\left(\vec{n}_A,R\left[\vec{n},\omega\right],\varepsilon\right)=\frac{\left\{\frac{1}{\varepsilon^2}+\frac{1}{2}\left(1-\sin^2\frac{\omega}{2}\cdot\left[\vec{n}_A,\vec{n}\right]^2\right)\right\}}{\left\{\left(1-\sin^2\frac{\omega}{2}\cdot\left[\vec{n}_A,\vec{n}\right]^2\right)\right\}^{3/2}}\exp\left\{-\frac{1}{\varepsilon^2}\frac{\sin^2\frac{\omega}{2}\cdot\left[\vec{n}_A,\vec{n}\right]^2}{1-\sin^2\frac{\omega}{2}\cdot\left[\vec{n}_A,\vec{n}\right]^2}\right\}, \quad (7)$$

In terms of rotation g , the approximating distribution (7) takes the form

$$f_{approx}^{axial}\left(\vec{n}_A,g,\varepsilon\right)=\frac{\left\{\frac{1}{\varepsilon^2}+\frac{1+\left(g\,\vec{n}_A,\vec{n}_A\right)}{4}\right\}}{\left\{\frac{1+\left(g\,\vec{n}_A,\vec{n}_A\right)}{2}\right\}^{3/2}}\exp\left\{-\frac{1}{\varepsilon^2}\frac{1-\left(g\,\vec{n}_A,\vec{n}_A\right)}{1+\left(g\,\vec{n}_A,\vec{n}_A\right)}\right\}, \quad (8)$$

If the axis of rotation $\vec{n}_A=\left(0,0,1\right)$, then $\left(g\,\vec{n}_A,\vec{n}_A\right)=\cos\beta$, thus

$$f_{approx}^{axial}\left((0,0,1),\ g,\ \varepsilon\right)=\frac{\left\{\frac{1}{\varepsilon^2}+\frac{1}{2}\cos^2\frac{\beta}{2}\right\}}{\cos^3\frac{\beta}{2}}\exp\left\{-\frac{1}{\varepsilon^2}\tan^2\frac{\beta}{2}\right\}. \quad (9)$$

4. Approximation of the axial component with non-center scattering

For the non-central distribution (4), the most interesting is the case, when $\vec{n}_A=\left(0,0,1\right)$. Averaging (4) over all possible rotations about this axis, we obtain

$$f_{approx}^{axial}\left((0,0,1),\ g,\ \vec{\varepsilon}\right)=\frac{1}{\varepsilon_1\varepsilon_2\varepsilon_3}\frac{\left\{1+\frac{1}{2A}+\frac{B^2}{A^2}\right\}}{\sqrt{A}\cos^4\frac{\beta}{2}}\exp\left\{-\frac{AC-B^2}{A}\right\}, \quad (10)$$

with

$$A=\left(\frac{\cos^2\gamma}{\varepsilon_1^2}+\frac{\sin^2\gamma}{\varepsilon_2^2}\right)\tan^2\frac{\beta}{2}+\frac{1}{\varepsilon_3^2},\ B=\left(\frac{\sin^2\gamma}{\varepsilon_1^2}+\frac{\cos^2\gamma}{\varepsilon_2^2}\right)\tan^2\frac{\beta}{2},\ C=\left(\frac{1}{\varepsilon_2^2}-\frac{1}{\varepsilon_1^2}\right)\sin\gamma\cos\gamma\tan^2\frac{\beta}{2}.$$

Note, that in case $\varepsilon_1=\varepsilon_2=\varepsilon_3=\varepsilon$, the distribution (10) goes into (9).

5. Conclusion

In the paper, distributions (7) – (10) approximating the axial component of the canonical normal distribution with the central and non-central scattering are obtained. They obey closed form and are convenient in calculations. These distributions can be recommended as standard functions for solving problems of quantitative texture analysis.

IOP Conf. Series: Journal of Physics: Conf. Series **1205** (2019) 012021 doi:10.1088/1742-6596/1205/1/012021

It should be noted that everywhere above it was assumed that the maximum of the distribution corresponds to zero rotation. The displacement of the maximum to the point g_0 in the orientation space can be achieved by replacing $g \to g_0^{-1}g$, and respectively, $R \to R_0^{-1}R$.

References

[1] Bunge H J 1982 *Texture Analysis in Material Science* (Butterworths:London)
[2] Nikolayev D I and Savyolova T I 1997 Normal distribution on the rotation group SO(3) , *Text. Microstruct.* **29** 201–33
[3] Bucharova T I and Savyolova T I 1993 Application of normal distribution on SO(3) and S2 for orientation distribution function approximation *Text. Microstruct.* **23** 161–176
[4] Ivanova T M and Savelova T I 2006 Robust method of approximating the orientation distribution function by canonical normal distributions *Phys. Met. Metallog.* 101 N2 114–118
[5] Ivanova T M and Serebryany V N 2017 Methods of recovering orientation distribution function using texture components with a circular normal distribution Indust. Lab. 83 I-1 43–47
[6] Ivanova T M Savyolova T I and Sypchenko M V 2010 The modified component method for calculation of orientation distribution function from pole figures *Inverse Problems in Science and Engineering* 18 N1 163–171
[7] Ivanova T M and Nikolayev D I 2001 New Standard Function for Quantitative Texture Analysis *Phisica Status Solidi* 228 N3 825–836

Scroll airmotor properties in comparison with piston and vane analogue[1]

V I Ivlev[1], S Yu Misyurin[12] and N Yu Nosova[12]

[1] Mechanical Engineering Research Institute of the Russian Academy of Sciences, M. Kharitonyevskiy Pereulok, 1, Moscow, 101990, Russia

[2] National Research Nuclear University MEPhI (Moscow Engineering Physics Institute), Kashirskoe shosse, 31, Moscow, 115409, Russia

E-mail: ssmmrr@mail.ru

Abstract. The experimental results of testing scroll machine in air motor mode, such as mechanical and flow performance; leakage coefficient; specific volume and weight power are presented. These results analyzed and compared with the same properties of radial – piston and vane air motors with equal power.

1. Introduction

Nowadays the scroll machines have been widely used as air or refrigeration compressors and also vacuum pumps. As compared with piston compressors, scroll compressors with the same capacity, have more energy efficiency (about 10-15 %), smaller dimension and weight (accordingly 25-40 % and 15-20 %), a low level of noise and vibrations. Now the actuator power range of scroll compressors is from few tens Watts [1] to tens kWts (Emerson ZP725K from "Copeland"). The volumetric rotary scroll machine is reversible and can function in air motor (expander) mode without any significant constructive modification and with preservation all advantages to piston air motors.

Today, quite a large number of works have been performed on using a scroll machine as a gas motor for various applications, especially for low-grade heat utilization systems based on Organic Rankine Cycle [2]. The "Air Squared Manufacturing"(USA) products specially designed scroll gas motors (working medium freon) for driving the generator in waste heat recovery systems [3]. Another application of scroll machine is associated with the creation of compressed air energy storage in solar and wind power stations [4, 5], as well as emergency power systems [6]. In [7] a scroll machine for energy storage system with two moving scrolls connected by synchronizing device is considered. This makes it possible to work in the compressor and motor modes. Increasing the energy efficiency of scroll air motor, by placing inside the scrolls insertions from rare earth magnets, is devoted to [8].

The investigation of the scroll machine in air motor mode began quite a long time ago [9, 10]. But at the present time, in technical literature there are no data allowing to comparing the

[1] This work was supported by the Russian Foundation for Basic Research, in the framework of a research project 16-29-04401 "Development of methods and algorithms for the synthesis of drive complexes (robots) working in different environments, taking into account the interaction of two or more robots."

characteristics of scroll air motor with the corresponding indices of the other types of air motors commercially produced by industry [11]. These data are usually given in the catalogues of manufacturers in the form of mechanical and air consumption characteristics, weight and dimensions. The most interesting is the comparison with the characteristics of such mass-produced types of air motors as vane (up to 75 – 80% of the market of air motors) and radial piston. Carrying out such a comparative analysis is the purpose of this paper.

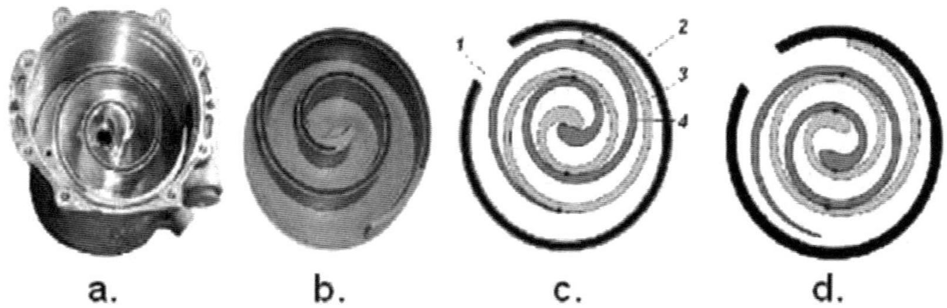

Figure 1. Appearance and mutual position of scrolls: a) fixed scroll, b) movable scroll, c), d) – two positions of the movable scroll with the angles of rotation of the output shaft differing by an angle of 180°. 1 — outlet port, 2 — housing, 3 — fixed scroll, 4 -– movable scroll

2. Test rig and experimental results

As a scroll air motor was used the converted compressor "Mitsubishi" model MSC90CAS from car air conditioner. In Fig. 1a, b show a fixed and movable scrolls respectively. As a counter-rotating device of moving scroll use the Oldheim coupling. In Fig.1c,d show the two positions of the movable scroll with the angles of rotation of the output shaft differing by an angle of 180°.

Figure 2. Test rig. 1 — scroll airmotor, 2 — bellows coupling, 3 — powder electromagnetic clutch, 4 — torque sensor.

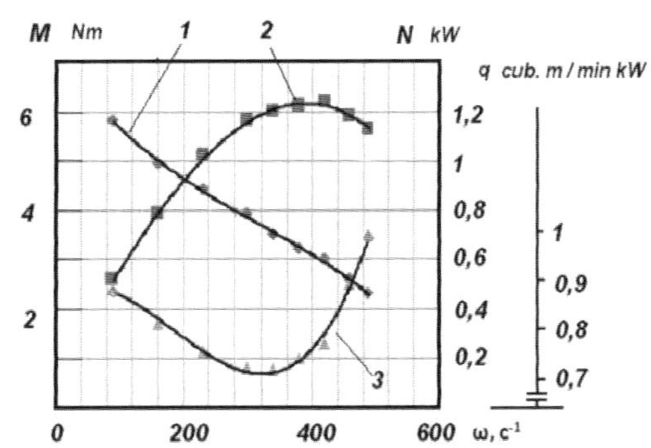

Figure 3. Mechanical and air consumption characteristics of scroll airmotor at 0,6 MPa. 1 — developed moment, 2 — power, 3 — specific flow consumption

To measure the mechanical and flow consumption characteristics of scroll air motor a test rig was assembled, part of which is shown in Fig. 2. In the compressed air preparation system a lubricator is installed to supply 5-7 mg/m^3 of oil having a reduced viscosity at low temperatures (up to -25°C). The loading moment was created by an electromagnetic powder clutch P12S (Poland), whose shaft was connected through a bellows coupling with a scroll air motor shaft. The clutch housing (stator) is connected to the shaft of the static torque sensor YDNF-100KC

(China). The steady angular velocity ω of scroll air motor shaft was measured by a laser tachometer "testo 460" (Germany). The compressed air flow consumption was measured at the air motor output using a "testo 6446" flowmeter, which was installed after the oil separator. The passage sections of the oil separator and the flowmeter were more than 5 times larger than the section of the exhaust port of air motor, which virtually eliminated the appearance of additional resistance on the exhaust.

Fig. 3 shows the mechanical and flow consumption characteristics of scroll air motor dependence of the steady – state (ω – const): moment, power and specific flow consumption at supply pressure $p_m = 0.6$ MPa. Here, the specific flow consumption pressed air, reduced to normal conditions is the consumption per unit of developed power. Mechanical and flow consumption characteristics for this scroll air motor at supply pressures of 0.5 and 0.3 MPa are given in [12]. Externally the curves in Fig. 3 look similar to the corresponding characteristics of other types of volumetric air motors. Pressure 0.6 or 0.63 MPa the most commonly used supply pressures, for which all characteristics are given in the catalogs of manufacturers. It should be noted that recalculation of experimental data given in [10] for scroll air motor with other parameters gives approximately the same value of specific flow consumption – 0.8 m^3/min kW at developing power of 0.6 kW and $p_m = 0.41$ MPa.

To determine leakage coefficient (the relation of flow consumption at $\omega = 0$ to flow consumption at maximum power) measurements were carried out at the braked rotor for two positions of the movable scroll shown in Fig. 1c, d. For the position corresponding to Fig. 1c, depending on the supply pressure, the flow consumption varies practically linearly from 0,03 m3/min at $p_m = 0.1$ MPa to 0.16 m^3/min at $p_m = 0.6$ MPa. At rotor position, as shown in Fig. 1d, the flow consumption is approximately 7 – 8% higher, which is explained by the decrease in pneumatic resistance formed by the tangential and radial gaps between the scrolls.

3. Comparison of characteristics with other types of air motors

It is necessary to compare the characteristics of different types of air motors with their approximately equal power developed at the same supply pressure because for air motors of one type, but different power, the specific characteristics may differ significantly. We have analyzed catalogs of more than ten leading world manufacturers – producers of piston and vane air motors. Some of these catalogs are given in [13–20]. For comparison, samples of air motors, which in our opinion have the best performance (except cost) were selected. So, among the vane air motors we take two models of "Atlas Copco": model LZB54A180-11 with a large ratio of the stator length to its diameter, which allows to obtain good volumetric and weight specific power, and model "heavy series" LZL05S, with smaller ratio stator length to its diameter. More details about these air motors can be found in [15]. Among the piston air motors were selected models of radial piston RM110 from "Globe" and axial piston A6 series from "Gardner – Denver".

Table 1 shows the main indicators of this air motors, as well as the performance of our scroll air motor. In Fig. 4 one graph shows the mechanical characteristics of the three types of air motors clearly illustrating their working areas.

First of all, pay attention to high energy efficiency of scroll air motor. This is due to the following factors.

Leakage effects. For radial -– piston air motors the main leakage occurs in a rotating distributor of compressed air. A small part of the compressed air from the supply line can fall right on the exhaust. For vane air motors tangential end leakage can lead to similar processes and significantly reduce energy efficiency. In scroll air motor the tangential and radial leakages from central chamber pass other chambers before come to atmosphere and take part in forming moving torque.

Harmful (or "dead") volume. Radial -– piston air motor has a harmful volume in the form of channels connecting the rotating distributor with the working chambers. Moreover, these

Table 1. Main indicators of air motors and scroll air motor

Model	Power kW	Supply pressure MPa	Specific flow rate m^3/min kW	Specific weight power kW/kg	Specific volume power kW/l^3	Leakage coefficient
Atlas Copco LZB54A180	1.2	0.63	1.13	0.51	2.12	0.22
Atlas Copco LZL05S	1.3	0.63	1.7	0.33	2.2	0.2
Globe RM-110	1.15	0.60	1.3	0.09	0.46	0.12
Gardner-Denver A6 series	1.42	0.62	1.24	0.12	0.45	–
Scroll airmotor	1.25	0.6	0.78	0.29	0.94	0.16

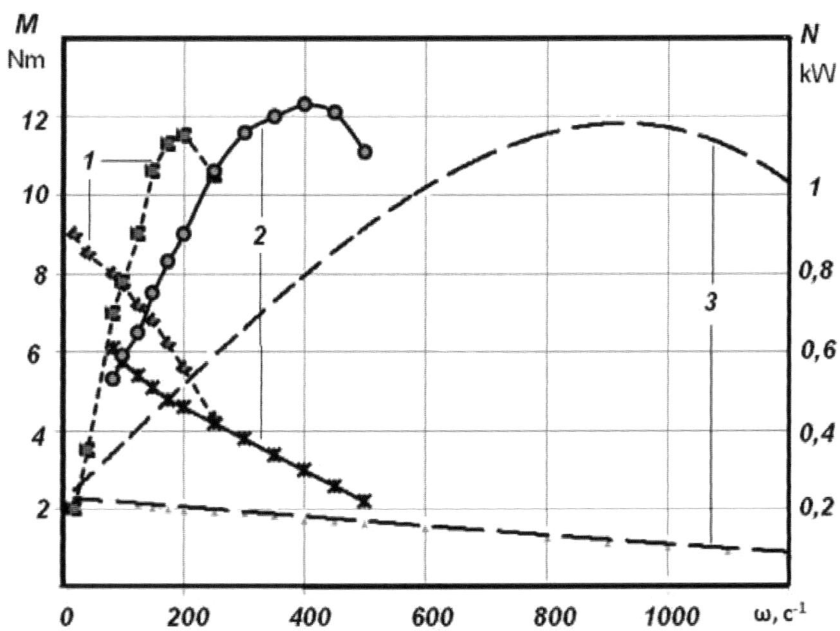

Figure 4. Mechanical characteristics: 1 — radial-piston airmotor RM110, 2 -– scroll airmotor, 3 -– vane airmotor LZB54A180

channels are used both for filling the working chambers with compressed air, and for emptying. This results in an unproductive consumption a portion of the compressed air. Therefore, radial - piston air motor have a "soft" torque characteristic (see Fig. 4), which indirectly makes it possible to judge about the perfection of the organization the processes of intake and exhaust the compressed air in the working chambers. Scroll and vane air motors have practically no harmful volume, and this is one of the reasons why they have more "hard" torque characteristics.

Built-in volume ratio. This indicator characterizes the ratio of the maximum to the minimum volume of the working chamber, when it is not communicating with the supply line and the atmosphere, and positive mechanical work is done only at the expansion of the compressed air in the chamber. Vane air motors are characterized by a small value of the built-in volume ratio – no more than 1.4 (for non-reversible models). For reversible models, this indicator is smaller. For piston air motors, the built-in volume ratio can reach 7-8, but in practice it is somewhat lower. The scroll air motor represented here has a built-in volume ratio of 2.95. This value was determined on the basis of the relationships for calculating the volumes of the scroll air motor chambers, given in [21]. Together, these factors provide high energy efficiency to scroll air motor.

An important operational indicator of air motor functioning quality is the noise level accompanying its operation. The noise effect consists of two components: from mechanical interaction of moving parts and pulsations of compressed air on the exhaust. In [22], noise data are reported for a 0.36 kW vane air motor (Atlas Copco), where, without a muffler on the exhaust, it is 94 dB, and with a muffler of 77 dB. For radial – piston P1V-P series of the company "Parker Pneumatic" with a power of up to 0.5 kW at idle, the noise level without a muffler is 95-100 dB, and with a muffler 75 – 80 dB at a supply pressure of 0.5 MPa. For a scroll air motor with a power of 10 kW [6], the noise level at a distance of 5 meters is less than 65 dB, and in [3] a 1 kW scroll motor in freon is 55 dB. As can be seen from the data presented, and according to this index, the scroll air motor has significant advantages over piston and vane air motors.

4. Concluding Remarks

From the characteristics of the scroll air motor considered here, it can be concluded that the latter is not inferior in most indices, but in terms of specific consumption it surpasses traditional piston and vane air motors of comparable power.

However, the list of characteristics considered here is not complete. In the future, it is necessary to determine such characteristics as the starting torque (its magnitude as a function of the initial angular position of the movable scroll), the effect of radial loads on the output shaft, the minimum stable revolutions, and the relative cost. These data will allow us to more reasonably judge the competitiveness of scroll air motor. It is necessary to compare the characteristics of different types of air motors with their approximately.

It should be noted that at a cost of pneumatic energy (the main line with $p_m = 0.8 - 1$ MPa) is 5 to 6 times higher than electricity, the energy efficiency index may in some cases be decisive. Also, the energy efficiency of air motors is important for autonomous systems with a tank power source, because this directly affects the working time of the system. In view of the foregoing, we can expect a wider use of scroll air motors in various applications where piston or vane air motors were previously used.

References

[1] Scroll Compressor P11H12N2.5. Copyright 2012 Air Squared Mfg. Inc. www.airsquared.com
[2] Lemort V, Quoilin S, Lebrun J 2010 Experimental study and modeling of an organic Rankine cycle using scroll expander *J. Applied Energy* Volume 87 Issue 4 pp. 1260—1268
[3] E15H22N4.25. 1kW Scroll Expander. www.airsquared.com
[4] Saiton T, Yamada N, Wakashima S 2007 Solar Rankine Cycle Using Scroll Expander *J. of Environment and Engineering* Volume 2 No 4 pp. 708–719
[5] Krupke C et al 2017 Modeling and experimental study of wind turbine system in hybrid connection with compressed air energy storage *J. IEEE Transactions on Energy Conversion* Volume 32 Issue 1 pp. 137-–145
[6] CAB 10kW The Smarter alternative to conventional batteries. www.flowbattery.co.uk
[7] Iglesias A, Favrat D Sept. 2014 Innovative isothermal oil-free co-rotation scroll compressor — expander for energy storage with first expander test *J. Energy Conversion and Management* Volume 85 pp. 565-–572

[8] Wang J, Shpanin L, Dooner M Nov.2017 A New Scroll — Type Air Motor with Magnetic Spirals *J. IEEE/ASME Transactions on Mechatronics* Volume 23 Issue 1 pp. 459-–468
[9] Zanelli R, Favrat D 1994 Experimental Investigation of a Hermetic Scroll Expander – Generator *Conf. Proc. 12-th International Compressor Engineering Perdue* p. 459—464
[10] Yanagisawa T, Fukuta M, Ogi Y 2001 Performance of an oil-free scroll — type air expander *Conf. Proc. of the ImagE Conference on Compressors and their systems* pp. 167-–174
[11] Ivlev V I., Misyurin S Yu. 2017 Calculated and Experimental Characteristics of a Scroll Machine Operating in the Air Motor Mode *J. Doklady Physics* Volume 62 No. 1 pp. 42-–45
[12] Ivlev V I, Bozrov V M, Voronov V A March 2015 Testing a Scroll Machine in Pneumatic Motor – Expander Modes *J. of Machinery Manufacture and Reliability* Volume 44 Issue 2 pp. 120—124
[13] TONSON Air Motors MFG. CORP. Catalog. www.tonson-motor.com
[14] Ingersoll Rand MVR and SM (select) Lub -– free reversible air motors www.ingersollrandprodacts.com
[15] Atlas Copco Air Motors. LZB, LZL vane motor. Catalog. 2016. 75 p. www.atlascopco.com
[16] Deprag Air Motors Customized drives solutions. 2012. 28 p. www.deprag.com
[17] GAST Air Motors / Gear Motors. Lubricated and Non-lubricated. ISO 9001. Catalog Final 2014. -– 41 p. www.gastmfg.com
[18] Gardner -– Denver Pneumatic Power Motors. Catalog 2011. www.cooperpowertools.com
[19] Globe Piston air motor. www.globe-benelux.nl
[20] Parker Pneumatic Air motors. Platform Catalog PDE2670TCUK. May 2014. 52 p. www.parker.com
[21] Wang J et al 2011 Mathematical Modeling Study of Scroll Air Motor and Energy Efficiency Analysis – Part I *J. IEEE/ASME Transactions on Mechatronics* Volume 16 Issue 1 pp. 112—121
[22] Pocket guide to Air Motors. Atlas Copco. 28 p. www.atlascopco.com

Determination of the right-hand side term in the degenerate parabolic equation with two variables

V L Kamynin[1] **and A B Kostin**[1]
[1]National Research Nuclear University MEPhI (Moscow Engineering Physics Institute), 31 Kashirskoe shosse, 115409 Moscow, Russia

E-mail: `vlkamynin2008@yandex.ru`, `abkostin@yandex.ru`

Abstract. We find two type conditions sufficient for unique solvability of inverse problem of source determination in degenerate parabolic equation with two independent variables.

1. Introduction

In the present work we study the unique solvability of the inverse problem of determination of a pair of functions $\{u(t,x);p(x)\}$ satisfying in the rectangle $Q \equiv [0,T] \times [0,l]$ the parabolic equation

$$\rho(t,x)u_t - u_{xx} - b(t,x)u_x - c(t,x)u = g(t,x)p(x) + r(t,x), \tag{1}$$

the initial and boundary conditions

$$u(0,x) = u_0(x),\ x \in [0,l], \qquad u(t,0) = u(t,l) = 0,\ t \in [0,T], \tag{2}$$

and the additional condition of integral observation

$$\mathbf{l}(u) \equiv \int_0^T u(t,x)\chi(t)\,dt = \varphi(x), \quad x \in [0,l]. \tag{3}$$

In what follows we use Lebesgue, Sobolev and Hölder spaces with corresponding norms in usual sense. For convenience we denote the space $L_2(0,l)$ by E and the norm in the space $L_q(0,l)$ by $\|\cdot\|_q$, $q \in [1,\infty]$. We will also use the notation $G(x) = \int_0^T g(t,x)\chi(t)\,dt \equiv \mathbf{l}(g)(x)$ and the well–known Poincaré – Steklov inequality, which, for $n = 1$ is in the form

$$\|z\|_2 \leqslant \frac{l}{\pi}\|z_x\|_2 \qquad \text{for all} \quad z \in \overset{\circ}{W}{}_2^1(0,l). \tag{4}$$

In what follows we assume that the functions appearing in the input data of the problem (1)–(3) are measurable and satisfy the following conditions:

$$0 \leqslant \rho(t,x) \leqslant \rho_1, \quad (t,x) \in Q; \qquad \frac{1}{\rho} \in L_q(Q), \quad q > 1, \quad \|\frac{1}{\rho}\|_{L_q(Q)} \leqslant \rho_2; \tag{A}$$

$$\begin{cases} b \in L_2(0,T;L_\infty(0,l)); & b^2/\rho,\ c^2/\rho \in L_\infty(Q); \\ \|b\|_{L_2(0,T;L_\infty(0,l))} \leqslant K_b; & \|b^2/\rho\|_{L_\infty(Q)} \leqslant K_{b,\rho},\ \|c^2/\rho\|_{L_\infty(Q)} \leqslant K_{c,\rho}; \end{cases} \tag{B}$$

$$u_0 \in \overset{\circ}{W}{}_2^1(0,l), \qquad \varphi \in W_2^2(0,l),\ \varphi(0)=\varphi(l)=0, \quad \|\varphi''\|_2 \leqslant K_\varphi^{**}; \tag{C}$$

$$\begin{cases} g\,,\ g^2/\rho \in L_1(0,T;L_\infty(0,l))\,; & \big\|g^2/\rho\big\|_{L_1(0,T;L_\infty(0,l))} \leqslant K_{g,\rho}; \\ r \in L_1(0,T;E)\,,\ r^2/\rho \in L_1(Q); & \|r\|_{L_1(o,T;L_2(0,l))} \leqslant K_r\,,\ \big\|r^2/\rho\big\|_{L_1(Q)} \leqslant K_{r,\rho}\,; \\ \chi \in L_\infty(0,T),\ \|\chi\|_{L_\infty(0,T)} \leqslant K_\chi\,, & c\chi,\ (\rho\chi)_t \in L_2(0,T;L_\infty(0,l)); \\ \Big|\int_0^T g(t,x)\chi(t)\,dt\Big| \geqslant g_0 > 0, & |\rho(T,x)\chi(T)| \leqslant \rho_3,\ |\rho(0,x)\chi(0)| \leqslant \rho_4, \\ \|c\chi\|_{L_2(0,T;L_\infty(0,l))} \leqslant K_{c,\chi}; & \|(\rho\chi)_t\|_{L_2(0,T;L_\infty(0,l))} \leqslant K_{\rho,\chi}\,; \end{cases} \tag{D}$$

$$\text{either} \quad \frac{\rho_t}{\rho} \in L_\infty(Q), \qquad \text{or} \quad \rho_t \in L_1(Q) \quad \text{and} \quad \rho_t(t,x) \leqslant 0, \quad (t,x) \in Q. \tag{E}$$

Here ρ_1, ρ_2, g_0, $K_{g,\rho}$, K_χ = const > 0, and ρ_3, ρ_4, K_b, $K_{b,\rho}$, $K_{c,\rho}$, $K_{\rho,\chi}$, $K_{c,\chi}$, K_r, $K_{r,\rho}$ = const $\geqslant 0$.

Definition 1. By the generalized solution of the problem (1)–(3) we mean the pair of functions $\{u(t,x);p(x)\}$ with

$$u \in C^{0,\alpha}(Q) \cap L_\infty(0,T;\overset{\circ}{W}{}_2^1(0,l)) \cap W_s^{1,2}(Q),\ s>1,\ \alpha \in (0,1), \quad p \in L_2(0,l) \equiv E,$$

such that this pair satisfies the equation (1) almost everywhere in Q and the function $u(t,x)$ satisfies the conditions (2), (3) in the classical sense.

By virtue of condition (A) the equation (1) is a degenerate parabolic equation. Inverse problems for degenerate parabolic equations are important in applications in engineering and financial mathematics (see, for example, [1, 2, 3]). Earlier the inverse problem (1)–(3) was investigated in [4] but with more stringent assumptions on the input data functions and in the different class than in the present one. Let us also note the papers [5, 6] where inverse problem for equation of the type (1) but with unknown function $p(t)$ and with another type of overdetermination condition was considered.

Let $p \in E$ be a known function and let us consider the direct problem (1), (2) with a given function $p(x)$ in the right-hand side of equation (1). For convenience, we introduce the notation

$$f(t,x) = g(t,x)p(x) + r(t,x), \qquad \text{so that} \quad f^2/\rho \in L_1(Q),$$

then the equation (1) takes the form

$$\rho(t,x)u_t - u_{xx} - b(t,x)u_x - c(t,x)u = f(t,x), \quad (t,x) \in Q. \tag{5}$$

In our investigation we need the following result from [4] concerning the direct problem (5), (2).

Theorem 1. *Let the conditions (A)–(E) hold and $p \in E$. Set*

$$q^* = \frac{2q}{q+1}, \qquad \lambda^* = 3\big(K_{b,\rho} + \frac{l^2}{\pi^2}K_{c,\rho}\big). \tag{6}$$

Then there exists a generalized solution $u(t,x)$ of the problem (5), (2) with $u \in W_{q^*}^{1,2}(Q)$. Moreover this solution is unique and it satisfies the estimates:

$$\sup_{0\leqslant t\leqslant T} \|u_x(t,\cdot)\|_2^2 + \|\rho u_t^2\|_{L_1(Q)} \leqslant e^{\lambda^* T}\big(\|u_0'\|_2^2 + 3\|f^2/\rho\|_{L_1(Q)}\big), \tag{7}$$

$$\|u\|^2_{W^{1,2}_{q^*}(Q)} \leqslant C_1, \tag{8}$$

$$|u(t_1,x_1)-u(t_2,x_2)| \leqslant C_2|x_1-x_2|^{1/2}+C_3|t_1-t_2|^{\frac{q-1}{6q}}, \quad \textit{for all} \quad (t_1,x_1),\,(t_2,x_2)\in Q, \tag{9}$$

where C_1, C_2, $C_3 = \text{const} > 0$ *depend only on* l, T, ρ_1, ρ_2, $\|u_0'\|_2$, $K_{b,\rho}$, $K_{c,\rho}$, $K_{g,\rho}$, $K_{r,\rho}$ *and on* $\|p\|_2$.

2. Unique solvability of inverse problem. The first variant of sufficient conditions

In this section we assume that the input data of the inverse problem (1)–(3) satisfy conditions (A) – (E), constants $q^* > 1$ and λ^* are defined in (6). Let us derive the operator equation for the unknown function $p(x)$.

Let pair of functions $\{u(t,x);p(x)\}$ be any generalized solution of the inverse problem (1)–(3). Let us multiply equation (1) by $\chi(t)$ and integrate over $[0,T]$. Taking into account conditions (2), (3) after integrating by parts we obtain the relation

$$p(x)=\frac{1}{G(x)}\Big[\rho(T,x)\chi(T)u(T,x)-\int_0^T\big(c(t,x)\chi(t)+(\rho(t,x)\chi(t))_t\big)u(t,x)\,dt-$$
$$-\int_0^T b(t,x)\chi(t)u_x(t,x)\,dt\Big]-\frac{1}{G(x)}\Big[\varphi''(x)+\mathbf{l}(r)(x)+\rho(0,x)\chi(0)u_0(x)\Big]. \tag{10}$$

In view of this relation let us introduce the operator $\mathcal{A}: E\to E$ by the formula

$$\mathcal{A}p=\frac{1}{G(x)}\Big[\rho(T,x)\chi(T)u(T,x)-\int_0^T\big(c(t,x)\chi(t)+(\rho(t,x)\chi(t))_t\big)u(t,x)\,dt-$$
$$-\int_0^T b(t,x)\chi(t)u_x(t,x)\,dt\Big]-\frac{1}{G(x)}\Big[\varphi''(x)+\mathbf{l}(r)(x)+\rho(0,x)\chi(0)u_0(x)\Big], \tag{11}$$

where $p(x)$ is an arbitrary function in E, and $u=u(t,x;p)$ is a solution of direct problem (1), (2) with given $p(x)$ in the right-hand side of equation (1). In view of theorem 1 solution $u(t,x;p)$ exists and is unique and the operator $\mathcal{A}$ is well–defined. By virtue (11) the relation (10) can be written as

$$p=\mathcal{A}p. \tag{12}$$

Lemma 1. *Let conditions* (A)–(E) *hold. Then the operator equation* (12) *is equivalent to the inverse problem* (1)–(3) *in the following sense. If pair* $\{u(t,x);p(x)\}$ *is a generalized solution of the inverse problem, then* $p(x)$ *satisfies* (12). *Conversely, if* $p\in E$ *is a solution of operator equation* (12), *and* $u=u(t,x;p)$ *is a solution of direct problem* (1), (2) *with this* p, *then the pair* $\{u(t,x);p(x)\}$ *is a generalized solution of inverse problem* (1)–(3).

The proof of this lemma is standard — see, for example, [4].

Theorem 2. *Let the conditions* (A)–(E) *hold, the numbers* q^* *and* λ^* *are defined in* (6). *Suppose that*

$$\beta\equiv\frac{12}{g_0^2}\exp(\lambda^*T)K_{g,\rho}\Big[\frac{l^2}{\pi^2}(\rho_3^2+TK^2_{c,\chi}+TK^2_{\rho,\chi})+TK^2_\chi K^2_b\Big]<1. \tag{13}$$

Then there exists a generalized solution of inverse problem (1)–(3) *and* $u\in W^{1,2}_{q^*}(Q)$. *Moreover, such a solution is unique and the following estimates*

$$\|p\|_2\leqslant\frac{1}{(1-\sqrt{\beta})g_0}\Big\{\exp(\lambda^*T/2)\Big[\frac{l^2}{\pi^2}(\rho_3^2+TK^2_{c,\chi}+TK^2_{\rho,\chi})+TK^2_\chi K^2_b\Big]^{1/2}(\|u_0'\|_2^2+3K_{r,\rho})^{1/2}+$$
$$+K^{**}_\varphi+K_\chi K_r T^{1/2}+\rho_4\|u_0\|_2\Big\}, \tag{14}$$

$$\sup_{0\leqslant t\leqslant T} \|u_x(t,\cdot)\|_2^2 \leqslant \exp(\lambda^* T)\Big[\|u_0'\|_2^2 + 6K_{g,\rho}\|p\|_2^2 + 6K_{r,\rho}\Big], \tag{15}$$

as well as (8) *and* (9) *hold.*

Proof. Suppose that $p^{(1)}$, $p^{(2)} \in E$ and $u^{(i)}(t,x) = u(t,x;p^{(i)})$ for $i = 1,2$ are solutions of corresponding direct problems (1), (2). Set $v(t,x) = u^{(1)}(t,x) - u^{(2)}(t,x)$ and $y(x) = p^{(1)}(x) - p^{(2)}(x)$. Then the pair $\{v(t,x);\, y(x)\}$ satisfies the relations

$$\rho(t,x)v_t - v_{xx} - b(t,x)v_x - c(t,x)v = g(t,x)y(x), \quad (t,x) \in Q, \tag{16}$$

$$v(0,x) = 0, \quad x \in (0,l); \qquad v(t,0) = v(t,l) = 0, \quad t \in [0,T]. \tag{17}$$

By virtue of (11), the conditions (*A*)–(*E*) and inequality (4) we have

$$\|\mathcal{A}p^{(1)} - \mathcal{A}p^{(2)}\|_2^2 \leqslant \frac{4}{g_0^2}\big[\rho_3^2\|v(T,\cdot)\|_2^2 + K_\chi^2 K_b^2\|v_x\|^2_{L_2(Q)} + \big(K_{c,\chi}^2 + K_{\rho,\chi}^2\big)\|v\|^2_{L_2(Q)}\big] \leqslant$$
$$\leqslant \frac{4}{g_0^2}\big[\frac{l^2}{\pi^2}(\rho_3^2 + TK_{c,\chi}^2 + TK_{\rho,\chi}^2) + TK_\chi^2 K_b^2\big] \sup_{0\leqslant t\leqslant T} \|v_x(t,\cdot)\|_2^2. \tag{18}$$

On the other hand due to the estimate(7) applied to the function $v(t,x)$ via relations (16), (17) we obtain that

$$\sup_{0\leqslant t\leqslant T} \|v_x(t,\cdot)\|_2^2 \leqslant 3\exp(\lambda^* T)\int_0^T \sup_{0\leqslant x\leqslant l}\frac{g^2(t,x)}{\rho(t,x)}\,dt \cdot \int_0^l y^2(x)\,dx \leqslant$$
$$\leqslant 3K_{g,\rho}\exp(\lambda^* T)\cdot\|p^{(1)} - p^{(2)}\|_2^2. \tag{19}$$

From (18), (19) taking into account assumption (13) we derive the contractibility of operator $\mathcal{A}$ and thus the operator equation (12) is uniquely solvable. Therefore in view of lemma 1 there exists a generalized solution of the inverse problem (1)–(3) and it is unique. The estimate (14) follows from the well–known error estimate of the first approximation in iterative method (see, for example, [7, p. 43]). Estimate (15), (8) and (9) were proved in theorem 1.

3. Solvability of inverse problem. The second variant of sufficient conditions

In the present section we assume that in the equation (1) $b(t,x) \equiv b(x)$ and the conditions (*A*)–(*E*) hold.

Performing the transformation of the right-hand side of equation (1)

$$g(t,x)p(x) \equiv \frac{g(t,x)}{G(x)}G(x)p(x) \equiv \tilde{g}(t,x)\tilde{p}(x),$$

we reduce our problem (1)–(3) to the equivalent problem of the same form, but for which the equality $\mathbf{l}(\tilde{g})(x) \equiv 1$, $x \in [0,l]$, holds. Thus, without loss of generality we assume that in the initial problem (1)–(3) $G(x) \equiv 1$, while we retain the previous notation for the input data and unknown functions.

Denote by $z^0(t,x)$ the solution of direct problem (1), (2) for $p(x) \equiv 0$ (so that $z^0(t,x) = u(t,x;0)$). Let us introduce the new unknown function $u(t,x) - z^0(t,x)$ for which we retain the notation $u(t,x)$ and we obtain the inverse problem of finding the pair $\{u(t,x); p(x)\}$ from the conditions

$$\rho(t,x)u_t - u_{xx} - b(x)u_x - c(t,x)u = g(t,x)p(x), \quad (t,x) \in [0,T]\times[0,l], \tag{20}$$

$$u(0,x) = 0, \; x \in [0,l]; \qquad u(t,0) = u(t,l) = 0, \; t \in [0,T], \tag{21}$$

$$\int_0^T u(t,x)\chi(t)\,dt = \varphi(x) - \mathbf{l}(z^0)(x) \equiv \tilde{\varphi}(x), \quad x \in [0,l]. \tag{22}$$

Obviously $\tilde{\varphi} \in W_2^2(0,l) \cap \overset{\circ}{W}{}_2^1(0,l)$. Further in present section we investigate just the problem (20)–(22). The results obtained for it, obviously, will also hold for the original inverse problem (1)–(3).

Let us consider in $E \equiv L_2(0,l)$ the cone of nonnegative functions

$$E_+ = \{p \in L_2(0,l) \mid p(x) \geqslant 0, \quad x \in [0,l]\}.$$

This cone is closed and reproducing (see, for example, [8]). We define the linear operator $\mathcal{B} : E \to E$ by the formula

$$\mathcal{B}p = \rho(T,x)\chi(T)u(T,x;p) - \int_0^T \big[(\rho(t,x)\chi(t))_t + c(t,x)\chi(t)\big]u(t,x;p)\,dt, \tag{23}$$

where $u = u(t,x;p)$ is a solution of direct problem (20), (21) with chosen function $p \in E$ in the right-hand side of (20). According to the theorem 1 the operator $\mathcal{B}$ is defined on the whole space E and is bounded, i.e. $\mathcal{B} \in \mathcal{L}(E)$ and following estimates of solutions of direct problem (20), (21)

$$u \in C(Q), \qquad \|u(t,\cdot)\|_2 \leqslant \text{const} \cdot \|p\|_2 \quad \text{for all} \quad t \in [0,T], \tag{24}$$

$$\|u_x(t,\cdot)\|_2 \leqslant \text{const} \cdot \|p\|_2 \quad \text{for all} \quad t \in [0,T]. \tag{25}$$

hold. Let us consider the operator equation of the second kind in the space E:

$$(I - \mathcal{B})p = \psi, \tag{26}$$

where $\psi(x) = -\tilde{\varphi}''(x) - b(x)\tilde{\varphi}'(x)$, $\psi \in E$ by virtue to conditions (B)–(D). As in [4] we prove the following lemma.

Lemma 2. *Let conditions* (A)–(E) *hold. Then the inverse problem* (20)–(22) *is equivalent to the operator equation* (26).

In addition to (A)–(E) we assume that the following conditions are satisfied:

$$\chi(t) \geqslant 0, \quad t \in [0,T]; \qquad g(t,x) \geqslant 0, \quad (\rho(t,x)\chi(t))_t + c(t,x)\chi(t) \leqslant 0, \quad (t,x) \in Q. \tag{27}$$

In the lemmas below we establish a number of properties for operator $\mathcal{B}$ from which the solvability of the equation (26) will follow.

Lemma 3. *Let conditions* (A)–(E) *and* (27) *hold. Then* $\mathcal{B}$ *is a positive operator.*

The proof is done in [4].

Lemma 4. *Let conditions* (A)–(E) *hold. Then* $\mathcal{B} \in \mathcal{L}(E)$ *is a compact operator.*

Proof. We represent $\mathcal{B}$ in the form

$$\mathcal{B}p = \mathcal{B}_1 p + \mathcal{B}_2 p, \quad \text{where} \quad \mathcal{B}_1 p = \rho(T,x)\chi(T)u(T,x;p),$$

$$\mathcal{B}_2 p = -\int_0^T \big[(\rho(t,x)\chi(t))_t + c(t,x)\chi(t)\big]u(t,x;p)\,dt \equiv \int_0^T R(t,x)u(t,x;p)\,dt.$$

By virtue of the estimate (25) the operator $\widetilde{\mathcal{B}}_1 p = u(T, x; p)$ transforms each bounded set of space E to a set bounded in $W_2^1(0, l)$ which is compact in E. Then the operator $\mathcal{B}_1$ is also compact, since it is a product of the operator bounded in E on the compact operator.

Let us consider the operator $\mathcal{B}_2$. Taking into account conditions (B), (D), it follows that $R(t, x) \equiv -(\rho\chi)_t - \sqrt{\rho}(c/\sqrt{\rho})\chi \in L_1(0, T; L_\infty(0, l))$. We extend $R(t, x)$ by zero for $t > T$ and introduce for it the mean Steklov functions

$$R^h(t, x) = \frac{1}{h}\int_t^{t+h} R(\tau, x)\, d\tau.$$

Using continuous in mean for functions from $L_1(0, T; L_\infty(0, l))$ (see [9, chapter 4, § 1]) it is easy to check that

$$\int_0^T \|R(t, \cdot) - R^h(t, \cdot)\|_\infty\, dt \to 0, \quad h \to 0+. \tag{28}$$

We define a family of operators $\mathcal{B}_2^h : E \to E$ by the formula

$$\mathcal{B}_2^h p = \int_0^T R^h(t, x) u(t, x; p)\, dt.$$

Since $R^h \in C(0, T; L_\infty(0, l))$, $u \in C(Q)$ and estimate (24) hold it is obvious that $\mathcal{B}_2^h \in \mathcal{L}(E)$. By virtue of the estimates (24), (25), the operators $\mathcal{R}^h p \equiv R^h(t, x)u(t, x; p)$ are compact in E for any $t \in [0, T]$, since they are the products of the bounded operators on the compact operator. Therefore the operators $\mathcal{B}_2^h$ are also compact as they are the limits of Riemann integral sums in the norm of $\mathcal{L}(E)$.

The simplest estimate shows that according to (28)

$$\|\mathcal{B}_2 p - \mathcal{B}_2^h p\|_E \leqslant \text{const} \cdot \|p\|_E \cdot \int_0^T \big\|R(t, \cdot) - R^h(t, \cdot)\big\|_\infty dt \to 0, \quad h \to 0+.$$

Therefore $\mathcal{B}_2^h \to \mathcal{B}_2$ in $\mathcal{L}(E)$, and hence $\mathcal{B}$ is compact operator in E (see [7, p. 191]). Lemma is proved.

From lemmas 2, 4 we obtain the following statement well–known in the theory of inverse problems for uniformly parabolic equations (see [10, 11], etc.).

Theorem 3. *Let conditions (A)–(E) hold. Then the inverse problem* (20)–(22) *is equivalent to the operator equation of second kind with compact operator in E, i.e. this inverse problem has a Fredholm property.*

The proof of following lemma is done in [4].

Lemma 5. *Let conditions (A)–(E) and* (27) *hold. Suppose that* $\chi(t) > 0$, $t \in [0, T]$. *Then the spectral radius* $r(\mathcal{B})$ *of the operator* $\mathcal{B}$ *is less then one, i.e.* $r(\mathcal{B}) < 1$.

Now we obtain a theorem on unique solvability of the inverse problem (20)–(22).

Theorem 4. *Let conditions (A)–(E),* (27) *hold. Suppose that* $\chi(t) > 0$, $t \in [0, T]$, *and function* $\tilde{\varphi} \equiv \varphi - \mathbf{l}(z^0) \in W_2^2(0, l) \cap \overset{\circ}{W}{}_2^1(0, l)$. *Then there exists a generalized solution of inverse problem* (1)–(3). *Moreover, such a solution is unique and the estimate*

$$\|p\|_E \leqslant C \cdot \|\psi\|_E, \tag{29}$$

where $\psi(x) = -\tilde{\varphi}''(x) - b(x)\tilde{\varphi}'(x)$, *as well as estimates* (8), (9) *hold.*

Proof. By virtue of lemma 5 $r(\mathcal{B}) < 1$, therefore the operator equation of second kind (26) has a unique solution $p \in E$ and the estimate of stability (29) holds. Then by virtue of lemma 2 the inverse problem (20)–(22) has a generalized solution which is unique. The estimates for u follow from (29) and the estimates for the solution of direct problem from theorem 1.

4. Examples

Let us present some examples of inverse problems for which the above-proved theorems hold.

Example 1. Consider the inverse problem

$$(T-t)^{2/3}\,u_t - u_{xx} = p(x)T, \quad (t,x)\in Q, \tag{30}$$

$$u(0,x)=u_0(x), \quad x\in[0,l]; \qquad u(0,t)=u(l,t)=0, \quad t\in[0,T], \tag{31}$$

$$\frac{1}{T}\int_0^T u(t,x)\,dt = \varphi(x), \quad x\in[0,l], \tag{32}$$

with arbitrary functions $u_0 \in \overset{\circ}{W}{}_2^1(0,l)$ and $\varphi \in W_2^2(0,l)$, $\varphi(0)=\varphi(l)=0$.

Obviously, for this problem all the conditions of the theorem 4 are satisfied. Therefore the inverse problem (30)–(32) is uniquely solvable.

Example 2. Consider the inverse problem for the equation

$$(T-t)^{2/3}(x+1)\,u_t - u_{xx} + (T-t)^{1/3}\,u_x + (T-t)^{1/3}\,u = p(x), \quad (t,x)\in Q, \tag{33}$$

with initial and boundary conditions (31) and additional condition (32). It is easy to verify that for this inverse problem the conditions (A)–(E) are fulfilled. The condition (13) can be written in the form

$$36l^2\exp\Big(\frac{3T(\pi^2+l^2)}{\pi^2}\Big)\Big[\frac{4l^2(l^2+3l+3)}{9\pi^2T^{1/3}} + \frac{3T(\pi^2+l^2)}{5\pi^2}\Big] < 1.$$

Thus the condition (13) of theorem 2 is certainly valid if T is fixed and l is sufficiently small. So in this case the inverse problem (33), (31), (32) has a generalized solution which is unique.

The authors were partially supported by the Program of competitiveness increase of the National Research Nuclear University MEPhI (Moscow Engineering Physics Institute); contract No 02.a03.21.0005, 27.08.2013.

References

[1] Bouchouev I and Isakov V 1999 Uniqueness, stability and numerical methods for the inverse problem that arises in financial markets *Inverse Problems* **15** N3 R 95–116
[2] Lishang J and Yourshan T 2001 Identifying the volatibility of underlying assets from option prices *Inverse Problems* **17** N1 137–155
[3] Lishang J, Qihong C and Lijun W 2003 A new well-posed algorithm to recover implied local volatibility *Quantitative Finance* **3** N6 451–457
[4] Prilepko A I, Kamynin V L and Kostin A B 2018 Inverse source problem for parabolic equation with the condition of integral observation in time *Journal Inverse and Ill-Posed problems* **26** N4 523–539
[5] Kamynin V L 2015 On the solvability of the inverse problem for determining the right-hand side of a degenerate parabolic equation with integral observation *Math. Notes* **98** N5 710–724
[6] Kamynin V L 2017 Inverse problem of determining the right-hand side in a degenerating parabolic equation with unbounded coefficients *Comp. Math. and Math. Physics* **57** N5 833-842
[7] Lyusternik L A and Sobolev V I 1982 *Kratkii Kurs Funktsional'nogo Analiza* (Moscow, Vyssh. Shkola)
[8] Birman M S et al 1972 *Funktsional'ny Analiz.* 2nd ed. (Moscow, Nauka)
[9] Gajewski H, Gröger K und Zacharias K 1974 *Nightlineare Operatorgleichungen und Operatordifferentialgleichungen* (Akademie–Verlag, Berlin)
[10] Prilepko A I and Solov'ev V V 1987 Solvability theorems and the Rothe method in inverse problems for an equation of parabolic type II. *Differential Equations* **23** 1341–1349
[11] Prilepko A I and Tkachenko D S 2003 Fredholm property and well-posedness of the inverse source problem with integral overdetermination, *Comput. Math. Math. Phys.* **43** 1138–1347

IOP Conf. Series: Journal of Physics: Conf. Series **1205** (2019) 012024 doi:10.1088/1742-6596/1205/1/012024

Dynamics of spatially distributed delay logistic equation

I Kashchenko

Yaroslavl State University, Yaroslavl, Russia

E-mail: ikashchenko@yandex.ru

Abstract. The local dynamics of spatially distributed logistic equation with delay is studied. It is shown, that the critical cases in problem of equilibrium's stability have infinite dimention. For each of the critical case special replacement is built, which reduce the original problem to a system of parabolic equations — quasi-normal form, the which solutions' behavior defines local dynamics.

1. Introduction

Logistic equation with delay

$$\frac{dN}{dt} = r[1 - N(t-h)]N \quad (r, h > 0)$$

appears in a number of applications. This equation plays an especial role in modelling of mathematical ecology problems [1, 2]. Consider more complex model for description of quantity $N = N(t, x)$, which depends also on spatial variable $x \in (-\infty, \infty)$:

$$\frac{\partial N}{\partial t} = r\left[1 - \int\limits_{-\infty}^{\infty} F(s)N(t-h, x+s)\,ds\right]N + \delta^2 \frac{\partial^2 N}{\partial x^2}. \tag{1}$$

Here as above $r, h > 0$. We will consider that $N(t, x)$ is periodic in second variable

$$N(t, x + 2\pi) \equiv N(t, x). \tag{2}$$

We will focus our attention below on one type of kernel functions $F(s)$, which have a specified biological meaning [2]

$$F(x) = \frac{\sigma}{\sqrt{\pi}} e^{-\sigma^2 x^2}, \quad (\sigma > 0).$$

The graph of the $F(x)$ is shown on Fig. 1. Note that

$$\int\limits_{-\infty}^{\infty} F(x)\,dx = 1,$$

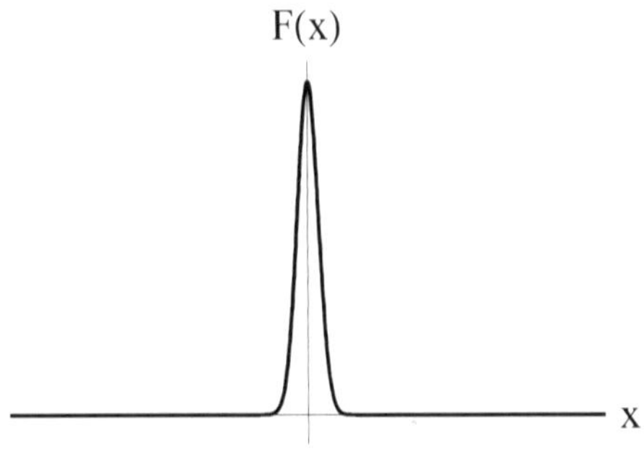

Figure 1. The graph of the function $F(x)$.

so problem (1), (2) have equilibrium state $N \equiv 1$. Let study the local dynamics.of (1), (2) in the neighborhood of this equilibrium state.

Main assumption that allows us using asymptotic methods of analysis is that the diffusion coefficient in (1) is a small parameter, and the values of $F(x)$ are small if x is not in the small neighborhood zero

$$\delta = \varepsilon d, \quad \sigma = \varepsilon^{-1} a, \quad 0 < \varepsilon \ll 1.$$

This means that (1), (2) is singular pertubed problem.

The method of research based on special method of quasynormal forms [3, 4]. The main idea of this method is to build special nonlinear equation, which dynamics describes behavior of initial equation's solutions in small neighbourhood of steady state.

2. Nonlinear analysis

The linearisation of the boundary value problem (1), (2) at the equilibrium $N \equiv 1$ gives a linear boundary value problem

$$\frac{\partial u}{\partial t} = -r \int\limits_{-\infty}^{\infty} F(s) u(t-h, x+s)\, ds + \varepsilon^2 d^2 \frac{\partial^2 u}{\partial x^2},$$

$$u(t, x+2\pi) \equiv u(t, x).$$

Using the eigenfunctions e^{ikx} of a periodic boundary value problem we obtain the following characteristic equation

$$\lambda = -r \exp(-\lambda h) \exp\left(-\frac{\varepsilon^2 k^2}{4a^2}\right) - d^2 \varepsilon^2 k^2, \quad k = 0, \pm 1, \pm 2, \ldots \tag{3}$$

The following lemma describes the root location for the equation (3).

Lemma. *Let $rh < \frac{\pi}{2}$, then all roots of the equation* (3) *have negative real parts and are bounded away from the imaginary axis for small values of ε. If $rh > \frac{\pi}{2}$, then the equation* (3) *has at least one root with a positive real part (and separated from 0 for $\varepsilon \to 0$)*

Let $rh = \frac{\pi}{2}$, then infinitely many roots of the equation (3) *tend to $i\frac{\pi}{2h}$ for $\varepsilon \to 0$.*

Thus (see [5]) for $rh < \frac{\pi}{2}$ the dynamics in a neighborhood of the equilibrium state $N \equiv 1$ is trivial: all solutions from its certain neighborhood tend to $N = 1$. For $rh > \frac{\pi}{2}$, the dynamics become nonlocal: there is no stable regime in the some small (but independent of ε neighborhood of equilibrium.

Let r_0 and h_0 be positive and such that $r_0 h_0 = \frac{\pi}{2}$. So the $r = r_0$, $h = h_0$ is the point of bifurcation. Consider

$$r = r_0 + \varepsilon^p r_1, \quad h = h_0 + \varepsilon^p h_1, \quad 0 < p \leq 2. \tag{4}$$

We consider only $0 < p \leq 2$, since for $p > 2$ all constructions are equivalent to the case $p = 2$ and $r_1 = h_1 = 0$. Note, that the case $p = 2$ is studied in [6].

From lemma it follows that under the condition (4) and for small enough ε we have the critical case of infinite dimension. Roots of (3), which tend to the imaginary axis, admit an asymptotic representation

$$\lambda_n = \frac{\pi}{2h_0} i + \frac{\varepsilon^p}{1 + i\pi/2}\left(\frac{r_0 h_1 \pi}{2h_0} + r_1 i - d^2 z^2 n^2 - \frac{r_0 z^2 n^2}{4a^2} i\right) + o(\varepsilon^p).$$

The critical spatial modes have the numbers $(z\varepsilon^{p/2-1} + \theta_z)n$, where $n \in \mathbb{Z}$, z is arbitrarily fixed, and $\theta_z = \theta_z(\varepsilon) \in [0, 1)$ such that $z\varepsilon^{p/2-1} + \theta_z$ takes integer values. However, if $p = 2$ then select $z = 1$ and $\theta_z = 0$.

Let us assume that in (1), (2)

$$N = 1 + \varepsilon^{p/2}\left(\exp(i\frac{\pi}{2h_0}t)u(\tau, y) + \exp(-i\frac{\pi}{2h_0}t)\overline{u}(\tau, y)\right) + \varepsilon^p u_2(t, \tau, y) + \varepsilon^{3p/2} u_3(t, \tau, y) + \ldots, \tag{5}$$

where $\tau = \varepsilon^p t$ – is a slow time, $y = (z\varepsilon^{p/2-1} + \theta_z)x$, the functions $u_j(t, \tau, x)$ are $4h_0$-periodic with respect to t. Put the formal series (5) into (1),(2) and collect coefficients at equal powers of ε. First, define $u_2(t, \tau, x)$, from the equation of coefficients at ε^p:

$$u_2 = \frac{2 - i}{5}\exp(i\frac{\pi}{h_0}t)u^2(\tau, y) + \frac{2 + i}{5}\exp(-i\frac{\pi}{h_0}t)\overline{u}^2(\tau, y).$$

Then, from the solvability condition of the relevant equation with respect to $u_3(t, \tau, x)$, we obtain a parabolic boundary value problem (without a small parameter) of the Ginzburg-Landau type

$$\frac{\partial u}{\partial \tau}(1 + \frac{\pi}{2}i) = z^2(d^2 + \frac{r_0}{4a^2}i)\frac{\partial^2 u}{\partial x^2} + (r_0\frac{\pi h_1}{2h_0} + r_1 i)u + \frac{r_0(1 - 3i)}{5}u|u|^2, \tag{6}$$

$$u(\tau, x) \equiv u(\tau, x + 2\pi). \tag{7}$$

The main result is the following. The problem (6), (7) is a quasinormal form of (1), (2).

Theorem. *To each periodic solution $u(\tau, y)$ of the problem (6), (7) for some z there corresponds solution (1), (2) with discrepancy up to $O(\varepsilon^{3p/2})$ uniformly with respect to $t \geq 0$, $x \in [0, 2\pi]$*

$$N(t, x) = 1 + \varepsilon^{p/2}\left(u(\varepsilon^p t, y)\exp(i\frac{\pi}{2h_0}t) + \varepsilon\overline{u}(\varepsilon^p t, y)\exp(-i\frac{\pi}{2h_0}t)\right) + $$
$$+ \varepsilon^p\left(\frac{2 - i}{5}\exp(i\frac{\pi}{h_0}t)u^2(\varepsilon^p t, y) + \frac{2 + i}{5}\exp(-i\frac{\pi}{h_0}t)\overline{u}^2(\varepsilon^p t, y)\right), \tag{8}$$

where $y = (z\varepsilon^{p/2-1} + \theta_z)x$.

Note, that if $p < 2$ then (8) is a family (depending on z) of solutions which fast oscillates in spatial variable around $N \equiv 1$.

3. Conclusion

We built special nonlinear parabolic Ginzburg-Landau type equations (that does not contain asymptotically small or large parameters) to find the main terms of asymptotic of initial equation's solutions. Existence of arbitrary parameter z in these equations leads to multistability. Constructing the main terms of asymptotic of solutions (1) can help to find solutions with any degree of accuracy.

Obtained results can be generalized to more complex equations — for example, to situation when spatial variable is two-dimentional [7].

Acknowledgments

This study is supported by the Russian Foundation for Basic Research (project no. 18-01-00672).

References

[1] Hutchinson G E 1948 *Ann. N.Y. Acad. Sci.* **50** 221–246
[2] Leven S and Segel L 1985 *SIAM Review* **27** 45–67
[3] Kaschenko S A 1989 *Differential equations* **25** 1448–1451
[4] Kashchenko I S 2008 *Computational Mathematics and Mathematical Physics* **48** 2172–2181
[5] Wu J 1996 *Theory and applications of partial functional differential equations* (New York: Springer-Verlag)
[6] Kashchenko I 2011 *Communications in Nonlinear Science and Numerical Simulation* **16** 3520–3524
[7] Kashchenko I S 2013 *Automatic Control and Computer Sciences* **47** 516–525

IOP Conf. Series: Journal of Physics: Conf. Series **1205** (2019) 012025 doi:10.1088/1742-6596/1205/1/012025

Developing mathematical method for multi alignment of DNA sequences with weak similarity

E V Korotkov[1,2] and M A Korotkova[2]

[1]Institute of Bioengineering, Research Center of Biotechnology of the Russian Academy of Sciences, Leninsky Ave. 33, bld. 2, 119071, Moscow, Russia
[2]National Research Nuclear University "MEPhI", Kashirskoe shosse, 31. Moscow 115409, Russia.

genekorotkov@gmail.com

Abstract. A new mathematical method is proposed for constructing multiple alignment of weakly similar amino acid or nucleotide sequences. The method uses a multiple alignment representation in the form of position-weight matrices (PWM) and global dynamic programming. An optimization procedure for PWM is developed with the aim of finding a multiple alignment with maximum statistical significance. The method allows to find multiple alignment of sequences with the number of nucleotide or amino acid substitutions more than 2.5. The developed approach is applied to obtain multiple alignment of promoter sequences of 600 DNA bases length.

1. Introduction

At present, a large number of DNA sequences from various species are known. Therefore, the problem of determining the functional value of accumulated sequences, that is, their annotation, is topical. The problem is put in this way: it is necessary to find some statistical regularities (*ST*) in sequences (*Seq*), which perform the same biological function (we denote it as *F*), show the specificity of *ST* only to *Seq*, and then develop a mathematical method for finding the revealed regularities in new sequences.
If *ST* are found somewhere, then we can say that with a certain probability we have found a sequence that has a biological function *F*. Usually in the role of *ST* appears a statistically significant multiple alignment of sequences that have the same biological function *F*. After constructing the multiple alignment, a hidden Markov model is developed, which then analyzes the new sequences for *ST* detection. It turns out that the effectiveness of the *ST* search depends on the quality of multiple alignment, ie, the efficiency of the annotation of sequences of different genomes.

2. Methods and algorithms for multiple alignment creation

The problem of multiple alignment of amino acid and nucleotide sequences is one of the central problems of bioinformatics. Because of this, it has attracted much attention over the last 25 years and many mathematical methods, algorithms and software have been developed to solve it. This is the so-called NP-complete problem and its direct solution by dynamic programming methods is impossible because of the lack of huge computer resources. It is unlikely that such resources will appear in the coming decades. Therefore, this problem has some approximate solutions. In these solutions, the so-called objective function is used to evaluate the quality of the created multiple alignment and an optimization procedure that allows improving the quality of the already constructed multiple alignment. Common algorithms for multiple leveling contain the following steps. 1. N sequences are input; 2. A pair comparison of the sequences is performed and the matrix of distances between the sequences is filled; 3. A guide tree is constructed based on the distance matrix; 4. Based on the distance matrix, a progressive multiple alignment is created; 5. For the created progressive alignment, an optimization procedure is applied; 6. There is a visualization of the created multi-leveling and final optimization. The objective function can be the value of the sum of pairs of similarity functions for pair alignments, which can be obtained by projecting the constructed multiple alignment into two-dimensional planes [1]. In total, $N(N-1)/2$ pair pairings are obtained. There are also alternative ways of specifying the objective function as a consensus [2], entropy [3] or a circular sum [4]. Let's briefly review some of the methods used to find multiple alignments.

1. Using n-dimensional dynamic programming. In this case, the space is trimmed to find the optimal alignment. This means that the similarity function is only populated for part of the space in n-dimensional dynamic programming. In the algorithm [4], pair alignments are performed for each pair of sequences by the dynamic programming method, and only the space next to the n-dimensional intersection of these found optimal pair alignments is used for n-dimensional alignment. In the DCA approach [5], the long multiple alignments are divided into shorter ones, where n-dimensional dynamic programming is possible. In practice, MSA and DCA methods are rarely used, since they are applicable only for a relatively small number of sequences (no more than 10-12).

2. Application of global optimization procedures in the construction of multiple alignment. A genetic algorithm is widely used to build multiple alignment. In this case, the multiple alignment or part of it is the organism for the genetic algorithm. To such multiple alignment operations of symbol replacement, deletion or insertion are applied. The developed programs include the programs SAGA, MAGA and PHGA. The shortcomings of the genetic algorithm include a long time of searching for optimal multiple alignment. And the less similar the sequence, the more evolution steps are needed to find the optimal alignment. In the case of weakly similar sequences, which are considered in this paper, the time becomes extremely large, which makes it impossible to find multiple alignment for weakly similar sequences.

3. Progressive alignment. This is one of the most popular methods for building multiple alignment. In the popular progressive alignment strategy [6], sequences that need to be aligned are attributed to individual leaves of a binary tree (known as the alignment tree). Further, the inner nodes of the tree are bottom-up, and a visit to each node is associated with multiple alignment of the sequences in the corresponding sub-tree. Further, internal nodes of the tree are analyzed from the bottom up, and each site visit is associated with multiple alignment of sequences in the corresponding sub-tree. At the end of the traversal, a multiple alignment is created, linked to the root node. Among the disadvantages of progressive alignment is that if a alignment is constructed for a certain group of sequences, then it is not already rebuilt. The created programs include DFALIGN, MULTAL, MAP, PCMA, PIMA, PRIME, PRRP, MULTALIN, CLUSTALW, MAFFT, MUSCLE, T-Coffee, KAlign, POA, PROBCONS and MUMMALS/PROMALS. These programs are commonly used in the search for distant similarities and for predicting the spatial structure of the amino acid sequence. The effectiveness and simplicity of progressive alignment algorithms has led to the widespread use of these algorithms and programs in modern molecular biology and bioinformatics. However, progressive alignment strategies can produce significant errors due to inaccuracies in the construction of the guide tree. Or there may be an accumulation of errors from the early stages of pairwise alignment. Therefore, a number of heuristic methods used in modern MSA programs were proposed to overcome the shortcomings of progressive alignment. Some of these methods try to improve the steering tree and make it similar to a tree of evolutionary distances. Other improvements concern the choice of the similarity measure between sequences. For example, MUSCLE, MAFFT, MUMMALS / PROMALS used a quick measure to find similarities between sequences, which consisted in counting k-words. Therefore, replacing this measure with a complete pairwise alignment of the sequences made it possible to improve substantially the alignment, but the software began to work much more slowly. It

also uses the iterative allocation of the directing tree to obtain a more accurate distance between pairs of sequences. This is used in the algorithms MAFFT, MUSCLE, PRIME, PRRP, and MUMMALS / PROMALS. A new guide tree is used to build a new multiple alignment and this process can be repeated. Also, the improvement of progressive multiple alignment went along the path of modification of the objective function. The idea is that if a pair alignment of the sequences *x* and *y* is performed and *i* is the position of the sequence *x* and *j*, the position of the sequence *y* are similar and they are similar to the *k* position of some multiple alignment of *Z*, this is strong evidence that in the final multiple alignment of position *i*, *j* and *k* must be in the same column. It turns out that the objective function should change from the kind of multiple alignment. This approach is implemented in algorithms for programs T-coffee, Dialign, Probcons, INF, Mummals, Promals and Align-m. The objective function for pair alignment is also improved by introducing position-specific penalties for the creation of inserts and deletions. The idea is to reduce the penalty for insertion or deletion at a certain position in pair alignment, if in the pair alignments of other sequences the deletion or insertion will be present at the same place. This applies to programs CLUSTALW, MUSCLE, MAFFIT. Another way to improve progressive alignment is to weight the sequences. In a plurality of sequences that undergo multiple equalization, there may be a group of very similar sequences. In this case, insertions and deletions of this group will be projected onto all the multiple alignment. This means that this group will be represented too much in the multiple alignment. Therefore, in the beginning, it is better to weigh the sequences to construct a progressive multiple alignment. There are many algorithms for how to weigh sequences in multiple alignment [7]. To improve progressive alignment, the so-called "post-processing" is also important, which is accomplished by iteratively rearranging the resulting alignment. The task here is to get rid of the errors that occur at the initial stages of building a multiple progressive alignment. The final quality of the alignment can be significantly improved. For example, it is possible to split the multiple alignment into two parts and these parts are re-aligned and then the final multiple alignment is constructed. This process is made iterative and the division into two groups is random. The process can stop when there is no significant improvement in the objective function. Programs ITERALIGN, TULLA, AMPS / AMULT, MULTAN, PRRP, PROBCONS, MUSCLE and MAFFIT include this iterative procedure.

3. Use of hidden Markov models. Hidden Markov models (HMM) are also used to construct multiple alignment. (Hughey & Krogh, 1996) However, Markov's models, including hidden ones, have the drawback that this "replicating" method is inherently. This means that you must already have a solution to the problem (that is, build in some way a multiple alignment) and determine the probabilities of generating insertions or deletions, the probability of symbol generation, and the probability of a transition between HMM states. In advance these probabilities are not known and therefore using only HMM to create the optimal (from the point of view of the objective function) multiple alignment is impossible. Therefore, the application of HMM to solve the problem of multiple alignment is accompanied by the application of a number of other optimization procedures and mathematical methods, often of an iterative nature. These include Baum-Welsh algorithm, simulation of "annealing", deterministic "annealing", gradient descent method and some other methods. After optimizing procedures, the optimal HMM is created and all sequences are aligned with this HMM using the Viterbi algorithm to find the best similarity between each sequence and the HMM.

4. Multiple alignment of sequences with a low level of similarity (less than 35% identity). For such sequences, it is advisable to use the programs T-coffee, Probcons, Maffit, MUMMALS, ProbAlign), Amap) as the most accurate ones for constructing multiple alignment of weakly similar sequences.

Mathematical algorithm and results. In this paper, a mathematical algorithm has been developed for carrying out multiple alignment for nucleotides for which there is no statistically significant similarity between pairs of sequences. This alignment can't be found by all previously developed algorithms. In this case, for aligned sequences, it is not possible to construct a dendrogram (guide tree), as well as finding statistically significant "seeds" or similarity between "words". We have also previously developed an optimization procedure that allows the use of randomly created position-weighted matrices (PVMs) to find multiple alignment of sequences using a genetic algorithm [8]. By this method, we could already detect the existence of multiple alignment for tandem sequences in the currency series for the euro / US $ exchange rate and the existence of tandem similar sequences in amino acid and nucleotide sequences [8]–[10]. In this paper, this approach has been applied to promoter sequences. Promoter sequences are sequences that are at the beginning of each gene and they regulate transcription processes. Promoters have a length of several hundred bases and multiple alignment for promoters from one genome has not been built. However, it is possible to have a statistically significant multiple alignment in the absence of pair alignment of the sequences. In this case, statistically significant alignment becomes a group property that is present only for a certain set of sequences with a number of sequences in a set greater than two. In this case, all previously developed methods for constructing multiple equalization can't find it at a statistically significant level. To build multiple sequence alignment, we used PWM, genetic algorithm, and two-dimensional dynamic programming [8]. The method was used to align more than 15,000 promoter sequences from the A.thaliana genome. Each promoter had a length of 600 bases of DNA. Alignment was constructed for 500 randomly selected promoter sequences from more than 15,000. To do this, 15,000 selected sequences were combined into one sequence S and for this sequence a two-dimensional alignment was constructed using random PWMs. The Niedelman-Wunsch algorithm [11] was used to build the alignment. Then, using a genetic algorithm, a PWM was constructed that has the greatest value of the similarity function when the sequence S is aligned. Then, from the resulting two-dimensional alignment, we restored the multidimensional alignment for 15,000 promoter sequences. After that, we used the best PWM to select the sequences that statistically enter into the created multiple alignment. For this, profile analysis was used. As a result, the created multiple alignment is statistically significant and contains only sequences that have a nonrandom similarity with the best PWM. We were able to obtain a statistically significant alignment of 75% of the promoter sequences (more than 15 thousand pieces) from the A.thaliana genome.

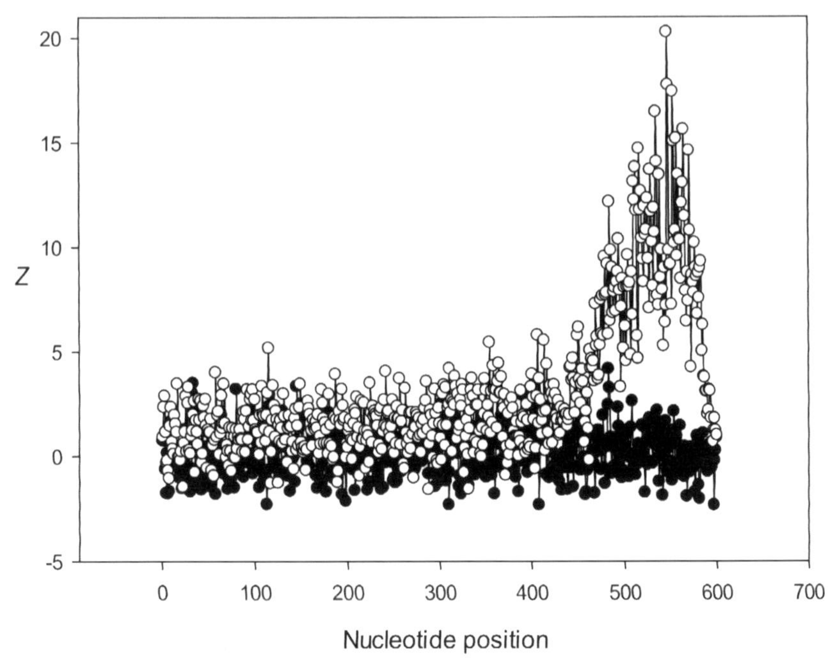

Fig.1 The safety of each position of promoters in multiple alignment is shown. Explanations are below in the text.

The safety of each position of the promoters in the multiple alignment is shown in Fig.1.The coordinate of each nucleotide is plotted along the abscissa axis. Position 501 corresponds to the first codon base in the gene. It turns out that positions 1-500 belong to non-coding regions of DNA, positions 501-600 belong to the coding regions of the gene. The ordinate shows the safety of each position as an argument of normal distribution. Using Z, we can calculate the probability $p=P(x> Z)$. This probability is the probability that the difference in the frequencies of nucleotides in position from the frequencies of nucleotides in the entire sequence is due to purely random factors. This means that the larger the Z, the less likely that the frequencies of the nucleotides in the position coincide with the averaged base frequencies for the entire promoter. Black circles are obtained for multiple alignment of the same number of random sequences. Light circles correspond to the multi alignment of promoter sequences. From Fig. 1 it can be seen that the most conservative sites fall on regions from 501 to 580 bases of DNA. You can also note the conservative section at a distance of -30 (470 base in Fig. 1), which corresponds to the TATA region. It is also evident that in the region from 1 to 400 bases there are many conservative positions. This shows that the promoter sequences of these positions are similar to each other. It also turned out that five classes of multi alignment can be created for approximately 15,000 promoter sequences. The results obtained can be used to search for promoter sequences in other eukaryotic genomes. We got the number of errors of the first kind when predicting promoter sequences no more than 0.1%. This will allow more accurate annotation of promoter sequences and genes in eukaryotic genomes.

References

[1] M. Murata, J. S. Richardson, and J. L. Sussman, "Simultaneous comparison of three protein sequences.," *Proc. Natl. Acad. Sci. U. S. A.*, vol. 82, no. 10, pp. 3073–7, May 1985.

[2] M. S. Waterman and R. Jones, "Consensus methods for DNA and protein sequence alignment.," *Methods Enzymol.*, vol. 183, pp. 221–37, 1990.

[3] R. Durbin, S. R. Eddy, A. Krogh, and G. Mitchison, *Biological Sequence Analysis: Probabilistic Models of Proteins and Nucleic Acids*. Cambridge University Press, 1998.

[4] G. H. Gonnet, C. Korostensky, and S. Benner, "Evaluation Measures of Multiple Sequence Alignments," *J. Comput. Biol.*, vol. 7, no. 1–2, pp. 261–276, Feb. 2000.

[5] A. Dress, G. Füllen, and S. Perrey, "A divide and conquer approach to multiple alignment.," *Proceedings. Int. Conf. Intell. Syst. Mol. Biol.*, vol. 3, pp. 107–13, 1995.

[6] D. F. Feng and R. F. Doolittle, "Progressive sequence alignment as a prerequisite to correct phylogenetic trees.," *J. Mol. Evol.*, vol. 25, no. 4, pp. 351–60, 1987.

[7] S. Henikoff and J. G. Henikoff, "Position-based sequence weights.," *J. Mol. Biol.*, vol. 243, no. 4, pp. 574–8, Nov. 1994.

[8] V. Pugacheva, A. Korotkov, and E. Korotkov, "Search of latent periodicity in amino acid sequences by means of genetic algorithm and dynamic programming," *Stat. Appl. Genet. Mol. Biol.*, vol. 15, no. 5, pp. 381–400, Jan. 2016.

[9] E. V. Korotkov and M. A. Korotkova, *Search of regions with periodicity using random position weight matrices in the genome of c. Elegans*, vol. 10209 LNCS. 2017.

[10] E. V. Korotkov and M. A. Korotkova, "Study of the periodicity in Euro-US Dollar exchange rates using local alignment and random matrices," *Algorithmic Financ.*, 2017.

[11] S. B. Needleman and C. D. Wunsch, "A general method applicable to the search for similarities in the amino acid sequence of two proteins.," *J. Mol. Biol.*, vol. 48, pp. 443–453, 1970.

Mathematical modeling of free convection problems in a gravity field in OpenFOAM

V K Kozlov[1] and M A Chmykhov[2]

National Research Nuclear University MEPhI (Moscow Engineering Physics Institute)

E-mail: [1]`viktorkozlovkonst@gmail.com`, [2]`MAChmykhov@mephi.ru`

Abstract. A mathematical model of natural convection in the gravity field, using Boussinesq approximation, has been presented. This model contains the continuity equation, where density variations are ignored, the Navier-Stokes equation and the equation for heat flow. Rayleigh-Benard-convection in a rectangular box, whith different types of border conditions, has been investigated. The equations solved numerically by an original solver with the use of object-oriented programming language OpenFOAM. Solver is based on PISO(Pressure-Implicit with Splitting of Operators) algorithm and finite volume method.

The Prandtl number, Grashof number and Rayleigh number has been examined. Rayleigh-Benard-convection between parallel planes of different temperatures, steady convection in horizontal fluid layer and natural convection flow in a square box, enclosed by non-isothermal wall has been used for solver verification. Results have been visualized with the use of open source application ParaView.

1. Introduction

Experiments of Benard (1900) initiated the study of convection. He observed the appearance of hexagonal honeycomb structures (Benard cells). Later Relay (1916) theoretically investigated the occurrence of convection in a horizontal fluid layer for two free boundaries. He established a transition from the thermal conductivity to the convection (the critical Rayleigh number). Further development was severely limited due to significant computational difficulties.

Lately, the study of convection processes developed very rapidly. Advances in the study of convection are associated with the progress of numerical calculations and with surviving interest in convection.

2. Mathematical model

A mathematical model of natural convection in the gravity field is determined by the continuity equation, where density variations are ignored, the Navier-Stokes equation and the equation for heat flow. In the Boussinesq approximation, they have the following form [1]

$$\begin{cases} div\vec{U} = 0, \\ \rho(\frac{\partial \vec{U}}{\partial t} + (\vec{U}\bigtriangledown)\vec{U}) - \nu\Delta\vec{U} = -\bigtriangledown p + g[\beta(T - T_0)], \\ \frac{\partial T}{\partial t} + \vec{U} \bigtriangledown T - \alpha\Delta T = 0, \end{cases} \tag{1}$$

where β — coefficient of thermal expansion of fluid, T_0 — temperature, α — the thermal diffusivity, ν — effective kinematic viscosity, ρ — the density of the fluid.

First boundary conditions :

$$\begin{cases} T(0,y,t) = T_H,\ T(l,y,t) = T_c,\ 0 \le y \le h,\ t > 0, \\ \frac{\partial T}{\partial y}|_{y=0} = \frac{\partial T}{\partial y}|_{y=h} = 0,\ 0 \le x \le l,\ t > 0, \\ \vec{U}|_{\partial\Omega} = 0, \\ p = f(T|_{\partial\Omega},\ \vec{U}|_{\partial\Omega}). \end{cases} \tag{2}$$

Second boundary conditions:

$$\begin{cases} T(x,0,t) = T_c,\ T(x,h,t) = T_H,\ 0 \le x \le l,\ t > 0, \\ \frac{\partial T}{\partial x}|_{x=0} = \frac{\partial T}{\partial x}|_{x=l} = 0,\ 0 \le y \le h,\ t > 0, \\ \vec{U}|_{\partial\Omega} = 0, \\ p = f(T|_{\partial\Omega},\ \vec{U}|_{\partial\Omega}). \end{cases} \tag{3}$$

Third boundary conditions:

$$\begin{cases} T(x,0,t) = T_H,\ T(x,h,t) = T_c,\ 0 \le x \le l,\ t > 0, \\ T(0,y,t) = T_c + (T_H - T_c)(h - y),\ 0 \le y \le h,\ t > 0, \\ T(l,y,t) = T_c,\ 0.05 \le y \le h,\ t > 0, \\ T(l,y,t) = T_c + (T_H - T_c)\frac{(0{,}05-y)}{0{,}05},\ 0 \le y \le 0.05,\ t > 0, \\ \vec{U}|_{\partial\Omega} = 0, \\ p = f(T|_{\partial\Omega},\ \vec{U}|_{\partial\Omega}). \end{cases} \tag{4}$$

Where f - Navier-Stokes equation solved with respect to pressure.

Initial condition:

$$\begin{cases} T(x,y,t)|_{t=0} = T_0,\ 0 \le x \le l,\ 0 \le y \le h, \\ \vec{U}(x,y,t)|_{t=0} = 0, \\ p(x,y,t)|_{t=0} = \rho g(h - y). \end{cases} \tag{5}$$

These equations solved numerically by an original solver with the use of object-oriented programming language OpenFOAM. To solve Navier-Stokes equation, PISO(Pressure-Implicit with Splitting of Operators) algorithm has been used [2]. It splits the operators into an implicit predictor and multiple explicit corrector steps, without iterations.

3. Solver verification

3.1. The case of a vertical fluid layer, task (1),(2),(5)

Solution of steady free convection in an infinite vertical fluid layer between parallel planes of different temperatures, has already been investigated [3]. Test case considered a rectangular enclosure of height h and length l, with different ratio ($\frac{h}{l} = 1; 2; 4; 8$). Results confirmed the validity of the solver as shown in Fig. 1.

3.2. The case of a horizontal fluid layer, task (1),(3),(5)

As for the horizontal fluid layer, the onset of steady convection has been calculated. It was determined by the critical value of the Rayleigh number. The Rayleigh number is defined as the product of the Grashof number and the Prandtl number. When the Rayleigh number is below

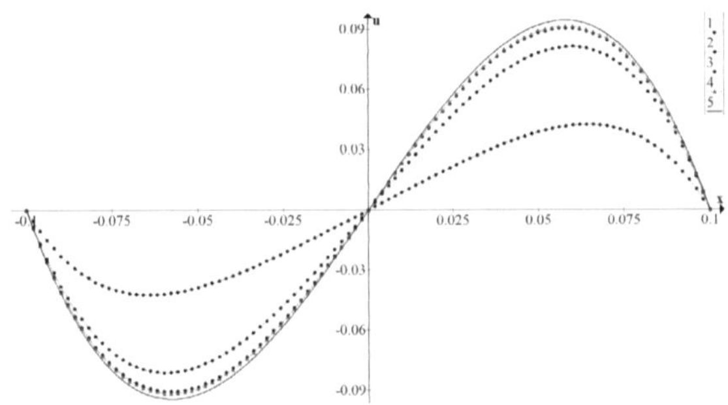

Figure 1. Comparison between numerical and exact solution, where 1 — numerical solution for $h/l = 1$, 2 — $h/l = 2$, 3 — $h/l = 4$, 4 — $h/l = 8$, 5 — exact solution and $l = 0{,}2$ m.

a critical value, heat transfer is primarily in the form of conduction, when it exceeds the critical value, heat transfer is primarily in the form of convection. Was established, that critical value of the Rayleigh number belongs to the range of values between 1949 and 1952 3 (theoretical value 1708, experimental value [5] 1820).

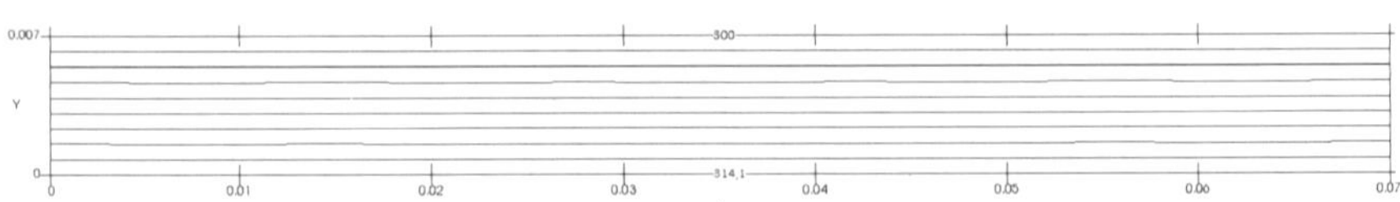

a) $Ra = 1949, \quad \Delta T = 14.1\ K, \quad U_{max} = 10^{-6}\ m^2/s$

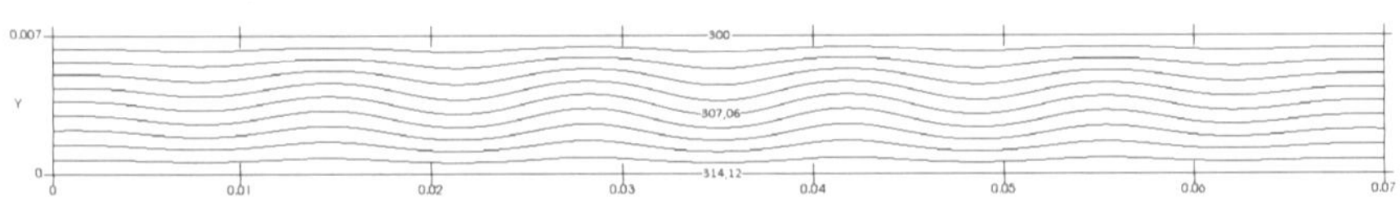

b) $Ra = 1952, \quad \Delta T = 14.12\ K, \quad U_{max} = 10^{-3}\ m^2/s$

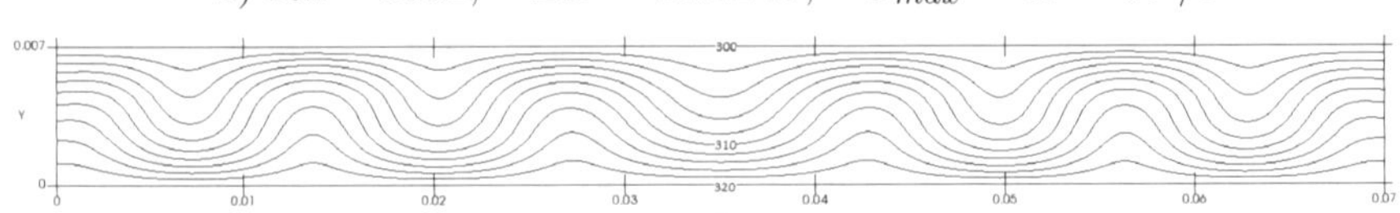

c) $Ra = 2756, \quad \Delta T = 20\ K, \quad U_{max} = 10^{-2}\ m^2/s$

Figure 2. Horizontal fluid layer, top wall cooler then the bottom one. Rayleigh number Ra, temperature difference ΔT and maximum speed U_{max} specified under the pictures.

3.3. The case of natural convection in square encloser, task (1),(4),(5)

The content of the article [6] was used for verification. The Navier-Stokes equation in the Boussinesq approximation in porous medium has the following form

$$\rho(\frac{\partial \vec{U}}{\partial t} + (\vec{U}\bigtriangledown)\vec{U}) - \nu\Delta\vec{U} = -\bigtriangledown p - \frac{\nu}{K}\vec{U} + g[\beta(T - T_0)]. \tag{6}$$

First, the value of the permeability coefficient was taken as infinity $K \approx \infty$. It was done to compare the content of the article [6] with results from the solver without adding modifications.

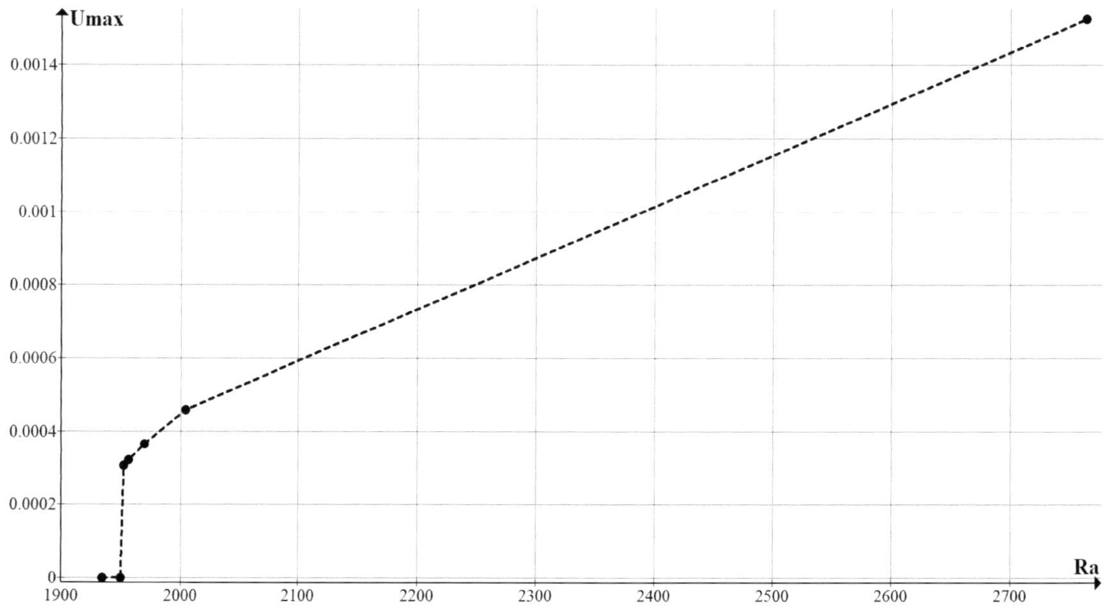

Figure 3. Variation of maximum velocity with Rayleigh number.

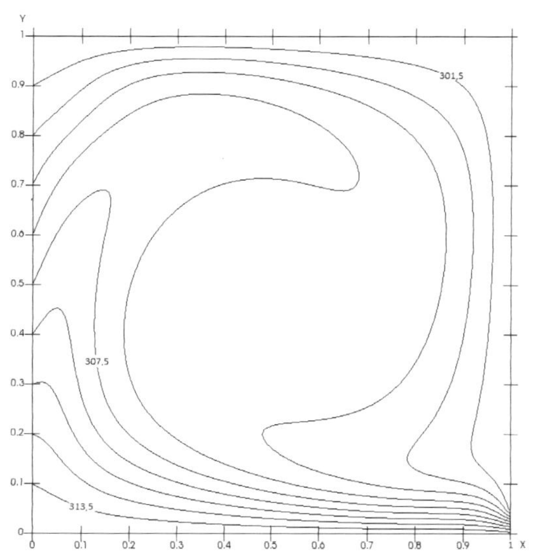

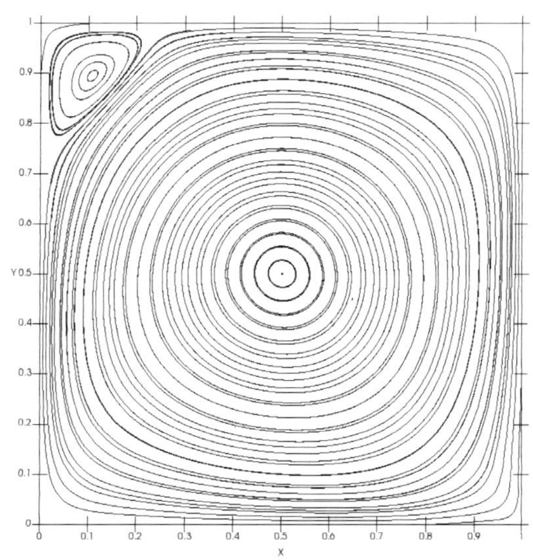

Figure 4. Dimensionless numbers $Pr = 0.7$, $Ra = 10^5$ Permability coefficient $K \approx \infty$, isotherms.

Figure 5. Dimensionless numbers $Pr = 0.7$, $Ra = 10^5$ Permability coefficient $K \approx \infty$, streamlines.

Second, the value of the permeability coefficient was taken as $K = 0.001$. And modified solver was used. Constants were taken with the values indicated in the Table 1.

In Fig. 4, 5 results of the task (1), (4), (5) with $K \approx \infty$. In Fig. 6, 7 results with $K = 0.001$. Two oppositely directed fluid flows are obtained, which converges with the simulation results taken for verification. In addition, the obtained isotherms have the form presented in the article [6], which allows to draw a conclusion about the correctness of the implemented solver.

Table 1. Values of constants and dimensionless numbers for the task (1),(4),(5)

$\nu = 7 \cdot 10^{-4}\ m^2/s$	$Pr = 0,7$
$a = 1 \cdot 10^{-3}\ m^2/s$	$Gr = 1,5 \cdot 10^5$
$\beta = 5 \cdot 10^{-4}\ K^{-1}$	$Ra = 1,05 \cdot 10^5$
$g = 9,81\ m/s^2$	$l = h = 1\ m$

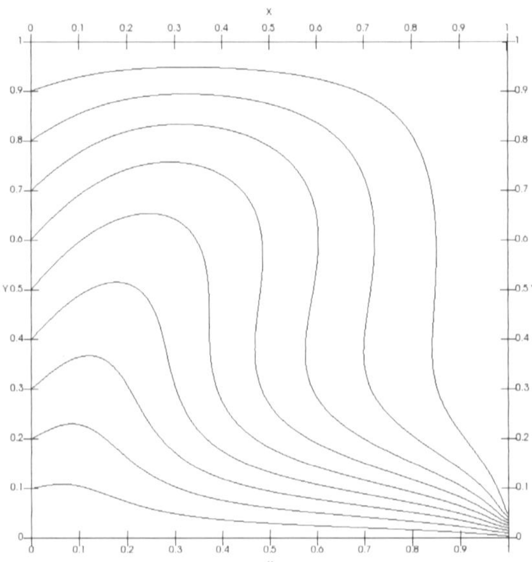

Figure 6. Dimensionless numbers $Pr = 0.7$, $Ra = 10^5$ Permability coefficient $K = 0.001$, isotherms.

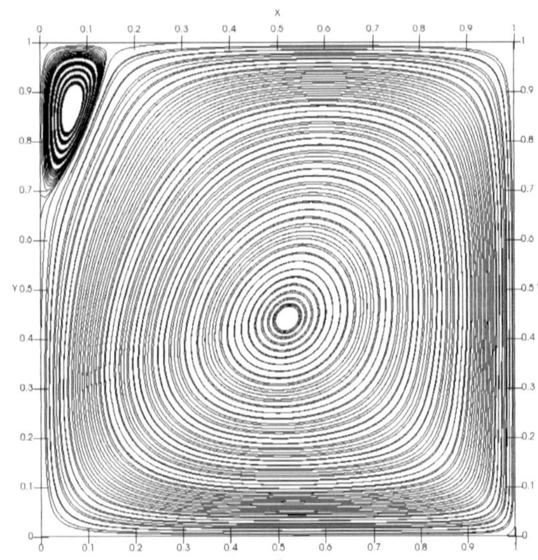

Figure 7. Dimensionless numbers $Pr = 0.7$, $Ra = 10^5$ Permability coefficient $K = 0.001$, streamlines.

4. Conclusion

The processes of natural convection in the field of gravity in the Boussinesq approximation has been examined. A mathematical model and numerical simulation has been made. Original solver and test cases were implemented using OpenFOAM. Simulation results for solver verification has been presented.

This work was supported by the Ministry of Higher Education and Science of the Russian Federation within the framework of the base part of state task no. 1.9746.2017/BP.

References

[1] Gershuni G.Z., Zhukhovitskii E.M. 1976 *Convective Stability of Incompressible Fluids.* Published by I.P.S.T. ISBN: 0706515625

[2] R.I.Issa 1986 Solution of the implicitly discretised fluid flow equations by operator-splitting. *Journal of Computational Physics* (ISSN 0021-9991), vol. 62, p. 40-65.

[3] Kozlov V. K., Kudryashov N. A., and Chmykhov M. A. Mathematical Modeling of the Convection in the Gravity Field, using Boussinesq Approximation in OpenFOAM - *Vestnik Natsional'nogo issledovatel'skogo yadernogo universiteta "MIFI"*, 2018, vol. 7, no. 4, pp. 347–355

[4] Issa R.I. 2001 An Improved Piso Algorithm for the Computation of Buoyancy-Driven Flows. *Numerical Heat Transfer: Part B: Fundamentals*, vol. 40, issue 6, pp. 473-493.

[5] Buhler K., Kirchartz K.R., Oertel H. 1979 Steady convection in a horizontal fluid layer *Acta Mechanica* 31 pp. 155

[6] Hossain M.A. and Wilson M. 2002 Natural convection flow in a fluid-saturated porous medium enclosed by non-isothermal walls with heat generation *International Journal of Thermal Sciences* 41 pp. 447–454

IOP Conf. Series: Journal of Physics: Conf. Series **1205** (2019) 012027 doi:10.1088/1742-6596/1205/1/012027

Modeling of acetylene detonation in a shock tube by the large particle method with TVD correction

I M Kozlov[1], **N I Misyuchenko**[1] **and A V Teterev**[1,2]

[1] Belarusian State University, Nezavisimosti Av., 4, 220030 Minsk, Belarus
[2] Department of Applied Mathematics, National Research Nuclear University MEPhI
31 Kashirskoe Shosse, 115409 Moscow, Russian Federation

teterev@bsu.by

Abstract. A one-dimensional TVD correction scheme is described, which can be successfully applied not only in 1D calculations, but also in 2D - 3D modeling [1-2]. The advantage of the proposed scheme is the lack of solving matrix equations. The implementation of the scheme in conjunction with the large particle method was tested on the problems of arbitrary discontinuity breakup and strong point explosion. The modeling of acetylene combustion in a shock tube and its transition to detonation was performed depending on the conditions set on the surface of the tube, and the parameters of the problem.

1. Introduction

The scheme of the large particle method [3] consists of several stages corresponding to the scheme of splitting into processes. We use the TVD scheme as the most last stage of this method to correct the solution obtained after the execution of the previous stages. Hence the name of the method includes – «TVD correction». The large particle method modified in this way allows to increase the accuracy of the approximation to the second order, as well as to get rid of the need for the use of artificial viscosity.

Verification of the joint operation of the large particle scheme and the TVD correction was performed on the problem of decay of an arbitrary discontinuity and the problem of a strong point explosion. Although these test problems are one-dimensional in nature, they were simulated using a two-dimensional axisymmetric formulation, which allowed validating the applicability of one-dimensional TVD scheme in 2D modeling.

Numerical simulation in a shock tube was performed for stoichiometric mixture of enriched air and acetylene represented by four elements: air – $2.5O_2 + 5.2N_2$; acetylene – C_2H_2; nitrogen – N_2; combustion products – H_2O+2CO_2. The proposed physical-chemical model uses tabular descriptions of the equilibrium thermodynamic state for each component of the gas mixture separately, but they do not have to be in a state of chemical equilibrium with respect to each other and can participate in mutual chemical conversion.

2. Mathematical model

The numerical method is based on the solution of a system of Euler equations written in the form of the mass, momentum, and energy conservation laws. The two-dimensional Euler equations can be represented as a system of conservation laws recorded in cylindrical coordinates (z,r) as follows:

$$\frac{\partial r\mathbf{U}}{\partial t}+\frac{\partial r\mathbf{F}(\mathbf{U})}{\partial z}+\frac{\partial r\mathbf{G}(\mathbf{U})}{\partial r}=\mathbf{S}(\mathbf{U}) \tag{1}$$

where $\mathbf{U}$ is the column vector of the conserved quantities, $\mathbf{F}$ and $\mathbf{G}$ are the column vectors of flows along the coordinates z and r, respectively, and $\mathbf{S}$ is the source vector. The number of components in the vectors is determined by the number of equations in system (1); this number, in turn, depends on the used physical model. When describing the processes of chemical conversion in combustible mixtures, the equations of conservation of the corresponding chemical components have to be added. Then the number of equations in (1) must be equal to $N+4$, where N is the number of components of the mixture. The vector quantities from the system (1) can be written as:

$$\mathbf{U}=\begin{bmatrix}\rho\\ \rho u\\ \rho v\\ \rho E\\ \rho Y\end{bmatrix},\quad \mathbf{F}=\begin{bmatrix}\rho u\\ \rho u^2+p\\ \rho uv\\ (\rho E+p)u\\ \rho uY\end{bmatrix},\quad \mathbf{G}=\begin{bmatrix}\rho v\\ \rho uv\\ \rho v^2+p\\ (\rho E+p)v\\ \rho vY\end{bmatrix},\quad \mathbf{S}=\begin{bmatrix}0\\ 0\\ p\\ -r\nabla\mathbf{q}\\ r\rho W\end{bmatrix} \tag{2}$$

where ρ is the density of the substance; u and v are the components of the velocity vector, respectively, in z and r coordinates; E is the total specific energy; p is the pressure; $Y\equiv\{Y_1,...,Y_N\}$ – the mass fraction vector of the chemical components; $W(T,\rho,Y)\equiv\{W_1,...,W_N\}$ – the vector of the chemical reaction rate; $\mathbf{q}$ – thermal conductivity flow. To close the system of equations (1)–(2), one should specify the equation of state or the system of equations by which the pressure in the form $p=p(T,\rho,Y)$ and its derivatives can be calculated:

$$\chi=\left(\frac{\partial p}{\partial\rho}\right)_{\rho\varepsilon,Y},\quad k=\left(\frac{\partial p}{\partial\rho\varepsilon}\right)_{\rho,Y},\quad \eta_i=\left(\frac{\partial p}{\partial Y_i}\right)_{\rho,\rho\varepsilon},\ \text{were}\ \varepsilon=E-\frac{u^2+v^2}{2} \tag{3}$$

The pressure derivatives determined in (3) are needed for calculating the TVD dissipation, in particular, of the local speed of sound:

$$c^2=\chi+k(\varepsilon+p/\rho) \tag{4}$$

The temperature used to calculate the partial pressure and the energy of the mixture components is determined based on the energy conservation equation for the mixture in a given calculation cell:

$$\varepsilon=\sum_i^N Y_i\varepsilon_i(T,Y_i\rho) \tag{5}$$

The numerical simulation used the functions $\varepsilon_i=\varepsilon_i(T,\rho)$ and $p_i=p_i(T,\rho)$ obtained in the form of tables, calculated with a given grid upon the parameters of temperature and density. A uniform temperature scale ranging from 200 K up to 6000 K with increments of 50 K and a density scale ranging from 10^{-8} up to $10^{2.5}$ kg/m^3 with a uniform logarithmic step of 0.25 were used. For a realistic description of the thermodynamic properties of the substance, equilibrium tables calculated by NASA CEA for specific components of the burning mixture considered in the simulation were used.

The combustion rate of acetylene in the air, determined in accordance with the Arrhenius law:

$$W(T,\rho,Y_1,Y_2)=\overline{Y}_{O2}B(T)\exp(-E_a/T)Y_1Y_2\rho \tag{6}$$

Were Y_1 – mass fraction of the enriched air; Y_2 – mass fraction of acetylene; $\overline{Y}_{O2}$ – fixed fraction of oxygen in the air (variable Y_1); $B(T)=2.3\cdot10^{11}$ m^{3}mol/kg^{2}s; E_a – activation energy expressed in units of temperature is a fitting parameter.

3. TVD correction

Initially, the scheme of Total Variation Diminishing (TVD) was developed for the conservation equations for the one-dimensional plane case [4]. However, as shown in [5], it can be successfully applied in both cylindrical and spherical geometry for one-dimensional gas dynamics problems. In the future, the TVD scheme has been widely used for the numerical solution of two-dimensional and three-dimensional gas dynamics problems [1-2]. Herewith, the same one-dimensional version of the scheme, applied alternately to different coordinate directions, was used. The possibility of using the TVD method in conjunction with the large particle method was described in [6] too.

Nonphysical oscillations are suppressed by adding the TVD correction after completion all of the large particle method stages. That is, after the values $\rho_{ij}^{n+1}, u_{ij}^{n+1}, v_{ij}^{n+1}, E_{ij}^{n+1}$, and Y_{ij}^{n+1} have been obtained on the ($n+1$)-th time layer, they are supplemented with the correction that plays the part of numerical dissipation. Corrections to each of the mentioned quantities have the form of the divergence of correcting fluxes calculated on the basis of the TVD scheme. Note that in solving of two-dimensional equations of hydrodynamics use is made of one-dimensional TVD schemes applied alternately along the respective coordinate lines. Since all of the operations for calculating these fluxes are performed on gas-dynamical quantities determined for the moment t^{n+1}, the index showing their relation to the time layer can be omitted in their designation. The change of gas-dynamic quantities included in the column vector $\mathbf{U}_{ij}$ from the definition (2) due to the considered dissipation can be presented in the following form:

$$\Delta\mathbf{U}_{ij} = \frac{\Delta t}{V_{ij}}\left[(\sigma\mathbf{f})_{i-1/2,j} - (\sigma\mathbf{f})_{i+1/2,j} + (\sigma\mathbf{f})_{i,j-1/2} - (\sigma\mathbf{f})_{i,j+1/2}\right] \tag{7}$$

Here $\mathbf{f}_{i+1/2,j}$ and $\mathbf{f}_{i,j+1/2}$ are the vector columns of correcting fluxes calculated from the formulas of the TVD scheme that yield a correction to the fluxes $\mathbf{F}_{i+1/2,j}$ and $\mathbf{G}_{i,j+1/2}$ of the components in the vector column $\mathbf{U}_{ij}$, respectively. Since the value of flux $\mathbf{f}$ is determined by a one-dimensional TVD scheme for an arbitrary fixed coordinate direction on the computational grid, we need to consider in detail only one direction, for example, cells with variable index $i(z)$ at a fixed value of index $j(r)$. Then, the fluxes $\mathbf{f}_{i+1/2}$ (omit the index j) are commonly calculated by expanding the variations $\delta\mathbf{U}_{i+1/2} = \mathbf{U}_{i+1} - \mathbf{U}_i$ at the edge of the cells by eigenvectors of Jacobi matrix $(\partial\mathbf{F}/\partial\mathbf{U})_{i+1/2}$ and applying the scalar TVD scheme for each component of the series. The matrix $(\partial\mathbf{G}/\partial\mathbf{U})_{j+1/2}$ is used for another direction. This approach is a version of the solution of the linearized Riemann problem [7].

However, with respect to the gas dynamics conservation equations, this decomposition problem can be solved in a different way, without the use of Jacobi matrices treatment. The way proposed in this work consists in the application of known expressions for Riemann invariants:

$$\delta J_+ = \delta u + \frac{\delta p}{\rho c}, \quad \delta J_- = \delta u - \frac{\delta p}{\rho c}, \tag{8}$$

respectively, for the characteristics with the eigenvalues equal to $u \pm c$, where c is the local velocity of sound. Variations of the quantities from (8) are the total differentials only for the isentropic flow. In the case of an arbitrary flow, the entropy variation differs from zero $\delta s \neq 0$ and propagates along the characteristic with the eigenvalue equaled to u, i.e., it moves with the mass of the matter. Thus, any perturbation of gas-dynamic quantities can be decomposed to independent parts of the mentioned three types that propagate along the corresponding characteristics.

We determine the variations of the quantities necessary for the described decomposition:

$$\begin{gathered}\delta u_{i+1/2} = u_{i+1} - u_i, \quad \delta v_{i+1/2} = v_{i+1} - v_i, \quad \delta\rho_{i+1/2} = \rho_{i+1} - \rho_i, \quad \delta\varepsilon_{i+1/2} = \varepsilon_{i+1} - \varepsilon_i, \\ \delta e^k_{i+1/2} = e^k_{i+1} - e^k_i, \quad \delta p_{i+1/2} = p_{i+1} - p_i, \quad \delta s_{i+1/2} = \delta\varepsilon_{i+1/2} - \frac{p_{i+1/2}}{\rho^2_{i+1/2}}\delta\rho_{i+1/2}.\end{gathered} \tag{9}$$

Next, we omit the spatial indices for the values defined at the cell boundaries. Then the decomposition vector α and the vector of its corresponding eigenvalues a will have the form:

$$\alpha\Delta z = \left[\frac{1}{c}\left(\delta u - \frac{\delta p}{\rho c} \right);\ \frac{1}{c}\left(\delta u + \frac{\delta p}{\rho c} \right);\ \frac{\delta v}{c};\ \frac{\delta s}{c^2};\ \delta Y \right] \tag{10}$$

$$a = \left[u - c;\ u + c;\ u;\ u;\ u \right] \tag{11}$$

For each characteristic solution $\alpha^l_{i+1/2}$, where $l = 1,2,...,N+4$, corrective flows $\Phi^l_{i+1/2}$ are calculated using formulas for the scalar counter-flow TVD scheme [4]:

$$\Phi^l_{i+1/2} = \frac{\Delta z_{i+1/2}}{2}\left[\frac{1}{2}\psi\left(a^l_{i+1/2}\right)(g^l_{i+1} + g^l_i) - \psi\left(a^l_{i+1/2} + \gamma^l_{i+1/2}\right)\alpha^l_{i+1/2} \right] \tag{12}$$

where

$$\psi(x) = \begin{cases} |x|, & |x|/c \geq \delta \\ (x^2 + \delta^2)/2\delta, & |x|/c < \delta \end{cases} \tag{13}$$

$$\gamma^l_{i+1/2} = \frac{1}{2}\psi\left(a^l_{i+1/2}\right)\begin{cases} (g^l_{i+1} - g^l_i)/\alpha^l_{i+1/2}, & \alpha^l_{i+1/2} \neq 0 \\ 0, & \alpha^l_{i+1/2} = 0 \end{cases} \tag{14}$$

$$g^l_i = \min\operatorname{mod}\left(\alpha^l_{i-1/2}, \alpha^l_{i+1/2}\right) \tag{15}$$

The minmod(x,y) function is the smallest argument in absolute value if all arguments have the same sign, or zero if the arguments have different signs.

Then, substituting the vector Φ into the expression (10) instead of the vector α, we can find the changes of flows for the values included in (10) in the form of variations. It should be borne in mind that Φ has the dimension of speed. Therefore, to avoid confusion, replace the symbol δ with the symbol Δ, and obtain a system of equations:

$$\Phi = \left[\frac{1}{c}\left(\Delta u - \frac{\Delta p}{\rho c} \right);\ \frac{1}{c}\left(\Delta u + \frac{\Delta p}{\rho c} \right);\ \frac{\Delta v}{c};\ \frac{\Delta s}{c^2};\ \Delta Y \right] \tag{16}$$

By solving this system of equations with respect to $\Delta u, \Delta p, \Delta v, \Delta s$ and ΔY one can find the correcting flows $\mathbf{f}$ for the corresponding flows from the vector $\mathbf{F(U)}$ with the help of formulas:

$$\begin{aligned} &\text{mass flow } - f_1 = \Delta\rho; \\ &\text{longitudinal pulse flux } - f_2 = u\Delta\rho + \rho\Delta u; \\ &\text{cross pulse flux} - f_3 = \rho\Delta v + v\Delta\rho; \\ &\text{flux of the total energy} - f_4 = (E + p/\rho)\Delta\rho + \rho\Delta s + \rho u\Delta u + \rho v\Delta v; \\ &\text{flux of the mass fraction} - f_{4+k} = \Delta Y_k; \end{aligned} \tag{17}$$

where $\Delta p = -\rho c^2(\Phi_1 - \Phi_2)/2 - \sum_k \eta_k\Phi_{4+k}$; $\Delta u = c(\Phi_1 + \Phi_2)/2$; $\Delta v = \Phi_3 c$; $\Delta s = \Phi_4 c^2$;

$\Delta\rho = (\Delta p - k\rho\Delta s)/c^2$; $\Delta Y_k = \Phi_{4+k}$. These expressions are derived explicitly from system (16).

4. Simulation results

Calculations were made for a long pipe with a diameter of about 0.04 m and a length of 1 m to 6 m, sealed on one side. Initially, the pipe was filled with the rest stoichiometric acetylene-air mixture enriched with oxygen at a temperature of about 300 K, with a reduced density of 0.18 kg/m^3 and a pressure of 0.142 Pa. Initiation of combustion was carried out by modeling Joule heat release from the electric spark of the spark plug. The energy of the spark plug stood out in a gas mixture with a constant power of 0.04 J per 1.5ms in a small volume at the axis adjacent to the closed end of the pipe.

Numerical simulation was performed by the modified large particle method with TVD correction in 2D axisymmetric coordinate system. The symmetric spatial difference scheme of the 2nd order without artificial viscosity was used. A homogeneous calculation grid with a cell size in both directions (axial and radial) equal to $9.77 \cdot 10^{-4}$ m was applied.

Simulation results for some variants were considered: variants differed in settings the conditions on the side walls – the slip or adhesion model, and differed in the value of activation energy (6) – low (53 kcal) or high (63.7 kcal).

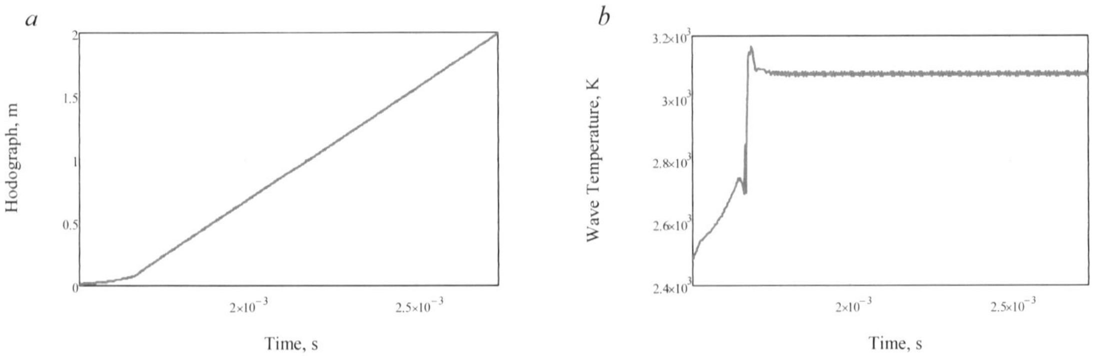

Figure 1. Combustion front dynamics at the axis for the adhesion and low activation energy: *a* – front location; *b* – front temperature.

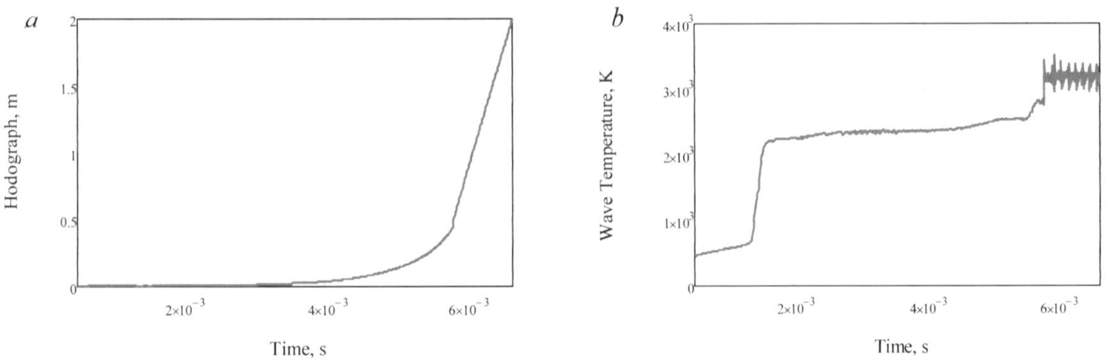

Figure 2. Combustion front dynamics at the axis for the adhesion and high activation energy: *a* – front location; *b* – front temperature.

In the variant with the slip and low activation energy, the slow combustion of the mixture almost immediately after ignition passes into the detonation mode with a stable flat combustion front. Such an option, but with high activation energy gives a stable plane combustion wave, never passing into detonation. In the case of an adhesion and a low activation energy pattern of propagation of the combustion slightly different from the similar option with the slip. However, the detonation front is less stable here, judging by the small temperature oscillation on the wave front in axis (figure 1). In the high-energy version of the activation detonation occurs much later and the temperature oscillation at the wave front have a much higher amplitude (figure 2).This suggests that the surface of the detonation wave front is not flat and stationary. That can be seen in figure 3 where color maps of pressure distribution are represented for some moments in the range of the oscillation period. Here is shown a fragment of the pipe located in the vicinity of the detonation front at three (a, b, c) sequential time points divided with time interval equaled to 0.01ms. The detonation wave moves from up to down in the figure.

Conclusion

The use of TVD correction in conjunction with the large particle method improves the accuracy and stability of the latter. TVD correction extends capabilities of large particle method relatively to legitimately including second-order processes such as thermal conductivity and molecular viscosity in the model. The simulation results showed the importance of taking into account the friction on the pipe walls. However, without taking into account viscous stresses, the effect of these boundary conditions disappears when the mesh cells decrease.

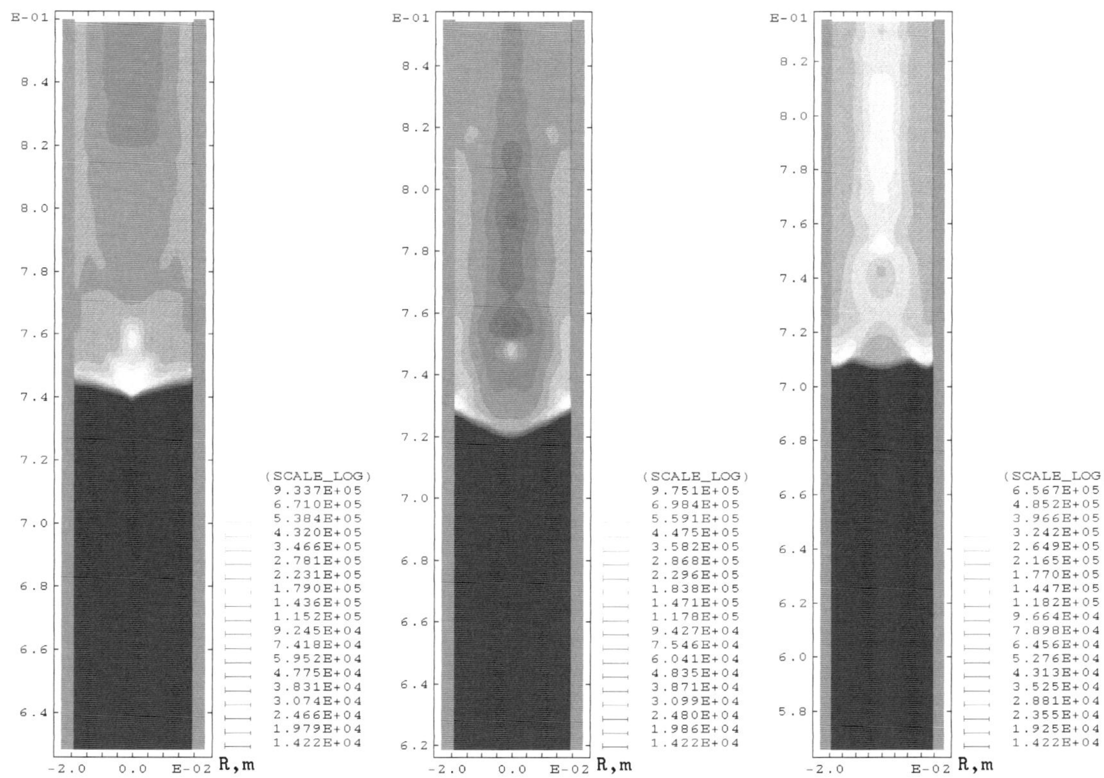

a – t_0 *b* – t_0+0.01ms *c* – t_0+0.02ms

Figure 3. Color maps of pressure distribution at three sequential time points.

References

[1] Moon J, Yee H C 1987 Numerical simulation by TVD schemes of complex shock reflections from airfoils at high angle of attack *AIAA Paper* **350** 1-17

[2] Yee H C 1986 Linearized form of implicit TVD schemes for multidimensional Euler and Navier-Stokes equations *Int. J. Comp. Math. Appl.* **12A** 413-32

[3] Belotserkovsky O M, Davydov Y M 1982 *Coarse particle (large particle) method in hydrodinamics* (Moscow, Nauka) p 392

[4] Harten A 1983 High Resolution Schemes for Hyperbolic Conservation Laws *Journal of Computational Physics* **49** 357-93

[5] Kozlov I M, Romanov G S, Suvorov A E 1993 Application of TVD-schemes for simulation of gas-dynamic flows with spherical or cylindrical symmetry *Journal of Engineering Physics and Thermophysics* **65** 296–301

[6] Davydov Y M, Davydova I M 2003 Modern difference schemes of increased accuracy with decreasing the total variation on the solution for the actual problems of machine building a // *Mathematical modeling of systems and processes* **11** 4-18

[7] Roe P L 1981 Approximate Riemann solvers, parameter vectors, and difference schemes *J. Comp. Phys.* **43** 357-72

IOP Conf. Series: Journal of Physics: Conf. Series **1205** (2019) 012028 doi:10.1088/1742-6596/1205/1/012028

Schemes for the formation of effective resource portfolios and areas of activity of organizations based on statistical forecasts

A V Kryanev[1,3], Yu G Pavlov[2], D E Sliva[1] and Yu A Ulyanin[2]

[1]National Research Nuclear University "MEPhI", Russia
[2]Tehsnabexport, Russia

[3]avkryanev@mephi.ru

Abstract. The problems of efficient allocation of resources in the areas of the company activity under conditions of uncertainty are considered. The indicators of effectiveness in the areas of the company activity are of an uncertain nature and are random variables with predetermined laws of probability distribution. To form an effective distribution of the company's activities in the areas of its activities, a scheme is used to form effective portfolios based on the probabilities of priority developed at the NRNU MEPhI. As estimates of the values of the efficiency index for the directions of the company's activities, their forecasts were taken as random variables with a uniform or normal probability distribution law.
Key words: resource allocation, activities of organizations, efficiency, uncertainty, portfolio, risk, effective portfolio formation schemes, probability of priority.

1. Introduction

At present the theory of portfolio investment is widely used in countries with developed market economies and is one of the main tools by which the efficiency of the use of material and financial resources is increased. The article gives an application of the scheme for the formation of effective portfolios, presented in [1], to determine the effective allocation of resources in the areas of activity of the organization. Unlike the well-known Markowitz schemes [2, 3] and the VaR schemes [4-6], the scheme presented in [1] is based on the use of priority indicators which provides the maximization of the opportunity to carry out large values of portfolio efficiency.

2. Scheme for the formation of effective portfolios of resources and company activities in accordance with Markowitz

Let's consider the problem of allocating resources or activities of organizations in the presence of uncertainty [1-6]. Uncertainty means that the magnitude of the effectiveness of the distributed resources in the selected object or the chosen direction of the organization's activities (direction) is random.

We number the objects by the index $i = 1, ..., n$. We denote by R_i the value of resource utilization efficiency in the i-th object (in the i-th direction). Mathematically, the assumption of uncertainty means that R_i are random variables.
The complete knowledge of a system of random variables $\boldsymbol{R} = (R_1, ..., R_n)^T$ (a random vector **R**) is determined by its distribution law. However, in order to solve the problem of the optimal allocation of the resources between the investment objects under consideration according to the Markowitz scheme, it is sufficient to know only two characteristics of the random efficiency vector - the vector of mathematical expectations $E(\boldsymbol{R}) = \boldsymbol{m} = (m_1, ..., m_n)^T, m_i = E(R_i)$, the expected average value of the i-th investment object efficiency and the covariance matrix W, $W_{ij} = cov(R_i, R_j), i, j - 1, ..., n$. In the formulation of Markowitz, the effectiveness of the investing resources in the object under consideration is a random value, the mathematical expectation of which is taken as the expected value of efficiency from investing capital in the object under consideration. As a numerical expression of the risk value, Markowitz proposed taking a measure of the deviation from the expected value - the standard deviation of efficiency as a random variable.
Such a choice of the mathematical expression for risk made it possible to carry out the principle known in economics in the portfolio investment scheme, expressed in the phrase "do not put eggs in one basket", i.e. the diversification of resources between several objects leads to a reduction in risk compared to the risk of investing resources in different facilities.
The estimations of the vector **m** and matrix W can be obtained depending on the specific economic content of the resources. After having received estimations $\widehat{m}_i$ и $\widehat{W}_{ij}$ it is possible to set and solve the tasks of optimizing portfolios of distributed resources. Let's denote x_i the share of the distributed resource in the object with the number i. Obviously the fractions x_i satisfy the natural conditions

$$0 \le x_i \le 1, \; i = 1, ..., n, \; \sum_{i=1}^{n} x_i = 1. \tag{1}$$

Under the portfolio of distributed resources we will understand the vector of shares $\boldsymbol{x} = (x_1, ..., x_n)^T$ of distributed resources in objects $i = 1, ..., n$. The task of choosing a portfolio is equivalent to the task of selecting a vector of shares **x**. For a fixed value of the vector $\vec{x}$, the value of the portfolio efficiency R_p is given by

$$R_p = \sum_{i=1}^{n} x_i \cdot R_i = (\boldsymbol{x}, \boldsymbol{R}). \tag{2}$$

Consequently, the expected average value m_p of efficiency and the variance of portfolio efficiency σ_p^2 are given by the equalities:

$$m_p = \sum_{i=1}^{n} x_i \cdot m_i = (\boldsymbol{x}, \boldsymbol{m}), \tag{3}$$

$$\sigma_p^2 = \sum_{i=1}^{n} \sum_{j=1}^{n} x_i \cdot x_j \cdot W_{ij} = (W\boldsymbol{x}, \; \boldsymbol{x}). \tag{4}$$

According to Markowitz, the risk of the portfolio is equal σ_p (sometimes as a risk in the Markowitz scheme choose σ_p^2). Thus, the portfolio is characterized by two criteria $m_p = m_p(\boldsymbol{x})$ - the expected average value of portfolio efficiency R_p and portfolio risk $\sigma_p = \sigma_p(\boldsymbol{x})$, both criteria depending on the selected portfolio composition. The larger the value of the criterion $m_p(\boldsymbol{x})$ and the smaller the criterion $\sigma_p(\boldsymbol{x})$, the better the composition of the portfolio (vector **x**) is chosen.
It is known that the optimal solution of multicriteria problems must be searched among the set of Pareto solutions possessing the property that the improvement of any criterion for the Pareto solution leads to the deterioration of other criteria. A set of Pareto solutions for a two-criteria problem can be found by fixing one of the criteria and optimizing the other.
According to the Markowitz scheme, in order to find the Pareto solutions of the two-criteria portfolio optimization problem, portfolio risk $\sigma_p(\boldsymbol{x})$ is minimized for fixed values of the expected mean value $m_p(\boldsymbol{x})$.
Thus, the mathematical model of portfolio optimization according to the Markowitz scheme has the form [2, 3]:

$$\sigma_p^2 = (W\boldsymbol{x}, \; \boldsymbol{x}) - \min \tag{5}$$

$0 \le x_i \le 1,\ i = 1, \dots, n,\ \sum_{i=1}^{n} x_i = 1,\ \ \sum_{i=1}^{n} x_i \cdot m_i = m_p.$.

Obviously, m_p belongs to the interval $[m_{min}, m_{max}]$, where $m_{min} = min\{m_1, \dots, m_n\}$, $m_{max} = max\{m_1, \dots, m_n\}$.

3. Formation of effective resource portfolios or company activities according to the VaR scheme

In the Markowitz scheme, the uncertainty expressed by the variance of efficiency is two-sided since it determines the possibility of deviating efficiency both at a lower side of the average expected value (which is undesirable and determines essentially the risk), and in the larger side (which, on the contrary, is desirable , the opposite of understanding the risk). Therefore, the concept of uncertainty used in the Markowitz scheme leads to the fact that a decrease in uncertainty σ_p^2 in the allocation of resources is possible only due to a large degree of diversification of resources. And the greatest decrease in uncertainty is achieved, as a rule, with a significant degree of uniform distribution of resources between the objects. Thus, the Markowitz scheme implements the rule that a greater decrease in uncertainty is due to a greater diversification of resources.

At first glance, this property of uncertainty in the Markowitz scheme does not raise objections. However, the consideration of specific practical problems of resource allocation in conditions of uncertainty shows that the rule of "large diversification of resources - a lower risk value" is not universal. Often, a greater degree of resource concentration and even a complete lack of diversification (when the entire resource is given to a single facility) can lead to a significant increase in the efficiency of its use while significantly reducing the risk of loss of efficiency.

Unfortunately, the concept of uncertainty concluded in the Markowitz scheme does not take into account the last circumstance. In fact, the main reason for the last lack of the concept of uncertainty in the Markowitz scheme is the inconsistency of this concept of uncertainty for the effectiveness of an individual object and the risk of a decrease for the total effectiveness of the entire distributed resource. When setting and solving the problem of optimal allocation of resources under uncertainty, the risk should reflect the possibility of a loss of resource use efficiency due to its improper diversification.

Thus, in order that the used concept of risk in setting up problems of optimizing resource allocation in conditions of uncertainty does not lead to undesirable results, the concept of risk should include the possibility of loss of efficiency due to improper diversification of resources. Unfortunately, the transfer of the concept of uncertainty in the Markowitz scheme does not provide the above account for the loss of fulfilled efficiency.

As mentioned above, the use of dispersion as a measure of risk includes a double focus of possible changes in efficiency, both to a smaller and larger extent. At the same time, in many applied fields of science and technology, a unidirectional risk concept is traditionally used associated with the probability of implementing unfavorable random events (probability of a catastrophe, probability of an accident, probability of failure of equipment, etc.). Therefore, in recent years, the economy has also begun to use the concept of risk based on the counting of adverse events of a kind $p(R < -R_V)$, , where R is efficiency and $R_V > 0$ determines the level of efficiency losses and is called Value-at-Risk (VaR) [5, 6].

Instead of posing Markowitz (5) for the problem of optimal allocation of resources under conditions of uncertainty, we propose the formulation of this problem on the basis of VaR:

$$\begin{gathered} p^* - min \\ R^* - max \\ 0 \le x_i \le 1,\ i = 1, \dots, n,\ \sum_{i=1}^{n} x_i = 1, \end{gathered} \tag{6}$$

where $p^* = p\left(R_p(\boldsymbol{x})\right)$, $R_p(\boldsymbol{x}) = \sum_{i=1}^{n} x_i \cdot R_i, R_i, i = 1, \dots, n$, R_i is efficiency of resource using in the *i*-th facility, R^* is level of efficiency; p^* is the probability level of the adverse event that the total resource allocation efficiency will be less than the level R^*.

By solving the problem of optimal allocation of resources in a VaR setting, we mean the Pareto solutions of the two-criteria problem (6) [4].

4. Algorithm for the numerical calculation of the composition of effective portfolios for the areas of company activity based on the probability of priority

Let's consider an algorithm for solving the problem of forming effective portfolios with the help of priority probabilities [1].
Let the possible values of the efficiency R_i belong to the interval $[r_{i\,min}, r_{i\,max}], i = 1, \dots, n$. This, in particular, means that the probability density values $\varphi_i(x), i = 1, \dots, n$ are zero outside the interval $[r_{i\,min}, r_{i\,max}], i = 1, \dots, n$. If among the intervals $[r_{i\,min}, r_{i\,max}], i = 1, \dots, n$ there are those for which the inequalities $r_{imax} < r_{jmin}$ are satisfied for some $j = 1, \dots, n$, then objects with such numbers i are excluded from the composition of the portfolio.
Let the remaining l objects be renumbered in descending order $r_{i\,max}$:

$$r_{1\,max} > r_{2\,max} > \cdots > r_{l\,max}. \tag{7}$$

We introduce the probability of priority of efficiency for objects according to equation

$$p(R_i \succ R_{i+1}) = \int_{r_{i+1\,max}}^{r_{i\,max}} \varphi_i(x)dx\,, \quad i = 1, \dots, l-1. \tag{8}$$

The formation of an effective portfolio is carried out according to the scheme of step-by-step increase of the shares of the distributed resource into objects, starting with the object with the number 1 with the largest value $r_{1\,max}$.
At the first stage, the proportion of the first object is assigned the value

$$x_1 = p(R_1 \succ R_2). \tag{9}$$

In the second stage, the percentage of the first object is assigned with the value

$$x_1 = x_1 + \left(\int_{r_{3\,max}}^{r_{2\,max}} \varphi_1(x)dx\right) \cdot (1 - x_1), \tag{10}$$

where x_1 in the right-hand side of (10) is taken from equation (9), and the fraction of the second object is assigned with the value

$$x_2 = p(R_2 \succ R_3) \cdot (1 - x_1), \tag{11}$$

where x_1 in the right-hand side of (11) is taken from (10).
Similarly, the assignment of shares for objects up to the object with the number l [1]. Table 1 presents the results of calculations according to scheme (7) - (11) of the composition of effective portfolios for the company activity based on the data of the forecast of performance indicators (the numerical values of the shares are given in percentages).

Table 1. The forecasted shares of company activities.

	2018	2019	2020	2021	2022	2023	2024	2025	2026	2027	2028	2029	2030
Area 1	50,0	50,0	50,0	50,0	50,0	50,0	50,0	49,9	46,3	43,3	38,7	35,6	33,0
Area 2	50,0	50,0	50,0	50,0	50,0	50,0	50,0	49,9	46,3	43,3	38,7	35,6	33,0
Area 3	0,0	0,0	0,0	0,0	0,0	0,0	0,0	0,0	0,0	0,0	6,4	7,3	8,7
Area 4	0,0	0,0	0,0	0,0	0,0	0,0	0,0	0,2	7,4	13,4	16,2	19,3	23,0
Area 5	0,0	0,0	0,0	0,0	0,0	0,0	0,0	0,0	0,0	0,0	0,0	0,0	0,0
Area 6	0,0	0,0	0,0	0,0	0,0	0,0	0,0	0,0	0,0	0,0	0,0	2,2	2,3

5. Conclusion

The article gives an application of schemes for the formation of effective portfolios, resource allocation, including a scheme based on the probabilities of priority. The presented scheme allows to determine the optimal allocation of resources or company activities from the use of these forecasts for indicators of the efficiency of resource development or the implementation of directions. The results of numerical calculations of effective portfolios for selected areas of the company are presented.

Acknowledgments
The work was supported by the Ministry of Education and Science of the Russian Federation (the basic part of the state assignment, project No. 1.9746.2017 / BC).

References

[1] Kryanev A V., Sliva D E and Udumyan D K 2018 Methods for the formation of effective portfolios with the constraints and algorithms for their implementation. Bulletin of the National Research Nuclear University "MEPhI" **7** №3 pp 289-293 (in Russian)

[2] Markowitz H M 1990 *Mean Variance Analysis in Portfolio Choice and Capital Markets* Basil Blackwell

[3] Sharpe U F, Alexander G Dg and Baily D V 2001 *Investment* Moscow: Infra-M

[4] Kryanev A V and Lukin G V 2001 On the formulation and solution of problems of optimization of investment portfolios Moscow (*Preprint MEPhI* 006-2001) (in Russian)

[5] Rockafellar R T and Uryasev S 2000 Optimization of conditional value-at-risk J. Risk **2**(3)

[6] Lim Churlzu, Sherali Hanif D and Uryasev S 2010 Portfolio optimization by minimizing conditional value-at-risk via non differentiable optimization Computational Optimization and Applications **46**(3)

Statistical features of plastic flow localization in materials

N A Kudryashov, R V Muratov and P N Ryabov

National Research Nuclear University MEPhI (Moscow Engineering Physics Institute), Kashirskoe shosse 31, Moscow, Russia

E-mail: nakudr@gmail.com, pnryabov31@gmail.com, rodyon-mur@yandex.ru

Abstract. We consider the processes of plastic flow localization in dipolar materials undergoing high speed shear deformations. The mathematical model of the processes of plastic flow localization is formulated taking into account dipolar effect. We introduce the numerical algorithm which is based on adaptive mesh refinement technique. We show that this algorithm allows to increase performance of computations. We also studied the statistical properties of shear bands formation. We show that dipolar effect changes average characteristics of the processes considered such as average temperature, stress and etc. Moreover this effect leads to increase in initiation time, changes the widths of localization zones and distances between them.

1. Mathematical formulation of the problem considered

We consider the infinite slab of incompressible thermoviscopastic material undergoing to simple shear deformation. The dipolar effects and effects of strain hardening of material are taken into account. The slab occupies the domain $0 \leqslant y \leqslant H$. The bottom surface $(y = 0)$ of slab is fixed and top surface $(y = H)$ is moving at the constant velocity $v_{up} = H\dot{\varepsilon}_0$. The direction of motionis parallel to the x axis. The mathematical model of the process of slab deformation in the above mentioned suggestions can be expressed by the following system of nonlinear equations [1, 2, 3]

$$v_t = \frac{1}{\rho}\left(s - \sigma_y\right)_y, \tag{1}$$

$$s_t = \mu\left(v_y - \dot{\varepsilon}\right), \tag{2}$$

$$\sigma_t = l\mu\left(v_{yy} - \frac{1}{l}\dot{d}\right), \tag{3}$$

$$\psi_t = \frac{s_e\dot{\varepsilon}_e}{\kappa(\psi)}, \tag{4}$$

$$C\rho T_t = \left(kT_y\right)_y + s_e\dot{\varepsilon}_e, \tag{5}$$

where s, σ are simple and dipolar stress respectively, v is the velocity, T is a temperature, $\dot{\varepsilon}$ is the plastic strain rate, ψ is a strain hardening variable, d is a dipolar strain, l is a characteristic length of material, ρ is a mass density. In Eqs. (1)-(5) we use the following notations

$$s_e = \sqrt{s^2 + \sigma^2}, \qquad \dot{\varepsilon}_e = \sqrt{\dot{\varepsilon}^2 + \dot{d}^2}, \qquad \varepsilon_e = \int_0^t \dot{\varepsilon}_e(\tau)d\tau, \tag{6}$$

Effective variables listed in (6) are connected by the plastic flow law $D(s_e, \psi, T) \geqslant 0$. This connection is given by relations

$$\dot{\varepsilon} = \frac{s}{s_e}\dot{\varepsilon}_e, \qquad \dot{d} = \frac{\sigma}{s_e}\dot{\varepsilon}_e, \qquad \dot{\varepsilon}_e = D(s_e, \psi, T). \tag{7}$$

To close the system of Eqs. (1) - (7) we use the following initial conditions

$$\begin{gathered} v(y,0) = y\dot{\varepsilon}_0, \quad s(y,0) = s_0(y), \quad T(y,0) = T_0(y), \\ \sigma(y,0) = \sigma_0(y), \quad \psi(y,0) = 0. \end{gathered} \tag{8}$$

In turn, boundary conditions have the form

$$\begin{gathered} v(0,t) = 0, \quad v(H,t) = v_{up}, \\ T_y(0,t) = T_y(H,t) = 0, \quad \sigma(0,t) = \sigma(H,t) = 0. \end{gathered} \tag{9}$$

We also use the Litonsky plastic flow to model the behavior of materials studied

$$s_e = \kappa_0 g(T) \left(1 + \left(\frac{\psi}{\psi_0}\right)^n\right) \left(1 + \frac{\dot{\varepsilon}_e}{\dot{\varepsilon}_y}\right)^m. \tag{10}$$

Here $g(T)$ is a thermal softening function, ψ_0 is a references strain hardening parameter, $\dot{\varepsilon}_y$ is a reference strain rate, κ_0 is a characteristic value of stress. The value of the above mentioned parameters can be found in [3].

2. Adaptive mesh refinement technique

Since the width of the slab in most numerical experiments equal to several millimeters in turn the area of localization zones usually equal to $10-100$ μm, there is no need to use homogeneous grids with high resolution on the whole computational domain. Also such approach is not effective in the case when we know the places of shear bands localization. To overcome this difficulty the adaptive mesh refinement algorithm can be used. It allows to reduce the number of computational nodes and helps to prove the efficiency of numerical algorithms.

In the present work we apply adaptive mesh refinement to numerical algorithm presented in [3]. We discretize the computational domain using the fixed number of nodes. At initial moment of time nodes distribution in computational domain is uniform. The main idea of the adaptive mesh refinement algorithm is to concentrate nodes in the areas with maximum strain-rate during the deformation process. So it allows us to use less number of nodes and as a result to reduce computational time. For description of the nodes motion function we use the following relation

$$p_i = \left(\frac{\dot{\varepsilon}_{i+1}^{n+1} - \dot{\varepsilon}_i^{n+1}}{y_{i+1}^n - y_i^n} + \frac{\dot{\varepsilon}_i^{n+1} - \dot{\varepsilon}_{i-1}^{n+1}}{y_i^n - y_{i-1}^n} \right) \Big/ \left(\frac{\dot{\varepsilon}_{i+1}^{n+1} + \dot{\varepsilon}_i^{n+1} + 1}{y_{i+1}^n - y_i^n} + \frac{\dot{\varepsilon}_i^{n+1} - \dot{\varepsilon}_{i-1}^{n+1} + 1}{y_i^n - y_{i-1}^n} \right), \tag{11}$$

where variable p_i has the meaning of strain-rate dimensionless gradient. In relation (11) number one in denominator allows us to avoid the problem caused by division by zero. This assumption can be done since in the localization areas value of $\dot{\varepsilon} >> 1$, in other regions p_i approximately equal to zero.

On the next step we compute velocity of the nodes

$$v_i = RSp_i, \tag{12}$$

where S is speed of a transverse acoustic wave in material and R is dimensionless speed regulation parameter (we use $R = 5$ in all experiments). On last step we calculate new coordinates of nodes using the following relation

$$y_i^{n+1} = y_i^n + \tau^n v_i, \tag{13}$$

and than transfer grid functions using simple linear interpolation. Also we use additional condition on node location which allows to control the process of nodes convergence and divergence relative to each other. So, the possible minimum distance between nodes equal to 1μm, and the maximum distance equal to 100μm. It should be pointed out that the above mentioned algorithm cannot be used in the tasks where randomly disturbed conditions are used.

3. Results of numerical simulation of shear bands formation using and adaptive grid method

Here we study the formation of multiple adiabatic shear bands in nonpolar copper ($l = 0$) specimen with height of $H = 4$ mm. We suppose that initially there are three maximum of temperature located at points $y = 1$, 2 and 3 mm. The value of temperature at these points equal to 100, 50 and 75 °C respectively. So, initial distribution of temperature in the specimen is given by relation

$$T(y,0) = 100\left(e^{-r(y-1)^2} + 0.5e^{-r(y-2)^2} + 0.75e^{-r(y-3)^2}\right). \tag{14}$$

Also we note that there are no initial perturbation of stress and strain.

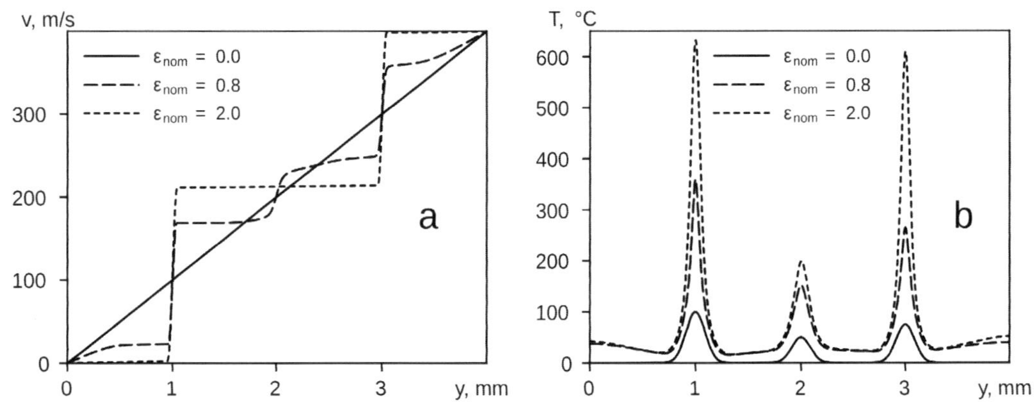

Figure 1. Distributions of velocity (a) and temperature (b) at different time moments.

In Fig. 1 we have illustrated the stages of shear bands formation from initial moment of time till the final stage of localization. From Fig. 1 we see that on the first stage of localization there is an increase of temperature at the whole computational domain especially at the points $y =$ 1, 2 and 3 mm. After a while, we see an abrupt jump of the temperature at the points $y = 1$ and 3 mm, where shear bands formed. In turn, the temperature at the center of the slab slowly increases during experiment. We also pointed that corresponding velocity gap disappears and plastic strain-rate equals to zero at this point after $\varepsilon_{\text{nom}} = 1.2$. Such behavior is caused not only by the difference of initial temperature at the points $y = 1, 3$ mm and $y = 2$ mm but also by existence of additional temperature gradients on the boundaries.

These results were checked using the uniform grid with high resolution and it was shown that they are identical to each other. We also investigate the influence of the value of spatial step on the final results. Numerical experiments show that if we increase the spatial step in simple uniform grid from $h = 1\,\mu$m to $h = 4\,\mu$m the behavior of temperature and velocity changes at the point $y = 2$ mm. However if use the adaptive mesh refinement technique we obtain the same results as in the case of uniform grid with high resolution even in the case $h = 4\,\mu$m. We also note, that this approach allows us to increase performance in five times.

Let's take a look at behavior of nodes. According to the algorithm on the first part of the experiment nodes must move to temperature maximum. On the second part nodes must move away from the center of block. In Fig. 2 we illustrate the trajectories of nodes corresponding to this problem. Form Fig. 2 we see that tracks of the nodes demonstrate expected behavior i.e.

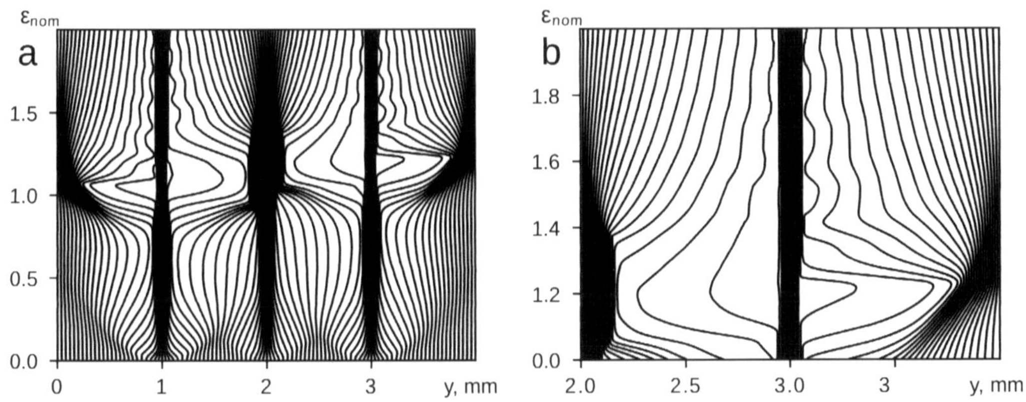

Figure 2. Tracks of nodes in the numerical experiment of three shear band formation.

on the first stage of localization nodes concentrated near the temperature maximum at points $y = 1, 2, 3$ mm. Than approximately at $\varepsilon_{\mathrm{nom}} = 1.2$ nodes started to move away form the point $y = 2$ mm(see Fig. 2b).

4. Statistical features of shear bands formation

In works [4, 5, 6, 7] it was shown that inhomogeneity in initial stress distribution leads to self-organization of adiabatic shear bands. Here we study the statistical properties of the multiple ASBs formation in two materials deformed at $\dot{\varepsilon}_0 = 10^5$ s^{-1} taking into account dipolar effect [3]. One of materials is steel HY-100 and another one is OFHC copper.

First we explore the behavior of the average values such as temperature and effective stress. It is shown that the average temperature tends to a constant which increases with an increase of dipolar parameter l. For steel this value changes in twice and for copper approximately in one and a half times if we increase l from 0 to 10 μm. Also it was found that dipolar effect significantly changes localization time. This fact also was mentioned in [1].

In work [1] authors noted that dipolar effects lead to an increase of shear bands width. This fact also was confirmed by our numerical experiments. So the question arise: is there any statistical law that describes the distribution of shear bands width depending on the value of dipolarity? Also it is interesting to study the distribution of distance between shear bands.

Numerical experiments show that the distribution of shear bands width is close to normal one. We see that the most probable value of distribution $w = w_m$ changes linearly with increasing the value of parameter l. In the case of steel we have $w_m = 11 + 2.4l$ μm, in the case of copper $w_m = 102 + 12l$ μm. However the distribution of distance between shear bands have more complex behavior. It turns out that if we increase the value of parameter l the distribution become multimodal. For steel we have bimodal and four-modal distribution in the case $l = 5\,\mu$m and $10\,\mu$m respectively, in the case of copper bimodal distribution is observed in the case $l = 10\,\mu$m. Moreover, for both materials all modes located on the same distance between each other. That allows us to find the characteristic distance between shear bands in steel and copper. In the case of steel we have $L_p = 1.03$ mm and $L_p = 0.68$ mm for copper.

IOP Conf. Series: Journal of Physics: Conf. Series **1205** (2019) 012029 doi:10.1088/1742-6596/1205/1/012029

Acknowledgment
This work was supported by the Grant for support of young Russian scientists MK-6044.2018.1 and by the Grant of Russian Science Foundation 18-11-00209.

References

[1] R C Batra and C H Kim 1990 *Int. J. Engng. Sci.* **28** 927
[2] R C Batra and C H Kim 1990 *Int. J. Plast.* **6** 127
[3] N A Kudryashov, R V Muratov and P N Ryabov 2018 *Appl. Math. Comput.* **338** 164
[4] F Zhou, T W Wright and K T Ramesh 2006 *J. Mech. Phys. Solids.* **54** 1376
[5] N A Kudryashov, P N Ryabov and A S Zakharchenko, 2015 *J. Mech. Phys. Solids* **76** 180
[6] P N Ryabov, N A Kudryashov and R V Muratov 2017 *J. Phys.: Conf. Ser.* **788** 012031
[7] P N Ryabov 2017 *AIP Conference Proceedings* **1863** 380007

IOP Conf. Series: Journal of Physics: Conf. Series **1205** (2019) 012030 doi:10.1088/1742-6596/1205/1/012030

The first integrals and exact solutions of a two-component Belousov–Zhabotinskii reaction system

Nikolay A Kudryashov

National Research Nuclear University MEPhI (Moscow Engineering Physics Institute), 31 Kashirskoe Shosse, 115409 Moscow, Russian Federation

E-mail: nakudr@gmail.com

Abstract. The popular Belousov–Zhabotinskii (BZ) system of equations for description of a two component reaction is considered. The Painlevé test is applied to determine integrability of this system. It is shown that the system of equations is nonintegrable in general case. The parameter values of the mathematical model are found for the case when the system of equations passes the Painlevé test. Simplest solutions of the system of equations are presented. Additional conditions are specified when the general solutions of the system can be found. These general solutions are found using the new generalized method for finding exact solutions and the first integrals. Two first integrals of the Belousov–Zhabotinskii reaction system are given at additional conditions on parameter values for the mathematical model.

1. Introduction

The reaction known as the Belousov–Zhabotinskii reaction was discovered by the russian biochemist Boris Belousov in 1951. He noted the first reaction with the Ce^{3+}/Ce^{4+} couple as a catalyst and citric acid as a reductant. He observed that the color of the reaction solution oscillated between colorless and yellow and found that the frequency of oscillations increased with the rise of temperature. Belousov made two attempts to publish his discovery, but he failed. His work has been finally published only eight years later.

The study of this reaction continued when professor Simon Shnol invited his graduate student Anatoly Zhabotinskii to investigate the reaction mechanism. Zhabotinskii's group conducted detailed studies of the reaction, including various options and the first mathematical model, which was able to show oscillatory behavior.

The first model of the reaction was obtained in 1967 by Zhabotinskii and Korzukhin. It was based on the selection of empirical relations correctly describing oscillations in the system. The mechanism proposed by Field and Noyes (Oregonator model) [1] is one of the simplest and at the same time is the most popular in the research works investigating behavior of the BZ reaction. This model shows simple oscillations similar to the experimentally observed, but it can not show more complex types of oscillations, such as complex-periodic and chaotic. The model of Showalter, Noyes and Bar-Eli (Modified Oregonator model) [2] was developed to simulate the complex-periodic and chaotic behavior of the BZ reaction.

A good and varied introduction to the reaction can be found in the book of articles edited by Field and Burger [3]. It also has several articles on chemical oscillators and wave phenomena.

Winfree [4] discusses some of the reaction's properties, both temporal and spatial. A brief review of the detailed reaction and its properties is given by Tyson [5]. Some other questions corresponding to the study of the Belousov–Zhabotinskii reaction system were considered in works [6, 7, 8, 9, 10, 11].

In this paper we follow a simplified reaction mechanism (based on the Noyes–Field model) suggested by Murray [12, 13, 14]. The key chemical elements in the BZ reaction are

$$X = \mathrm{HBrO_2}, \quad Y = \mathrm{Br^-}, \quad A = \mathrm{BrO_3^-}, \quad P = \mathrm{HBrO}, \tag{1}$$

and the model reactions can be approximated by the sequence

$$X + Y \to X + P, \quad X + Y \to 2\,P, \quad A + X \to 2\,X, \quad 2\,X \to P + A \tag{2}$$

with the known rate constants k_1, k_2, k_3, k_4.

So, applying the law of mass action to reactions (2) we get the second-order system of kinetics equations for the concentrations of the bromous acid and bromide ion, denoted by U and V [13], [14], [15]

$$\frac{dU}{dt} = k_1\,V - k_2\,U\,V + k_3\,U - k_4\,U^2, \tag{3}$$

$$\frac{dV}{dt} = -k_1\,V - k_2\,U\,V, \tag{4}$$

where t is the time and k_i's are rate constants. We assume that reactants do not diffuse in the media. The compound HBrO does not appear in our analysis and the concentration A ($\mathrm{BrO_3^-}$) is constant.

We use the variables and parameters further in the form

$$U = \frac{k_3}{k_4}\,u, \quad V = \frac{k_3}{k_2}\,v, \quad t = \frac{t'}{k_3}, \quad \alpha = \frac{k_1}{k_2}, \quad \beta = \frac{k_2}{k_4}, \quad \delta = \frac{k_1}{k_3}. \tag{5}$$

For simplicity we omit the primes and obtain the following system

$$\frac{du}{dt} = \alpha\,v + u\,(1 - u - \beta\,v), \tag{6}$$

$$\frac{dv}{dt} = -\delta\,v - \beta\,u\,v. \tag{7}$$

The rest of this work is organized as follows. In Section 2 we apply the Painlevé test to understand analytical properties of system (6). In Section 3 we found two partial cases of the general solution for the Belousov–Zhabotinskii system at $\beta = 0$ and $\beta = 1$. Using the standard simplest equation method we present one-parameter exact solutions of the Belousov–Zhabotinskii system in Section 4. The key moment of this Letter is given in Section 5, where we present a new generalized approach for finding exact solutions and first integrals of nonlinear ordinary differential equations. In this Section we find the general solutions of the Belousov–Zhabotinskii system for some parameter values of mathematical model and obtain the first integrals of this system.

2. An application of the Painlevé test to the Belousov–Zhabotinskii reaction system

Detailed investigation of the integrability and the Painlevé property for low-dimensional systems was suggested in paper [16]. Here we apply the Painlevé test for understanding the integrability of the Belousov–Zhabotinskii reaction system of equations because to the best of our knowledge the Painlevé test was not used for consideration of this system. To apply the Painlevé test to (6) we transform the system to one equation. From the last equation of system (6) we get

$$u = -\frac{\delta}{\beta} - \frac{v_t}{\beta\, v}, \qquad (\beta \neq 0). \tag{8}$$

(The case $\beta = 0$ is considered in the next Section). Substituting (8) into the first equation of system (6) we have the second-order differential equation in the form

$$v\, v_{tt} - \left(1 + \frac{1}{\beta}\right) v_t^2 - \left(1 + \frac{2\,\delta}{\beta}\right) v\, v_t + + \beta\, v^2\, v_t + \beta\,(\alpha + \delta)\, v^3 - \left(\delta + \frac{\delta^2}{\beta}\right) v^2 = 0. \tag{9}$$

Let us apply the Painlevé test to study the integrability of equation (9) using three steps of the Kovalevsky algorithm [17, 18].

At the first step we look for the first member of the general solution expansion in the Laurent series substituting

$$u = \frac{a_0}{t^p}. \tag{10}$$

into equation (9). Note that (9) is autonomous and we keep in mind that we can change $t \to t - t_0$ in all results (where t_0 is an arbitrary constant).

Substituting (10) into the equation with the leading members

$$v\, v_{tt} - \left(1 + \frac{1}{\beta}\right) v_t^2 + \beta\, v^2\, v_t = 0 \tag{11}$$

we get $p = 1$ and $a_0 = \frac{\beta - 1}{\beta^2}$.

At the second step of the Painlevé test we substitute the expression in the form

$$u = \frac{\beta - 1}{\beta^2\, t} + u_j\, t^{1-j}, \tag{12}$$

into the equation with the leading members (11) and look for linear expression with respect to u_j. Here the coefficient u_j is one of coefficients for the general solution expansion in the Laurent series. As a result we obtain the Fuchs indices which are determined by formulas

$$j_1 = -1, \qquad j_2 = \frac{\beta - 1}{\beta}. \tag{13}$$

Taking into account the values of the Fuchs indices we see that equation (9) does not pass the Painlevé test in the general case. However it can have the Painlevé property in case of

$$\beta = \frac{1}{1 - N}, \quad (N \neq 1). \tag{14}$$

In this case the second Fuchs index will be integer.

Substituting the expansion in the form

$$u(t) = \frac{\beta - 1}{\beta^2\, t} + a_1 + a_2\, t + a_3\, t^2 + \ldots. \tag{15}$$

into equation (9) and equating the expressions at different powers t to zero, we get

$$a_1 = -\frac{\alpha\,\beta^2 + \beta^2 + \delta\,\beta^2 - 2\,\alpha\,\beta - \beta - \delta + \alpha}{2\,\beta^2}. \tag{16}$$

At $\beta = -1$ the coefficient $a_1 = -2\,\alpha - 1$ but the coefficient a_2 is not determined. Equation (9) passes the Painlevé test only in two cases

$$\delta_1 = -2\,\alpha, \qquad \delta_2 = -2\,\alpha - 1. \tag{17}$$

In case of $\beta = -\frac{1}{2}$ the coefficients a_1 and a_2 are determined. They take the form

$$a_1^{(1)} = \frac{3}{2}\,\delta - \frac{9}{2}\,\alpha - \frac{3}{2}, a_2^{(1)} = -\frac{3}{4}\,\delta - \frac{9}{8}\,\delta^2 - \frac{21}{4}\,\alpha\,\delta - \frac{15}{4}\,\alpha - \frac{81}{8}\,\alpha^2 - \frac{1}{8}. \tag{18}$$

However the coefficient a_3 can not be found. The equation passes the Painlevé test when δ is equal to

$$\delta_3 = -3\,\alpha, \qquad \delta_4 = -3\,\alpha - 1. \tag{19}$$

We can hope that this regularity is carried out for the next members of the expansion for the general solution.

One can see that it is difficult to determine the arbitrary constant a_j at large j because we need to calculate all previous coefficients. So we suggest checking the arbitrary constant at the third step of the Painlevé test to use the possible form of the first integral.

We know that the value of the Fuchs index obtained at the second step by Painlevé test corresponds to the order of poles for leading members of the first integral [19], [20]. Moreover it is well known that the Painlevé test is a necessary condition for integrability of nonlinear differential equations. And if we find the first integrals of equation (9) we will obtain the sufficient condition of integrability for this equation. We discuss this property in Section 5.

3. Solutions of the Belousov–Zhabotiskii system at $\beta = 0$ and $\beta = 1$

The second equation in the Belousov–Zhabotinskii system (6) at $\beta = 0$ can be easily solved. In this case we can write the equation for u(t) in the form

$$\frac{\mathrm{d\,u}}{\mathrm{d\,t}} = \mathrm{u} - \mathrm{u}^2 - \frac{\mathrm{C}_1^2\,\delta^2}{4}\,\mathrm{e}^{-\delta\,\mathrm{t}}, \tag{20}$$

where an arbitrary constant C_1 is taken in this form for convenience of calculations.

The general solution of equation (20) is expressed by formula

$$\mathrm{u} = -\frac{\psi_{\mathrm{t}}}{\psi}, \tag{21}$$

where $\psi(t)$ is expressed via the Bessel function in the form

$$\psi(\mathrm{t}) = \mathrm{e}^{-\frac{\mathrm{t}}{2}}\left[\mathrm{C}_2\,\mathrm{J}_{-\frac{1}{\delta}}\left(\mathrm{C}_1\,\mathrm{e}^{\frac{\delta\,\mathrm{t}}{2}}\right) + \mathrm{C}_3\,\mathrm{Y}_{-\frac{1}{\delta}}\left(\mathrm{C}_1\,\mathrm{e}^{\frac{\delta\,\mathrm{t}}{2}}\right)\right]. \tag{22}$$

In case of $\beta = 1$ from (9) we obtain the following equation

$$\mathrm{v\,v_{tt}} - 2\,\mathrm{v_t^2} - (1 + 2\,\delta)\,\mathrm{v\,v_t} + \mathrm{v^2\,v_t} + (\alpha + \delta)\,\mathrm{v}^3 - \left(\delta + \delta^2\right)\,\mathrm{v}^2 = 0. \tag{23}$$

From 23 at $\delta = 0$ or $\delta = -1$ we obtain the first integral of equation 9 in the form

$$\mathrm{v_t} + (1 + 2\,\delta)\,\mathrm{v} + \mathrm{v}^2 \ln(\mathrm{v}) + (\alpha + \delta)\,\mathrm{v}^2\,\mathrm{t} = \mathrm{C}_4\,\mathrm{v}^2, \tag{24}$$

Additionally in case of $\alpha = -\delta$ we obtain the solution in the form of integral

$$\int \frac{\mathrm{dv}}{\mathrm{C}_4\,\mathrm{v}^2 - (1 + 2\,\delta)\,\mathrm{v} - \mathrm{v}^2\,\ln\mathrm{v}} = \mathrm{t} + \mathrm{C}_5, \tag{25}$$

where C_5 is an arbitrary constant.

4. One–parameter exact solutions of the Belousov–Zhabotinskii reaction system

At the present time there are a lot of approaches for finding exact solutions of nonlinear differential equations (see, for example, papers [22], [23], [24], [25], [26], [27], [28], [29], [30], [31], [32]). However, their application to the Belousov–Zhabotinskii equation can give only one–parameter exact solutions. To find these exact solutions of equation (9) let us use the simplest equation method [22].

The general solution of equation (9) has the first order of pole and we can look for the exact solution of this equation in the form

$$v(t) = a_1\, Q(t), \tag{26}$$

where $Q(t)$ is a solution of the Riccati equation in the form

$$Q_t - A\,Q + Q^2 = 0. \tag{27}$$

Formula (26) was taken because the general solution of equation (27) also has the first order. Differentiating equation (27) in t we get that the solution of equation (27) also satisfies the following equation

$$Q_{tt} - A^2\,Q + 3\,A\,Q^2 - 2\,Q^3 = 0. \tag{28}$$

Substituting (26) into equation (9) and taking into account the derivatives $\mathrm{Q_t}$ and $\mathrm{Q_{tt}}$ from equations (27) and (28) we obtain the polynomial in Q(t) in place of differential equation (9). Equating expressions in this polynomial at the same degrees of Q(t) to zero, we have the algebraic equations with respect to parameters $\mathrm{a_1}$, A, α, β and δ.

Solving these equations we obtain

$$\mathrm{a_1} = \frac{\beta - 1}{\beta^2}. \tag{29}$$

We also get two cases. One of them is

$$\delta_1 = \frac{\alpha}{\beta} - \alpha - 1, \quad \mathrm{A_1} = 1 + \alpha - \frac{\alpha}{\beta} \tag{30}$$

and the second one is the following

$$\delta_2 = \frac{\alpha}{\beta} - \alpha, \quad \mathrm{A_2} = \alpha - \beta - \frac{\alpha}{\beta}. \tag{31}$$

Substituting these values of parameters into equation (27) and into (26) we have exact solutions of equation (9).

In the first case at $\delta = \delta_1 = \frac{\alpha}{\beta} - \alpha - 1$ we have the solution of equation (9) in the form

$$\mathrm{v_1(t)} = \frac{(\beta - 1)(\alpha\,\beta + \beta - \alpha)}{\beta^3 + \mathrm{C_1}\,\beta^2\,(\alpha\,\beta + \beta - \alpha)\,\mathrm{e}^{\frac{(\alpha - \beta - \alpha\,\beta)\,\mathrm{t}}{\beta}}} \tag{32}$$

and in the case of $\delta = \delta_2 = \frac{\alpha}{\beta} - \alpha$ the solution takes the form

$$\mathrm{v_2(t)} = \frac{(\beta - 1)(\alpha\,\beta - \beta^2 - \alpha)}{\beta^3 + \mathrm{C_1}\,\beta^2\,(\alpha\,\beta - \beta^2 - \alpha)\,\mathrm{e}^{\frac{(\alpha + \beta^2 - \alpha\,\beta)\,\mathrm{t}}{\beta}}}. \tag{33}$$

Here C_1 is an arbitrary constant.

Solutions (32) and (33) are kinks in time.

Other solutions of equation (9) can be found in cases of $A_1 = 0$ ($\alpha = \frac{\beta}{1-\beta}$, $\delta_1 = 0$) and at $A_2 = 0$ ($\alpha = \frac{\beta^2}{\beta-1}$, $\delta_2 = -\beta$). These rational solutions of equations (27) and (9) take the form

$$Q(t) = \frac{1}{t + C_1}, \qquad v(t) = \frac{\beta - 1}{\beta\,(t + C_1)}. \tag{34}$$

Solutions (32), (33) and (34) have only one arbitrary constant of the second–order ordinary differential equation and are not the general solutions of equation (9).

5. First integrals and the general solutions of the Belousov–Zhabotiskii reaction system

Let us consider the generalization of the simplest equation that allows us to obtain the first integrals and the general solution of equation (9). First of all let us note that the leading members of equation (27) have the second order of pole. Moreover the Fuchs index for the general solution of equation (9) at $\beta = -1$ is equal to $N = 2$. It is known that the arbitrary coefficient in the expansion of the general solution in the Laurent series corresponds to the leading members of the first integral [19], [20]. Using this observation let us look for the first integral of equation (9) at $\beta = -1$ assuming [21]

$$v(t) = b_1\,Q(t), \tag{35}$$

where Q(t) satisfies the equation in the form

$$Q_t = A\,Q - Q^2 + S(t). \tag{36}$$

As this takes place S(t) is an unknown function of t that can be found.

Differentiating (36) in t we have

$$Q_{tt} = A^2\,Q - 3\,A\,Q^2 + A\,S + 2\,Q^3 - 2\,Q\,S + S_t. \tag{37}$$

Substituting (35) into equation (9) and taking into account (36) and (37), after equating expressions at the same degrees of Q(t) to zero we have the algebraic system of equation. Solving this system of equations we get

$$b_1 = -2, \quad A = 2\,\alpha + 1, \quad \delta_1 = -2\,\alpha - 1, \quad \delta_2 = -2\,\alpha. \tag{38}$$

and

$$S_1 = C_1\,e^{(2\,\alpha+1)\,t}, \qquad S_2 = C_1\,e^{2\,\alpha\,t}. \tag{39}$$

The first integral at $\beta = -1$ and at $\delta = \delta_1 = -2\,\alpha - 1$ is expressed by formula

$$v_t - (2\,\alpha + 1)\,v - \frac{1}{2}\,v^2 = C_1\,e^{(2\,\alpha+1)\,t}. \tag{40}$$

and the first integral at $\beta = -1$, $\delta = \delta_2 = -2\,\alpha$ takes the form

$$v_t - (2\,\alpha + 1)\,v - \frac{1}{2}\,v^2 = C_1\,e^{2\,\alpha\,t}. \tag{41}$$

Equations (40) and (41) are reduced to the linear equations of the second order using the transformation

$$v(t) = \frac{\psi_z}{\psi}. \tag{42}$$

In fact the general solutions of equation (9) can be reduced to the nonlinear second-order differential equation which was studied by Painlevé school more than century ago [33], [34], [35], [36], so this example has an illustrative purpose.

Let us note that equation (9) can be transformed to the Abel equation of the second order that can be solved at conditions (38).

Let us also demonstrate the application of our approach for the value of the Fuchs index $N = -3$. In this case $\beta = \frac{1}{4}$ and we search for the first integral assuming $v(t) = b_1 Q$, where $Q(t)$ satisfies the equation in the form

$$Q^{-5} Q_t - AQ^{-4} + Q^{-3} = S(t). \qquad (43)$$

From (43) we get

$$Q_t = A\,Q - Q^2 + Q^5 S(t). \qquad (44)$$

Differentiating (44) in t we obtain the consequence of equation (44) in the form

$$Q_{tt} = 5\,S^2\,Q^9 - 7\,S\,Q^6 + (6\,A\,S + S_t)\,Q^5 + 2\,Q^3 - 3\,A\,Q^2 + A^2\,Q. \qquad (45)$$

Substituting (35) into equation (9) and using equations (44) and (45) we obtain a polynomial in $Q(t)$ again. Equating expressions at the same degrees $Q(t)$ to zero, after solving the algebraic system of equations we get

$$b_1 = -12, \quad \delta_1 = 3\,\alpha - 1, \quad \delta_2 = 3\,\alpha, \quad A_1 = 1 - 3\,\alpha, \quad A_2 = -3\,\alpha - \frac{1}{4}, \qquad (46)$$

and functions $S_1(t)$ and $S_2(t)$ in the form

$$S_1(t) = C_1\,e^{(12\,\alpha - 3)\,t}, \qquad S_2(t) = C_1\,e^{12\,\alpha\,t},. \qquad (47)$$

We have found that at $N = -3$ equation (9) passes the Painlevé test and, what is more, we get the sufficient condition for the Painlevé property at values of parameters (46). As a result we have the two first integrals for equation (9) at $\beta = \frac{1}{4}$.

The first integral at $\delta_1 = 3\,\alpha - 1$ takes the form

$$v_t - (3\,\alpha - 1)\,v - \frac{1}{12}\,v^2 = C_2\,v^5\,e^{(12\,\alpha - 3)\,t}, \qquad (48)$$

and at $\delta_2 = 3\,\alpha$ the first integral can be written as

$$v_t - \left(3\,\alpha + \frac{1}{4}\right)\,v - \frac{1}{12}\,v^2 = C_2\,v^5\,e^{12\,\alpha\,t}, \qquad (49)$$

where C_2 is an arbitrary constant. We can use this approach to check the third step of the Painlevé test for equation (9) in case of other values of the Fuchs index j_2.

Taking into consideration the observations mentioned above we can look for the first integral of equation (9) in the form

$$v^{-\frac{\beta+1}{\beta}}\,v_t + A\,v^{-\frac{1}{\beta}} + B\,v^{\frac{\beta-1}{\beta}} = S_n(t), \qquad (50)$$

where the parameters A and B are unknown. We can find them if we substitute derivatives v_t and v_{tt} into equation (9).

We obtain at $B = \frac{\beta^2}{\beta-1}$ the following values of parameters δ and A

$$\delta_1 = \frac{\alpha}{\beta} - 1 - \alpha, \quad A_1 = 1 + \alpha - \frac{\alpha}{\beta}, \quad \delta_2 = \frac{\alpha}{\beta} - \alpha, \quad A_2 = \alpha - \beta - \frac{\alpha}{\beta}. \qquad (51)$$

Using the values of parameters (51) and the form of first integrals (50) we get the two first integrals of equation (9) for two values δ.

At $\delta_1 = \frac{\alpha}{\beta} - 1 - \alpha$ we have the first integral in form

$$v^{-\frac{\beta+1}{\beta}} v_t + \left(1 + \alpha - \frac{\alpha}{\beta}\right) v^{-\frac{1}{\beta}} + \frac{\beta^2}{\beta - 1} v^{\frac{\beta-1}{\beta}} = I_1 \, e^{\frac{(\alpha-\beta)(\beta-1)}{\beta^2} t}, \quad (\beta \neq 0;\ 1) \tag{52}$$

and at $\delta_2 = \frac{\alpha}{\beta} - \alpha$ we get in the form

$$v^{-\frac{\beta+1}{\beta}} v_t + \left(\alpha - \beta - \frac{\alpha}{\beta}\right) v^{-\frac{1}{\beta}} + \frac{\beta^2}{\beta - 1} v^{\frac{\beta-1}{\beta}} = I_2 \, e^{\frac{(\beta-1)\,\alpha}{\beta^2} t}, \quad (\beta \neq 0;\ 1) \tag{53}$$

where I_1 and I_2 are arbitrary constants.

These first integrals can also be written in the another form.

At $\delta = \delta_1 = \frac{\alpha}{\beta} - \alpha - 1$ the first integral takes the form

$$v_t + \left(1 + \alpha - \frac{\alpha}{\beta}\right) v + \frac{\beta^2}{\beta - 1} v^2 = I_1 \, v^{\frac{\beta+1}{\beta}} \, e^{\frac{(\beta-1)(\beta-\alpha)}{\beta^2} t}, \quad (\beta \neq 0;\ 1) \tag{54}$$

and at $\delta = \delta_2 = \frac{\alpha}{\beta} - \alpha$ we get the first integral in the form

$$v_t + \left(\alpha - \beta - \frac{\alpha}{\beta}\right) v + \frac{\beta^2}{\beta - 1} v^2 = I_2 \, v^{\frac{\beta+1}{\beta}} \, e^{\frac{(\beta-1)\,\alpha}{\beta^2} t}, \quad (\beta \neq 0;\ 1). \tag{55}$$

From the first integrals (54) and (55) we see that at $\delta = \delta_1$ in case of $\alpha < 1$ and $\alpha < \beta < 1$ we have a dissipative process for the first integral of the Belousov–Zhabotinskii reaction system. In case of $\delta = \delta_2$ we observe the dissipation in the first integral at $\alpha > 0$ and $\beta < 1$ or at $\alpha < 0$ and $\beta > 1$.

In case of $\delta = \delta_1$ and $\alpha = \beta$ (or $\delta = \delta_2$ and $\alpha = 0$) the general solution of equation (9) is found from the first integrals (54) and (55) and is determined by the formula

$$\mathrm{t} + \mathrm{C}_6 = \int \frac{(\beta - 1)\,\mathrm{dv}}{\mathrm{I}_1\,(1 - \beta)\,\mathrm{v}^{\frac{1+\beta}{\beta}} + \beta^2\,\mathrm{v}^2 + \beta\,(1 - \beta)\,\mathrm{v}} \tag{56}$$

where C_6 is an arbitrary constant.

Table 1. Exact solutions of the Belousov–Zhabotinskii system at given parameters

n	α	β	δ	$v(t)$
1	α	0	δ	(18)
2	α	1	0	(23)
3	α	1	-1	(23)
4	α	β	$\frac{\alpha}{\beta} - \alpha - 1$	(30)
5	α	β	$\frac{\alpha}{\beta} - \alpha$	(31)
6	$\frac{\beta}{1-\beta}$	β	0	(32)
7	$\frac{\beta^2}{\beta-1}$	β	$-\beta$	(32)
8	α	α	$-\alpha$	(54)
9	0	β	0	(54)

Exact solutions of the two–component Belousov–Zhabotinskii equation (9) are given at Table 1.

6. Conclusion

In this paper we have considered the integrability of the Belousov–Zhabotinskii reaction system using the Painlevé approach. We have found that this system could not be integrable in general case, and we tried to look for conditions on parameters of this system when it can be integrable. Using the Painlevé test we have obtained conditions on parameter β of the mathematical model when the system can pass the Painlevé test. Taking into account the standard simplest equation method for finding exact solutions we have obtained two exact solutions of the considered system. Using the link between Fuchs indices and first integrals of nonlinear differential equations we suggested the modification in the third step of the Painlevé test for nonlinear differential equations. We have also found the first integrals of the Belousov–Zhabotinskii reaction system for some values of parameters δ taking into account the values of the Fuchs indices. It is worth to note that we used the Painlevé approach to find the first integrals of the Belousov–Zhabotinckii reaction system. But we obtained that there are the first integrals for values of parameters when equation (9) does not pass the Painlevé test.

Acknowledgments

This research was supported by Russian Science Foundation Grant No 18-11-00209 "Development of methods for investigation of nonlinear mathematical models".

References

[1] Field R.J., Noyes R.M., 1974, *J. Chem. Phys.* **60**, 1877.
[2] Showalter K., Noyes R.M., Bar-Eli K., 1978, *J. Chem. Phys.* **69**, 2514.
[3] Field R.J., Burger M., 1985 *Oscillations and Travelling Waves in Chemical Systems*, (Wiley, New-York,).
[4] Winfree A.T., 2000 *The Geometry of Biological Time*, (Springer–Verlag, Berlin–Heidelberg–New York, 2nd edition).
[5] Tyson J.J., 1994, *What everyone should know about the Belousov–Zhabotinsky reaction. In Frontiers in Mathematical Biology, vol. 100 of Lect. Notes in Biomathematics*, (Springer–Verlag, Berlin-Heidelberg-New York) 569.
[6] Tyson J.J., 1977 *J. Chem. Phys.* **66 (3)**, 905.
[7] Ye Q.X., Wang M.X., 1987, *Nonl. Analysis*, **11(11)**, 1289.
[8] 2001 Shiwang M., *Journal of Differential Equations*, **171**, 294.
[9] 2013 Trofimchuk E., Pinto M., Trofimchuk S. , 2013 *Journal of Differential Equations*, **254(9)**, 3690.
[10] Troy W.C., 1980, *Journal of Differential Equations*, **36(1)**, 89.
[11] Zhibin L., Shi He, 1996, *Appl. Math.*, **11B**, 1.
[12] Murray J.D., 1976 *Journal of Theoretical Biology*, **56(2)**, 329.
[13] Murray J.D., 2001, *Mathematical biology. I. An introduction*, (Springer-Verlag, Berlin).
[14] Murray J.D., 2003 *Mathematical biology. II. Spatial models and biomedical applications*, (Springer-Verlag, Berlin).
[15] Kudryashov N.A., Zakharchenko A.S. 2014 *Chaos, Soliton Fract.*, **65**, 111.
[16] Ramani A., Dorizzi B., Grammaticos B., Bountis T., 1984 *J. Math.Phys.* **25(4)**, 878.
[17] Ablowitz M.J., Ramani A., Segur H., 1984 *Nuovo Cim.*, **23**, 878.
[18] Ablowitz M.J., Ramani A., Segur H., 1980. *J. Math.Phys.* **21**, 715; 1006.
[19] Yoshida H., 1983 *Celest. Mech.*, 31, 363;381.
[20] Kudryashov N.A., 2005 *Chaos, Solitons and Fractals*, **26(2)**, 591.
[21] Kudryashov N.A., 2018 *Regular and Chaotic Dynamics*, **23(4)**, 471.
[22] Kudryashov N.A., 2005 *Chaos Soliton Frac.*, **24**, 1217.
[23] Kudryashov N.A., 2012 *Commun. Nonlinear Sci. Numer. Simulat.* 17, 2248.
[24] Biswas A., 2009 *Appl. Math. Lett.*, **22**, 208.
[25] Parkes E.J., Duffy B.R. , 1996 *Comput. Phys. Commun.*, **98**, 288.
[26] A.D. Polyanin, V.F. Zaitsev, 2012 *Handbook of Nonlinear Partial Differential Equations*, 2nd ed. (Chapman and Hall/CRC, Boca Raton).
[27] Malfliet W., Hereman W., 1996, *Phys. Scripta*, **54**, 563.
[28] Vitanov N.K., 2010, *Commun. Nonlinear Sci. Numer. Simulat.*, **15**, 2050.
[29] Kudryashov N.A., 2008, *Regular and Chaotic Dynamics*, **13(3)**, 234.
[30] Eslami M., Neirameh N., 2014, *Eur. Phys. J. Plus*, **129**, 54.

[31] M.B. Hubert, N.A. Kudryashov, M.Justin, S. Abbagari, G, Betchewe, S.Y. Doka, 2018 *Eur. Phys. J. Plus*, **133**, 108.
[32] Zayed E.M.E., Al-Nowehy A.G., 2017, *Eur. Phys. J. Plus*, **132**, 475.
[33] Ince E.L., 1956, *Ordinary differential equations*, (Dover, New York)
[34] P. Painleve, 1902, *Acta Math.*, **25(1)**, 1.
[35] B. Gambier, 1910, *Acta Math.*, 1.
[36] A.D. Polyanin, V.F. Zaitsev, 2003 *Handbook of Exact Solutions for Ordinary Differential Equations*, (Boca Ration, Chapman and Hall/CRC).

IOP Conf. Series: Journal of Physics: Conf. Series **1205** (2019) 012031 doi:10.1088/1742-6596/1205/1/012031

Remarks on the Fuchs indices and the first integrals for nonlinear ordinary differential equations

Nikolay A Kudryashov

National Research Nuclear University MEPhI (Moscow Engineering Physics Institute), 31 Kashirskoe Shosse, 115409 Moscow, Russian Federation

E-mail: nakudr@gmail.com

Abstract. The Painlevé analysis of nonlinear ordinary differential equations is used to construct the first integrals. The connection between the Fuchs indices and the first integrals of nonlinear differential equations is discussed. Some simple prepositions are presented. Some first integrals for nonlinear ordinary differential are found taking into account the values of the Fuchs indexes.

1. Introduction

It is well known that many processes in physics, biology, economics and in other disciplines are described by nonlinear differential equations and there is the important problem for finding the analytical solutions of them. Recently many methods were suggested for finding exact solutions of nonlinear differential equations.

At the present we have a long list of nonlinear partial differential equations for which the Cauchy problem can be solved by the Inverse scattering transform. However many processes are described by nonlinear nonintegrable partial differential equations and we need to have exact solutions for these differential equations too.

In fact, the first attempt of finding exact solutions for nonlinear nonintegrable differential equations were made in works [1, 2, 3] almost thirty years ago and later many methods have been suggested to find exact solutions of various nonlinear differential equations (see, for example papers [4, 5, 6, 7, 8, 9, 10, 11, 12, 13]). However we should note here that some of the methods are not effective and some of them lead to many redundant exact solutions (see papers [14, 15, 16, 17, 18]).

In this paper we study the connection between analytical properties of nonlinear ordinary differential equations and first integrals for them. This approach allows us to look for the general solutions for some nonlinear differential equations. Let us consider the nonlinear ordinary differential equation in the general form

$$E(w, w_z, w_{zz}, \ldots, w_{n,z}, z) = 0, \qquad w_{n,z} = \frac{d^n w}{dz^n}. \tag{1}$$

Let us assume that the general solution w(z) has pole p. It means that w(z) takes the form

$$w(z) = \frac{a_0}{(z - z_0)^p} \tag{2}$$

or has the expansion in the Laurent series in the form

$$w(z) = \frac{a_0}{(z - z_0)^p} + \ldots, \tag{3}$$

where z_0 is a movable singular point.

There are some simple prepositions.

Preposition 1. The derivative $w_{k,z}$ in z of order k has the order of pole $k + p$.

Proof. The proof follows from the following calculation

$$w_{k,z} = \frac{(-1)^p\, p\,(p-1)\ldots(p-k)\, a_0}{(z - z_0)^{p+k}} + \ldots, \tag{4}$$

Preposition 2. The power m of function w(z) has the order of pole $m\,p$.

Proof. The proof follows from the formula

$$w^m = \frac{a_0^m}{(z - z_0)^{m\,p}} + \ldots, \tag{5}$$

2. The Fuchs indices and the first integrals for nonlinear ordinary differential equations

For nonlinear ordinary differential equations one can suggest some additional properties. To understand these properties let us consider the nonlinear ordinary differential equation in the form

$$E_k = w_{k,z} - b\, w^m = 0. \tag{6}$$

Preposition 3. The order of pole for the general solution of equation (6) is determined by formula

$$p = \frac{k}{m-1}. \tag{7}$$

Proof. The proof follows from the formula

$$p + k = m\,p. \tag{8}$$

Preposition 4. The derivative of E_k in z does not change the order of pole for the general solution of equation (6).

Proof. The differentiation of equation (6) leads to multiplying formula (8) on integer.

Example 1. Let us take the first order equation in the from

$$E_1 = w_z + w^2 = 0. \tag{9}$$

The general solution of equation (9) takes the pole of the first order $p = 1$ and takes the form

$$w = \frac{1}{z - z_0}. \tag{10}$$

Differentiating (9) in z we obtain the second-order differential equation in the form

$$E_{1a} = w_{zz} + 2\, w\, w_z = 0. \tag{11}$$

We can see that the order of the pole for the general solution of equation (11) is not changed and solution (10) satisfies equation (11).

IOP Conf. Series: Journal of Physics: Conf. Series **1205** (2019) 012031 doi:10.1088/1742-6596/1205/1/012031

Example 2. Let us consider the second-order equation in the form

$$E_2 = w_{zz} - 6\,w^2 = 0. \tag{12}$$

The general solution of equation (12) has the second order pole $p = 2$ and is expressed via the Weierstrass elliptic function

$$w = \wp(z - z_0, 0, g_3), \tag{13}$$

where z_0 and g_3 are arbitrary constants.

At investigation of the analytical properties for nonlinear differential equation appears the important notion – the Fuchs indices. The discussion this notion can be found in many books. For equation (9) we obtain the one Fuchs index which is equal to $j_1 = -1$. For equation (12) we have two Fuchs indices $j_1 = -1$ and $j_2 = 6$.

Remark 1. It is obviously that differentiation of equation leads to the change of the general solution of equation.

We can see that the equation of the second order (11) that is found by differentiation of equation (8) in z has the general solution with the first order of pole in the form

$$w(z) = \sqrt{C_1}\tanh\Big(\sqrt{C_1}(z - z_0)\Big). \tag{14}$$

because equation (11) has the first integral with arbitrary constant C_1 in the form

$$E_{1a} = w_z + w^2 - C_1 = 0. \tag{15}$$

It is well known that the quantity of the Fuchs indices is equal to the order of nonlinear differential equation. So, differentiating equation (11) in z we have the third-order equation

$$E_{3a} = w_{zzz} + 2\,w\,w_{zz} + 2\,w_z^2 = 0 \tag{16}$$

which have three Fuchs indices in the form

$$j_1 = -1, \quad j_2 = 2, \quad j_3 = 3. \tag{17}$$

Equation (17) has the general solution that was found from the irreducible equation in the form

$$E_{1aa} = w_z + w^2 - C_1\,z - C_2 = 0 \tag{18}$$

and that is expressed via the Airy functions.

Let us note that the the order of pole for leading members of equation (18) w_z and w^2 is equal to 2 and correspond to the Fuchs index $j_2 = 2$. Arbitrary constant C_2 also corresponds to the Fuchs index $j_2 = 2$. At the same time the arbitrary constant C_1 in equation (18) correspond to the Fuchs index $j_3 = 3$ because multiplying on z leads to the second order of pole.

So, the values of the Fuchs indexes point out us the leading members of the First integrals in the polynomial form for nonlinear ordinary differential equations. These ideas can be used for finding first integrals of nonlinear differential equations. Let us demonstrate our approach taking into account three examples.

Example 3. Consider the nonlinear ordinary differential equation in the form

$$y_{zzz} + \delta\,y_{zz} + \varepsilon\,y_z + \beta\,y\,y_z + \alpha\,y^2 - C_0\,y + C_1 = 0. \tag{19}$$

Equation (19) can be obtained from the nonlinear evolution equation for description of nonlinear waves in a convecting fluid

$$u_t + \alpha\,u\,u_x + \beta\,(u\,u_x)_x + \varepsilon\,u_{xx} + \delta\,u_{xxx} + u_{xxxx} = 0 \tag{20}$$

if we look for exact solution using the traveling wave

$$u(x,t) = y(z), \quad z = x - C_0\, t. \tag{21}$$

Equation (20) is quite popular and was studied many times. It was obtained in works [19, 20] for description of long shallow wave evolution in a convecting fluid in case when the Rayleigh number slightly exceeds its critical value.

Nonlinear waves described by equation (20) were studied in paper [21]. Some exact solutions of the nonlinear dissipative equation (20) for surface waves in a convecting liquid layer were obtained in works [22, 23, 24, 25]. The study of equation (20) was presented in [26] including symmetries (nonlocal and contact transformations), similarity reductions and the application of the Ablowitz–Ramani–Segur algorithm to the reductions.

Various methods were applied to solve equation (20) in [27]. Some explicit traveling wave solutions of equation (20) were obtained in paper [28] through factorization techniques when coefficients of the equation fulfill a certain condition. Some exact solutions of equation (20) were found in paper [29] using the tanh-expansion method. Using some kinds of standard and nonstandard truncation approaches for extended singular manifolds to an equation of convecting fluid in paper [30] solitary wave structures were obtained.

Equation (19) has the following Fuchs indices at the expansion of the general solution in the Laurent series

$$y_1 = -1, \quad j_2 = 4, \quad j_3 = 6. \tag{22}$$

Taking into account the Fuchs index $j_2 = 4$ we can look for the First integral in the form

$$y_{zz} + a\,y^2 + b\,y + c\,y_z - S(z) = 0, \tag{23}$$

where expressions y_{zz} and y^2 are taken because they has the fourth order of poles, parameters a, b and the function S(z) can be found.

Substituting y_{zz} from (23) into equation (19) and equating various expressions of powers y_z and y to zero we obtain the following conditions

$$a = -\frac{\beta}{2}, \quad b = \varepsilon - \frac{2\,\alpha\,\delta}{\beta} - \frac{4\,\alpha^2}{\beta^2}, \quad c = \delta + \frac{2\,\alpha}{\beta}, \tag{24}$$

$$C_0 = \frac{2\,\alpha\,\epsilon}{\beta} - \frac{8\,\alpha^3}{\beta^3} - \frac{4\,\alpha^2\delta}{\beta^2}, \tag{25}$$

$$S(z) = \frac{C_1\,\beta}{2\,\alpha} + C_2\,e^{-2\,\frac{\alpha\,z}{\beta}}. \tag{26}$$

Using conditions (24) and 26 we have that nonlinear ordinary differential equation (19) at C_0 determined by formula (25) has the first integral in the form

$$y_{zz} - \frac{\beta}{2}\,y^2 + \left(\varepsilon - \frac{2\,\alpha\,\delta}{\beta} - \frac{4\,\alpha^2}{\beta^2}\right)y + \left(\delta + \frac{2\,\alpha}{\beta}\right)y_z - \frac{C_1\,\beta}{2\,\alpha} + C_2\,e^{\frac{-2\,\alpha\,z}{\beta}} = 0, \tag{27}$$

where C_2 is an arbitrary constant.

Taking into consideration the first integral (27) we can look for exact solutions of equation (19).

Example 4. As the second example of nonlinear differential equation let us consider the nonlinear ordinary differential equation

$$y_{zzz} - \mu\,y_{zz} + \alpha\,y\,y_z - \beta\,y^2 - C_0\,y_z - \varepsilon\,y = 0. \tag{28}$$

IOP Conf. Series: Journal of Physics: Conf. Series **1205** (2019) 012031 doi:10.1088/1742-6596/1205/1/012031

Equation (28) is found from the generalized Korteweg-de Vries equation with source in the form

$$u_t + \alpha\, u\, u_x + u_{xxx} = \mu\, u_{xx} + \beta\, u^2 + \varepsilon\, y \tag{29}$$

if we look for exact solution using the traveling wave (21).

Equation (28) has the following Fuchs indices at the expansion of the general solution in the Laurent series

$$y_1 = -1, \quad j_2 = 4, \quad j_3 = 6. \tag{30}$$

Taking into account the Fuchs index $j_2 = 4$ we can look for the First integral again in the form

$$y_{zz} - a\, y^2 + b\, y_z + d\, y - S(z) = 0, \tag{31}$$

where parameters a, b, d and the function S(z) can be found.

Substituting y_{zz} from (31) into equation (28) and equating various expressions of powers y_z and y to zero we obtain

$$a = -\frac{\alpha}{2}, \quad b = \mu - \frac{2\,\beta}{\alpha}, \quad d = -\frac{\alpha\,\varepsilon}{2\,\beta}, \tag{32}$$

$$C_0 = \frac{4\,\beta^2}{\alpha^2} - \frac{2\,\beta\,\mu}{\alpha} - \frac{\alpha\,\varepsilon}{2\,\beta}, \tag{33}$$

$$S(z) = C_1\, e^{-\frac{2\,\beta\, z}{\alpha}}. \tag{34}$$

Using conditions (32) and (34) we obtain that nonlinear ordinary differential equation (28) at (33) has the first integral in the form

$$y_{zz} + \frac{\alpha}{2}\, y^2 - \left(\mu - \frac{2\,\beta}{\alpha}\right) y_z + \frac{\alpha\,\varepsilon}{2\,\beta}\, y + C_1\, e^{-\frac{2\,\beta\, z}{\alpha}} = 0, \tag{35}$$

where C_1 is an arbitrary constant.

Example 5. As the third example of nonlinear differential let us consider the equation in the form

$$y_{zzz} - \mu\, y_{zz} + \alpha\, y^2\, y_z - \beta\, y^3 - C_0\, y_z - \varepsilon\, y = 0. \tag{36}$$

Equation (36) can be found from the generalized modified Korteweg-de Vries equation with source in the form

$$u_t + \alpha\, u^2\, u_x + u_{xxx} = \mu\, u_{xx} + \beta\, u^3 + \varepsilon\, y \tag{37}$$

if we look for exact solution using the traveling wave (21).

Equation (36) has the following Fuchs indices at the expansion of the general solution in the Laurent series

$$y_1 = -1, \quad j_2 = 3, \quad j_3 = 4. \tag{38}$$

Taking into account the Fuchs index $j_2 = 3$ we can look for the First integral in the form

$$y_{zz} - a\, y_z - b\, y^3 - h\, y + S(z) = 0, \tag{39}$$

where parameters a, b, h and the function S(z) can be found.

Substituting y_{zz} from (39) into equation (36) and equating various expressions of powers y_z and y to zero we obtain

$$a = \mu - \frac{3\,\beta}{\alpha}, \quad b = -\frac{\alpha}{3}, \quad h = -\frac{\alpha\,\epsilon}{3\,\beta}, \tag{40}$$

$$C_0 = \frac{9\,\beta^2}{\alpha^2} - \frac{3\,\mu\,\beta}{\alpha} - \frac{\alpha\,\varepsilon}{3\,\beta}, \tag{41}$$

$$S(z) = C_1\, e^{\frac{3\beta z}{\alpha}}. \tag{42}$$

Using conditions (40) and (42) we obtain that nonlinear ordinary differential equation (36) at condition (41) has the first integral in the form

$$y_{zz} - \frac{\alpha}{3}\, y^3 - \left(\mu - \frac{3\beta}{\alpha}\right) y_z - \frac{\alpha\,\varepsilon}{3\,\beta}\, y - C_1\, e^{\frac{3\beta z}{\alpha}} = 0, \tag{43}$$

where C_1 is an arbitrary constant.

These examples illustrate that the values of the Fuchs indices allows us to see the form of the first integrals for nonlinear ordinary differential equations. The open question is the connection of the Fuchs indices with the first integrals for the Painlevé equations and higher-order Painlevé analogies [32, 33, 34]. We know that these equations have arbitrary constants at expansion of the general solution in the Laurent series but all these equations do not have any first integrals in the polynomial form.

3. Acknowledgments

This research was supported by Russian Science Foundation Grant No 18-11-00209 "Development of methods for investigation of nonlinear mathematical models".

References

[1] Kudryashov N.A., 1988 *Journal of Applied Mathematics and Mechanics*, **52(3)**, 360
[2] Kudryashov N.A., 1990 *Phys. Lett. A.* **147** 287.
[3] Kudryashov N.A., 1991 *Phys.Lett. A.* **155** 269.
[4] Parkes E.J., Duffy B.R., 1996 *Computer Physics Communications*, **98** 288.
[5] Malfliet W., Hereman W., 1996 *Phys. Scripta* **54** 563.
[6] Fan E., 2000 *Physics Letters A*, **227(4-5)** 212.
[7] Polynin A.D., Zaitsev V.F. 2003 *Handbook of Exact Solutions for Ordinary Differential Equations*(Chapman and Hall/CRC, Boca Ration–London).
[8] Kudryashov N.A., 2005 *Chaos Soliton Fractals* **24** 1217.
[9] Kudryashov N.A., 2005 *Physics Letters A.*, **342**, 99.
[10] Kudryashov N.A. 2010, *Appl. Math. Comput.* **217**, 1755.
[11] Kudryashov N.A., Loguinova N.B., 2008 *Appl Math Comput* **205**396.
[12] Kudryashov N.A., 2012 *Commun. Nonlinear Sci. Numer. Simulat.* **17** (2012) 2248.
[13] Kudryashov N.A., Loguinova N.B., 2009 *Commun Nonlinear Sci Numer Simulat* **14** 1881.
[14] Kudryashov N.A., 2009 *Commun Nonlinear Sci. Numer. Simulat.* **14** 1891.
[15] Kudryashov N.A., 2009 *Commun Nonlinear Sci Numer Simulat.* **14** 3507.
[16] Kudryashov N.A., 2018 *Regular and Chaotic dynamics*, **23(4)**
[17] Kudryashov N.A. 2018 *Regular and Chaotic Dynamics*, **23** 152.
[18] Polyanin A.D. and Zaitsev V.F., 2018 *Handbook of Ordinary Differential Equations: Exact Solutions, Methods, and Problems*, (CRC Press, Boca Raton–London).
[19] Alfaro C.M., Depassier M.C., 1989 *Physical Revieew Letters*, **62(22)** 2597.
[20] Aspe H., Depassier M.C., 1990 *Physical Revieew A*, **22(19) 008** 4135.
[21] Garazo A.N.,Velarde M.G., 1991 *Physics of Fluids A: Fluid Dynamics*, **3** 2295.
[22] Kliakhander I.L., Porubov A.V., Velarde M.G., 2000, *Physical Review E.*, **62(4)** 4959.
[23] Porubov A.V. , 1993 *J. of Phys. A.: Math. Gen.* **26** L797.
[24] Porubov A.V., 1996 *Physics Letters A* **221** 391.
[25] Kudryashov N.A., 2015 *Commun Nonlinear Sci Numer Simulat* **28** 1.
[26] Cerveró J.M., Zurròn O., 1996 *Jurnal of Nonlinear Mathematical Physics*, **3(1-2)**, 1.
[27] Abourabia Aly Maher, Soliman Ibrahim Mohammed, 2015 *Eur. Phys. J. Plus*, **130**130.
[28] Conejo–Pérez Octavio, Rosu Haret C., 2010 *Cent. Eur. J. Phys.* **9(4)** 523.
[29] Lou Sen-yue, Huang Guo-xiang and Ruan Hang-yu, 1991 *J. Phys. A. Math. Gen.* **24** L587.
[30] Lou Sen-yue, Zhu Yue Jin and Wang Lie Yan, 1999 *Commun. Theor. Phys.* **32** 563.
[31] Kudryashov N.A., 2008 *Regular and Chaotic Dynamics*, **13 (3)**, 234.
[32] Kudryashov N.A., 1998 *Journal of Physics A. Math. Gen.*, **31**, L129.
[33] Kudryashov N.A., 1999 *Journal of Physics A. Math. Gen.*, **32**, 999.
[34] Kudryashov N.A., 2002 *J. Phys. A.: Math. Gen.* **35** (2002) 4617.

Modelling of some mechanism of metal electroplasticity under pulsed high-energy electromagnetic field action

K V Kukudzhanov* **and A L Levitin****

Ishlinsky Institute for Problems in Mechanics of the Russian Academy of Sciences, Moscow, Russia

E-mail: *kconstantin@mail.ru, **alex_lev@ipmnet.ru

Abstract. The processes of evolution of microcracks and micropores occurring in the material during processing of metal specimens by short-term pulses of high-energy electromagnetic field (HEEMF) are considered. The study is done numerically on the basis of a coupled model of the action an intense electromagnetic field on a previously damaged thermoelastoplastic material with an ordered system of defects, which takes into account the melting and evaporation of the metal and the dependence of all its physical and mechanical properties on temperature. Simulation has shown that, under certain conditions, microcracks can be almost completely healed. There is welding of the cracks by simultaneously decreasing the length of the microcrack, ejecting a jet of molten metal from the crack tip into the crack and closing its edges. The influence of the geometry and orientation of microdefects in the process of their healing is investigated. Based on the simulation results, simple approximate dependencies of the metal damage under the influence of the HEEMF on its initial damage, the characteristic "length" of the microdefects and their slope are obtained.

1. Introduction

Electroplasticity phenomenon is the material yield strength decrease and the ultimate plastic strain increase under a short electromagnetic pulses during or after plastic deformation. Plastic strain is accompanied by the appearance of microscopic discontinuity of the structure (microdamages or microdefects) and an increase in their number and size. Microdefects with linear dimensions of 3–10 μm are the most common defect size in the polycrystalline metals. The assumption on possibility of healing (transformation) of intergranular and intragranular microdefects in metals under the action of short pulses of a high-energy electromagnetic field (HEEMF) was done by researchers [1–5] initially to explain the phenomenon of electroplasticity of polycrystalline metals. In these works, the hypothesis of healing was supported by analytical and numerical solutions of model problems. Healing was understood as the barriers creation (due to HEEMF action) for the further cracks propagation: the occurrence of compress stresses in the tips of microcracks and the approaching of their edges, accompanied by the melting of craters (pores) in the tips. Later, in [6], on the basis of thermodynamic considerations it was shown that when we apply the current over certain threshold value, the "length" of the elliptical crack can decrease.

Since then, the hypothesis of the defects healing has received serious experimental confirmation [7–11]. However, in these experiments not the creation of barriers to the microcracks propagation was observed, but a change of defect shape and the metals structure continuity recovery, accompanied by the microdefects volume fraction change. Thus, the experiments showed a decrease in the damage (porosity) of the material. Meanwhile, the proposed in [3–6] mathematical models did not explain damage decrease due to HEEMF action.

For a mathematical description of the physical processes taking place in the vicinity of microcracks under the action of HEEMF pulses, a model of the effect of a pulsed HEEMF on a pre-damaged material with defects was proposed [12–14]. With this model it was possible to reproduce the experimentally observed process of microdefects transformation and metal damage decrease. In [15–17] the authors showed that the shape and mutual arrangement of microdefects practically do not influence on the dependence of healing or damage of the metal on time, but depends only on the initial damage (with the same initial "length" and orientation of the microcracks).

In the present work, we consider the influence of the geometry and orientation of microcracks and micropores on changes in the healing and damage in a metal under pulsed HEEMF action.

2. Electro-thermo-mechanical model

A damaged conductive material with microdefects (microcracks or micropores) with sizes $l_0 \times h_0$ is considered. All microdefects have uniform shape, size and spatial orientation (Fig. 1). We consider a plane strain and the size of the defect along the z-axis is assumed to be infinite. We assume that the defects in the material are arranged at the nodes of the rectangular lattice (Fig. 1b). In this case, separation of a representative cell is easy possible (Fig. 1a).

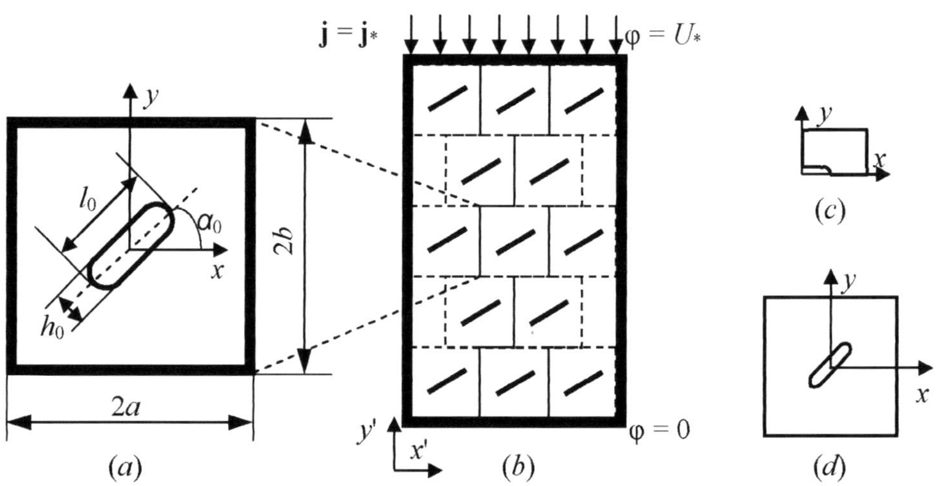

Figure 1. (a) Representative cell with a micro-defect; (b) the ordered structure of micro-defects, here the arrows show the direction of the electric current density vector at the specimen boundary; (c–d) domains of integration used in the calculations.

Material is under the action of a pulsed HEEMF (potential difference at top and bottom specimen boundaries) that causes the electric current in the specimen with a density $10^8 - 10^{11}$ A/m^2 and duration $10^{-5} - 10^{-4}$ s. The specimens sizes are much larger than microdefect size.

The solution is obtained in the integration cells shown in Fig. 1c and Fig. 1d and containing either one quarter (for $\alpha_0 = 0$) or one whole representative cell (for $\alpha_0 \neq 0$) [14–16].

The basic assumptions, equations of the electro-thermo-mechanical model, the initial, boundary, contact conditions and conditions on the interphases (solid–melt, melt–gas) boundaries are described in detail in [12, 13, 15, 16].

In this paper we consider the parameters of the damage (porosity) $f(t)$ and healing $\chi(t)$:

$$f(t) = \frac{V(t)}{V_{\mathrm{re}}}, \qquad \chi(t) = \frac{V(0) - V(t)}{V(0)} \tag{1}$$

where $V(t)$ is the volume of a single microdefect, $V_0 = V(0)$ is the initial volume of the defect, V_{re} is the volume of the representative cell in which microdefect is located. The initial material damage at time $t = 0$ is $f_0 = f(0) = V_0/V_{\mathrm{re}}$. The volume of the microdefect under the action of the HEEMF decreases [12, 13] and during pulse action the damage (porosity) will decrease, and the healing will increase.

3. Results of numerical simulation

The coupled equations of the model are solved numerically together with boundary, contact, interphase and initial conditions. The computations were performed for the plane strain using linear four-node isoparametric and three-node finite elements. The modeling was performed for zinc specimens while physical and mechanical properties of the material were dependent on temperature [18].

The representative cell sizes (a and b) varied in the range of 15–180 μm. The influence of the shape and size of representative cells during the processes was studied in [14, 15, 17].

The microcracks with two parallel edges and rounded tips (Fig. 2) were considered. In the case $h_0 = 2r_0$ (where r_0 is the curvature radius at the crack tip) the microcrack degenerated into a circular micropore. All defects had the same initial area (in the xy-plane), but with different initial length l_0 (size along the x-axis). For all microdefects the initial damage was the same (for identical sizes of the representative cell). The initial lengths of microcracks in the calculations varied in the range 3.53 μm $\leq l_0 \leq$ 12.5 μm.

Figure 2. Shapes of microdefects (the areas are the same)

The slope angle of the microcrack axis was also varied in the range $0° \leq \alpha_0 \leq 75°$ (only for microcracks $l_0 = 10$ μm).

The potentials difference (per unit length) used in the calculations was 534.3 mV/mm in the defect–free material that corresponds to the electric current density 8.95 kA/mm^2. The potentials difference was constant during pulse time $\tau_0 = 90$ μm.

Fig. 3 shows the healed microcracks ($l_0 = 10$ μm, $\alpha_0 = 0°$) and the temperature fields near it after pulse action for different initial damage. During continue HEEMF action the microcracks edges close and the molten metal jet from the crack tip clampes. Thus, the microcrack is welded.

Dependence of the damage f on pulse time t, initial damage f_0, initial "length" of the microdefect l_0 and initial slope angle α_0 are shown in Fig. 4.

Dependencies given in Fig. 4 are well approximated by a piecewise linear function:

$$f(t) = \begin{cases} f_0, & t < t_0, \\ f_0 - C\,(t - t_0)\,, & t \geq t_0, \end{cases} \tag{2}$$

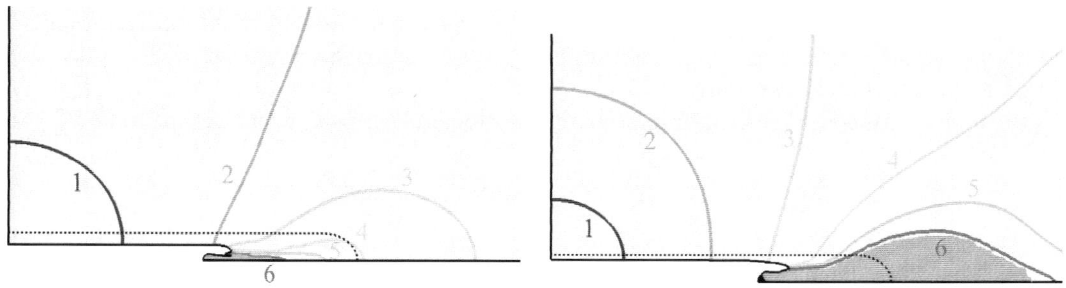

(*a*) (*b*)

Figure 3. Microcrack healing (dashed line is the initial crack boundary at $t = 0$) and the temperature contours (1 is 25°C, 2 is 50°C, 3 is 100°C, 4 is 200°C, 5 is 300°C, 6 is 400°C), dark gray color is melting area ($T \geq 419$°C), black is evaporation area ($T = 906$°C).
(*a*) at time $t = 22.1$ µs with initial damage $f_0 = 0.273\%$,
(*b*) at time $t = 76.6$ µs at initial damage $f_0 = 2.45\%$

where threshold time $t_0 = 9.63$ µm and coefficient $C = C(l_0, \alpha_0, h_0)$ is a function which does not depend on time, but depend on initial microdefect geometry and orientation.

Note that microdefects length increase leads to more slow healing (Fig. 4*a*), and slope angle increase leads to faster microdefects healing (Fig. 4*b*). This is explained by the fact that microdefects length increase leads to increase in the effective material resistivity, and current density decrease causing slower heating in the vicinity of the microdefect. Increasing the slope angle leads to the effective material resistivity decrease, and current density increase causes faster heating in the vicinity of the microdefect.

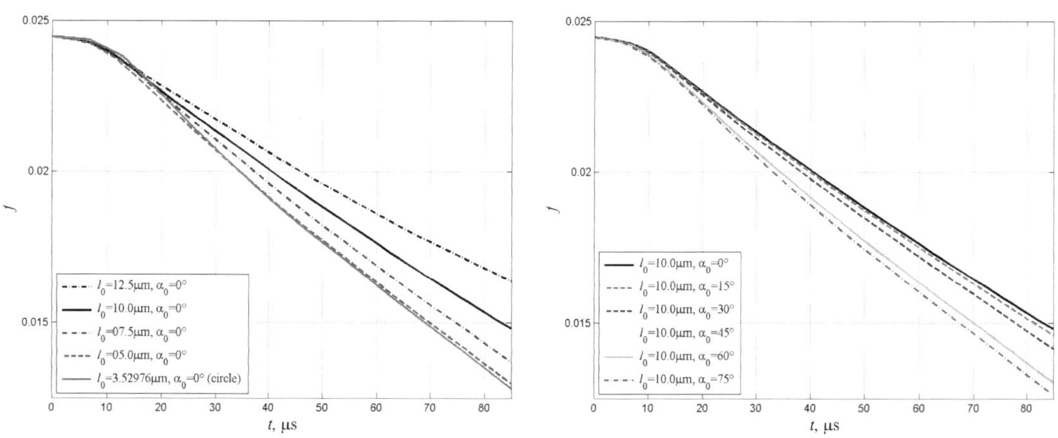

(*a*) (*b*)

Figure 4. Dependence of the damage f on time t (µs) for various initial
(*a*) microdefect lengths l_0, (*b*) slope angles α_0 (all for initial damage $f_0 = 2.45\%$)

Calculations also show that in the investigated range of lengths and inclinations of microcracks, the curves of damage $f(t)$ practically coincide with each other if the projections of the initial lengths of microdefects on the x-axis are equal. Fig. 5*a* shows the coincidence of $f(t)$ curves for microcracks with $l_0 = 10$ µm, $\alpha_0 = 45°$ and $l_0 = 10$ µm, $\alpha_0 = 60°$, respectively with similar curves for microdefects with $l_0 = \frac{\sqrt{2}}{2}10$ µm, $\alpha_0 = 0°$ and $l_0 = \frac{1}{2}10$ µm, $\alpha_0 = 0°$.

Therefore, it was assumed that the function C in the first approximation depends only on the projection of the initial length of the microdefect on the x axis: $l_{0x} = l_0 \cos\alpha_0$.

Fig. 5b shows dependence of $C(l_{0x})$ (for $t \geq t_0$). We conclude from this graph that C is practically linear on l_{0x} in the range 3.53 μm $\leq l_0 \leq$ 12.5 μm, $0° \leq \alpha_0 \leq 60°$ and has the form

$$C = B - A \cdot l_0 \cos\alpha_0 \tag{3}$$

where A and B are defined graphical coefficients fit as

$$A = 6.95 \cdot 10^6 \ (\mathrm{m \cdot s})^{-1}, \qquad B = 2.01 \cdot 10^2 \ \mathrm{s}^{-1}$$

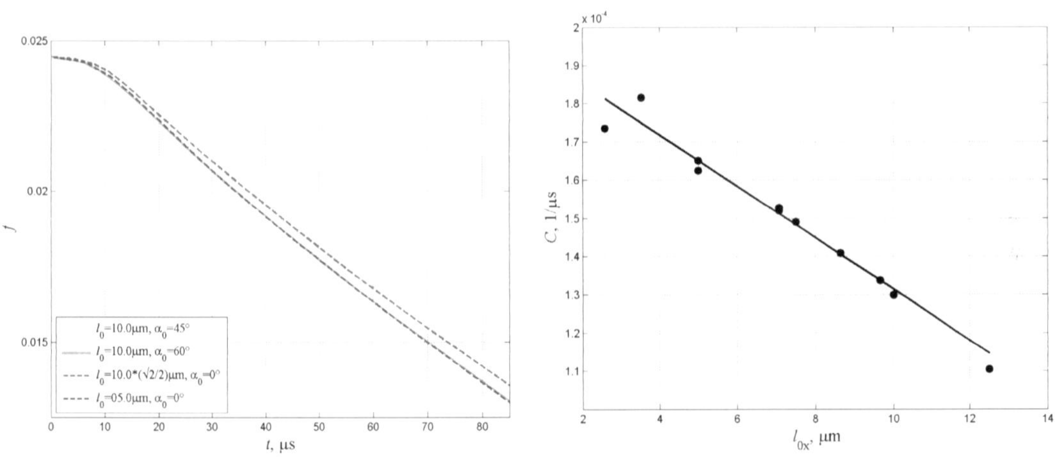

(a) (b)

Figure 5. Dependencies of (a) the damage $f(t)$ for identical projections l_0 and (b) tangent of the slope of the damage curve $f(t)$ on l_{0x} (for $f_0 = 2.45\%$)

Calculations show that (2)–(3) occur in a wide range of initial damage from 0.2% to 5%.

Thus, for the model it possible to obtain a simple approximate dependence $f(t, f_0, l_0, \alpha_0)$.

Experiments [1, 2, 7–11] have confirmed that the continuity of the material structure is restored, accompanied by a change in the volume fraction of microdefects (up to the complete healing of certain microdefects). The obtained results are in qualitative agreement with these experiments.

Conclusions

Metal damage decrease during HEEMF action leads to improvement of the plastic properties and increase ultimate plastic strain. Understanding the damage dependencies will allow determine constitutive equations for modelling electroplastic deformation and fracture processes. With the same initial material damage f_0, for changing material porosity during electric current pulse action the determining factors are not the geometry and microdefect orientation, but a generalized parameter equal to the projection of the initial "length" of the microdefect on the line (plane) perpendicular to the current density vector $l_{0x} = l_0 \cos\alpha_0$.

Acknowledgments

This work was partially supported by Russian State Assignment under contract No. AAAA-A17-117021310380-1 and partially supported by the Russian Foundation for Basic Research (RFBR) under Grant No. 18-08-00958.

References

[1] Beklemishev N N, Koryagin N I, and Shapiro G C 1984 Influence of locally inhomogeneous pulse electric field on plasticity and strength of conducting materials *Izv. AN SSSR. Metals* (4) 184–7

[2] Beklemishev N N, Gorbunov N M, Koryagin N I, Kukudzhanov V N, Mitin B S, Naumov N M, and Porokhov V A 1989 Plasticity and strength of metallic materials with the pulse action of a high-energy electromagnetic field taken into account *Preprint No. 372. IPM AN SSSR, Moscow* [in Russian]

[3] Klyushnikov V D and Ovchinnikov I V 1988 Plane problem of effect of an instantaneous point heat source *Mech. Solids* **23** (4) 113–7

[4] Ovchinnikov I V 1993 Determination of the reserve of plasticity under the action of a current *Strength of Materials* **25** (6) 437–41

[5] Kukudzhanov V N and Kolomiets-Romanenko A V 2010 Study of the influence of electric current dynamical action on mechanical properties of materials with ordered structure of defects *Mech. Solids* **45** (3) 465–75 doi: 10.3103/S0025654410030167

[6] Qin R S and Su S X 2002 Thermodynamics of crack healing under electropulsing *J. Mater. Res.* **17** (8) 2048–52 doi: 10.1557/JMR.2002.0303

[7] Conrad H 1989 A study into the mechanism(s) for the electroplastic effect in metals and its application to metalworking, processing and fatigue (Final Report ARO Proposal Number 23090-MS, ARO Funding Document DAAL03-86-K-0015, U.S. Army Research Office, North Carolina State University, Raleigh, NC 27695)

[8] Zuev L B, Sosnin O V, Chirakadze D Z, Gromov V E, and Murav'ev V V 1998 Acoustic evaluation of the endurance of steel specimens and recovery of their serviceability *J. Appl. Mech. Techn. Phys.* **39** (4) 639–41 doi: 10.1007/BF02471262

[9] Song Hui, Wang Zhong-jin, and Gao Tie-jun 2007 Effect of high density electropulsing treatment on formability of TC4 titanium alloy sheet *Trans. Nonferrous Soc. China* **17** (1) 87–92 doi: 10.1016/S1003-6326(07)60053-3

[10] Troitskii O A, Baranov Yu V, Avraamov Y S, and Shlyapin A D 2004 *Physical Fundamentals and Technologies of Processing Advanced Materials (Theory, Technology, Structure, and Properties)* vol 1 (Moscow-Izhevsk: Inst. Komp. Issled.) [in Russian]

[11] Yu J, Zhang H, Deng D, Hao S, and Iqbal A 2014 Numerical calculation and experimental research on crack arrest by detour effect and joule heating of high pulsed current in remanufacturing *Chinese J. Mech. Engng* **27** (4) 745–53 doi: 10.3901/CJME.2014.0414.075

[12] Kukudzhanov K V 2015 Modeling the treatment of high-energy pulsed electromagnetic field of the micro-cracks in a polycrystalline metal *PNRPU Mechanics Bulletin* (4) 138–58 doi: 10.15593/perm.mech/2015.4.09

[13] Kukudzhanov K V and Levitin A L 2015 Modeling the healing of microcracks in metal stimulated by a pulsed high-energy electromagnetic field. Part I *Nanomech. Sci. Technol.: Int. J.* **6** (3) 233–50 doi: 10.1615/NanomechanicsSciTechnolIntJ.v6.i3.60

[14] Kukudzhanov K V and Levitin A L 2016 Modeling the healing of microcracks in metal stimulated by a pulsed high-energy electromagnetic field. Part II *Nanomech. Sci. Technol.: Int. J.* **7** (2) 123–48 doi: 10.1615/NanomechanicsSciTechnolIntJ.v7.i2.30

[15] Kukudzhanov K V 2017 On healing metal damages using high-energy pulsed electromagnetic field *PNRPU Mechanics Bulletin* (2) 99–124 doi: 10.15593/perm.mech/2017.2.06

[16] Kukudzhanov K V and Levitin A L 2017 Phase Transformations in Metals Stimulated by a Pulsed High-Energy Electromagnetic Field *Procedia IUTAM* **23** 84–100 doi: 10.1016/j.piutam.2017.06.008

[17] Kukudzhanov K V and Levitin A L 2018 Healing of damaged metal by a pulsed high-energy electromagnetic field *J. Phys.: Conf. Ser.* **991** 012049 doi: 10.1088/1742-6596/991/1/012049

[18] Pikunov M V 1997 *Metal Smelting. Alloy Crystallization. The Solidification of Castings* (Moscow: MISiS) [in Russian]

IOP Conf. Series: Journal of Physics: Conf. Series **1205** (2019) 012033 doi:10.1088/1742-6596/1205/1/012033

The generalized Solow model

D A Kulikov

Faculty of Mathematics, Yaroslavl State University, Sovetskaya str., 14, Yaroslavl, 150003, Russia

E-mail: kulikov_d_a@mail.ru

Abstract.

A modification of the known mathematical macroeconomics model of macroeconomics are studied in which a delay factor is presumed. This led to the replacement of the ordinary differential equation, which cannot exhibit periodic cycles on the equations with a deviating argument (functional-differential equations). It was possible to show the existence of periodic solutions that can and are intended to describe the periodic cycles in the market economy.

The mathematical portion is based on the application of the modern theory of dynamical systems with an infinite-dimensional space of initial conditions. This will allow us to apply the Andronov-Hopf Theorem for equations with a deviating argument in such a form that the parameters of the cycles are located. We will also apply the well-known Krylov-Bogolyubov algorithm that is extended to infinite-dimensional dynamical systems that is used and reduces the problem to the analysis of the finite-dimensional system of ordinary differential equations-the normal Poincare-Dulac form.

1. Introduction

In macroeconomics, there are a large number of mathematical models that are traditionally considered classical and basic. Traditionally among them, a mathematical model proposed by R.M. Solow [1,2]. This model is usually called the Solow model or the Solow-Swan model. It is designed to describe the dynamics of the labor-power ratio of labor (see also [3-8])

$$p' = -\alpha p + \beta p^k, \tag{1.1}$$

where α, β are positive constants, $p = p(\tau)$ is the value of funds at the time τ. After replacements (normalizations)

$$\tau = \gamma_0 t,\ p = \gamma_1 x,\ \gamma_0\alpha = 1,\ \beta\gamma_0\gamma_1^{k-1} = 1,\ \gamma_0 = \frac{1}{\alpha},\ \gamma_1 = \left(\frac{\alpha}{\beta}\right)^{1/(k-1)},\ \beta, \gamma_0, \gamma_1 > 0.$$

Equation (1.1) acquires the following form

$$\dot{x} = -x + x^k,\ x = x(t). \tag{1.2}$$

The first terms on the right-hand side of equations (1.1), (1.2) are responsible for the depreciation of the funds. The second terms are proportional to the investments. Next we will consider equation (1.1) for the normalized value of the funds $x(t) > 0$, via Equation (1.2).

Observe that Equation (1.2) has a unique positive equilibrium state $x(t) = 1$ if $k \neq 1$. Then for $k \in (0, 1)$, this equilibrium state is globally asymptotically stable (all solutions of the Equation (1.2) with positive initial conditions converge to a given equilibrium position).

In this particular model, it is usually assumed that $k \in (0, 1)$ is a stable economic equilibrium $x = 1$. Notice that the positive equilibrium state $x = 1$ is unstable, when $k \in (1, \infty)$. Equation (1.1) cannot have stationary solutions with the exception of the equilibrium states. Hence this mathematical model does not quite adequately describe the real economic processes, for which, of course, periods of ups and downs are the characteristic.

In this paper, we will show that the assumption of the delay factor in the Solow Model leads to a meaningful change in the dynamics of the solutions and makes it possible to isolate the range of the parameters of the problem under which the periodic cycles exist. It is possible that the delay is introduced into both terms, which produces the following equation

$$\dot{x} = -y + y^k, \tag{1.3}$$

where as previously $y = x(t - h)$. Naturally we see that $h > 0$.

If we supplement equations (1.3) with the initial condition

$$x(t) = \varphi(t), \tag{1.4}$$

where the given function $x(t) \in C[-h, 0]$ to the space of continuous functions defined on $[-h, 0]$, then we obtain Cauchy problem: (1.3), (1.4). This problem generates a local semiflow.

Equation (1.3) has an equilibrium state $x(t) = 1$. In the next section, we will address the stability question. Equation (1.3) can be rewritten in the following form

$$\dot{u} = -v + kv + \frac{k(k-1)}{2}v^2 + \frac{k(k-1)(k-2)}{6}v^3 + o(v^3), \tag{1.5}$$

after replacement which is acquired from Equation (1.3) we get

$$x = 1 + u, y = 1 + v, u = u(t), v = v(t) = u(t - h),$$

and applying the Taylors Formula. Notice that all three equations have a zero equilibrium state corresponding to the equilibrium states $x = 1$ of the original equation (1.3). Now we will consider the equation (1.5) for which the structure of the neighborhood of the zero equilibrium state will be studied.

2. Analysis of linearized equation

We will first address the stability question of the zero equilibrium state of Equation (1.5). To study this question, the analysis of the linearized Equation (1.5) will be necessary. In this case we obtain the following equation

$$\dot{u} = -v + kv. \tag{2.1}$$

It is well known [9] that stability of solutions of the linear differential equation (2.1) can be reduced to an analysis of the characteristic equation

$$\lambda = (k - 1)\exp(-\lambda h). \tag{2.2}$$

Observe when $h = 0$ then we have the root $\lambda = k - 1 < 0$. To find the positive $min\, h = H$ for which roots with $Re\lambda = 0$ emerge in Equation (2.2). The case when $\lambda = 0$ is not possible for any h.

Consequently, the critical case in the stability problem is possible. For the corresponding value h, the stability spectrum (the set of roots of the characteristic equation (2.2)) contains the pair of purely imaginary roots $\pm i\sigma$, where $\sigma > 0$.

The corresponding pairs (h, σ) are defined as solutions of the system

$$(k-1)\cos\sigma h = 0,\ \sigma = (1-k)\sin\sigma h\,(k-1\neq 0).$$

Therefore, we finally obtain a system for determining H and σ already of the following form

$$\cos\sigma h = 0,\ \sigma h = (1-k)h\sin\sigma h \tag{2.3}$$

or after relabeling $\omega = \sigma h$, the system (2.3) can be written in the form

$$\cos\omega = 0,\ (1-k)h\sin\omega = \omega. \tag{2.4}$$

Note that system (2.4) has the following set of solutions

$$\omega_m = \frac{\pi}{2} + \pi m,\ h_m = \frac{\omega_m}{(1-k)\sin\omega_m},\ m = 0, \pm 1, \pm 2, \ldots$$

The smallest $H = \min\{h_m\} > 0$ is possible if $\omega = \dfrac{\pi}{2}$, $H = \dfrac{\pi}{2(1-k)}$, $\sigma = 1-k$. Therefore for $H = \dfrac{\pi}{2(1-k)}$, the stability spectrum contains a pair of pure imaginary roots $\lambda_{1,2} = \pm i\sigma$ and for the remaining λ_k the inequality $Re\lambda_k < 0$ is valid.

Now let $h = H(1+\gamma\varepsilon)$. Then from the characteristic equation (2.2) we obtain $\lambda_0' = \dfrac{d\lambda(\varepsilon)}{d\varepsilon}|_{\varepsilon=0} = \tau_0' + i\sigma_0'$, and the equality $\lambda_0' = -i\sigma H\lambda_0' + H\sigma^2\gamma$, which allows us to determine λ_0'. In our case we acquire the following equality: $\tau_0' = 2\pi(1-k)\gamma/(4+\pi^2)$. Note that if we choose the constant $\gamma = 1$, then the inequality $\tau_0' > 0$ will hold true. In this case, as h increases, the roots $\pm i\sigma$ go to the right half-plane of the complex plane. Consequently, the zero solution of the auxiliary equation (2.1) loses its stability; in particular, the existence of a stable cycle is possible. The last question will be studied in the next section.

3. Periodic Solutions

Our intents of this section is to study the existence and stability of cycles of Equation (1.5). In Equation (1.5) we set:

$$t = \frac{h(\varepsilon)}{H}\Theta,\ h(\varepsilon) = H(1+\gamma\varepsilon), \gamma \in R, \varepsilon \in (0, \varepsilon_0), 0 < \varepsilon_0 << 1, H = \frac{\pi}{2(1-k)}.$$

Notice that Θ is a new time. As a result Equation (1.5) is rewritten in the form

$$u' = (1+\gamma\varepsilon)[-v + kv + \frac{k(k-1)}{2}v^2 + \frac{k(k-1)(k-2)}{6}v^3 + o(v^3)], \tag{3.1}$$

where $v = u(\Theta - H)$. In the neighborhood of the zero equilibrium state Equation (3.1) has a two-dimensional smooth invariant manifold $M_2(\varepsilon)$ [10]. In this case, all solutions of Equation (3.1) approach it with the velocity of the exponent with time if their initial conditions are small. The dynamics of the solutions of Equation (3.1) is restored after analyzing the system of two ordinary differential equations - normal form (NF). In the assumed case, NF can be written in complex form [6-8]

$$z' = (\alpha + i\beta)z + (d + ic)z|z|^2 + O(\varepsilon), \tag{3.2}$$

where $\alpha, \beta, d, c \in R$. These coefficients can be written out in an explicit form, which will be done below after the implementation of the algorithm for constructing the defining equation, which is commonly called NF. These coefficients depend on the parameters of Equation (3.1). In our case, these are k and H. For this purpose, recently it has been customary to use the adaptation of the Krylov-Bogolyubov algorithm to infinite-dimensional dynamical systems [6-8]. In the NF (3.2) $z = z(s)$ is a complex function, and $s = \varepsilon\Theta, \varepsilon \in (0, \varepsilon_0)$. If we assume that the first Lyapunov value d is non-zero a priori ($d \neq 0$), then the solution of Equations (3.1) with initial conditions in a small neighborhood of the zero solution can be expediently sought in the following form [6-8]

$$u(\Theta, \varepsilon) = \varepsilon^{1/2}u_1(\Theta, z) + \varepsilon u_2(\Theta, z) + \varepsilon^{3/2}u_3(\Theta, z) + O(\varepsilon^2), \tag{3.3}$$

where $u_1(\Theta, z) = z(s)\exp(i\sigma\Theta) + \overline{z}(s)\exp(-i\sigma\Theta)$, $z(s)$ is one of the solutions of the NF. The sufficiently smooth functions $u_2(\Theta, z), u_3(\Theta, z)$ with respect to the variable Θ have period $2\pi/\sigma$ and

$$M_{\pm}(u_m) = \frac{\sigma}{2\pi}\int_0^{2\pi/\sigma} u_m(\Theta, z)\exp(\pm i\sigma\Theta)d\Theta = 0,\ m = 2, 3.$$

We substitute the sum (3.3) into Equation (3.1) and equate the coefficients for the powers $\varepsilon, \varepsilon^{3/2}$. Hence we obtain two linear nonhomogeneous delayed differential equations. Thus, to determine $u_2(\Theta, z)$, we obtain the following equation

$$\frac{\partial u_2}{\partial\Theta} + (1-k)v_2 = \Phi_2(\Theta, z), \tag{3.4}$$

where $\Phi_2(\Theta, z) = \dfrac{k(k-1)}{2}v_1^2, v_2 = u_2(\Theta - H, z), v_1(\Theta) = u_1(\Theta - H)$. When deriving the equation for $u_3(\Theta, z)$, one can note that $\dfrac{d}{d\Theta}\psi(\Theta, s) = \dfrac{\partial\psi}{\partial\Theta} + \dfrac{\partial\psi}{\partial s}\varepsilon$. Therefore, we obtain the equation

$$\frac{\partial u_3}{\partial\Theta} + v_3 - kv_3 = \Phi_3(\Theta, z), \tag{3.5}$$

$$\Phi_3(\Theta, z) = k(k-1)v_1v_2 + \frac{k(k-1)(k-2)}{6}v_1^3 + \gamma(k-1)v_1 -$$
$$-z'i\frac{\pi}{2}\exp(i\sigma\Theta) + \overline{z}'i\frac{\pi}{2}\exp(-i\sigma\Theta) - z'\exp(i\sigma\Theta) - \overline{z}'\exp(-i\sigma\Theta).$$

Comment. *Linear Differential Equation with a deviating argument*

$$\frac{du}{d\Theta} + (1-k)u(t-H) = \Phi(\Theta),\ H = \frac{\pi}{2(1-k)},$$

where $\Phi(\Theta)$ *is a* $2\pi/\sigma$ *periodic function, has periodic solutions with the same period if*

$$M_{\pm}(\Phi(\Theta)) = \frac{\sigma}{2\pi}\int_0^{2\pi/\sigma}\Phi(\Theta)\exp(\pm i\sigma\Theta)d\Theta = 0.$$

The equalities of $M_{\pm}(u) = 0$ *single one such solution.*

Therefore, in the case of Equation (3.4), when

$$\Phi_2(\Theta, z) = \frac{k(1-k)}{2}[z^2\exp(2i\sigma\Theta) - 2|z|^2 + \overline{z}^2\exp(-2i\sigma\Theta)],$$

the solution of this equation should be selected in the following form

$$u_2(\Theta, z) = \eta_2 z^2 \exp(2i\sigma\Theta) + \eta_0 |z|^2 + \overline{\eta}_2 \overline{z}^2 \exp(-2i\sigma\Theta),$$

where $\eta_2 \in C$, $\eta_0 \in R$. After substituting $u_2(\Theta, z)$ in the selected form in the corresponding equation we determine that $\eta_0 = -k$, $\eta_2 = -\frac{k}{10}(1+2i)$. We now analyze the nonhomogeneous differential equation (3.5). From the solvability conditions for this equation in the class $2\pi/\sigma$ of periodic functions, we obtain the following equation for $z(s)$

$$z' = (\alpha + i\beta)z + (d + ic)z|z|^2. \tag{3.6}$$

In our case we get

$$\alpha = 2\pi\gamma\frac{1-k}{4+\pi^2},\ \beta = 4\gamma\frac{1-k}{4+\pi^2},\ d = -\frac{4k(1-k)}{5(4+\pi^2)}[(\frac{3\pi}{2} - 1)k + \frac{5\pi}{2}],\ c = -\frac{4k(1-k)}{5(4+\pi^2)}[5 + (3 + \frac{\pi}{2})k].$$

We emphasize that Equation (3.6) is a "shorter" version of the NF. Note that for all k under the assumption of the first Lyapunov value $d < 0$.

Lemma. *Equation (3.6) has a periodic solution*

$$z(s) = \rho\exp(i\nu s),$$

where $\rho = \sqrt{\gamma\frac{5\pi}{2k}[(\frac{3\pi}{2} - 1)k + \frac{5\pi}{2}]^{-1}}$, $\nu = \beta + c\rho^2$, *if* $\gamma > 0$. *Moreover, this periodic solution is stable while the zero solution is unstable. For* $\gamma < 0$ *this equation has no non-trivial periodic solutions and the zero solution of the NF is asymptotically stable.*

Now by substituting $z(s) = \rho\exp(i\nu s)$ into Equation (3.6) leads to a system of algebraic equations for determining ρ and ν

$$\alpha\rho + d\rho^3 = 0,\ \nu = \beta + c\rho^2.$$

The first equation of the given system has a non-trivial solution $\rho > 0$, if α and d of different signs and only the zero solution if the signs α, d coincide. The stability analysis of a non-trivial periodic solution is described by the standard scheme.

Now we set $z = \rho\exp(i\nu s)(1 + w)$. The linearized version for the vector-valued function $\eta = (w_1, w_2)$ $(w = w_1 + iw_2)$ has the following form

$$\eta' = B\eta, B = \begin{pmatrix} d\rho^2 & 0 \\ d\rho^2 & 0 \end{pmatrix}.$$

One eigenvalue of the matrix is $\lambda_1 = d\rho^2 < 0$, if $d < 0\,(\alpha > 0)$ and this is an eigenvalue of $\lambda_1 > 0$, if $d > 0\,(\alpha < 0)$, and the second eigenvalue of the matrix is $\lambda_2 = 0$. The Andronov-Witt theorem implies the validity of the lemmas assertion. The results from [10-13] imply the validity of the assertion.

Theorem. *There exists* $\varepsilon_0 > 0$, *such that for all* $\varepsilon \in (0, \varepsilon_0)$ *Equation (3.1) for* $\gamma > 0$ *has a stable (orbitally asymptotically stable) limit cycle corresponding to the NF cycle with the following asymptotic formula*

$$u_*(\Theta, \varepsilon) = \varepsilon^{1/2}\rho[\exp(i(\sigma + \varepsilon\nu)\Theta + i\varphi) + \exp(-i(\sigma + \varepsilon\nu)\Theta - i\varphi)] +$$
$$+ \varepsilon\rho^2[\eta_2\exp(2i(\sigma + \varepsilon\nu)\Theta + 2i\varphi) + \eta_0 + \overline{\eta}_2\exp(-2i(\sigma + \varepsilon\nu)\Theta - 2i\varphi)] + o(\varepsilon),$$

where $\varphi \in R$ *and the constants* $\rho, \sigma, \nu, \eta_0, \eta_2$ *were indicated earlier.*

These substitutions allow us to transfer the results of the theorem 1 to Equation (1.5). Now let $h = H(1+\varepsilon)$. Then the periodic solution $u_*(\Theta, \varepsilon)$ corresponds to the periodic solution of Equation (1.5)

$$x_*(t, \varepsilon) = 1 + u_*(\frac{t}{1+\varepsilon}, \varepsilon)\,(\gamma = 1).$$

Naturally, the solution $x_*(t, \varepsilon)$ is stable in the sense of A.M. Lyapunov.

Also notice that the periodic solutions depend on the selection of the parameter k. For instance when $k = 1 - \delta$ and δ is sufficiently small, we obtain periodic solutions with long periods that can be interpreted as "long waves " N.D. Kondratiev (see also [14]).

4. Conclusions

In this paper we have shown that the delay effects can significantly change the dynamics of solutions in the classical Solow model. At least in the presented form of the correspondingly modified Solow Equation may appear to fluctuate prices, which is typical for pricing within a market economy.

A similar effect occurs if a delay is introduced into another classical model of macroeconomics - the "supply-demand" model (market model). In this classical version, there are no oscillatory solutions but the assumption of the delay showed that under certain variations periodic cycles may exist. The selection of the parameters of the problem has a modified version of the equation that already has a periodic cycle ([7]).

In macroeconomics, a similar situation is reproduced for the one that occurred in mathematical ecology. If we consider the well-known Verhulst Equation

$$\dot{N} = \alpha N(1 - N),$$

then of course it does not exhibit periodic solutions. Notice that $\alpha > 0, N = N(t)$ resembles the number of species. This is a significant contribution by Hutchinson [15], who proposed to study the following equation

$$\dot{N} = \alpha N(1 - N(t-h)),\ h > 0,$$

which is already capable of describing fluctuations in the population size in a single-species bio-sinopsis.

The study was conducted with the financial support of the Russian Foundation for Basic Research in the framework of the scientific project No 18-01-00672.

5. References

[1] Solow R M 1956 *The Quarterly Journal of Economics* **70** 65
[2] Swan T W 1956 *Economic Record* **32** 334
[3] Zhang W B 1983 *Synergetic Economics: Time and Change in Nonlinear Rconomics*(Berlin-New-York: Springer-Verlag)
[4] Puu T 1997 *Nonlinear Economic Dynamics*(Berlin-New-York: Springer-Verlag)
[5] Ferrara M, Guerini L and Mavilla R 2013 Applied Mathematical Sciences **7** 4249
[6] Kulikov A N and Kulikov D A 2015 Taurida Journal of Computer Science Theory and Mathematics **2** 87
[7] Kulikov D A 2017 IOP Conf. Series: Journal of Physics: Conf. Series **788** 6 p.
[8] Kulikov D A 2018 Zhurnal SVMO **20** 225
[9] Hale J 1977 *Theory of functional differential equations*(Berlin-New-York: Springer-Verlag)
[10] Kulikov 1976 Studies of stability and the theory of oscillation 67
[11] Kolesov Yu S 1974 Vestnik Yaroslavl. univ. **7** 3
[12] Kulikov Yu S and Svitra D 1976 Differential equations and applications **16** 41
[13] Kolesov A Yu, Kulikov A N and Rozov N Kh 2003 Differ. Equ. **39** 614
[14] Kondratiev N D 1993 Special opinion. Book 2.(Moscow: Nauka(in Russian))
[15] Hutchinson G E 1948 Annals of the New York Academy of Sciences **50** 221

IOP Conf. Series: Journal of Physics: Conf. Series **1205** (2019) 012034 doi:10.1088/1742-6596/1205/1/012034

Numerical analysis of dynamical systems: unstable periodic orbits, hidden transient chaotic sets, hidden attractors, and finite-time Lyapunov dimension

N V Kuznetsov

Faculty of Mathematics and Mechanics, St. Petersburg State University, Peterhof, St. Petersburg, Russia
Department of Mathematical Information Technology, University of Jyväskylä, Jyväskylä, Finland
Institute of Problems of Mechanical Engineering RAS, Russia

E-mail: nkuznetsov239@gmail.com

T N Mokaev

Faculty of Mathematics and Mechanics, St. Petersburg State University, Peterhof, St. Petersburg, Russia

E-mail: t.mokaev@spbu.ru

Abstract. In this article, on the example of the known low-order dynamical models, namely Lorenz, Rössler and Vallis systems, the difficulties of reliable numerical analysis of chaotic dynamical systems are discussed. For the Lorenz system, the problems of existence of hidden chaotic attractors and hidden transient chaotic sets and their numerical investigation are considered. The problems of the numerical characterization of a chaotic attractor by calculating finite-time time Lyapunov exponents and finite-time Lyapunov dimension along one trajectory are demonstrated using the example of computing unstable periodic orbits in the Rössler system. Using the example of the Vallis system describing the El Ninõ-Southern Oscillation it is demonstrated an analytical approach for localization of self-excited and hidden attractors, which allows to obtain the exact formulas or estimates of their Lyapunov dimensions.

1. Introduction

History of the turbulence phenomena study is associated with the consideration of various models, which include the Navier-Stokes equations, their Galerkin approximations, and the development of the theory of chaos [1, 2, 3, 4]. Here let us note the significant results by D. Ruelle, F. Takens [3], and S. Smale [4], who proposed a chaotic attractor as a mathematical prototype describing the onset of turbulence, and by O. Ladyzhenskaya, who studied the case when the two-dimensional Navier-Stokes equation generates a dynamical system and proved the finite dimensionality of its attractor [5]. The first vivid example of chaotic attractor in a hydrodynamic system was obtained by E. Lorenz [6]. Using the Galerkin method he derived a crude three-dimensional mathematical model for Rayleigh-Bénard convective flow, which has the following form:

$$\begin{cases} \dot{x} = -\sigma(x-y), \\ \dot{y} = rx - y - xz, \\ \dot{z} = -bz + xy, \end{cases} \tag{1}$$

where r, σ, b are positive parameters. For $0 < r < 1$, there is one globally asymptotically stable equilibrium $S_0 = (0,0,0)$. For $r > 1$, equilibrium S_0 is a saddle, and a pair of symmetric equilibria $S_\pm = (\pm\sqrt{b(r-1)}, \pm\sqrt{b(r-1)}, r-1)$ appears.

For the parameters $r = 28$, $\sigma = 10$, $b = 8/3$ in system (1) E. Lorenz numerically found a chaotic attractor in the model. In general, for numerical localization of attractor, it is necessary to explore its basin of attraction and choose an initial point in it. If for a particular attractor its basin of attraction is connected with the unstable manifold of unstable equilibrium, then the localization procedure is quite simple. From this perspective, the following classification of attractors is suggested [7, 8, 9, 10]: an attractor is called a *self-excited attractor* if its basin of attraction intersects an arbitrarily small open neighborhood of an equilibrium; otherwise, it is called a *hidden attractor*. Numerical localization of hidden attractors is much more challenging and requires the development of special methods. The classical Lorenz attractor is a self-excited one with respect to all equilibria S_0, $S_\pm$, and it is an open question [10, p. 14] whether for some parameters there exists a hidden Lorenz attractor. This question is related to the "chaotic" generalization [11] of the second part of Hilbert's 16th problem *on the number and mutual disposition of attractors and repellers in the chaotic multidimensional dynamical systems and, in particular, their dependence on the degree of polynomials in the model*; see corresponding discussion, e.g. in [12, 13]. There are a number of physical dynamical models which possess hidden chaotic attractors, e.g, the Rabinovich system (describes the interaction of plasma waves) [14, 15], the Glukhovsky-Dolghansky system (describes convective fluid motion in a rotating cavity) [16, 9] and others [17, 18, 19, 20].

We note that the Lorenz system (1) with parameters $r = 28$, $\sigma = 10$, $b = 8/3$ is dissipative in the sense of Levinson, and for any initial data (except for equilibria) the trajectory tends to the attractor. Thus, system (1) generates a *dynamical system* $\big(\{\varphi^t\}_{t\geq 0}, (U \subseteq \mathbb{R}^3, ||\cdot||)\big)$.

2. Hidden transient chaotic sets in Lorenz system

In numerical computation of a trajectory over a finite-time interval it is difficult to distinguish a *sustained chaos* from a *transient chaos* (a transient chaotic set in the phase space, which can nevertheless persist for a long time) [21, 22], thus it is reasonable to give a similar classification for transient chaotic sets [23, 15]: a *transient chaotic set* is called a *hidden transient chaotic set* if it does not involve and attract trajectories from a small neighborhood of equilibria; otherwise, it is called *self-excited*. In order to distinguish an attracting chaotic set (attractor) from a transient chaotic set in numerical experiments, one can consider a grid of points in a small neighborhood of the set and check the attraction of corresponding trajectories towards the set. There one can reveal a subset of points for which the trajectories leave the transient set.

For the Lorenz system (1), suppose that $\sigma = 10$, $b = \frac{8}{3}$ are fixed and r varies. If $r = 28$, then all three equilibria S_0, $S_\pm$ are unstable and in the phase space there exists a self-excited attractor with respect to these equilibria. For $r < \sigma(\frac{\sigma+b+3}{\sigma-b-1}) \approx 24.7368$, the equilibria $S_\pm$ become stable, and for $24.06 < r < 24.7368$, there exists a self-excited attractor with respect to equilibrium S_0. Near the point $r \approx 24.06$ it is possible to observe a long living transient chaotic set, which is hidden since it's basin of attraction does not intersect with the small vicinities of equilibrium S_0. For example, for $r = 24$ a hidden transient chaotic set can be visualized[1] [24] from the initial

[1] In this work, we use MATLAB's standard procedure `ode45` with default parameters.

point $(2, 2, 2)$ (see Fig. 1). In [25], hidden transient chaotic set was obtained in system (1) with $r = 29$, $\sigma = 4$, $b = 2$.

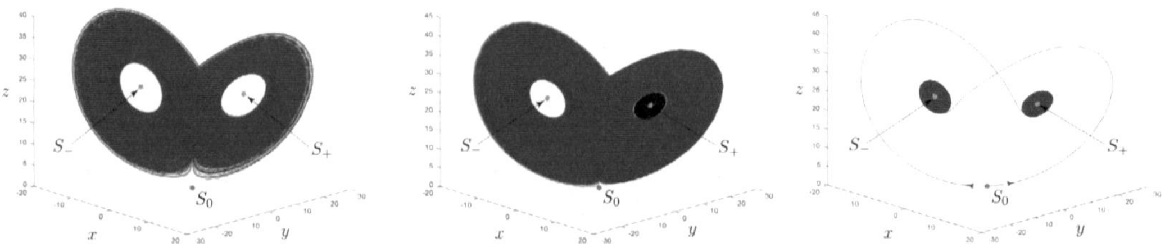

(a) $u_0 = (2, 2, 2)$, $t \in [0, 1000]$ (b) $u_0 = (2, 2, 2)$, $t \in [0, 2.3 \cdot 10^5]$ (c) u_0 in vicinity of S_0

Figure 1: Visualization of the hidden transient chaotic set in system (1), $r = 24$, $\sigma = 10$, $b = \frac{8}{3}$.

The time of the transient process in this case depends strongly on the choice of the initial data, which complicates the task of distinguishing an attracting chaotic set (attractor) from a transient chaotic set in numerical experiments. E.g., for system (1) with parameters $r = 24$, $\sigma = 10$, $b = 8/3$ and for initial point $u_0 = (20, 20, 20)$ a transient chaotic behavior is observed[2] on the time interval $[0,\ 1.8{\cdot}10^4]$, for initial point $u_0 = (-7, 8, 22)$ — on the time interval $[0,\ 7.2{\cdot}10^4]$, for initial point $u_0 = (2, 2, 2)$ — on the time interval $[0,\ 2.26 \cdot 10^5]$, and for initial point $u_0 = (0, -0.5, 0.5)$ a transient chaotic behavior continues over a time interval of more than $[0,\ 1 \cdot 10^7]$. In order to distinguish an attracting chaotic set from a transient chaotic set by computing trajectories on a reasonable time interval, one can consider a grid of points in a small neighborhood of the set and check the attraction of corresponding trajectories towards the set. There one can reveal a subset of points for which the trajectories leave the transient set.

Next, on the example of the Lorenz system (1) we will demonstrate difficulties in the reliable numerical computation of the finite-time Lyapunov exponents and finite-time Lyapunov dimension.

3. Finite-time Lyapunov dimension of a transient chaotic set

Consider system (1) with parameters $r = 24$, $\sigma = 10$, $b = 8/3$ and integrate numerically the trajectory with initial data $u_0 = (20,\ 20,\ 20)$. We numerically approximate the finite-time Lyapunov exponents and finite-time Lyapunov dimension (see corresponding definitions, e.g. in [26, 14]). Integration with $t > T_1 \approx 1.8 \cdot 10^4$ leads to the collapse of the "*attractor*", i.e. the "*attractor*" turns out to be a transient chaotic set. However, on the time interval $t \in [0,\ T_3 \approx 507883]$ we have $\mathrm{LE}_1(t,\ u_{\mathrm{init}}) > 0$ and, thus, may conclude that the behavior is chaotic, and for the time interval $t \in [0,\ T_2 \approx 262954]$ we have $d_{\mathrm{L}}^{\mathrm{KY}}(\{\mathrm{LE}_i(t,\ u_{\mathrm{init}})\}_{i=1}^3) > 2$. This effect is due to the fact that the finite-time Lyapunov exponents and finite-time Lyapunov dimension are the values averaged over the considered time interval. Since the lifetime of transient chaotic process can be extremely long and taking into account the limitations of reliable integration of chaotic ODEs, even long-time computation of the finite-time Lyapunov exponents and the finite-time Lyapunov dimension does not necessary lead to a relevant approximation of the Lyapunov exponents and the Lyapunov dimension.

On the one hand, computational errors (caused by a finite precision arithmetic and numerical integration of differential equations) and sensitivity to initial data allow one to get a more

[2] The time of transient chaotic behavior is often estimated approximately by analyzing the sign of the largest Lyapunov exponent. For simplicity, here we approximate the transient behavior time by the time of entering a small ball with the center at the points $S_\pm$.

complete visualization of chaotic attractor (pseudo-attractor) by one pseudo-trajectory computed for a sufficiently large time interval. On the other hand, there arises a question of the reliability of calculating the trajectory itself and its various characteristics, such as finite-time Lyapunov exponents (FTLEs) and finite-time Lyapunov dimension (FTLD), over a long time interval. In [27] for the Lorenz system (1) the time interval for reliable computation with 16 significant digits and error 10^{-4} is estimated as $[0, 36]$, with error 10^{-8} is estimated as $[0, 26]$, and reliable computation for a longer time interval, e.g. $[0, 10000]$ in [28], is a challenging task.

For two different long-time pseudo-trajectories $\tilde{u}(t, u_0^1)$ and $\tilde{u}(t, u_0^2)$ visualizing the same attractor, the corresponding FTLEs can be, within the specified accuracy, similar due to averaging over time and similar sets of points $\{\tilde{u}(t, u_0^1)\}_{t\geq 0}$ and $\{\tilde{u}(t, u_0^2)\}_{t\geq 0}$. At the same time, one of the corresponding real trajectories $u(t, u_0^{1,2})$ may correspond to an unstable periodic orbit (UPO) which is embedded in the attractor and does not allow one to visualize it. The limitations of the possibilities of numerical integration procedures demonstrates [29] an example of the Rössler system [30]:

$$\begin{cases} \dot{x} = -y - z, \\ \dot{y} = x + 0.2y, \\ \dot{z} = 0.2 - 5.7z + xz. \end{cases} \tag{2}$$

For system (2) it is possible to stabilize an unstable periodic orbit (UPO) u^{upo_1} with period $\tau \approx 5.8811$ embedded into attractor. The corresponding computations by the standard MATLAB numerical integration procedure with and without application of the Pyragas' correction control [31] (see also [32]) of the largest FTLE, $\mathrm{LE}_1(t, u_0^{\mathrm{upo}_1})$, and FTLD $\dim_{\mathrm{L}}(t, u_0^{\mathrm{upo}_1})$ along a trajectory with initial data $u_0^{\mathrm{upo}_1} \in u^{\mathrm{upo}_1}$ over the time interval $[0, 500]$ give us the following results. On the initial part of the time interval, one can indicate the coincidence of these values with a sufficiently high accuracy. For the UPO and for the unstabilized trajectory $\mathrm{LE}_1(t, u_0^{\mathrm{upo}_1})$ coincide up to the 5th decimal place inclusive on the interval $[0, 30.4]$, up to the 4th decimal place inclusive on the interval $[0, 53.8]$, up to the 3rd decimal place inclusive on the interval $[0, 71.5]$. After $t > 71.5$ the difference in values becomes significant and the corresponding graphics diverge in such a way that the part of the graph corresponding to the unstabilized trajectory is lower than the part of the graph corresponding to the UPO (see Fig. 2b). Thus, the application of the Pyragas' procedure makes it possible to compensate round-off errors and to trace the UPO numerically[3]. Note that other UPOs could be revealed, e.g., by various evolutionary algorithms [34].

4. Analytical localization of attractors via Lyapunov functions

In order to simplify the numerical search for attractors, one can apply an analytical localization approach related to the dissipativity in the sense of Levinson [9]. Also this approach may help to obtain an exact formula or to estimate Lyapunov dimension in the entire phase space. As an example of the effectiveness of such approach, we consider the localization of attractors in the Vallis system describing the El Niño-Southern Oscillation (ENSO) phenomenon of irregular, anomalous, Christmas time warming of the coastal waters of Peru and Ecuador about every 3–6 years that affected weather on a global scale. A low-order model for the ENSO phenomenon was suggested by G.K. Vallis [35] and has the following form:

$$\begin{cases} \dot{x} = By - C(x + p), \\ \dot{y} = xz - y, \\ \dot{z} = -xy - z + 1, \end{cases} \tag{3}$$

[3] There are well-known cases when the accumulation of errors in the computer representation of real data led to catastrophes (see, e.g. [33])

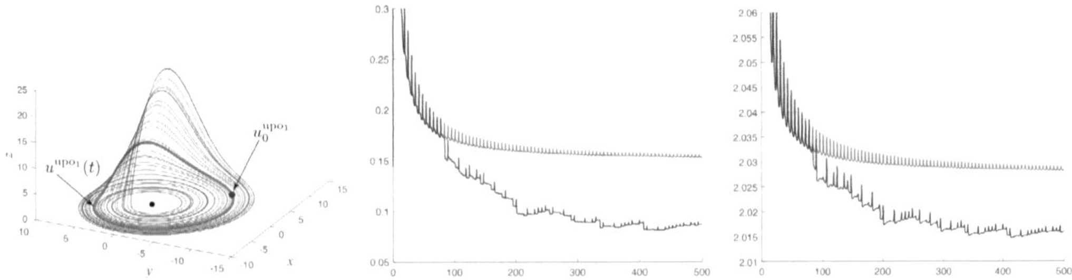

(a) (b) (c)

Figure 2: Numerical computations of trajectory (a), largest finite-time Lyapunov exponent $\mathrm{LE}_1(t, u_0^{\mathrm{upo}_1})$ (b), and finite-time local Lyapunov dimension $\dim_{\mathrm{L}}(t, u_0^{\mathrm{upo}_1})$ (c) with (red) and without (blue) application of Pyragas' correction control on the time interval $t \in [0, 500]$. Initial point $u_0^{\mathrm{upo}_1} = (6.491, -7.0078, 0.1155)$ (dark red) is chosen at the computed UPO u^{upo_1}.

with parameters $B, C > 0$, and real p. If $p = 0$, then system (3) can be transformed to the Lorenz system (1). The number of equilibria in system (1) depends on the sign of the discriminant

$$D = \tfrac{Q^2}{4} + \tfrac{P^3}{27}, \quad Q = \tfrac{p}{3}\left(\tfrac{2p^2}{9} + \tfrac{B}{C} + 2\right), \quad P = -\left(\tfrac{p^2}{3} + \tfrac{B}{C} - 1\right)$$

of the cubic equation

$$x^3 + px^2 + \left(1 - \tfrac{B}{C}\right)x + p = 0. \tag{4}$$

If $B < C$, then for any real p system (3) has only one equilibrium, Otherwise, if $B > C$, and also $p \in (-p^*, p^*)$, $\quad p^* = \sqrt{\left(\frac{B^2}{8C^2} - \frac{5B}{2C} - 1\right) + \frac{1}{8}\sqrt{\frac{B}{C}\left(\frac{B}{C} + 8\right)^3}}$ we get $D < 0$ and system (3) has three equilibria $O_j = (x_j, y_j, z_j)$, $j = 1, 2, 3$, where x_j are the solutions of (4), and $y_j = \frac{C(x_j+p)}{B}$, $z_j = \frac{C(x_j+p)}{Bx_j}$. One can express x_j as follows

$$x_j = -\frac{p}{3} + \xi^{j-1}\sqrt[3]{-\tfrac{Q}{2} + \sqrt{D}} + \xi^{2(j-1)}\sqrt[3]{-\tfrac{Q}{2} - \sqrt{D}}, \quad j = 1, 2, 3, \quad \text{where} \quad \xi = -\tfrac{1}{2} + \tfrac{\sqrt{3}}{2}i. \tag{5}$$

Using the direct Lyapunov method, we can prove the dissipativity of system (1) for $C \geq \frac{1}{2}$ and obtain the following ellipsoidal absorbing set:

$$\mathcal{B} = \left\{(x, y, z) \in \mathbb{R}^3 \mid \tfrac{1}{B}x^2 + y^2 + (z+1)^2 \leq 4 + \tfrac{(Cp)^2}{(2C-1)B}\right\}. \tag{6}$$

Let $u = (x, y, z) \in U = \mathbb{R}^3$, and the dynamical system $\{\varphi_{\mathrm{V}}^t\}_{t\geq 0}$, is generated by the Vallis system (3) with positive parameters B, C and real parameter p, and $\mathcal{A}_{\mathrm{V}} \subset \mathbb{R}^3$ is a nonempty closed bounded set, which is invariant with respect to the dynamical system $\{\varphi_{\mathrm{V}}^t\}_{t\geq 0}$. i.e. $\varphi_{\mathrm{V}}^t(\mathcal{A}_{\mathrm{V}}) = \mathcal{A}_{\mathrm{V}}$ for all $t \geq 0$. Using an effective analytical approach, proposed by Leonov [36, 26], which is based on a combination of the Douady-Oesterlé approach with the direct Lyapunov method we obtain the upper estimate of the Lyapunov dimension for the global attractor in system (3).

Theorem 1. *For the Vallis dynamical system $\{\varphi_{\mathrm{V}}^t\}_{t\geq 0}$, generated by system (3) with $B, C > 0$, and real p we have the following estimate for the Lyapunov dimension of it global B-attractor*

$$\dim_{\mathrm{L}} \mathcal{A}_{\mathrm{V}} \leq 3 - \frac{2(C+2)}{C+1+\sqrt{(C-1)^2 + \frac{25}{4}B}}. \tag{7}$$

Let us compare the obtained estimate (7) with the values of local Lyapunov dimension at the equilibria O_1, O_2, O_3. E.g., for $B = 102$, $C = 3$, $p = 0.83$ we get the following local Lyapunov dimensions

$$\dim_{\rm L} O_1 = 2.58834,\ \dim_{\rm L} O_2 = 2.00368,\ \dim_{\rm L} O_3 = 2.07175.$$

and the corresponding estimate (7) is in accordance with the values of local Lyapunov dimensions

$$\dim_{\rm L} O_2 < \dim_{\rm L} O_3 < \dim_{\rm L} O_1 \le \dim_{\rm L} \mathcal{A} \le 2.65903.$$

Acknowledgments

The work is supported by the Russian Science Foundation project (14-21-00041).

References

[1] Landau L D 1944 On the problem of turbulence *Dokl. Akad. Nauk SSSR* vol 44 pp 339–349
[2] Hopf E 1948 *Communications on Pure and Applied Mathematics* **1** 303–322
[3] Ruelle D and Takens F 1971 *Communications in Mathematical Physics* **20** 167–192
[4] Smale S 1967 *Bulletin of the American mathematical Society* **73** 747–817
[5] Ladyzhenskaya O A 1982 *Zapiski Nauchnykh Seminarov POMI* **115** 137–155
[6] Lorenz E 1963 *J. Atmos. Sci.* **20** 130–141
[7] Leonov G, Kuznetsov N and Vagaitsev V 2011 *Physics Letters A* **375** 2230–2233
[8] Leonov G and Kuznetsov N 2013 *International Journal of Bifurcation and Chaos* **23** art. no. 1330002
[9] Leonov G, Kuznetsov N and Mokaev T 2015 *The European Physical Journal Special Topics* **224** 1421–1458
[10] Kuznetsov N 2016 *Lecture Notes in Electrical Engineering* **371** 13–25 (Plenary lecture at International Conference on Advanced Engineering Theory and Applications 2015)
[11] Leonov G and Kuznetsov N 2015 *Applied Mathematics and Computation* **256** 334–343
[12] Sprott J C, Jafari S, Khalaf A and Kapitaniak T 2017 *The European Physical Journal Special Topics* **226** 1979–1985
[13] Zhang X and Chen G 2017 *Chaos: An Interdisciplinary Journal of Nonlinear Science* **27** art. num. 071101
[14] Kuznetsov N, Leonov G, Mokaev T, Prasad A and Shrimali M 2018 *Nonlinear Dynamics* **92** 267–285
[15] Chen G, Kuznetsov N, Leonov G and Mokaev T 2017 *International Journal of Bifurcation and Chaos* **27** art. num. 1750115
[16] Leonov G, Kuznetsov N and Mokaev T 2015 *Communications in Nonlinear Science and Numerical Simulation* **28** 166–174
[17] Danca M F, Kuznetsov N and Chen G 2017 *Nonlinear Dynamics* **88**(1) 791–805
[18] Dudkowski D, Jafari S, Kapitaniak T, Kuznetsov N, Leonov G and Prasad A 2016 *Physics Reports* **637** 1–50
[19] Garashchuk I, Sinelshchikov D and Kudryashov N 2018 *EPJ Web of Conferences* **173** art. num. 06006
[20] Garashchuk I, Sinelshchikov D and Kudryashov N 2018 *Regular and Chaotic Dynamics* **23** 257–272
[21] Grebogi C, Ott E and Yorke J 1983 *Physical Review Letters* **50** 935–938
[22] Lai Y and Tel T 2011 *Transient Chaos: Complex Dynamics on Finite Time Scales* (New York: Springer)
[23] Danca M F and Kuznetsov N 2017 *Chaos, Solitons & Fractals* **103** 144–150
[24] Yuan Q, Yang F Y and Wang L 2017 *International Journal of Nonlinear Sciences and Numerical Simulation* **18** 427–434
[25] Munmuangsaen B and Srisuchinwong B 2018 *Chaos, Solitons & Fractals* **107** 61 – 66
[26] Kuznetsov N 2016 *Physics Letters A* **380** 2142–2149
[27] Kehlet B and Logg A 2013 Quantifying the computability of the Lorenz system using a posteriori analysis *Proceedings of the VI Int. conf. on Adaptive Modeling and Simulation (ADMOS 2013)*
[28] Liao S and Wang P 2014 *Science China Physics, Mechanics and Astronomy* **57** 330–335
[29] Kuznetsov N and Mokaev T 2018 *arXiv preprint arXiv:1807.00235*
[30] Rossler O E 1976 *Physics Letters A* **57** 397–398
[31] Pyragas K 1992 *Phys. Lett. A.* **170** 421–428
[32] Kuznetsov N, Leonov G and Shumafov M 2015 *IFAC-PapersOnLine* **48** 706 – 709
[33] Skeel R 1992 *SIAM News* **25** 11
[34] Zelinka I 2015 *Swarm and Evolutionary Computation* **25** 2–14
[35] Vallis G K 1986 *Science* **232** 243–245
[36] Leonov G 1991 *Vestnik St. Petersburg University: Mathematics* **24** 38–41 [Transl. from Russian: Vestnik Leningradskogo Universiteta. Mathematika, 24(3), 1991, pp. 41-44]

IOP Conf. Series: Journal of Physics: Conf. Series **1205** (2019) 012035 doi:10.1088/1742-6596/1205/1/012035

On some properties of the coupled Fitzhugh-Nagumo equations

S F Lavrova[1], **N A Kudryashov**[2] **and D I Sinelshchikov**[3]

Department of Applied Mathematics, National Research Nuclear University MEPHI, Kashirskoe highway, 31, 115409, Moscow, Russia

E-mail: [1]infuriatedot@gmail.com, [2]nakudr@gmail.com, [3]disine@gmail.com

Abstract. We consider the FitzHugh-Nagumo model describing two neurons electrically coupled via ion flow through gap juctions between them. This model is a simple example of a neural network, which has a vast amount of periodic behaviors. It is shown that system of equations describing this model does not pass the Painlevé test. Analysis of stability of system's trivial stationary point is carried out. It is shown that this equilibrium point is not always stable. For some parameter regions where solution oscillates bifurcation diagrams are plotted and Lyapunov exponents are calculated. It is shown that analyzed non-stationary solutions are quasiperiodic.

1. Introduction

FitzHugh-Nagumo model is a simple model for the description of impulse propagation from one neuron to another [1]. It is described by the following system of ordinary differential equations

$$\frac{dv}{dt} = v(v-a)(1-v) - R,$$
$$\frac{dR}{dt} = \epsilon(v - \beta R),$$

where a, β, ϵ, R are positive parameters, v is neuron's membrane potential, R is its recovery variable. This model has been studied a lot in periodical literature. Some recent results can be found in [2–4].

In this work we study a system of two FitzHugh-Nagumo neurons coupled electrically via ion flow through gap junctions between them. This model is described by the following system of equations [5]

$$\begin{aligned} \frac{dv_1}{dt} &= v_1(v_1-a)(1-v_1) - R_1 + g(v_1 - v_2),\\ \frac{dR_1}{dt} &= \epsilon(v_1 - \beta R_1),\\ \frac{dv_2}{dt} &= v_2(v_2-a)(1-v_2) - R_2 + g(v_2 - v_1),\\ \frac{dR_2}{dt} &= \epsilon(v_2 - \beta R_2), \end{aligned} \tag{1}$$

IOP Conf. Series: Journal of Physics: Conf. Series **1205** (2019) 012035 doi:10.1088/1742-6596/1205/1/012035

where v_1, v_2 are membrane potentials of considered neurons, R_1, R_2 are their recovery variables. The neurons are paired through the voltage equations [5].

It is important to study periodic sequences of neural impulses to control dynamic functions of the body. Understanding the mechanism of neural networks which have a wide variety of periodic activities is highly significant [6]. FitzHugh-Nagumo model is one of the simplest examples of such network.

2. Painlevé analysis of the equations studied

System of eqs. (1) be reduced to a system of two second-order equations

$$\begin{aligned} &v_{1tt} + (3{v_1}^2 - 2v_1(a+1) + \beta\epsilon + a - g)v_{1t} + \epsilon\beta {v_1}^3 - \epsilon\beta(a+1){v_1}^2 + \\ &+ \epsilon(1+\beta a - \beta g)v_1 + g(v_{2t} + \epsilon\beta v_2) = 0, \\ &v_{2tt} + (3{v_2}^2 - 2v_2(a+1) + \beta\epsilon + a - g)v_{2t} + \epsilon\beta {v_2}^3 - \epsilon\beta(a+1){v_2}^2 + \\ &+ \epsilon(1+\beta a - \beta g)v_2 + g(v_{1t} + \epsilon\beta v_1) = 0. \end{aligned} \tag{2}$$

Substituting $v_1 = \sqrt{u_1}$, $v_2 = \sqrt{u_2}$ into the system of equations (2) we get

$$\begin{aligned} &u_1 u_{1tt} - \frac{1}{2}{u_{1t}}^2 + (3{u_1}^2 - 2(a+1){u_1}^{\frac{3}{2}} + (\epsilon\beta + a - g)u_1)u_{1t} + 2\epsilon\beta {u_1}^3 - \\ &- 2\epsilon\beta(a+1){u_1}^{\frac{5}{2}} + 2\epsilon(\beta a - \beta g + 1){u_1}^2 + 2{u_1}^{\frac{3}{2}} g(\epsilon\beta\sqrt{u_2} + \frac{u_{2t}}{\sqrt{u_2}}) = 0, \\ &u_2 u_{2tt} - \frac{1}{2}{u_{2t}}^2 + (3{u_2}^2 - 2(a+1){u_2}^{\frac{3}{2}} + (\epsilon\beta + a - g)u_2)u_{2t} + 2\epsilon\beta {u_2}^3 - \\ &- 2\epsilon\beta(a+1){u_2}^{\frac{5}{2}} + 2\epsilon(\beta a - \beta g + 1){u_2}^2 + 2{u_2}^{\frac{3}{2}} g(\epsilon\beta\sqrt{u_1} + \frac{u_{1t}}{\sqrt{u_1}}) = 0. \end{aligned} \tag{3}$$

We begin by applying the Painlev'e test to eqs. (3)(see e.g. [7]). Leading terms of this system are

$$\begin{aligned} &u_1 u_{1tt} + 3{u_1}^2 u_{1t} - \frac{1}{2}{u_{1t}}^2 = 0, \\ &u_2 u_{2tt} + 3{u_2}^2 u_{2t} - \frac{1}{2}{u_{2t}}^2 = 0. \end{aligned} \tag{4}$$

Subst ituting $u_1 = a_0 t^{-n}$, $u_2 = b_0 t^{-m}$ into eqs. (4), we find first coefficients of the solution's expansion into Laurent series $(a_0, n) = (b_0, m) = (1/2, 1)$.

To find Fuchs indices we substitute

$$\begin{aligned} u_1 &= \frac{1}{2t} + \beta_1 t^{r-1}, \\ u_2 &= \frac{1}{2t} + \beta_2 t^{r-1} \end{aligned}$$

into (4). Equating to zero terms linear in β_1 and β_2 we get

$$\begin{pmatrix} \frac{1}{2}r^2 - \frac{1}{4}r - \frac{3}{4} & 0 \\ 0 & \frac{1}{2}r^2 - \frac{1}{4}r - \frac{3}{4} \end{pmatrix} \begin{pmatrix} \beta_1 \\ \beta_2 \end{pmatrix} = \begin{pmatrix} 0 \\ 0 \end{pmatrix}.$$

Accordingly, the Fuchs indices are $r = -1, \frac{3}{2}, -1, \frac{3}{2}$. We can see that system of equations (3) does not pass the Painlevé test, since it does not have three non-negative integer Fuchs indices. None of the local solutions of (3) is general [7].

However, if we look for the expansions of the solution of eqs. (2) in the Puiseux series

$$\begin{aligned} v_1(t) &= \frac{a_0}{\sqrt{t-t_0}} + a_1 + a_2\sqrt{t-t_0} + a_3(t-t_0) + a_4(t-t_0)^{\frac{3}{2}} + a_5(t-t_0)^2 + \ldots, \\ v_2(t) &= \frac{b_0}{\sqrt{t-t_0}} + b_1 + b_2\sqrt{t-t_0} + b_3(t-t_0) + b_4(t-t_0)^{\frac{3}{2}} + b_5(t-t_0)^2 + \ldots. \end{aligned} \tag{5}$$

We get the following values of expansion's coefficients

$$\begin{aligned} &a_0 = b_0 = \pm\frac{1}{\sqrt{2}}, \quad a_1 = b_1 = \frac{a}{3} + \frac{1}{3}, \quad a_2 = b_2 = \frac{\sqrt{2}}{12}(a^2 - a + 1), \\ &a_4 = b_4 = \frac{1}{432}(a^4 - 2a^3 + 3a^2 - 2a + 144\epsilon + 1). \end{aligned} \tag{6}$$

There are arbitrary constants (t_0, a_3) in the first expansion and (t_0, b_3) in the second. They correspond to Fuchs indices and are arbitrary. So, we have obtained expansion of the solution of system of eq.(2) into the Puiseux series.

3. Stability analysis of the equations studied

Stationary points of (1) can be found from

$$\begin{aligned} &v_1(1-v_1)(v_1-a) - R_1 + g(v_1 - v_2) = 0, \\ &\epsilon(v_1 - \beta R_1) = 0, \\ &v_2(1-v_2)(v_2-a) - R_2 + g(v_2 - v_1) = 0, \\ &\epsilon(v_2 - \beta R_2) = 0. \end{aligned} \tag{7}$$

This system of equations can have up to nine real solutions. We denote them as $(v_{1i}, v_{2i}, u_{1i}, u_{2i})$, $i = \overline{1,9}$. Clearly, $(v_1, R_1, v_2, R_2) = (0,0,0,0)$ is a stationary point. We denote it as $(0,0,0,0) = O$.

Linearization of eqs. (1) about the steady states is

$$\begin{aligned} \frac{d\xi_1}{dt} &= (-3{v_{1i}}^2 + 2(a+1)v_{1i} - a + g)\xi_1 - \eta_1 - g\xi_2, \\ \frac{d\eta_1}{dt} &= \epsilon(\xi_1 - \beta\eta_1), \\ \frac{d\xi_2}{dt} &= (-3{v_{2i}}^2 + 2(a+1)v_{2i} - a + g)\xi_2 - \eta_2 - g\xi_1, \\ \frac{d\eta_2}{dt} &= \epsilon(\xi_2 - \beta\eta_2). \end{aligned} \tag{8}$$

Jacobian matrix of the stationary state is

$$\begin{pmatrix} -3{v_{1i}}^2 + 2(a+1)v_{1i} - a + g & -1 & -g & 0 \\ \epsilon & -\epsilon\beta & 0 & 0 \\ -g & 0 & -3{v_{2i}}^2 + 2(a+1)v_{2i} - a + g & -1 \\ 0 & 0 & \epsilon & -\epsilon\beta \end{pmatrix}. \tag{9}$$

We will examine stability of the state O. Jacobian matrix (9) at this point is

$$\begin{pmatrix} -a+g & -1 & -g & 0 \\ \epsilon & -\epsilon\beta & 0 & 0 \\ -g & 0 & -a+g & -1 \\ 0 & 0 & \epsilon & -\epsilon\beta \end{pmatrix}, \tag{10}$$

and the corresponding characteristic equation is

$$\lambda^4 + 2(\beta\epsilon + 2a - 2g)\lambda^3 + (\beta^2\epsilon^2 + 4a\beta\epsilon - 4\beta\epsilon g + a^2 - 2ag + \epsilon)\lambda^2 +$$
$$+(2a\beta^2\epsilon^2 - 2\beta^2\epsilon^2 g + 2a^2\beta\epsilon - 4a\beta\epsilon g + \beta\epsilon^2 + a\epsilon - \epsilon g)\lambda + a^2\beta^2\epsilon^2 - 2a\beta^2\epsilon^2 g + a\beta\epsilon^2 - \beta\epsilon^2 - \beta\epsilon^2 g = 0. \quad (11)$$

Eigenalues of (10) are

$$\lambda_{1,2} = \frac{1}{2}(-\beta\epsilon - a \pm \sqrt{(\beta\epsilon - a)^2 - 4\epsilon}),$$
$$\lambda_{3,4} = \frac{1}{2}(-\beta\epsilon - a + 2g \pm \sqrt{(\beta\epsilon - a + 2g)^2 - 4\epsilon}). \quad (12)$$

For all admissible parameter values $Re\lambda_{1,2} < 0$, because ϵ, β and a are positive parameters. Stationary point O loses stability when $g = 1/2(\beta\epsilon + a)$, since $Re\lambda_{3,4} > 0$ when $-\beta\epsilon - a + 2g > 0$. Let us fix $a = 0.3, \beta = 0.1, \epsilon = 0.01$.

In case of phase-repulsive coupling ($g > 0$) and fixed parameter values trivial stationary point is unstable when $g > \frac{1}{2}(a + \epsilon\beta) = 0.1505$. It is an unstalbe node if $(\beta\epsilon - a + 2g)^2 - 4\epsilon > 0$ and an unstable focus otherwise. For example, when $g = 0.2$ eigenvalues of (10): $\lambda_1 = -0.039, \lambda_2 = -0.2616, \lambda_{3,4} = 0.049 \pm 0.086i$. This rest state is an unstable focus. When $g = 0.5$ eigenvalues of (10): $\lambda_1 = -0.039, \lambda_2 = -0.2616, \lambda_3 = 0.6854, \lambda_4 = 0.014$. Stationary point is an unstable node. Dependences of neurons' membrane potentials on time when $g > 0$ are presented on fig.3 (a),(b).

In case of excitable coupling ($g < 0$) neuron potentials quickly synchronize after excitement. When $g = -0.5$ eigenvalues of (10) are $\lambda_1 = -0.2616, \lambda_2 = -0.039, \lambda_3 = -1.2923, \lambda_4 = -0.009$. This rest state is a stable node. Dependence of neurons' membrane potentials on time when $g > 0$ is depicted on fig. 3 (c).

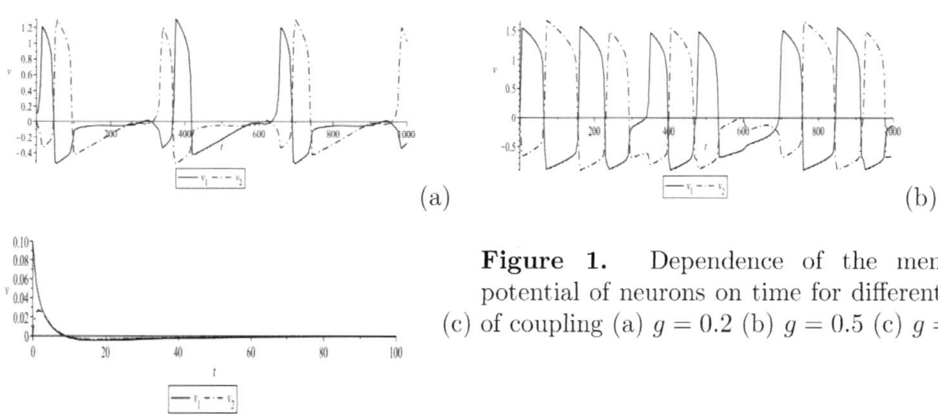

Figure 1. Dependence of the membrane potential of neurons on time for different types of coupling (a) $g = 0.2$ (b) $g = 0.5$ (c) $g = -0.5$

To determine whether solutions for $g > 0$ have periodic or chaotic nature we plot bifurcation diagrams and find Lyapunov exponents of a system. To compute maximal Lyapunov exponents we use the standard algorithm by Bennetin [8].

Sectioning the solution by a plane $R_2 = const$ in a positive direction gives one, two or three points in the Poincare section for all g except for $g \in (0.5, 0.54)$ (fig.2). Fig.3 shows that for all g the maximal Lyapunov exponent is close to zero. For studied values of parameter g computations show that the second largest Lyapunov exponent is close to zero as well. For example for $g = 0.507$ Lyapunov spectrum is $(\lambda_1, \lambda_2, \lambda_3, \lambda_4) = (-6{\cdot}10^{-5}, -9{\cdot}10^{-4}, -1.09, -4.16)$. Accordingly, all investigated oscillating solutions for $g > 0$ are quasiperiodic.

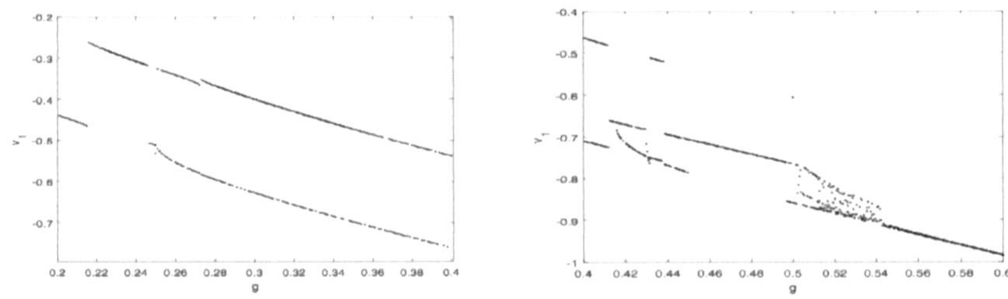

Figure 2. Bifurcation diagrams when $a = 0.3, \beta = 0.1, \epsilon = 0.01$; section by the plane $R_2 = 0.3$ on the left, section by the plane $R_2 = 0.5$ on the right

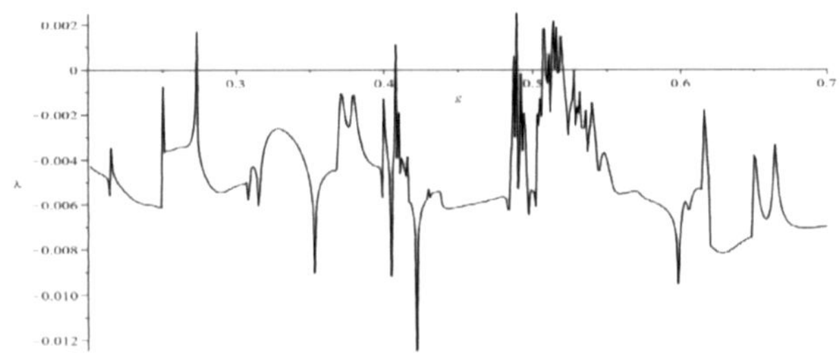

Figure 3. Dependence of the maximal Lyapunov exponent on coupling strength

4. Conclusion

FitzNagumo model describing impulse propagation between two coupled neurons is considered in this work.

For studied system of equations Painlevé test is performed. We obtain that we cannot find an expansion of a general solution of the investigated system in a Laurent or Puasieux series. Stability of system's equilibrium points is investigated. It is shown that all of the examined non-stationary solutions are quasiperiodic.

This work was supported by the Russian Science Foundation, grant number 18-11-000209.

References

[1] FitzHugh R 1961 *Biophys. J.* **1** 445
[2] Postnikov E B and Titkova O V 2016 *Eur. Phys. J. Plus* **131** 411
[3] Kudryashov N A, Rybka R B and Sboev A G 2018 *Appl. Math. Lett.* **76** 142
[4] Kudryashov N A 2018 *Regul. Chaotic. Dyn.* **23** 152
[5] Skinner F K, Zhang L, Velazquez J P and Carlen P 1999 *J. Neurophysiol.* **81** 1274
[6] Pearson K 2000 *Curr. Opin. Neurobiol.* **10** 649
[7] Ablowitz M J, Ramani A and Segur H 1980 J. *Math. Phys.* **21** 715
[8] Benettin G, Galgani L, Giorgilli A and Strelcyn M 1980 *Meccanica* **15** 9

Application of M.Riesz potentials for solving a 3D inverse problem of acoustic sounding

A S Leonov[1]

[1]National Research Nuclear University MEPhI (Moscow Engineering Physics Institute), Moscow, 115409 Russia

E-mail: asleonov@mephi.ru

Abstract. An inverse coefficient problem for time-dependent 3D wave equation is under consideration. We recover a spatially varying coefficient of this equation knowing special time integrals of the wave field in an observation domain. The inverse problem has applications to the acoustic sounding, medical imaging, etc. We reduce the inverse problem to a new linear 3D Fredholm integral equation of the first kind in which the integral operator has the form of well-known M.Riesz potentials. The equation has a unique solution for a considered class of coefficients. Assuming a special scheme for recording the data of the inverse problem, we present and substantiate a numerical algorithm of solving this integral equation. The algorithm does not require significant computational resources and a long solution time. It is based on the use of fast Fourier transform. Typical results of solving 3D inverse problem in question on a personal computer for simulated data demonstrate high capabilities of the proposed algorithm.

1. Introduction

We consider the wave equation with homogeneous initial conditions:

$$\begin{cases} \frac{1}{c^2(\mathbf{x})} u_{tt}(\mathbf{x},t) = \Delta u(\mathbf{x},t) - g(t)\varphi(\mathbf{x}), \ \ \mathbf{x} \in \mathbb{R}^3, \, t > 0 \\ u(\mathbf{x},0) = u_t(\mathbf{x},0) = 0, \ \ \mathbf{x} = (x_1, x_2, x_3) \in \mathbb{R}^3. \end{cases} \tag{1}$$

For this equation, the coefficient inverse problem is usually formulated as follows: knowing the wave field $u(\mathbf{x},t)$ in a domain $Y \subset \mathbb{R}^3$ and the source functions $g(t)$, $\varphi(\mathbf{x}), \mathbf{x} \in D$, to find the function $c(\mathbf{x})$ in a known domain X, $X \cap Y = \varnothing$, $X \cap D = \varnothing$, if $c(\mathbf{x})$ is a known constant c_0 outside the region X. A detailed statement of the inverse problem with an indication of the spaces used is given below. Such a task, as established in a number of works (see, e.g. [1, 2, 3]), requires for its solution significant computing resources (supercomputer, long calculation time, etc.). However, there are alternative statements of such a problem, in which special integral functionals of the function $u(\mathbf{x},t)$ are used as data. For example, in [4], a three-dimensional linear integral equation of the first kind was obtained for finding the function $\xi(\mathbf{x}) = \frac{1}{c_0^2} - \frac{1}{c^2(\mathbf{x})}$ associated with $c(\mathbf{x})$. The right-hand side of this equation contains the function $V_2(\mathbf{x}) = \int_0^\infty t^2 u(\mathbf{x},t)dt$, $\mathbf{x} \in Y$ that can be calculated from the scattered wave field or directly measured. However, the solution of the integral equation may not be unique, and it was suggested in [4] to seek the normal solution. In the present report, the uniqueness problem is eliminated by considering a new integral equation.

2. A new equation for finding the function $c(\mathbf{x})$

Our goal is to obtain the basic integral equation with a unique solution for finding the function $c(\mathbf{x})$. We assume that the following conditions are satisfied. 1) $u(\mathbf{x},t) \in C^{2,2}(\mathbb{R}^3 \times [0,\infty))$; 2) the integrals $V_0(\mathbf{x}) = \int_0^\infty u(\mathbf{x},t)dt$, $V_2(\mathbf{x}) = \int_0^\infty t^2 u(\mathbf{x},t)dt$ converge for all $\mathbf{x} \in \mathbb{R}^3$, and the functions $V_0(\mathbf{x})$, $V_2(\mathbf{x})$ are regular at infinity ($|\mathbf{x}| \to \infty$); 3) the equalities

$$\int_0^\infty \Delta u(\mathbf{x},t)dt = \Delta\left(\int_0^\infty u(\mathbf{x},t)dt\right),\ \int_0^\infty t^2\Delta u(\mathbf{x},t)dt = \Delta\left(\int_0^\infty t^2 u(\mathbf{x},t)dt\right),$$

are valid; 4) $\lim\limits_{t\to\infty} t\,u(\mathbf{x},t) = \lim\limits_{t\to\infty} t^2 u_t(\mathbf{x},t) = 0$, $\forall \mathbf{x} \in \mathbb{R}^3$; 5) the auxiliary function $\xi(\mathbf{x}) = \frac{1}{c_0^2} - \frac{1}{c^2(\mathbf{x})}$ is continuous in $\mathbb{R}^3$ and compactly supported in the bounded domain X; the function $\varphi(\mathbf{x}) \in C^1(\mathbb{R}^3)$ is positive and compactly supported in a bounded domain D; $g(t) \in C[0,+\infty)$; the integrals $A_0 = \int_0^\infty g(t)dt$, $A_2 = \int_0^\infty t^2 g(t)dt$ converge, and $A_0 \neq 0$.

Similar conditions were used in the article [4], in which their realizability is discussed.

Integrating the equation (1) with respect to the argument t under conditions 1) – 5), we find: $\Delta V_0(\mathbf{x}) = A_0\varphi(\mathbf{x})$, and therefore the regular at infinity solution of this equation, that is the function

$$V_0(\mathbf{x}) = -\frac{A_0}{4\pi}\int_D \frac{\varphi(\mathbf{x}')d\mathbf{x}'}{|\mathbf{x}-\mathbf{x}'|} \in C^2(\mathbb{R}^3),$$

can be precomputed. Thereafter, it is assumed to be known. Moreover, we further assume that the function $V_2(\mathbf{x}) = \int_0^\infty t^2 u(\mathbf{x},t)dt$ and its analogue $V_2^{(0)}(\mathbf{x}) = \int_0^\infty t^2 u^{(0)}(\mathbf{x},t)dt$ for the case when there are no scatterers in the region X can be measured or calculated in the region Y.

Introducing the finite function $\zeta(\mathbf{x}) = V_0(\mathbf{x})(\frac{1}{c_0^2} - \frac{1}{c^2(\mathbf{x})})$ associated with $c(\mathbf{x})$, we can further obtain from (1) the equality

$$\int_X \frac{\zeta(\mathbf{x}')d\mathbf{x}'}{|\mathbf{x}-\mathbf{x}'|^\lambda} = w(\mathbf{x}),\ \mathbf{x} \in Y, \tag{2}$$

with $w(\mathbf{x}) = \frac{1}{2}b(\lambda)\left\{(-\Delta)^{\frac{\lambda-1}{2}}(V_2(\mathbf{x}) - V_2^{(0)}(\mathbf{x}))\right\}$, $b(\lambda) = 2^{3-\lambda}\pi^{3/2}\Gamma\left(\frac{3-\lambda}{2}\right)\Gamma^{-1}\left(\frac{\lambda}{2}\right)$. The function $w(\mathbf{x}) \in L_2(Y)$ can be found by using the functions $V_2(\mathbf{x}), V_2^{(0)}(\mathbf{x}) \in L_2(X)$. How to do this will be discussed in Sec.3. Then the relation (2) is the required integral equation for $\zeta(\mathbf{x})$. To the left in this equation, there is an integral, which is called the M.Riss potential. As shown by M. Riss in 1938 [5], the solution of such an equation is unique for $1 < \lambda < 3$ (and for some other λ). Especially convenient is the case $\lambda = 2$, in which

$$\int_X \frac{\zeta(\mathbf{x}')d\mathbf{x}'}{|\mathbf{x}-\mathbf{x}'|^2} = w(\mathbf{x}),\ \mathbf{x} \in Y;\ w(\mathbf{x}) = \pi^2(-\Delta)^{\frac{1}{2}}[V_2(\mathbf{x}) - V_2^{(0)}(\mathbf{x})]. \tag{3}$$

Below we will solve this particular integral equation.

3. Data for the inverse problem

The initial data for the inverse problem (3) are the function $V_2(\mathbf{x}) = \int_0^\infty t^2 u(\mathbf{x},t)dt$, $x \in Y$, and its analogue $V_2^{(0)}(\mathbf{x})$. For the inverse problem of acoustic sounding, their finding is associated with a special accumulation of information about the sound pressure $u(\mathbf{x},t)$ at the registration points. To solve the equation (3), preliminary data processing is required, namely, the calculation of the function $w(\mathbf{x}) = \pi^2(-\Delta)^{\frac{1}{2}}[V_2(\mathbf{x}) - V_2^{(0)}(\mathbf{x})]$. Now we describe a special algorithm for calculating it. The algorithm uses the system of eigenfunctions $\{\Psi_k(\mathbf{x})\}$ and the eigenvalues $\{\lambda_k\}$ of the Laplace operator for the domain Y that can be found from the boundary value problem

$$\begin{cases} \Delta\Psi(\mathbf{x}) + \lambda\Psi(\mathbf{x}) = 0,\ \mathbf{x} \in Y \\ \Psi(\mathbf{x})|_{\partial Y} = 0. \end{cases}$$

Algorithm 1 (for computing the function $w(\mathbf{x})$ in the domain Y).

1) Find a function $\bar{u}(\mathbf{x}) \in L_2(\mathbb{R}^3)$ satisfying the condition $\bar{u}(\mathbf{x})|_{\partial Y} = u(\mathbf{x})|_{\partial Y}$ and "sufficiently fast" decreasing as $|\mathbf{x}| \to \infty$.

2) Calculate the function $\bar{U}(\mathbf{x}) = (-\Delta)^{\frac{1}{2}}\bar{u}(\mathbf{x}) = F^{-1}(|\omega|F(\bar{u})(\omega))$ by the use of the direct and inverse Fourier transforms $F(\cdot)(\omega)$, $F^{-1}(\cdot)(\mathbf{x})$.

3) Find the expansion of the function $v(\mathbf{x}) = u(\mathbf{x}) - \bar{u}(\mathbf{x})$ with respect to the basis $\{\Psi_k(\mathbf{x})\}$: $v(\mathbf{x}) = \sum_k C_k \Psi_k(\mathbf{x})$ in the domain Y.

4) Calculate the function $W(\mathbf{x}) = (-\Delta)^{\frac{1}{2}}v(\mathbf{x}) = \sum_k C_k \sqrt{\lambda_k}\Psi_k(\mathbf{x})$, $\mathbf{x} \in Y$. Here, the regularization of this ill-posed problem may be necessary.

5) Find the function $w(\mathbf{x}) = W(\mathbf{x}) + \bar{U}(\mathbf{x})$, $\mathbf{x} \in Y$.

4. Scheme for obtaining data of the inverse problem and a special form of the basic equation

In what follows, we take for the coordinates $\mathbf{x} = (x_1, x_2, x_3)$ the notation x, y, z. Then for the geometric scheme of solving the inverse problem in a plane layer (see Fig.1), that is for $X \subset \mathbb{R}^2_{xy} \times [h, H]$, $Y = \mathbb{R}^2_{xy} \times [l, L]$, $H < l$, the equation (3) can be written in the form

$$\int_h^H dz' \iint_{\mathbb{R}^2_{xy}} K(x-x', y-y', z-z')\zeta(x', y', z')dx'dy' = w(x, y, z),$$

$(x, y) \in \mathbb{R}^2_{xy}$, $z \in [l, L]$, with the kernel $K(x, y, z) = \left(x^2 + y^2 + z^2\right)^{-2}$. From the assumptions about the function $c(x, y, z)$ and the form of the functions K, w, the inclusions $\zeta \in L_2(X)$, $w \in L_2(Y)$ can be derived, as well as the inclusions $\zeta(x', y', z') \in L_2(\mathbb{R}^2_{xy})$, $w(x, y, z) \in L_2(\mathbb{R}^2_{xy})$ for every admissible z, z'. The following relations also hold: $||K(x-x', y-y', z-z')||_{L_2(Y\times R)} < \infty$, $||K(x, y, z-z')||_{L_2(\mathbb{R}^2_{xy})} < \infty$, $\forall z, z'$, $z \in [l, L]$, $z' \in [h, H]$. Then, using the two-dimensional Fourier transforms of the functions K, ζ and w with respect to the variables (x, y), i.e. the values $\tilde{K}(\omega_1, \omega_2, z)$, $\tilde{\zeta}(\omega_1, \omega_2, z)$,$\tilde{w}(\omega_1, \omega_2, z)$, we obtain by the convolution theorem the family of one-dimensional integral equations of the first kind:

$$\int_h^H \tilde{K}(\omega_1, \omega_2, z-z')\tilde{\zeta}(\omega_1, \omega_2, z')dz' = \tilde{w}(\omega_1, \omega_2, z), \;\; z \in [l, L]. \tag{4}$$

To solve these equations, we first need to find $w(x)$ using Algorithm 1, and then calculate $\tilde{w}(\omega_1, \omega_2, z)$.

5. Method for solving the integral equations (4)

Equations (4) can be represented in the operator form $A_0(\omega_1, \omega_2)\tilde{\zeta} = \tilde{w}(\omega_1, \omega_2, z)$ with linear bounded integral operators $A_0(\omega_1, \omega_2)[\cdot] = \int_h^H \tilde{K}(\omega_1, \omega_2, z-z')[\cdot]dz'$ acting from $L_2[h, H]$ to $L_2[l, L]$. These equations belong to the class of linear operator equations $A_0\varphi = \psi$, where $\varphi \in Z$ is an unknown, $\psi \in U$ is data, and Z, U are Hilbert spaces. At present, many stable methods (regularizing algorithms, RAs) are worked out for finding solutions of linear operator equations in Hilbert spaces (see, e.g. [6, 8]). Suppose that a parametric family of operators $R_\alpha(A_0, \psi) : U \to Z$ represents one of these RAs. We assume that we have at our disposal approximate data of the problem $A_0\varphi = \psi$ with a known level of error δ, i.e. instead of ψ the element $\psi_\delta \in U$ is given for which the estimate $||\psi - \psi_\delta||_U \leq \delta$ holds. Then, with the proper choice of the parameter $\alpha = \alpha(\delta)$, the convergence $||\varphi_\delta - \varphi||_Z \to 0$ as $\delta \to 0$ is provided for approximate solutions $\varphi_\delta = R_{\alpha(\delta)}(A, \psi_\delta)$. Sufficient conditions for such convergence (see [6]) are the regularity conditions for elements φ_δ: $\overline{\lim}_{\delta\to 0}||\varphi_\delta||_Z \leq ||\varphi||_Z$, $\lim_{\delta\to 0}||A\varphi_\delta - \psi_\delta||_U = 0$.

Now we formulate the **Algorithm 2 of solving the inverse problem**, assuming that $w(\mathbf{x})$ is already found by Algorithm 1.
1) Computation of two-dimensional Fourier transforms $\tilde{K}(\omega_1, \omega_2, z)$ for each $z \in [h, H]$, and $\tilde{w}(\omega_1, \omega_2, z)$ for each $z \in [l, L]$.
2) Finding the approximate solution of the integral equation (4), $\tilde{\zeta}_\delta(\omega_1, \omega_2, z)$, for each (ω_1, ω_2) with the aid of an RA that ensures the fulfillment of the regularity conditions.
3) Calculation of the approximate solution $\zeta_\delta(x, y, z)$ for each $z \in [h, H]$ using the two-dimensional inverse Fourier transforms F^{-1}: $\zeta_\delta(x, y, z) = F^{-1}[\tilde{\zeta}_\delta(\omega_1, \omega_2, z)](x, y)$.
4) Finding the function $\xi_\delta(x, y, z) = \zeta_\delta(x, y, z)/V_0(x, y, z)$ that is an approximation to $\xi(x, y, z)$, and, further, an approximate determination of the coefficient $c(x, y, z)$ from the equality $\xi(x, y, z) = \frac{1}{c_0^2} - \frac{1}{c^2(x,y,z)}$.

We can prove the following assertion.

Theorem 1. *Algorithm 2 ensures the convergence of* $\left\|\zeta_\delta(x, y, z) - \bar{\zeta}(x, y, z)\right\|_{L_2(X)} \to 0$ *of approximate solutions to the exact solution of the inverse problem,* $\bar{\zeta}(x, y, z)$, *as* $\delta \to 0$.

The regularity conditions required in Sec.5) for the used RA are fulfilled, for example, for Tikhonov regularization with a choice of the parameter by the discrepancy principle or by the generalized discrepancy principle [6]. Similar property has well-known TSVD method (see, e.g. [7, 8]) with the appropriate choice of the regularization parameter.

6. The finite-dimensional approximation of the problem and the numerical implementation of Algorithm 2

We replace the spaces $\mathbb{R}^3$ and $\mathbb{R}^2_{xy}$ by the domains $\Pi = [-r, r] \times [-r, r] \times [-r, r]$ and $\Pi_{xy} = [-r, r] \times [-r, r]$ respectively with r large enough and carry out an approximation of the equations (4) in Π by a finite-difference method. To this end, we introduce uniform grids with respect to $x, \omega_1 \in [-r, r]$ and $y, \omega_2 \in [-r, r]$ of size N, as well as grids for z, z': $\{z_i\} \in [l, L]$, $\left\{z'_j\right\} \in [h, H]$ of size M and M', accordingly.

Using the fast Fourier transform in part 1 of the *Algorithm 2* for calculating the discrete analogs of the functions $\tilde{K}(\omega_1, \omega_2, z)$, $\tilde{w}(\omega_1, \omega_2, z)$, we obtain N^2 systems of linear algebraic equations (SLAEs) for $\tilde{\zeta}$:

$$A_0^{(m)} \tilde{\zeta}^{(m)} = \tilde{w}^{(m)}, \ m = 1, ..., N^2, \tag{5}$$

with the matrices $A_0^{(m)} = \left[\nu_{ij} \tilde{K}(\omega_1^{(m)}, \omega_2^{(m)}, z_i - z'_j)\right]$ and the right-hand sides $\tilde{w}^{(m)} = \tilde{w}(\omega_1^{(m)}, \omega_2^{(m)}, z_i)$. Here $\left(\omega_1^{(m)}, \omega_2^{(m)}\right)$ are points of the grid with respect to (ω_1, ω_2) enumerated by a single index m, and ν_{ij} are quadrature coefficients.

Systems (5) should be solved with the help of RAs that guarantee the regularity conditions. For these purposes, we used several regularizers: Tikhonov and iterated Tikhonov algorithms, the TSVD method and other. Numerical experiments showed that the best results when solving systems (5) gives the TSVD method. These results are presented in the next section.

7. Numerical experiments

For the experimental scheme shown in Fig.1, we define the model solution in the form

$$\xi(x, y, z) = a_1 \exp\left(-x^2 - 2y^2\right) + a_2 \exp[-3(x+4)^2 - (y-5)^2 + \\ + (x+4)(y-5)] + a_3 \exp\{-0.9\left[(x-4)^2 - (y+4)^2 + (x-4)(y+4)\right]\},$$

with $a_1 = 16 \max\left\{0.16 - (z - 1.5)^2, 0\right\}$, $a_2 = 13 \max\left\{0.04 - (z - 1.3)^2, 0\right\}$, $a_3 = 8 \max\left\{0.04 - (z - 1.7)^2, 0\right\}$ and $(x, y, z) \in X = [-10, 10] \times [-10, 10] \times [1, 2]$. The region of registration of the scattered wave field is $Y = [-10, 10] \times [-10, 10] \times [6, 7]$. We use a δ-like source

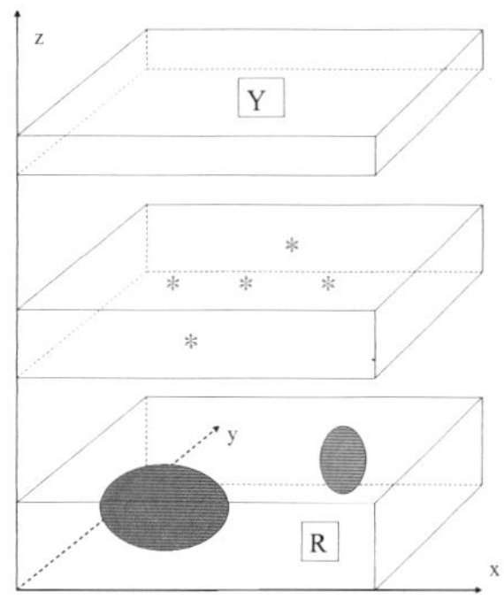

Figure 1. Scheme for data measuring 'in a layer' Y with $X = R$. Asterisks show conditional positions of the field sources.

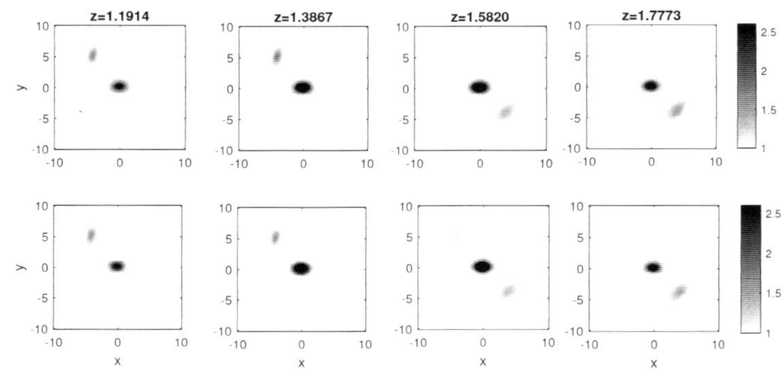

Figure 2. A qualitative comparison of the exact contrasts μ_{exact} (above) and found approximations μ_{appr} (below) for unperturbed data.

$g(t)\varphi(\mathbf{x})$ with position $(0,0,5)$ assuming also that $g(t) = \exp(-t)$ and $c_0 = 1$. Next, we calculate the model functions $w(\mathbf{x})$ and $u(\mathbf{x})$ from the equalities

$$w(\mathbf{x}) = \int_X \frac{\xi(\mathbf{x}')V_0(x')d\mathbf{x}'}{|\mathbf{x}-\mathbf{x}'|^2},\ u(\mathbf{x}) = \frac{1}{2\pi^4}\int_{\mathbb{R}^3} \frac{w(\mathbf{x}')d\mathbf{x}'}{|\mathbf{x}-\mathbf{x}'|^2},\ \mathbf{x} \in Y,$$

by applying the discrete Fourier transform (algorithmically, fast Fourier transform) on fine grids $N = M = M' = 2048$. Then we form the matrices $A_0^{(m)}$ of the systems of linear equations (5) and the right-hand sides $\tilde{w}^{(m)}$ on coarser grids ($N = 128 - 512, M = M' = 51 - 100$). We solve the systems (5) by the proposed algorithm and find the function $\xi(\mathbf{x}) = \frac{\zeta(x)}{V_0(\mathbf{x})}$. Finally, we compute from the approximately found function $\xi(\mathbf{x})$ the so-called contrast $\mu = \frac{c_0^2}{c^2(\mathbf{x})} = 1 - c_0^2\xi(\mathbf{x})$.

In Fig.2, the exact contrasts $\mu_{exact} = \frac{c_0^2}{c^2(\mathbf{x})}$ (above) and found approximations μ_{appr} (below) are shown in pairs for qualitative comparison for several values of $z \in [1,2]$. For quantitative

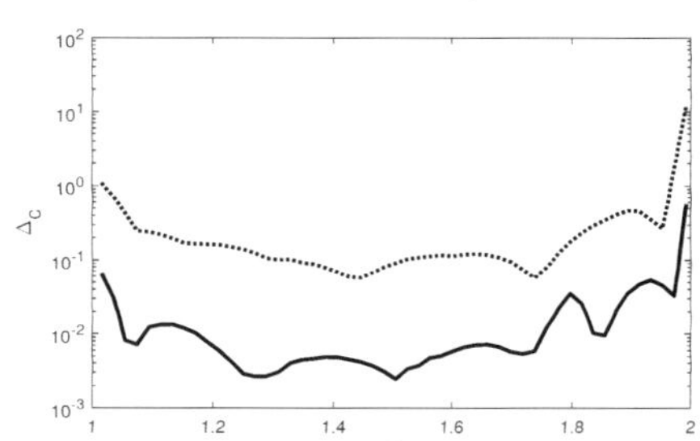

Figure 3. Relative accuracy $\Delta_C(z)$ of the approximate solutions for various z.

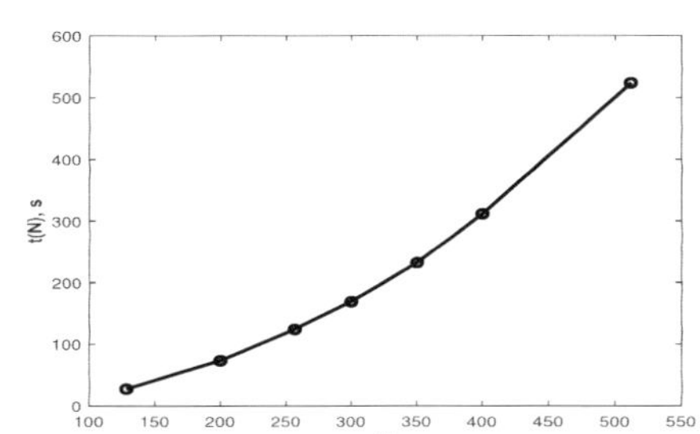

Figure 4. The solving time vs N.

comparison of the contrasts μ_{exact} and μ_{appr}, we use the relative error for each $z \in [1,2]$,

$$\Delta_C(z) = \frac{\|\mu_{appr}(x,y,z) - \mu_{exact}(x,y,z)\|_{C(\mathbb{R}^2_{xy})}}{\|\mu_{exact}(x,y,z)\|_{C(\mathbb{R}^2_{xy})}}.$$

This dependence is shown in Fig.3 for exact and perturbed data. The solid line represents the value $\Delta_C(z)$ for the unperturbed data $u(\mathbf{x})$, and the dotted line depicts the same for data with random point-wise perturbation of the order 10^{-8}. Calculations were carried out in MATLAB on a PC with an Intel (R) Core (TM) i7-7700 CPU 3.60 GHz, 16 GB RAM (without parallelization).

8. The speed of the algorithm

Since we are talking about the creation of a fast method for solving our inverse problem, we point out some temporal characteristics of proposed algorithms. The dimensions of grids in the variables z, z', that is, the numbers M and M', determine the speed of solving the one-dimensional integral equation (4) at fixed ω_1, ω_2, or what is the same, solving the SLAE (5) for a fixed m. The corresponding solution time, $t_0(M, M')$, varies a little in passing from one of equations (5) to another. This time is controlled by the desired resolution of the algorithm in the variable z. Actually, we have the estimate $t(N, M, M') \approx t_0(M, M') \cdot N^2$ for full solution time of the inverse problem for chosen grids, and the number N is controlled here by the required resolution in x, y. Figure 4 presents the dependence $t(N) = t(N, M, M')$ calculated on the same computer for fixed $M = 51, M' = 51$ and for different N. In particular, the time to solve the three-dimensional inverse problem for $N = 512$ is less than 10 minutes. Data preparation, that is calculation of $w(\mathbf{x})$, takes about 3 minutes.

9. Conclusions

1) The 3D inverse coefficient problem for the wave equation that arises in the modeling of acoustic sounding can be solved numerically faster if its data are not the time dependencies of the scattered field $u(\mathbf{x}, t)$ in a region of space Y, but some integrals of these dependencies in time. Possible integrals of such a kind are the functions $V_2(\mathbf{x}), V_2^{(0)}(\mathbf{x})$ or their partial derivatives.
2) For this type of data recorded in a plane layer, a numerical algorithm is proposed that allows to solve the inverse problem on a personal computer for sufficiently fine grids in a relatively short time, without the use of supercomputer systems. A similar algorithm can be proposed for other schemes of data acquisition.

Acknowledgments

This work was supported by the Russian Foundation for Basic Research (project no.16-01-00039a) and the Programme of Competitiveness Increase of the National Research Nuclear University MEPhI (Moscow Engineering Physics Institute); contract no. 02.a03.21.0005, 27.08.2013.

References

[1] Beilina L and Klibanov M V 2012 *Approximate Global Convergence and Adaptivity for Coefficient Inverse Problems* (New York: Springer)
[2] Bakushinsky A B and Kokurin M Yu 2004 *Iterative methods for approximate solution of inverse problems* (Dordrecht: Springer)
[3] Goncharsky A V and Romanov S Yu 2012 *Comp. Math. Math. Phys.* **52** 263
[4] Bakushinsky A B and Leonov A S 2017 *J. Inv. Ill-Posed Probl.* aop (DOI: 10.1515/jiip-2017-0041)
[5] Riesz M 1938 *Acta Litt. Acad. Sci. Szeged* **9** 1
[6] Tikhonov A N, Leonov A S and Yagola A G 1998 *Non-Linear Ill-Posed Problems* vol 1,2 (London: Chapmen & Hall)
[7] Hanson R J 1971 *SIAM J. Numer. Anal.* **8** 616
[8] Engl H W, Hanke M and Neubauer A 1996 *Regularization of Inverse Problems* (Dordrecht: Kluwer)

IOP Conf. Series: Journal of Physics: Conf. Series **1205** (2019) 012037 doi:10.1088/1742-6596/1205/1/012037

Plane problem of contact interaction for foundation with multilayer nonuniform coating

A V Manzhirov[1] **and K E Kazakov**[1,2*]

[1]Ishlinsky Institute for Problems in Mechanics of the Russian Academy of Sciences, Moscow, Russia

[2]Bauman Moscow State Technical University, Moscow, Russia

E-mail: *kazakov-ke@yandex.ru

Abstract.
Present article is devoted to solving the problem of contact interaction between a rigid punch and a viscoelastic aging foundation with a thin coating, which consists of several elastic layers. The nonuniformities of layer depend on longitudial coordinate. They can be described be discontinuous or rapidly changing unctions. The basic mixed integral equation and integral additional conditions for this problem are presented. The analytic solutions for all versions of the problem are presented.

1. Introduction

Solutions for contact problems for foundations with nonuniform coatings were constructed in [1, 2]. It were the problems for viscoelastic foundation with one surface nonuniform coating (onelayer coating). This paper devoted to the plane contact problem for foundation with coating, which consist of several different longitudially nonuniform layers.

2. Statement of the plane contact problem

Viscoelastic aging layer of a thickness H with a thin elastic multilayer coating lies on a rigid basis (number of layers equal to N). All layers have different thickness h_k $(k = 1, 2, \dots, N)$. These layers have longitudial nonuniformity, i.e. elastic characteristics of each layer depends on longitudial coordinate, but constant in depth. These nonuniformities can be described even by discontinuous functions. We also assume that the rigidities of coating layers are less than the rigidity of the lower layer or they are of the same order of magnitude. There is smooth contact or perfect contact between layers and between the lower layer and the rigid base. We consider the case of plane strain.

At time $\tau_0 \geq \tau_{\text{lower}}$ (τ_{lower} is a time of lower layer production), the force $P(t)$ starts to indent a rigid smooth punch into the surface of such a foundation. This punch width equal to $2a$ and this width much more than the total thickness of the coating layers, i.e. $2a \gg h_1 + h_2 + \ldots + h_N$. The function $g(x)$ describes distance between contact surfaces in nondeformable state and called backlash function. This function describes punch base form in this problem. We assume that we apply such force that the contact area coincides with the punch width, i.e. equal to $2a$. It is a problem with known contact area.

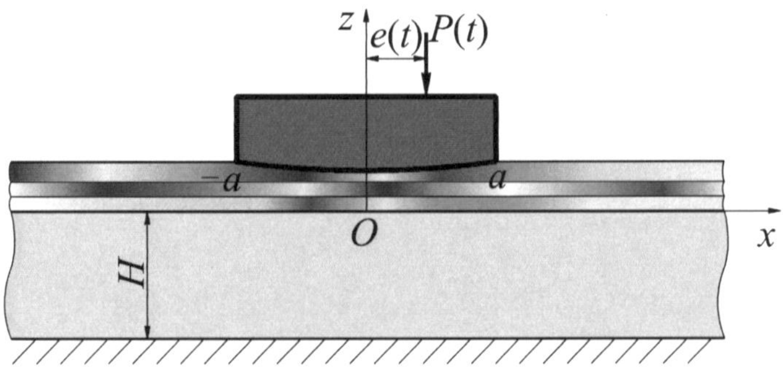

Figure 1. Plane contact interaction

As a result of this interaction, the punch is immersed in the base by the quantity $\delta(t)$ and rotated on the angle $\alpha(t)$. The contact interaction scheme is shown in figure 1.

To construct the mathematical model for this problem we should use the analytic solutions for layer (for the strip), presented in [3] and modify it for viscoelastic aging materials (see, for example, [4–6]). Then we assume the integral equation and additional conditions in the form

$$\begin{gathered}\sum_{k=1}^{N}\frac{q(x,t)h_k}{R_k(x)}+\frac{2(1-\nu_{\text{lower}}^2)}{\pi}\left[\int_{-a}^{a}k_{\text{pl}}\left(\frac{x-\xi}{H}\right)\frac{q(\xi,t)}{E_{\text{lower}}(t-\tau_{\text{lower}})}\,d\xi\right.\\ \left.+\int_{\tau_0}^{t}K(t-\tau_{\text{lower}},\tau-\tau_{\text{lower}})\int_{-a}^{a}k_{\text{pl}}\left(\frac{x-\xi}{H}\right)\frac{q(\xi,\tau)}{E_{\text{lower}}(\tau-\tau_{\text{lower}})}\,d\xi\,d\tau\right]\\ =\delta(t)+\alpha(t)x-g(x),\\ \int_{-a}^{a}q(\xi,t)\,d\xi=P(t),\quad \int_{-a}^{a}q(\xi,t)\xi\,d\xi=P(t)e(t),\quad -a\le x\le a,\quad t\ge\tau_0,\end{gathered}\tag{1}$$

where $q(x,t)$ is contact pressure under the punch, $R_k(x)$ are contact rigidities of the coating layers depend on Young moduli and Poisson's ratios of this layers [3, 6]: in the case of a smooth contact between layers k and $k-1$ (or between first layer and upper viscoelastic layer if $k=1$), we have $R_k(x)=E_k(x)/[1-\nu_k^2(x)]$, and in the case of an perfect contact, $R_k(x)=E_k(x)[1-\nu_k(x)]/[1-\nu_k(x)-2\nu_k^2(x)]$, $E_k(x)$ and $\nu_k(x)$ are the Young modulus and Poisson's ratio of the kth coating layer; $E_{\text{lower}}(t)$ and ν_{lower} are the Young modulus and Poisson's ratio of lower layer; $k_{\text{pl}}[(x-\xi)/H]$ is known kernel of the plane contact problem, which has the form $k_{\text{pl}}(s)=\int_0^\infty[L(u)/u]\cos(su)\,du$, where function $L(u)$ depend on conditions between upper layer and rigid base: in the case of smooth contact $L(u)=[\cosh(2u)-1]/[\sinh(2u)+2u]$, and in the case of perfect contact $L(u)=[2\kappa\sinh(2u)-4u]/[2\kappa\cosh(2u)+4u^2+1+\kappa^2]$ ($\kappa=3-4\nu_{\text{lower}}$); $K(t,\tau)$ is creep kernel which has a form (see [4–6]) $K(t,\tau)=E_{\text{lower}}(\tau)\frac{\partial}{\partial\tau}[1/E_{\text{lower}}(\tau)+C^{\text{lower}}(t,\tau)]$, $C^{\text{lower}}(t,\tau)$ is the tensile creep functions; $e(t)$ is force eccentricity.

There are exist 4 different versions of the problem: 1) the settlement and tilt angle of the punch are given (i.e., the right-hand side of the equation is given); 2) the punch settlement and the moment of the load application are given; 3) the tilt angle of the punch and the force of the load application are given; 4) the force and the moment of the load application are given.

3. Dimensionless form and connected problems

Make a change of variables in (1) according to the formulas

$$
\begin{aligned}
&x^* = \frac{x}{a}, \quad \xi^* = \frac{\xi}{a}, \quad t^* = \frac{t}{\tau_0}, \quad \tau^* = \frac{\tau}{\tau_0}, \quad \tau^*_{\text{lower}} = \frac{\tau_{\text{lower}}}{\tau_0}, \quad \lambda = \frac{H}{a},\\
&\delta^*(t^*) = \frac{\delta(t)}{a}, \quad \alpha^*(t^*) = \alpha(t), \quad g^*(x^*) = \frac{g(x)}{a}, \quad c^*(t^*) = \frac{E_{\text{lower}}(t-\tau_{\text{lower}})}{E_0},\\
&m^*(x^*) = \frac{E_0}{2a(1-\nu^2_{\text{lower}})}\sum_{k=1}^{N}\frac{h_k}{R_k(x)}, \quad q^*(x^*,t^*) = \frac{2(1-\nu^2_{\text{lower}})q(x,t)}{E_{\text{lower}}(t-\tau_{\text{lower}})},\\
&P^*(t^*) = \frac{2P(t)(1-\nu^2_{\text{lower}})}{E_{\text{lower}}(t-\tau_{\text{lower}})a}, \quad M^*(t^*) = \frac{2P(t)e(t)(1-\nu^2_{\text{lower}})}{E_{\text{lower}}(t-\tau_{\text{lower}})a^2},\\
&\mathbf{V}^*f(t^*) = \int_1^{t^*} K^*(t^*,\tau^*)f(\tau^*)\,d\tau^*, \quad K^*(t^*,\tau^*) = K(t-\tau_{\text{lower}},\tau-\tau_{\text{lower}})\tau_0,\\
&\mathbf{F}^*f(x^*) = \int_{-1}^{1} k^*_{\text{pl}}(x^*,\xi^*)f(\xi^*)\,d\xi^*, \quad k^*_{\text{pl}}(x^*,\xi^*) = \frac{1}{\pi}k_{\text{pl}}\Big(\frac{x-\xi}{H}\Big) = \frac{1}{\pi}k_{\text{pl}}\Big(\frac{x^*-\xi^*}{\lambda}\Big).
\end{aligned}
$$

If we omit the asterisks in formulas we obtain the following mixed integral equation with the additional condition in dimensionless form:

$$
\begin{aligned}
&c(t)m(x)q(x,t) + (\mathbf{I}-\mathbf{V})\mathbf{F}q(x,t) = \delta(t) + \alpha(t)x - g(x),\\
&\int_{-1}^{1} q(\xi,t)\,d\xi = P(t), \quad \int_{-1}^{1} q(\xi,t)\xi\,d\xi = M(t), \quad x\in[-1,1], \quad t\geq 1.
\end{aligned} \tag{2}
$$

These equations are similar to dimensionless equations of the problem for foundation with one nonuniform coating [1, 2, 8]. The construction of their analytical solutions is based on special orthogonal projection method [7]. So, the solution for our problem will have similar form. We will present only final formulas for all versions of the problems.

4. Solutions

4.1. Solution for known punch settlement and tilt angle

Final formulas for the case of known right side have a form

$$
\begin{aligned}
&q(x,t) = \frac{1}{m(x)}\sum_{k=0}^{\infty} z_k(t)\Phi_k(x),\\
&P(t) = \sum_{k=0}^{\infty} z_k(t)\delta_k, \quad M(t) = \sum_{k=0}^{\infty} z_k(t)\alpha_k,
\end{aligned} \tag{3}
$$

where

$$
\begin{aligned}
&\Phi_k(x) = \sum_{m=0}^{\infty}\psi_{km}p^*_m(x), \quad z_k(t) = (\mathbf{I}+\mathbf{W}_k)\frac{\delta_k\delta(t)+\alpha_k\alpha(t)-g_k}{c(t)+\gamma_k},\\
&\delta_k = \sqrt{J_0}\psi_{k0}, \quad \alpha_k = \frac{J_1}{\sqrt{J_0}}\psi_{k0} + \sqrt{\frac{J_0J_2-J_1^2}{J_0}}, \quad g_k = \int_{-1}^{1}\frac{g(\xi)\Phi_k(\xi)}{m(\xi)}\,d\xi,\\
&\mathbf{W}_kf(t) = \int_1^t R_k(t,\tau)f(\tau)\,d\tau, \quad k=0,1,2,\dots,
\end{aligned}
$$

polynomials $p_m^*(x)$ and constants J_0, J_1, J_2 can be determined from

$$J_m = \int_{-1}^{1} \frac{\xi^m d\xi}{m(\xi)}, \quad d_{-1} = 1, \quad d_m = \begin{vmatrix} J_0 & \cdots & J_m \\ \vdots & \ddots & \vdots \\ J_m & \cdots & J_{2m} \end{vmatrix},$$
$$p_m^*(x) = \frac{1}{d_{m-1}d_m} \begin{vmatrix} J_0 & J_1 & \cdots & J_m \\ \vdots & \vdots & \ddots & \vdots \\ J_{m-1} & J_m & \cdots & J_{2m-1} \\ 1 & x & \cdots & x^m \end{vmatrix}, \quad m = 0, 1, 2, \ldots, \tag{4}$$

coefficients ψ_{km} and γ_k can be derived from spectral problem $\sum\limits_{l=0}^{\infty} K_{ml}\psi_{kl} = \gamma_k \psi_{km}$ $(k, m = 0, 1, 2, \ldots)$, coefficients K_{ml} determined by ratios

$$K_{ml} = \int_{-1}^{1}\int_{-1}^{1} \frac{p_m^*(x) k_{\mathrm{pl}}(x,\xi) p_l^*(\xi)}{m(x)m(\xi)} dx\, d\xi, \quad m, l = 0, 1, 2, \ldots, \tag{5}$$

and kernels $R_k(t,\tau)$ are resolvents for the kernels $\gamma_k K(t,\tau)/[c(t)+\gamma_k]$ $(k = 0, 1, 2, \ldots)$.

4.2. Solution for version with known indenting force and moment

Solution of equations (2) for the case with known indenting force and moment has a form

$$q(x,t) = \frac{1}{m(x)}\left[z_0(t)p_0^*(x) + z_1(t)p_1^*(x) + \sum_{k=2}^{\infty} z_k(t)\Phi_k(x)\right],$$
$$\alpha(t) = \sqrt{\frac{J_0}{J_0 J_2 - J_1^2}}\left\{g_1 + c(t)z_1(t) + (\mathbf{I}-\mathbf{V})\left[K_{10}z_0(t) + K_{11}z_1(t) + \sum_{k=2}^{\infty} K_k^{(1)} z_k(t)\right]\right\},$$
$$\delta(t) = \frac{1}{\sqrt{J_0}}\left\{g_0 + c(t)z_0(t) + (\mathbf{I}-\mathbf{V})\left[K_{00}z_0(t) + K_{01}z_1(t) + \sum_{k=2}^{\infty} K_k^{(0)} z_k(t)\right]\right\} - \alpha(t)\frac{J_1}{J_0},$$

where

$$z_0(t) = \frac{P(t)}{\sqrt{J_0}}, \quad z_1(t) = \frac{J_0 M(t) - J_1 P(t)}{\sqrt{J_0(J_0J_2 - J_1^2)}}, \quad \Phi_k(x) = \sum_{m=2}^{\infty} \psi_{km} p_m^*(x),$$
$$z_k(t) = (\mathbf{I} + \mathbf{W}_k)\frac{-(\mathbf{I}-\mathbf{V})[z_0(t)K_k^{(0)} + z_1(t)K_k^{(1)}] - g_k}{c(t)+\gamma_k}, \quad \mathbf{W}_k f(t) = \int_1^t R_k(t,\tau) f(\tau)\, d\tau,$$
$$g_0 = \int_{-1}^{1} \frac{g(\xi)p_0^*(\xi)}{m(\xi)}\, d\xi, \quad g_1 = \int_{-1}^{1} \frac{g(\xi)p_1^*(\xi)}{m(\xi)}\, d\xi, \quad g_k = \int_{-1}^{1} \frac{g(\xi)\Phi_k(\xi)}{m(\xi)}\, d\xi,$$
$$K_k^{(0)} = \sum_{m=2}^{\infty} K_{0m}\psi_{km}, \quad K_k^{(1)} = \sum_{m=2}^{\infty} K_{1m}\psi_{km}, \quad k = 2, 3, 4, \ldots,$$

polynomials $p_m^*(x)$ and constants J_0, J_1, J_2 can be determined from (4), coefficients ψ_{km} and γ_k can be derived from spectral problem $\sum\limits_{l=2}^{\infty} K_{ml}\psi_{kl} = \gamma_k\psi_{km}$ $(m, l = 2, 3, 4, \ldots)$, coefficients K_{ml} determined by ratios (5), and kernels $R_k(t,\tau)$ are resolvents for the kernels $\gamma_k K(t,\tau)/[c(t)+\gamma_k]$ $(k = 2, 3, 4, \ldots)$.

4.3. Solution for the case with known indenting force and punch tilt angle

Solution for this case has a form

$$
\begin{aligned}
q(x,t) &= \frac{1}{m(x)}\left[z_0(t)p_0^*(x) + \sum_{k=1}^{\infty} z_k(t)\Phi_k(x)\right],\\
\delta(t) &= \frac{1}{\sqrt{J_0}}\left\{g_0 + c(t)z_0(t) + (\mathbf{I}-\mathbf{V})\left[K_{00}z_0(t) + \sum_{k=1}^{\infty} K_k^{(0)} z_k(t)\right]\right\} - \alpha(t)\frac{J_1}{J_0},\\
M(t) &= z_0(t)\frac{J_1}{\sqrt{J_0}} + \sum_{k=1}^{\infty} z_k(t)\alpha_{1k},
\end{aligned}
$$

where

$$
z_0(t) = \frac{P(t)}{\sqrt{J_0}}, \quad \Phi_k(x) = \sum_{m=1}^{\infty}\psi_{km}p_m^*(x), \quad z_k(t) = (\mathbf{I}+\mathbf{W}_k)\frac{\alpha_{1k}\alpha(t) - (\mathbf{I}-\mathbf{V})z_0(t)K_k^{(0)} - g_k}{c(t)+\gamma_k},
$$

$$
g_0 = \int_{-1}^{1}\frac{g(\xi)p_0^*(\xi)}{m(\xi)}\,d\xi, \quad g_k = \int_{-1}^{1}\frac{g(\xi)\Phi_k(\xi)}{m(\xi)}\,d\xi, \quad \alpha_{1k} = \sqrt{\frac{J_0J_2 - J_1^2}{J_0}}\psi_{k1},
$$

$$
K_k^{(0)} = \sum_{n=1}^{\infty} K_{0n}\psi_{km}, \quad \mathbf{W}_k f(t) = \int_1^t R_k(t,\tau)f(\tau)\,d\tau, \quad k = 1,2,3,\ldots,
$$

polynomials $p_m^*(x)$ and constants J_0, J_1, J_2 can be determined from (4), coefficients ψ_{km} and γ_k can be derived from spectral problem $\sum_{l=1}^{\infty} K_{ml}\psi_{kl} = \gamma_k\psi_{km}$ $(m, l = 1, 2, 3, \ldots)$, coefficients K_{ml} determined by ratios (5), and kernels $R_k(t,\tau)$ are resolvents for the kernels $\gamma_k K(t,\tau)/[c(t)+\gamma_k]$ $(k = 1, 2, 3, \ldots)$.

4.4. Solution for the case with known moment and punch settlement

Final formulas are

$$
\begin{aligned}
q(x,t) &= \frac{1}{m(x)}\left[z_0(t)\hat{p}_0^*(x) + \sum_{k=1}^{\infty} z_k(t)\Phi_k(x)\right],\\
\alpha(t) &= \frac{1}{\sqrt{J_2}}\left\{g_0 + c(t)z_0(t) + (\mathbf{I}-\mathbf{V})\left[\hat{K}_{00}z_0(t) + \sum_{k=1}^{\infty} K_k^{(0)} z_k(t)\right]\right\} - \delta(t)\frac{J_1}{J_2},\\
P(t) &= z_0(t)\frac{J_1}{\sqrt{J_2}} + \sum_{k=1}^{\infty} z_k(t)\hat{\delta}_{1k},
\end{aligned}
$$

where

$$
z_0(t) = \frac{M(t)}{\sqrt{J_2}}, \quad \Phi_k(x) = \sum_{m=1}^{\infty}\psi_{km}\hat{p}_m^*(x), \quad z_k(t) = (\mathbf{I}+\mathbf{W}_k)\frac{\hat{\delta}_{1k}\delta(t) - (\mathbf{I}-\mathbf{V})z_0(t)K_k^{(0)} - g_k}{c(t)+\gamma_k},
$$

$$
g_0 = \int_{-1}^{1}\frac{g(\xi)\hat{p}_0^*(\xi)}{m(\xi)}\,d\xi, \quad g_k = \int_{-1}^{1}\frac{g(\xi)\Phi_k(\xi)}{m(\xi)}\,d\xi, \quad \hat{\delta}_{1k} = \sqrt{\frac{J_0J_2 - J_1^2}{J_2}}\psi_{k1},
$$

$$
K_k^{(0)} = \sum_{m=1}^{\infty} \hat{K}_{0m}\psi_{km}, \quad \mathbf{W}_k f(t) = \int_1^t R_k(t,\tau)f(\tau)\,d\tau, \quad k = 1,2,3,\ldots,
$$

polynomials $\hat{p}_m^*(x)$ can be determined from

$$\hat{p}_0^*(x) = \frac{J_1}{\sqrt{J_0 J_2}} p_0^*(x) + \sqrt{\frac{J_0 J_2 - J_1^2}{J_0 J_2}} p_1^*(x), \quad \hat{p}_1^*(x) = \sqrt{\frac{J_0 J_2 - J_1^2}{J_0 J_2}} p_0^*(x) - \frac{J_1}{\sqrt{J_0 J_2}} p_1^*(x),$$
$$\hat{p}_l^*(x) = p_l^*(x), \quad l = 2, 3, 4, \ldots,$$

polynomials $p_m^*(x)$ and constants J_0, J_1, J_2 determines by (4), coefficients ψ_{km} and γ_k can be derived from spectral problem $\sum_{l=0}^{\infty} \hat{K}_{ml} \psi_{kl} = \gamma_k \psi_{km}$ $(k, m = 1, 2, 3, \ldots)$, coefficients $\hat{K}_{ml}$ determined by ratios

$$\hat{K}_{ml} = \int_{-1}^{1} \int_{-1}^{1} \frac{\hat{p}_m^*(x) k_{\mathrm{pl}}(x, \xi) \hat{p}_l^*(\xi)}{m(x) m(\xi)} dx\, d\xi, \quad m, l = 0, 1, 2, \ldots,$$

and kernels $R_k(t, \tau)$ are resolvents for the kernels $\gamma_k K(t, \tau)/[c(t) + \gamma_k]$ $(k = 1, 2, 3, \ldots)$.

Main Results and Conclusions

- We pose and solve plane contact problem for viscoelastic aging base with multilayer nonuniform coatings and rigid punches in the case, where nonuniformities are described by different rapidly changing functions.
- The solutions of all versions of the problem are obtained analytically. In the expressions for the contact stresses, function $m(x)$ and hence contact rigidities of the coating layers is distinguished explicitly. It is allows one to perform effective computations for actual nonuniformities of coating layers. Other known methods don't allow to do this.

Acknowledgments

The present study is financially supported by the Ministry of Science and Higher Education (State Registration Number AAAA-A17-117021310381-8) and, in part, by the Russian Foundation for Basic Research (under grants Nos. 18-51-05012-Arm_a, 18-01-00770-a).

References

[1] Kazakov K E 2007 Contact problems for foundations with coatings *Vestnik Samar. Gos. Univ. Estestvennonauchn. Ser.* No. 4(54) 176–96
[2] Kazakov K E 2009 Modeling of contact interaction for solids with inhomogeneous coatings *J. Phys.: Conf. Ser.* **181** 012013
[3] Alexandrov V M and Chebakov M I 2007 *Introduction into Mechanics of Contact Interaction* (Rostov-on-Don: Izdat. "TsVVR") [in Russian]
[4] Arutyunyan N Kh 1952 *Several Problems of the Theory of Creep* (Moscow-Leningrad: Gostekhizdat) [in Russian]
[5] Arutyunyan N Kh and Kolmanovskii V B 1983 *Creep Theory of Inhomogeneous Solids* (Moscow: Nauka) [in Russian]
[6] Arutyunyan N Kh and Manzhirov A V 1999 *Contact Problems of Theory of Creep* (Erevan: Izdat. Inst. Mekaniki NAN RA) [in Russian]
[7] Manzhirov A V 2016 A mixed integral equation of mechanics and a generalized projection method of its solution *Dokl. Phys.* **61** (10) 489–93
[8] Kazakov K E 2007 *Contact Problems of Deformation and Wear for Elastic and Viscoelastic Bodies with Complex Properties and Surface Shape*, Candidate's Dissertation in Physics and Mathematics (Moscow) [in Russian]

Integrated control of a robotic group with partial dominance of decision variants

S Yu Misyurin[1], M A Potapov[2] and A P Nelyubin[3]

[1] National Research Nuclear University MEPhI, 31 Kashirskoe shosse, Moscow 115409, Russia
[2] Institute of Computer Aided Design RAS, 18/19 2nd Brestskaya, Moscow 123056, Russia
[3] Mechanical Engineering Research Institute RAS, 4 Malyi Kharitonievski pereulok, Moscow 101990, Russia

nelubin@gmail.com

Abstract. We consider the problem of multicriteria design of a robotic group, consisting of a group of robots, with the aim of achieving specified targets. A new approach to the organization of control of such robotic systems based on the partial dominance of decisions over target characteristics is proposed.

1. Introduction

In recent years, there have been many promising developments in the world in the field of robotic systems, consisting of groups of robots that interact with each other. Such systems include groups of mobile robots designed to carry out search, explore or research work in hard-to-reach or completely inaccessible areas (in space, under water, underground, with high radiation, etc.) [1-4]. Groups of unmanned aerial vehicles (UAVs) are increasingly used in the military field [5]. Other promising areas of application are security systems, smart home technologies [6], and surgery with microscopic robots.

In practice, there are many situations where there are goal system characteristics (or target indicators) that need to be achieved or surpassed. As such a goal can be the characteristics of the enemy system in military confrontation or a rival in another type of competition. It may also be the characteristics necessary for a robotic system to perform a task. But the available resource and other constraints often do not allow us to achieve or dominate the goal with feasible solution variants. We propose an approach to solving such problems through the use of integrated and coordinated control of a group of robots.

In [7–8], the authors laid the foundations for a new multicriteria approach to designing complex robotic systems capable of adapting to dynamically changing operating conditions. In particular, on the examples of groups of mobile robots. This work is a continuation of the developing multicriteria approach.

2. Problem formulation

For the formulation of a multicriteria problem of designing a robotic system, we will use the following mathematical model:

$$< X, K, Z, G >.$$

IOP Conf. Series: Journal of Physics: Conf. Series **1205** (2019) 012038 doi:10.1088/1742-6596/1205/1/012038

Here X is the set of feasible solutions (designs); $K = (K_1, \ldots, K_m)$ is the vector criterion consisting of $m \geq 2$ partial criteria; Z is the vector criterion K range or scale; $G = (G_1, \ldots, G_m)$ is the goal (target) vector in the criteria space Z.

The solution variant $x \in X$ is a set of parameter values that describe the design of a robotic system. The design of a separate mechanism (robot) includes technical parameters, the values of which can be determined (selected, adjusted) in a certain area of possible values. In the case of a system consisting of many individual mechanisms (a group of robots), the design description also includes information on the composition and number of different mechanisms, on the systems of group control and interaction between them.

Each partial criterion K_i is a separate characteristic (performance indicator) of the solution $x \in X$, measured or evaluated on its scale Z_i, i.e. K_i: $X \to Z_i$, $i = 1, \ldots, m$. The scales Z_i can be ordinal or more advanced. For convenience, we assume that preferences increase along the scales Z_i. Then the non-strict Pareto preference relation R^0 can be introduced on the set of feasible solutions X [9]:

$$\forall x', x'' \in X: \quad x'R^0x'' \quad \Leftrightarrow \quad K_i(x') \geq K_i(x''), \quad i = 1, \ldots, m,$$

as well as the strict Pareto preference (dominance) relation P^0:

$$\forall x', x'' \in X: \quad x'P^0x'' \quad \Leftrightarrow \quad x'R^0x'' \text{ и } K(x') \neq K(x'').$$

The set of Pareto optimal solutions form X is denoted by X^0. These solutions $x^0 \in X$ are not dominated by any other solution from X, i.e. not $\exists x \in X$: xPx^0.

The problem of designing a robotic system considered in this paper is to find such a solution $x^* \in X$, which will provide the fulfillment of specified target characteristics:

$$K_i(x^*) \geq G_i, \; i = 1, \ldots, m. \qquad (1)$$

The problem formulation (1) differs from the classical problem of goal programming. First, the problem (1) is not an optimization problem, but consists in finding feasible solutions. However, one can further optimize one of the criteria K_i or add a new one, for example, the time or cost of manufacturing the system. Secondly, the problem (1) does not allow a solution close to goal G, but not reaching it, which can be considered as optimal when solving a goal programming problem. This key difference is due to the fact that in a number of practical situations in robotics it is not enough to get as close as possible to the goal, and it is necessary to achieve or exceed it.

3. The proposed approach to solving the problem

The problem of designing a robotic system is formulated by us in a general form, without specifying the type of constraints on the set X and the type of functions of the criteria K_i. Therefore, the specific method for estimating the range of feasible values of the vector function $K(X)$ in the space Z is not considered in this paper. For these purposes, you can use the methods of global multicriteria optimization (such as genetic algorithms [10]), parameter space investigation methods [11-13], methods for approximating the Edgeworth-Pareto hull [14]. Further, we assume that the region of feasible values of the vector function $K(X)$ is constructed (estimated) and the boundary of the Pareto optimal solutions $K(X^0)$ is picked out on it.

Let us estimate the relative position of the region $K(X^0)$ and the goal G. If some of the variants from X^0 dominate the goal G with respect to Pareto relation, then we can choose any of these variants as the solution to problem (1). In Fig. 1 this case is illustrated by the example of two criteria.

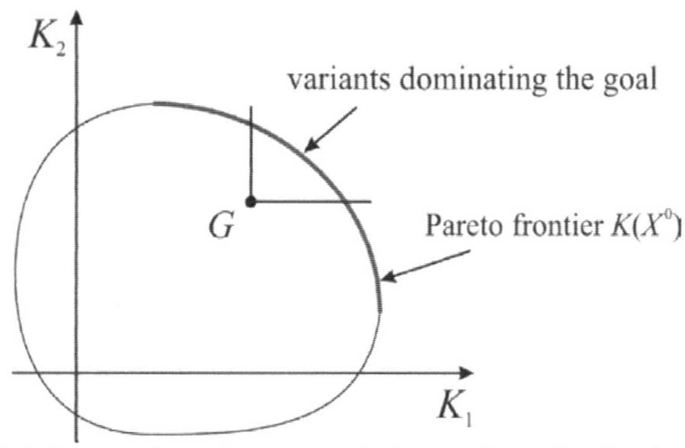

Fig. 1. Trivial case where there are solution variants dominating the goal.

In the general case, the Pareto frontier $K(X^0)$ is divided into zones (regions, sets) of partial dominance and a zone of complete dominance by a goal (see Fig. 2). In the zones of partial dominance, the solutions we have provide an advantage in terms of some criteria over the corresponding criteria of the goal. As a rule, the measure of the set of complete dominance of a goal is substantially less than the sum of sets of partial dominance.

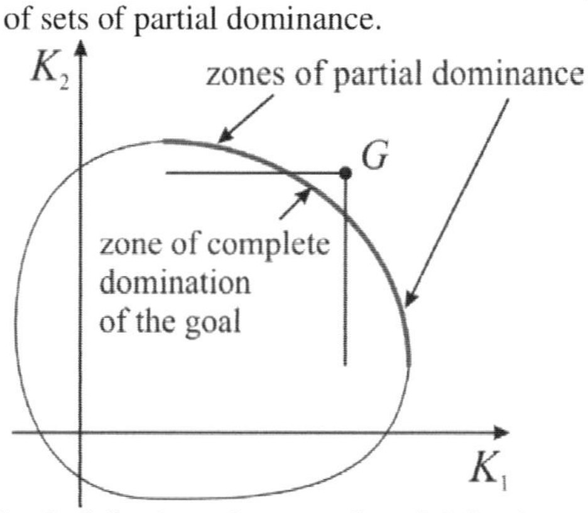

Fig. 2. Selection of zones of partial dominance.

Solution variants in the field of complete domination by goal are inferior by all the criteria. However, as a rule, it is in this area that the optimal solutions to the problem of goal programming are found. The proposed approach is based on the use of partial dominance zones depending on the operating conditions of the designed system, determining the relative importance of the criteria K_i. Under some conditions, one group of criteria will be more important, and it will be preferable to choose solutions that dominate the goal for this group of criteria. Under other conditions, the second group of criteria will be more important, and other solutions from another zone of partial dominance will be preferable. Thus, the problem (1) can be solved by organizing a mechanism for switching between different zones of partial dominance depending on the operating conditions of the system.

For robotic systems consisting of a group of robots, such a switching mechanism may be the transfer of control between heterogeneous subgroups of robots with different characteristics. By organizing such an integrated (coordinated) functioning of a group of robots and ensuring the transfer of control to that part of the robots that at the moment partially dominates the goal, it is possible to ensure the mode of complete domination of the group of robots in the vast majority of operating space. A single integrated control provides, in general, complete domination over the goal with limited resource capabilities to increase the ranges of characteristics of each robot through the most effective control of a group of robots.

4. Conclusion

We have proposed a new approach to organizing the control of robotic systems consisting of groups of robots, based on the partial dominance of variants over target indicators. The integrated control provides, in general, complete domination over the goal with limited resource capabilities to increase the ranges of characteristics of each robot.

The study was financially supported by the RFBR project No. 16-29-04401.

References

[1] Wuhui Chen, et al. 2018 A Study of Robotic Cooperation in Cloud Robotics: Architecture and Challenges. *IEEE Access* vol 6 pp 36662-36682
[2] Rubenstein M, Cornejo A and Nagpal R 2014 Programmable self-assembly in a thousand-robot swarm. *Science* vol 345 issue 6198 pp 795–799
[3] Lopatina N B, Frolov D V 2017 Analysis of prospects for the development of ground-based robotic systems for conducting radiation, chemical and biological intelligence in the United States. *Robotics and Technical cybernetics* vol 4(17) pp 3-5
[4] Vasiliev I A 2017 The use of mobile robots for rescue missions in the sea. *Robotics and Technical cybernetics* vol 4(17) pp 6-9
[5] Antohina E A, Panasenko N N, Panov V V 2017 Experience and effectiveness of the use of unmanned aerial vehicles for solving functional problems of test sites. *Robotics and Technical cybernetics* vol 3(16) pp 53-58
[6] Essa I A 2000 Ubiquitous sensing for smart and aware environments. *IEEE Personal Communication* vol 7(5) pp 47–49
[7] Misyurin S Yu, Nelyubin A P, Ivlev V I 2017 Multicriteria adaptation of robotic groups to dynamically changing conditions. *IOP Conf. Series: Journal of Physics* vol 788(1) 012027
[8] Misyurin S Yu, Nelyubin A P 2017 Multicriteria adaptation principle on example of groups of mobile robots. *IOP Conf. Series: Journal of Physics* vol 937(1) 012034
[9] Podinovski V V, Nogin V D. Pareto-optimal solutions of multicriteria problems. The second edition. Moscow: PhysMatLit, 2007
[10] Mitchell M. An Introduction to Genetic Algorithms. Cambridge, MA: MIT Press, 1996.
[11] Sobol I M, Statnikov R B. Optimal parameters choice in the problems with multiple criteria. Moscow: Nauka, 1981
[12] Nelyubin A P, Galkin T P, Galaev A A, Popov D D, Misyurin S Yu, Pilyugin V V 2017 Usage of Visualization in the Solution of Multicriteria Choice Problems. *Scientific Visualization* vol 9(5) pp 59–70
[13] Galkin T P, Nelubin A P, Galaev A A, Popov D D, Pilyugin V V, Misyurin S Yu 2018 Visual analytics in the case of multicriteria optimization. *Scientific Visualization* vol 10(2) pp 29–47
[14] Lotov A V, Bushenkov V A, Kamenev G K. Interactive Decision Maps. Approximation and Visualization of Pareto Frontier. Boston: Kluwer Academic Publishers, 2004
[15] Kudryashov N A 2008 Solitary and periodic solutions of the generalized Kuramoto-Sivashinsky equation. *Regular & Chaotic Dynamics* vol 13(3) pp 234-238
[16] Kudryashov N A 2013 Polynomials in logistic function and solitary waves of nonlinear differential equations. *Applied Mathematics and Computation* vol 219(17) pp 9245-9253

Determination of paleotemperature for the Elbrus glacier based on the inverse problem solution

O V Nagornov[1], S A Tyuflin[1], V N Mikhalenko[2] and G A Chernyakov[2]

[1] National Research Nuclear University MEPhI (Moscow Engineering Physics Institute), 31 Kashirskoe shosse, 115409 Moscow, Russia

[2] Institute of Geography, Russian Academy of Sciences, Moscow, Russia

E-mail: nagornov@yandex.ru

Abstract. The surface temperature reconstruction for the Elbrus Plato is built. It takes into account new data on the annual layers in the glacier. The dendrochronological data are used as an additional information that allows us to improve accuracy of calculations and stability of the solutions.

1. Introduction

Previous surface temperature reconstructions for the Elbrus Plato had big discrepancies between each other [1]. It was connected with absence of data on the advection rate in glacier. The new dating of the annual glacier layers [2] allows us to find out the proper solution.

2. Mathematical model

The underground temperature distribution is mainly determined by two types of processes [3]. The first is the surface temperature changes and the second is the heat flux from the Earth that is subjected to the long-time geological processes. The mathematical statement of the inverse problem consists of the thermal conductivity equation that takes into account the vertical advection term, the initial condition, the boundary condition at the bottom of glacier and the re-determination condition. The measured-temperature-depth profile is used as the re-determination condition, $\chi(z)$, where z is vertical coordinate. Then the inverse problem to find the temperature in the past is the solution of the following one-dimensional problem:

$$\begin{aligned} \rho(z)C(z)T_t = (k(z)T_z)_z - \rho(z)C(z)w(z)T_z, &\qquad (t,z) \in Q \equiv [0,t_f] \times [0,H], \\ T(z,0) = U(z), &\qquad z \in [0,H], \\ T(0,t) = U_s + \mu(t), &\qquad t \in [0,t_f], \\ -k(H)T_z(H,t) = q, &\qquad t \in [0,t_f], \\ T(z,t_f) = \chi(z), &\qquad z \in [0,H], \end{aligned} \tag{1}$$

where H is the ice sheet thickness, $\rho(z)$, $C(z)$, and $k(z)$ are the density, specific heat, and thermal conductivity of ice, $w(z)$ is the vertical ice velocity, q is the geothermal heat flux, $U(z)$ is the steady-state temperature profile associated with this flux. U_s – is the initial temperature on the surface, which characterizes the average temperature that was on the surface in the past before the beginning of sharp temperature variations on the surface. $\mu(t)$ is temperature variations on the surface in time with respect to its initial value U_s from the moment $t = 0$ ($\mu(0) = 0$) to the time of measurements of the borehole temperature profile t_f.

The solution of the direct problem (1) can be represented in the form of the operator relation $T(z, t_f) = R\{\mu(t)\}$. In terms of the measured temperature profile $\chi(z)$, the solution of the inverse problem can be represented in the form $\mu(t) = R^{-1}\{\chi(z)\}$. This equation has no exact solution for the element $\chi(z)$ on the set of the histories of surface temperature variations $F(\mu(t) \in F)$, because the measured temperature profile $\chi(z)$ includes temperature perturbations owing to which $\chi(z) \notin G$, where $G = RF$ is the set of images of the mapping specified by the operator R. These temperature perturbations appear because measurement errors exist and the mathematical model does not include all the possible processes affecting the temperature distribution in the glacier. Moreover, the operator R is not mutually continuous; i.e., the solution of the inverse problem is unstable with respect to "small" variations in the profile $T(z, t_f)$ on the set of images G. Therefore, the problem of the reconstruction of variations in the surface temperature from the borehole temperature measurements is an ill-posed problem.

The Tikhonov method is applied to determine the past surface temperatures [4]. The Tikhonov regularization method is the determination of the boundary temperature minimizing a smoothing functional consisting of the difference and stabilizer:

$$\Psi = \frac{1}{2}\int_0^H [R\{\mu(t)\} - \chi(z)]^2 dz + \alpha\Omega\{\mu(t)\}, \tag{2}$$

where α is the regularization parameter matched with the accuracy of the input data. The functional $\Omega\{\mu(t)\}$ is called stabilizing functional or stabilizer:

$$\Omega\{\mu(t)\} = \int_0^{t_f} \sum_{j=0}^{r} q_j \left(\frac{d^j \mu(t)}{dt^j}\right)^2 dt. \tag{3}$$

Where r is the stabilizer order, $q_j \geq 0$, and $q_r > 0$. The procedure of the minimization of the smoothing functional Ψ can be performed by means of the gradient method and is an iteration procedure. The iteration procedure is carried out until the functional Ψ reaches a minimum with a given accuracy, which corresponds to the optimal solution of the inverse problem.

3. Reconstruction

The temperature reconstruction for the Elbrus glacier is based on the measured temperature profile in the borehole taking into account new data on the annual layers dating [2]. We assume that the reconstructed temperature corresponds to the temperature at the depth of the activity layer equal approximately to $10m$. The temperature changes at this depth arise due to long-term temperature changes at the surface. The seasonal temperature variations do not penetrate at this depth. New data on the annual layers dating allows us to determine the advection rate as a function of depth. The advection rate is approximated by quadratic parabola so that it is equal to zero at the bottom. The other parameters were determined early [2]. The steady-state temperature in the borehole and the geothermal heat flux can be found out by eq. (1). As a rule the borehole temperature near the glacier bottom is very close to the steady-state. Thus, the stady-state temperature profile and the geothermal flux were found out $Q = 0.3\ Wt/m^2$. We assume that the geothermal flux is not changed for the time of consideration. The calculated steady-state temperature profile is shown in Fig. 1.

Let us look for the surface temperature by the finite set of trigonometric Fourier series

$$\mu(t) = \frac{a_0}{2} + \sum_{n=1}^{N} a_m \cos(\frac{2\pi t}{T_n}) + b_m \sin(\frac{2\pi t}{T_n}) \tag{4}$$

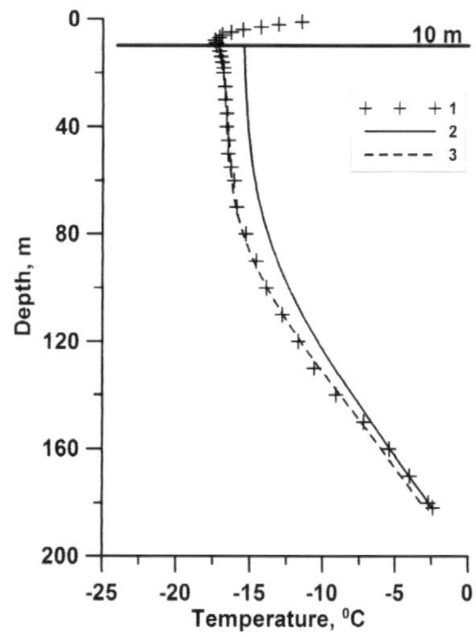

Figure 1. The temperature profiles in the borehole: measured (1), steady-state (2) and calculated (3) by the reconstructed surface temperature.

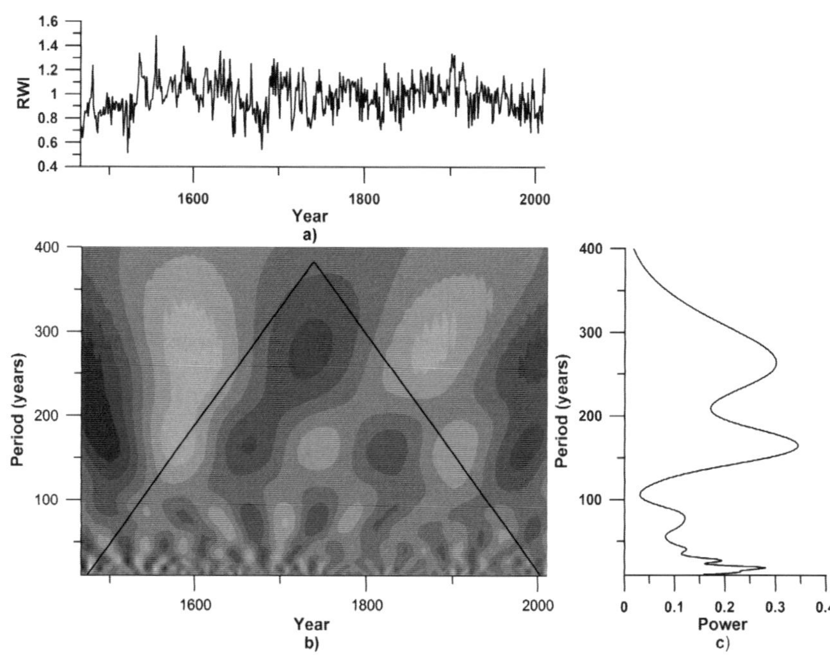

Figure 2. Tree-ring chronology (a); wavelet power spectrum (b) and global wavelet (c).

The periods of the Fourier series terms are defined by the indirect data based on the high resolution climate sources for the studied region. It is the tree-ring chronology (Fig. 2, a). Such approach enhances the accuracy of the reconstruction. We use the wavelet analysis to retrieve the characteristic periods. The wavelet analysis results are shown in Fig. 2 (b,c). This is spectrum of the wavelet coefficients and the global spectrum of energy. The analysis shows that there are periods of ~ 264, ~ 165, ~ 78, ~ 41, ~ 27 and ~ 19 years.

The reconstruction for this region is shown in Fig. 3. The temperature profile in the

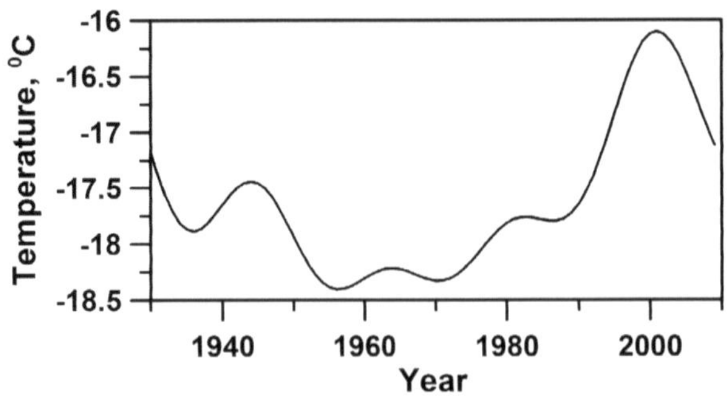

Figure 3. Past surface temperature reconstruction.

borehole corresponding to the determined surface temperature histore is shown in Fig. 1. The reconstruction correlates enough well with the meteo-data of the Teberda station situated near the glacier. The correlation coefficient is 0.65.

Acknowledgment. The authors acknowledge financial support of the Russian Foundation for basic research, grant # 18-29-10039.

References

[1] Mikhalenko et al. 2011 Stratigraphic structure and thermal regime of the infiltration layer at the Elbrus west plate, in Extreme natural phenomena and disasters, vol. 2: Uranium geology, geo-ecology, glaciology, Moscow: IFZ RAN, 180–188

[2] Mikhalenko et al. 2015 Investigation of a deep ice core from the Elbrus western plateau, the Caucasus, Russia. The Cryosphere 9, 2253–2270

[3] V.M. Kotlyakov et al. 2004 Deep drilling of glaciers in Eurasian Arctic as a source of paleoclimatic records. Quaternary Science Reviews 23, 1371–1390

[4] O.V. Nagornov and S.A. Tyuflin 2017 Inverse Problem for Paleo-Temperature Reconstruction Based on the Tree-Ring Width and Glacier-Borehole Data. Lecture Notes in Computer Science 10187, 508–516

To mathematical modeling of deformation of micropolar thin bodies with two small sizes

M Nikabadze[1], A Ulukhanyan[2] and G Sakhvadze[3]

[1]Lomonosov Moscow State University, Russia
[1,2]Bauman Moscow State Technical University, Russia
[3]Blagonravov Mechanical Engineering Research Institute of the RAS, Russia

E-mail: [1]munikabadze@yandex.ru, [2]armine_msu@mail.ru, [3]sakhvadze@mail.ru

Abstract. We consider some problems of modeling the deformation of micropolar thin bodies with two small sizes. Using the three-dimensional equations of motion, the constitutive relations and the boundary conditions of the micropolar elasticity theory [1,2], we got the equations of motion, the constitutive relations and the boundary conditions of the micropolar theory of thin bodies under the parametrization of the thin body domain with two small sizes [3]. The boundary conditions and various representations of the system of equations of motion and the constitutive relations of physical content in moments with respect to the Legendre polynomials are obtained. Note that using this method of construction the thin bodies theory with two small sizes, we get an infinite system of ordinary differential equations. This system contains quantities which depends on one variable, namely, depends on the parameter of the base line. Thus, decreasing the number of independent variables from three to one we increase the number of equations to infinity, which, of course, has its obvious practical inconveniences. In this regard, we reduce an infinite system to the finite system. The initial-boundary value problems are formulated. To satisfy the boundary conditions on the front surfaces we constructed correcting terms [4]. As a special case, we considered a prismatic body. We used the tensor calculus to do this research [5–8].

1. To parametrization of a thin body domain with two small sizes

Let the thin body have a cross-section in the form of a rectangle, which sides are much smaller than its length (third dimension). Let the considered domain does not have the line of symmetry.

The position vector of an arbitrary point of the thin body domain is represented as [3]

$$\begin{aligned}\hat{\mathbf{r}}(x', x^3) &= \mathbf{r}(x^3) + \sum_{K=1}^{2} \left[\bar{h}_K(x^3) + x^K h_K(x^3)\right]\mathbf{e}_K(x^3) = \\ &= \mathbf{r}(x^3) + \sum_{K=1}^{2} h_K^{-1}\left[\bar{h}_K(x^3) + x^K h_K(x^3)\right]\mathbf{r}_K(x^3), \quad -1 \le x^I \le 1,\end{aligned} \tag{1}$$

where $\mathbf{r} = \mathbf{r}(x^3)$ is the vector parametric equation of the baseline, $x^3 \equiv s$ is a natural parameter, $\bar{h}_I = (\overset{(+)}{h}_I - \overset{(-)}{h}_I)/2$, $h_I = (\overset{(+)}{h}_I + \overset{(-)}{h}_I)/2$, $\mathbf{r}_I = h_I \mathbf{e}_I$, $< I = 1,2 >$, $\mathbf{r}_3 = \partial_3 \mathbf{r}$ is the unit tangent vector to the baseline, $\mathbf{e}_1$ and $\mathbf{e}_2$ are the unit vectors of the principal normal and binormal to the baseline, respectively. So, $\mathbf{e}_1, \mathbf{e}_2, \mathbf{r}_3$ is the natural trihedral. Obviously, the ratio (1) for $-1 \leq x^I \leq 1$ and $x^3 \in [0, l)$, where $l \leq \infty$, is the vector parametric equation of a thin body with two small sizes. Moreover, if $l < \infty$, then we have a thin body of finite length, and if $l = \infty$, then we have a thin body of infinite length.

Since the number of pages is limited, we will not dwell on the issues of parameterization of the thin body domain. All calculations can be found in [3]. We also will not discuss problems of the theory of moments of (m,n)-th order. We only give a definition of this moment with respect to the system of Legendre polynomials. The calculation of the moments of some functions under this parametrization can be found in [3].

Let $\{u_k\}_{k=0}^{\infty}$ be some orthogonal system of the Legendre polynomials on the segment $[-1, 1]$, and $\mathbb{F}(x^1, x^2, x^3)$ is a tensor field.

Definition *Moment* $\overset{(m,n)}{\mathbb{M}}(\mathbb{F})$ *of* (m, n) *order of the quantity* $\mathbb{F}(x', x^3)$ *with respect to the system of polynomials* $\{u_k\}_{k=0}^{\infty}$ *is the integral*

$$\overset{(m,n)}{\mathbb{M}}(\mathbb{F}) = \frac{2m+1}{2}\frac{2n+1}{2}\int_{-1}^{1}\int_{-1}^{1} \mathbb{F}(x^1, x^2, x^3)P_m(x^1)P_n(x^2)dx^1dx^2. \tag{2}$$

Thereafter, we will call $\overset{(m,n)}{\mathbb{M}}(\mathbb{F})$ the operator of moments of (m, n)-th order with respect to the Legendre polynomials. In this case, the first index m refers to the variable x^1, and the second index n refers to the variable x^2. Taking into account the orthogonality of the system of Legendre polynomials and the definition of (2), it is easy to see that we will have

$$\mathbb{F}(x^1, x^2, x^3) = \sum_{m=0}^{\infty}\sum_{n=0}^{\infty}\overset{(m,n)}{\mathbb{M}}(\mathbb{F})P_m(x^1)P_n(x^2) = \sum_{m=0}^{\infty}\overset{(m,)}{\mathbb{M}}(\mathbb{F})P_m(x^1) = \sum_{n=0}^{\infty}\overset{(,n)}{\mathbb{M}}(\mathbb{F})P_n(x^2). \tag{3}$$

2. Various representations of the system of equations of motion in moments with respect to the system of Legendre polynomials

Taking into account representations of gradient and divergence under this parametrization (see [3]) we can write the equations of motion of the moment theory in the following forms:

$$g_3^{\hat{3}} N_3 \mathbf{P}^3 + \partial_P \mathbf{P}^P + \rho\mathbf{F} = \rho\partial_t^2\mathbf{u}, \quad g_3^{\hat{3}} N_3 \boldsymbol{\mu}^3 + \partial_P \boldsymbol{\mu}^P + \underset{\approx}{\mathbf{C}} \overset{2}{\otimes} \underset{\sim}{\mathbf{P}} + \rho\mathbf{m} = \underset{\sim}{\mathbf{J}} \cdot \partial_t^2\boldsymbol{\varphi}, \tag{4}$$

$$\nabla_3 \mathbf{P}^{\hat{3}} + \nabla_P \mathbf{P}^{\hat{P}} + \rho\mathbf{F} = \rho\partial_t^2\mathbf{u}, \quad \partial_3 \boldsymbol{\mu}^{\hat{3}} + \nabla_P \boldsymbol{\mu}^{\hat{P}} + \underset{\approx}{\mathbf{C}} \overset{2}{\otimes} \underset{\sim}{\mathbf{P}} + \rho\mathbf{m} = \underset{\sim}{\mathbf{J}} \cdot \partial_t^2\boldsymbol{\varphi}, \tag{5}$$

where $\underset{\approx}{\mathbf{C}}$ is the 3d rank discriminant tensor, $\mathbf{P}^q$ and $\boldsymbol{\mu}^q$ are the components of stress tensor and couple stress tensor, respectively, $\mathbf{u}$ is the displacement vector, $\boldsymbol{\varphi}$ is the rotation vector, ρ is the density, $\underset{\sim}{\mathbf{J}}$ is the 2d rank inertia tensor, $\overset{2}{\otimes}$ is the inner 2-product [3, 5, 7, 9].

Applying the moment operator of (m, n)th-order to (5), we get

$$\begin{gathered}
\nabla_3 \overset{(m,n)}{\mathbf{M}}(\mathbf{P}^{\hat{3}}) + \overset{(m',n)}{\mathbf{M}}(\mathbf{P}^{\hat{1}}) + \overset{(m,n')}{\mathbf{M}}(\mathbf{P}^{\hat{2}}) + \rho\overset{(m,n)}{\mathbf{M}}(\mathbf{F}) = \rho\overset{(m,n)}{\mathbf{M}}(\partial_t^2\mathbf{u}),\\
\nabla_3 \overset{(m,n)}{\mathbf{M}}(\boldsymbol{\mu}^{\hat{3}}) + \overset{(m',n)}{\mathbf{M}}(\boldsymbol{\mu}^{\hat{1}}) + \overset{(m,n')}{\mathbf{M}}(\boldsymbol{\mu}^{\hat{2}}) + \underset{\approx}{\mathbf{C}} \overset{2}{\otimes} \underset{\sim}{\overset{(m,n)}{\mathbf{M}}}(\underset{\sim}{\mathbf{P}}) + \rho\overset{(m,n)}{\mathbf{M}}(\mathbf{m}) = \underset{\sim}{\mathbf{J}} \cdot \overset{(m,n)}{\mathbf{M}}(\partial_t^2\boldsymbol{\varphi}),\\
-(\delta_1^I + \delta_2^I) \leq x^I \leq \delta_1^I + \delta_2^I, \quad m, n \in \mathbb{N}_0.
\end{gathered} \tag{6}$$

3. On the boundary conditions under the parametrization with an arbitrary base line of a thin body domain with two small sizes

Let $\overset{(-)}{S}_1$ ($\overset{(+)}{S}_1$) and $\overset{(-)}{S}_2$ ($\overset{(+)}{S}_2$) be the front surfaces of a thin body with two small sizes, which are determined by (1) when $x^1 = -1$ ($x^1 = 1$) and by arbitrary x^2 and x^3, and $x^2 = -1$ ($x^2 = 1$) and arbitrary x^1 and x^3, respectively. Let S_1 (S_2) be the left (right) element of a covering determined as (1) when $x^3 = x^3_1 = const$ ($x^3 = x^3_2 = const$ ($x^3_2 > x^3_1$)) and $-1 \le x^I \le 1$. $\overset{(-)}{\mathbf{P}}_{(1)}(x^2, x^3, t)$ and $\overset{(-)}{\boldsymbol{\mu}}_{(1)}(x^2, x^3, t)$ ($\overset{(+)}{\mathbf{P}}_{(1)}(x^2, x^2, t)$ and $\overset{(+)}{\boldsymbol{\mu}}(x^2, x^3, t)$) are known vectors of stress and couple stress on the surfaces $\overset{(-)}{S}_1$ ($\overset{(+)}{S}_1$); $\overset{(-)}{\mathbf{P}}_{(2)}(x^1, x^3, t)$ and $\overset{(-)}{\boldsymbol{\mu}}_{(2)}(x^1, x^3, t)$ $\big(\overset{(+)}{\mathbf{P}}_{(2)}(x^1, x^3, t)$ and $\overset{(+)}{\boldsymbol{\mu}}_{(2)}(x^1, x^3, t)\big)$ are known vectors of stress and couple stress on the surfaces $\overset{(-)}{S}_2$ $\big(\overset{(+)}{S}_2\big)$. Besides, let us introduce $\mathbf{P}_{(1)}(x^1, x^2, t)$ and $\boldsymbol{\mu}_{(1)}(x^1, x^2, t)$, $\big(\mathbf{P}_{(2)}(x^1, x^2, t)$ and $\boldsymbol{\mu}_{(2)}(x^1, x^2, t)\big)$ as known vectors of stress and couple stress on the end-faces S_1 $\big(S_2\big)$.

The boundary conditions of physical content on the front surfaces of the thin body can be represented as

$$\begin{aligned}
&\overset{(-)}{\mathbf{P}}{}^1_{(1)} + h_1^{-1}\big(\partial_3 \overset{(-)}{h}_1 - k_2\bar{h}_2\big)\overset{(-)}{g}{}^{\hat{3}}_{3}\overset{(-)}{\mathbf{P}}{}^3_{(1)} - h_1^{-1}h_2k_2\big(x^2\overset{(-)}{g}{}^{\hat{3}}_{3}\overset{(-)}{\mathbf{P}}{}^3_{(1)}\big) = -\overset{(-)}{\mathbf{P}}{}^{\vee}_{(1)}(x^2, x^3, t),\\
&\overset{(+)}{\mathbf{P}}{}^1_{(1)} - h_1^{-1}\big(\partial_3 \overset{(+)}{h}_1 + k_2\bar{h}_2\big)\overset{(+)}{g}{}^{\hat{3}}_{3}\overset{(+)}{\mathbf{P}}{}^3_{(1)} - h_1^{-1}h_2k_2\big(x^2\overset{(+)}{g}{}^{\hat{3}}_{3}\overset{(+)}{\mathbf{P}}{}^3_{(1)}\big) = \overset{(+)}{\mathbf{P}}{}^{\vee}_{(1)}(x^2, x^3, t),\\
&\overset{(-)}{\mathbf{P}}{}^2_{(2)} + h_2^{-1}\big(\partial_3 \overset{(-)}{h}_2 + k_2\bar{h}_1\big)g^{\hat{3}}_{3}\overset{(-)}{\mathbf{P}}{}^3_{(2)} + h_2^{-1}h_1k_2\big(x^1 g^{\hat{3}}_{3}\overset{(-)}{\mathbf{P}}{}^3_{(2)}\big) = -\overset{(-)}{\mathbf{P}}{}^{\vee}_{(2)}(x^1, x^3, t),\\
&\overset{(+)}{\mathbf{P}}{}^2_{(2)} - h_2^{-1}\big(\partial_3 \overset{(+)}{h}_2 - k_2\bar{h}_1\big)g^{\hat{3}}_{3}\overset{(+)}{\mathbf{P}}{}^3_{(2)} + h_2^{-1}h_1k_2\big(x^1 g^{\hat{3}}_{3}\overset{(+)}{\mathbf{P}}{}^3_{(2)}\big) = \overset{(+)}{\mathbf{P}}{}^{\vee}_{(2)}(x^1, x^3, t), \quad \mathbf{P} \to \boldsymbol{\mu}.
\end{aligned} \tag{7}$$

Here we introduced notations

$$\overset{(-)}{\mathbf{P}}{}^{\vee}_{(\alpha)} = \sqrt{g^{\hat{\alpha}\hat{\alpha}}}\Big|_{x^\alpha=-1}\overset{(-)}{\mathbf{P}}_{(\alpha)}, \quad \overset{(+)}{\mathbf{P}}{}^{\vee}_{(\alpha)} = \sqrt{g^{\hat{\alpha}\hat{\alpha}}}\Big|_{x^\alpha=1}\overset{(+)}{\mathbf{P}}_{(\alpha)}, \quad \mathbf{P} \to \boldsymbol{\mu}. \tag{8}$$

We can see that based on (7) the boundary conditions of physical content on the front surfaces in moments will be written in the form

$$\begin{aligned}
&\overset{(-,n)}{\mathbf{M}}(\mathbf{P}^1) + h_1^{-1}\big(\partial_3 \overset{(-)}{h}_1 - k_2\bar{h}_2\big)\overset{(-,n)}{\mathbf{M}}(\mathbf{P}^{\hat{3}}) - h_1^{-1}h_2k_2\overset{(-,n)}{\mathbf{M}}\big(x^2\mathbf{P}^{\hat{3}}\big) = -\overset{(,n)}{\mathbf{M}}\big(\overset{(-)}{\mathbf{P}}{}^{\vee}_{(1)}\big),\\
&\overset{(+,n)}{\mathbf{M}}(\mathbf{P}^1) - h_1^{-1}\big(\partial_3 \overset{(+)}{h}_1 + k_2\bar{h}_2\big)\overset{(+,n)}{\mathbf{M}}(\mathbf{P}^{\hat{3}}) - h_1^{-1}h_2k_2\overset{(+,n)}{\mathbf{M}}\big(x^2\mathbf{P}^{\hat{3}}\big) = \overset{(,n)}{\mathbf{M}}\big(\overset{(+)}{\mathbf{P}}{}^{\vee}_{(1)}\big),\\
&\overset{(m,-)}{\mathbf{M}}(\mathbf{P}^2) + h_2^{-1}\big(\partial_3 \overset{(-)}{h}_2 + k_2\bar{h}_1\big)\overset{(m,-)}{\mathbf{M}}(\mathbf{P}^{\hat{3}}) + h_2^{-1}h_1k_2\overset{(m,-)}{\mathbf{M}}\big(x^1\mathbf{P}^{\hat{3}}\big) = -\overset{(m,)}{\mathbf{M}}\big(\overset{(-)}{\mathbf{P}}{}^{\vee}_{(2)}\big),\\
&\overset{(m,+)}{\mathbf{M}}(\mathbf{P}^2) - h_2^{-1}\big(\partial_3 \overset{(+)}{h}_2 - k_2\bar{h}_1\big)\overset{(m,+)}{\mathbf{M}}(\mathbf{P}^{\hat{3}}) + h_2^{-1}h_1k_2\overset{(m,+)}{\mathbf{M}}\big(x^1\mathbf{P}^{\hat{3}}\big) = \overset{(m,)}{\mathbf{M}}\big(\overset{(+)}{\mathbf{P}}{}^{\vee}_{(2)}\big), \quad \mathbf{P} \to \boldsymbol{\mu}.
\end{aligned} \tag{9}$$

Applying the moment operator of (m, n)th-order to boundary conditions, we will obtain the required boundary conditions of the physical content in moments on the end-faces of a thin body in the form

$$\overset{(m,n)}{\mathbf{P}}{}^3_{(1)} = -\overset{(m,n)}{\mathbf{M}}\big(\hat{\vartheta}\mathbf{P}_{(1)}\big) \text{ when } x^3 = x^3_1; \quad \overset{(m,n)}{\mathbf{P}}{}^3_{(2)} = -\overset{(m,n)}{\mathbf{M}}\big(\hat{\vartheta}\mathbf{P}_{(2)}\big), \text{ when } x^3 = x^3_2; \ \mathbf{P} \to \boldsymbol{\mu}. \tag{10}$$

Note that we can also present (10) in the form

$$\overset{(m,n)}{\mathbf{M}}\big(\mathbf{P}^{\hat{3}}_{(1)}\big) = -\overset{(m,n)}{\mathbf{P}}_{(1)} \text{ when } x^3 = x^3_1, \quad \overset{(m,n)}{\mathbf{M}}\big(\mathbf{P}^{\hat{3}}_{(2)}\big) = \overset{(m,n)}{\mathbf{P}}_{(2)} \text{ when } x^3 = x^3_2; \ \mathbf{P} \to \boldsymbol{\mu}. \tag{11}$$

The boundary conditions of the thermal content can be considered in the same way as (9)–(11).

Now it is not difficult to write down some of the required representations of the system of motion equations in moments with respect to the system of Legendre polynomials. Indeed, from (6) we get

$$
\begin{aligned}
&\Big\{\nabla_3 \overset{(m,n)}{\mathbf{M}}(\mathbf{P}^{\hat{3}}) + \frac{2m+1}{2}\sum_{p=m}^{\infty}\big[1-(-1)^{m+p}\big]\overset{(p,n)}{\mathbf{M}}(\mathbf{P}^{1}) + h_1^{-1}\partial_3 h_1 (m+1)\overset{(m,n)}{\mathbf{M}}(\mathbf{P}^{\hat{3}}) -\\
&-\frac{2m+1}{2}\sum_{p=m}^{\infty} h_1^{-1}\Big[\big(\partial_3 \overset{(+)}{h}_1 + k_2\bar{h}_2\big) + (-1)^{m+p}\big(\partial_3 \overset{(-)}{h}_1 - k_2\bar{h}_2\big)\Big]\overset{(p,n)}{\mathbf{M}}(\mathbf{P}^{\hat{3}}) -\\
&-\frac{2m+1}{2}h_1^{-1}h_2k_2\sum_{p=m}^{\infty}\big[1-(-1)^{m+p}\big]\Big[\frac{n}{2n-1}\overset{(p,n-1)}{\mathbf{M}}(\mathbf{P}^{\hat{3}}) + \frac{n+1}{2n+3}\overset{(p,n+1)}{\mathbf{M}}(\mathbf{P}^{\hat{3}})\Big] +\\
&+\frac{2n+1}{2}\sum_{p=n}^{\infty}\big[1-(-1)^{n+p}\big]\overset{(m,p)}{\mathbf{M}}(\mathbf{P}^{2}) + h_2^{-1}\partial_3 h_2 (n+1)\overset{(m,n)}{\mathbf{M}}(\mathbf{P}^{\hat{3}}) -\\
&-\frac{2n+1}{2}\sum_{p=n}^{\infty} h_2^{-1}\Big[\big(\partial_3 \overset{(+)}{h}_2 - k_2\bar{h}_1\big) + (-1)^{n+p}\big(\partial_3 \overset{(-)}{h}_2 + k_2\bar{h}_1\big)\Big]\overset{(m,p)}{\mathbf{M}}(\mathbf{P}^{\hat{3}}) +\\
&+\frac{2n+1}{2}h_2^{-1}h_1k_2\sum_{p=n}^{\infty}\big[1-(-1)^{n+p}\big]\Big[\frac{m}{2m-1}\overset{(m-1,p)}{\mathbf{M}}(\mathbf{P}^{\hat{3}}) + \frac{m+1}{2m+3}\overset{(m+1,p)}{\mathbf{M}}(\mathbf{P}^{\hat{3}})\Big]\Big\} +\\
&\qquad\qquad\qquad\qquad\qquad\qquad + \rho\overset{(m,n)}{\mathbf{M}}(\mathbf{F}) = \rho\overset{(m,n)}{\mathbf{M}}(\partial_t^2\mathbf{u}),\\
&\{\mathbf{P} \to \boldsymbol{\mu}\} + \underset{\approx}{\mathbf{C}} \overset{2}{\otimes} \overset{(m,n)}{\underset{\sim}{\mathbf{M}}}(\underset{\sim}{\mathbf{P}}) + \rho\overset{(m,n)}{\mathbf{M}}(\mathbf{m}) = \underset{\sim}{\mathbf{J}}\cdot\overset{(m,n)}{\mathbf{M}}(\partial_t^2\boldsymbol{\varphi}), \quad m,n\in\mathbb{N}_0,
\end{aligned}
\tag{12}
$$

Further, from (6) we will have the following required representation of the system of motion equations in moments:

$$
\begin{aligned}
&\Big\{\nabla_3 \overset{(m,n)}{\mathbf{M}}(\mathbf{P}^{\hat{3}}) - \frac{2m+1}{2}\sum_{p=0}^{m}\big[1-(-1)^{m+p}\big]\overset{(p,n)}{\mathbf{M}}(\mathbf{P}^{1}) - h_1^{-1}\partial_3 h_1 m\overset{(m,n)}{\mathbf{M}}(\mathbf{P}^{\hat{3}}) +\\
&+\frac{2m+1}{2}\sum_{p=0}^{m} h_1^{-1}\Big[\big(\partial_3 \overset{(+)}{h}_1 + k_2\bar{h}_2\big) + (-1)^{m+p}\big(\partial_3 \overset{(-)}{h}_1 - k_2\bar{h}_2\big)\Big]\overset{(p,n)}{\mathbf{M}}(\mathbf{P}^{\hat{3}}) +\\
&+\frac{2m+1}{2}h_1^{-1}h_2k_2\sum_{p=0}^{m}\big[1-(-1)^{m+p}\big]\Big[\frac{n}{2n-1}\overset{(p,n-1)}{\mathbf{M}}(\mathbf{P}^{\hat{3}}) + \frac{n+1}{2n+3}\overset{(p,n+1)}{\mathbf{M}}(\mathbf{P}^{\hat{3}})\Big] -\\
&-\frac{2n+1}{2}\sum_{p=0}^{n}\big[1-(-1)^{n+p}\big]\overset{(m,p)}{\mathbf{M}}(\mathbf{P}^{2}) - h_2^{-1}\partial_3 h_2 n\overset{(m,n)}{\mathbf{M}}(\mathbf{P}^{\hat{3}}) +\\
&+\frac{2n+1}{2}\sum_{p=0}^{n} h_2^{-1}\Big[\big(\partial_3 \overset{(+)}{h}_2 - k_2\bar{h}_1\big) + (-1)^{n+p}\big(\partial_3 \overset{(-)}{h}_2 + k_2\bar{h}_1\big)\Big]\overset{(m,p)}{\mathbf{M}}(\mathbf{P}^{\hat{3}}) -\\
&-\frac{2n+1}{2}h_2^{-1}h_1k_2\sum_{p=0}^{n}\big[1-(-1)^{n+p}\big]\Big[\frac{m}{2m-1}\overset{(m-1,p)}{\mathbf{M}}(\mathbf{P}^{\hat{3}}) + \frac{m+1}{2m+3}\overset{(m+1,p)}{\mathbf{M}}(\mathbf{P}^{\hat{3}})\Big]\Big\} +\\
&\qquad\qquad\qquad\qquad\qquad\qquad + \overset{(m,n)}{\boldsymbol{\Phi}} = \rho\overset{(m,n)}{\mathbf{M}}(\partial_t^2\mathbf{u}),\\
&\{\mathbf{P} \to \boldsymbol{\mu}\} + \underset{\approx}{\mathbf{C}} \overset{2}{\otimes} \overset{(m,n)}{\underset{\sim}{\mathbf{M}}}(\underset{\sim}{\mathbf{P}}) + \overset{(m,n)}{\mathbf{M}} = \underset{\sim}{\mathbf{J}}\cdot\overset{(m,n)}{\mathbf{M}}(\partial_t^2\boldsymbol{\varphi}), \quad m,n\in\mathbb{N}_0,
\end{aligned}
\tag{13}
$$

where we introduced notations

$$
\begin{aligned}
\overset{(m,n)}{\boldsymbol{\Phi}}(x^3,t) = \Big\{&\frac{2m+1}{2}\Big[\overset{(,n)}{\mathbf{M}}\big(\overset{(+)}{\mathbf{P}}{}^{\vee}_{(1)}\big) + (-1)^m\overset{(,n)}{\mathbf{M}}\big(\overset{(-)}{\mathbf{P}}{}^{\vee}_{(1)}\big)\Big] +\\
&+\frac{2n+1}{2}\Big[\overset{(m,)}{\mathbf{M}}\big(\overset{(+)}{\mathbf{P}}{}^{\vee}_{(2)}\big) + (-1)^n\overset{(m,)}{\mathbf{M}}\big(\overset{(-)}{\mathbf{P}}{}^{\vee}_{(2)}\big)\Big]\Big\} + \rho\overset{(m,n)}{\mathbf{M}}(\mathbf{F}),\\
\overset{(m,n)}{\mathbf{M}}(x^3,t) = \{&\mathbf{P} \to \boldsymbol{\mu}\} + \rho\overset{(m,n)}{\mathbf{M}}(\mathbf{m}).
\end{aligned}
$$

The relations (12) – (13) are different representations of the system of motion equations in moments with respect to the system of Legendre polynomials of the thin body with two small sizes. Using (4) we can get other representations of the system of equations in the moments [3,10]. But we will not dwell on this.

Note that each equation (without taking into account the boundary conditions) (12), with fixed values of m and n, contains an infinite number of terms. And each equation (using the boundary conditions) (13) contains the finite number of terms. Moreover, each equation contains moments of the contravariant component $\mathbf{P}^{\hat{3}} = g_3^{\hat{3}}\mathbf{P}^3$ $(\boldsymbol{\mu}^{\hat{3}} = g_3^{\hat{3}}\boldsymbol{\mu}^3)$ of the stress tensor (couple stress tensor). Introducing $\underset{(r)}{g}{}_3^{\hat{3}} = \sum\limits_{s=0}^{r}[k_1(\bar{h}_1 + x^1 h_1)]^s$ and replacing $\mathbf{P}^{\hat{3}}$ and $\boldsymbol{\mu}^{\hat{3}}$ with $\mathbf{P}^{\hat{3}}_{(r)}$ and $\boldsymbol{\mu}^{\hat{3}}_{(r)}$ into (12) and (13) respectively, where

$$\mathbf{P}^{\hat{3}}_{(r)} = \underset{(r)}{g}{}_3^{\hat{3}}\mathbf{P}^3, \quad \boldsymbol{\mu}^{\hat{3}}_{(r)} = \underset{(r)}{g}{}_3^{\hat{3}}\boldsymbol{\mu}^3, \tag{14}$$

we will obtain the representations of the system of motion equations in moments of rth approximation. In addition, let us fix some non-negative integers M and N. Then, choose the first $(M+1)(N+1)$ equations ($m = \overline{0,M}$, $n = \overline{0,N}$) from the system of equations. In equations containing an infinite number of terms, we neglect the moments of unknown quantities whose order is greater than M (by the first index) and N (by the second index). Thus, we obtain representations of the system of motion equations in moments of (M,N)th approximation. If the representations of the system of motion equations in moments are obtained from the corresponding representations of the rth approximation, then such representations are called representations of the system of motion equations in moments of (r,M,N)th approximation. Above mentioned applies to the representations of the heat inflow equation and the constitutive relations.

4. Presentations of constitutive relations of micropolar theory of thermoelasticity in moments for a thin body with two small sizes

The constitutive relations (CR) for non-isothermal processes [1,3,11,12] have the form

$$\underset{\sim}{\mathbf{P}} = \underset{\approx}{\mathbf{A}} \overset{2}{\otimes} \hat{\nabla}\mathbf{u} + \underset{\approx}{\mathbf{B}} \overset{2}{\otimes} \hat{\nabla}\boldsymbol{\varphi} - \underset{\approx}{\mathbf{A}} \overset{2}{\otimes} \underset{\simeq}{\mathbf{C}} \cdot \boldsymbol{\varphi} - \underset{\sim}{\mathbf{b}}\vartheta, \quad \underset{\sim}{\boldsymbol{\mu}} = \underset{\approx}{\mathbf{C}} \overset{2}{\otimes} \hat{\nabla}\mathbf{u} + \underset{\approx}{\mathbf{D}} \overset{2}{\otimes} \hat{\nabla}\boldsymbol{\varphi} - \underset{\approx}{\mathbf{C}} \overset{2}{\otimes} \underset{\simeq}{\mathbf{C}} \cdot \boldsymbol{\varphi} - \underset{\sim}{\boldsymbol{\beta}}\vartheta, \tag{15}$$

where $\underset{\approx}{\mathbf{A}}$, $\underset{\approx}{\mathbf{B}}$, $\underset{\approx}{\mathbf{C}}$, $\underset{\approx}{\mathbf{D}}$ are the fourth-rank tensors, called tensors of elastic moduli, $\underset{\sim}{\mathbf{b}} = \underset{\approx}{\mathbf{A}} \overset{2}{\otimes} \underset{\sim}{\mathbf{a}} + \underset{\approx}{\mathbf{B}} \overset{2}{\otimes} \underset{\sim}{\mathbf{d}}$, $\underset{\sim}{\boldsymbol{\beta}} = \underset{\approx}{\mathbf{C}} \overset{2}{\otimes} \underset{\sim}{\mathbf{a}} + \underset{\approx}{\mathbf{D}} \overset{2}{\otimes} \underset{\sim}{\mathbf{d}}$ are the tensors of thermomechanical properties, $\mathbf{u}$ is the displacement vector, $\boldsymbol{\varphi}$ is the internal rotation vector, $\vartheta = T - T_0$ is temperature drop, $\underset{\sim}{\mathbf{a}}$ and $\underset{\sim}{\mathbf{d}}$ are the tensors of thermal expansion. Particular cases of CR can be viewed in [2,13].

Therefore, the relations (15) can be written in the appropriate form and then presented in moments. From (15) it is clear that to represent them in moments it is sufficient to write down the gradient of a vector in moments. If the body is homogeneous with respect to x^I, $I = 1,2$, then applying the (m,n)th-order moment operator to (15), we will have

$$\begin{aligned}
\overset{(m,n)}{\underset{\sim}{\mathbf{P}}} &= \underset{\approx}{\mathbf{A}} \overset{2}{\otimes} \overset{(m,n)}{\underset{\sim}{\mathbf{M}}}(\hat{\nabla}\mathbf{u}) + \underset{\approx}{\mathbf{B}} \overset{2}{\otimes} \overset{(m,n)}{\underset{\sim}{\mathbf{M}}}(\hat{\nabla}\boldsymbol{\varphi}) - \underset{\approx}{\mathbf{A}} \overset{2}{\otimes} \underset{\simeq}{\mathbf{C}} \cdot \overset{(m,n)}{\boldsymbol{\varphi}} - \underset{\sim}{\mathbf{b}} \overset{(m,n)}{\vartheta}, \quad -1 \le x^I \le 1, \\
\overset{(m,n)}{\underset{\sim}{\boldsymbol{\mu}}} &= \underset{\approx}{\mathbf{C}} \overset{2}{\otimes} \overset{(m,n)}{\underset{\sim}{\mathbf{M}}}(\hat{\nabla}\mathbf{u}) + \underset{\approx}{\mathbf{D}} \overset{2}{\otimes} \overset{(m,n)}{\underset{\sim}{\mathbf{M}}}(\hat{\nabla}\boldsymbol{\varphi}) - \underset{\approx}{\mathbf{C}} \overset{2}{\otimes} \underset{\simeq}{\mathbf{C}} \cdot \overset{(m,n)}{\boldsymbol{\varphi}} - \underset{\sim}{\boldsymbol{\beta}} \overset{(m,n)}{\vartheta}, \quad m,n \in \mathbb{N}_0.
\end{aligned} \tag{16}$$

Note that (16) are valid for any system of polynomials (Legendre, Chebyshev) on given segment.

Solving, for example, some boundary-value problem of $(0, M, N)$ approximation, we will obtain approximate expressions for the fields of displacement vector and stress tensor, respectively, in the following form [3]

$$\mathbf{u}_{(M,N)} = \sum_{m=0}^{M}\sum_{n=0}^{N} \overset{(m,n)}{\mathbf{u}}_{(M,N)} P_m(x^1)P_n(x^2), \quad \underset{\sim}{\mathbf{P}}_{(0,M,N)} = \sum_{m=0}^{M}\sum_{n=0}^{N} \overset{(m,n)}{\underset{\sim}{\mathbf{P}}}_{(0,M,N)} P_m(x^1)P_n(x^2). \quad (17)$$

Consequently, the expressions (17) satisfy the boundary conditions on the end-faces. For example, if the displacement field is given on the end-faces, then the constructed approximate solution (the first relation (17)) is consistent with the kinematic boundary conditions $\overset{(m,n)}{\mathbf{u}} = \overset{(m,n)}{f}_{(1)}$ when $x^3 = x^3_1$ and $\overset{(m,n)}{\mathbf{u}} = \overset{(m,n)}{f}_{(2)}$ when $x^3 = x^3_2$, where $\overset{(m,n)}{f}_{(I)}$, $I = 1, 2$, are moments of given vector fields on the end-faces S_1 and S_2, respectively.

Obviously, the issue arises how accurately the relations (17) satisfy the boundary conditions of physical content on the front surfaces. Generally, these relations will not be fulfilled with the required accuracy. Thus, we add the term $\mathbf{U}_0(x', x^3, t)$ to the approximate expression for the displacement vector $\mathbf{u}_{(M,N)}(x', x^3, t)$ (corrective factor) for satisfying the boundary conditions of physical content on the front surfaces. These problems are considered in detail in [3].

Note that, similarly to the above, we can consider various representations of the heat influx equation, the Fourier thermal conductivity law, and the boundary conditions for the thermal content in moments. We also note that different methods for reducing the infinite system to finite in case of thin bodies with one small size can be viewed in [3,11], and for thin bodies with two small sizes can be seen in [3,10,14]. It is not difficult to obtain initial conditions for the kinematic and thermal contents in moments, and to formulate the statements of problems in moments [3,10]. In order to shorten the text we will not dwell on these issues.

We also note that applying the canonical representations of material tensors and tensor-block matrices, the above relations can be represented in terms of eigenvalues and eigenvectors (eigentensor columns) of material tensors (tensor-block matrices). The application of eigenvalue problems of tensors and tensor-block matrices in mechanics is an upcoming trend, which deserves the attention of researchers. Some issues related to this direction are presented in the papers [7,9,15–17]. Finally, we note that from the point of practice the next works [18–21] have merited attention. Note also, that the authors studied the dynamics and design of the power unit with hydraulic piston drive in last two papers.

Conclusion. Some problems of modeling the deformation of thin bodies with two small sizes are considered. The equations of motion and the constitutive relations, and also the boundary conditions are obtained. Definitions of the moment (m, n)th order of a certain quantity are given with respect to the system of Legendre polynomials. The boundary conditions and the various representations of the system of equations of motion and the constitutive relations in moments are obtained.

Acknowledgements. The work is financially supported by the Ministry of Education and Sciences of the Russian Federation within the federal target program "Studies and Developments of Promising Trends in Russia's Science and Technology Sector for 2014-2020." Subsidy agreement no. 14.607.21.0191 of September 26, 2017, project RFMEFI60714X0191.

References

[1] Eringen A C 1999, *Microcontinuum field theories. 1. Foundation and solids* (N.Y.: Springer-Verlag)
[2] Kupradze V D, Gegelia T G, Basheleishvili M O and T. V. Burchuladze T V 1976, *Three-dimensional problems of the mathematical theory of elasticity and thermoelasticity* (Moscow: Nauka) (in Russian)
[3] Nikabadze M U 2014, *Development of the method of orthogonal polynomials in the classical and micropolar mechanics of elastic thin bodies* (M., Publ. House of the Board of Trustees mech.-math. facul. of MSU) 515 p (in Russian)

http://istina.msu.ru/media/publications/book/707/ea1/6738800/Monographiya.pdf

[4] Vekua I N 1985, *Shell theory, general methods of construction* Pitman Advanced Pub. Program.

[5] Vekua I N 1978, *Fundamentals of tensor analysis and covariant theory* (Moscow: Nauka) (in Russian).

[6] Pobedrya B E 1986, *Lectures in tensor analysis* (Moscow: Moscow State Univ.) (in Russian).

[7] Nikabadze M U 2017, Topics on tensor calculus with applications to mechanics *J. Math. Sci.* **225**:1 194 p DOI: 10.1007/s10958-017-3467-4

[8] Lur'e A I 1980, *Nonlinear elasticity* (Moscow: Nauka) (in Russian).

[9] Nikabadze M U 2009, On some Problems of Tensor Calculus. I *J. Math. Sci.* **161**:5 668–697.

[10] Nikabadze M U 2008, *Mathematical modeling of elastic thin bodies with two small sizes using systems of orthogonal polynomials* (Manuscript in VINITI Ross. Akad. Nauk) 722 – B2008 107 p (in Russian).

[11] Nikabadze M U 2008, *The application of systems of Legendre and Chebyshev polynomials in modeling elastic thin bodies with one small size* (Manuscript in VINITI Ross. Akad. Nauk) 720 – B2008 287 p (in Russian).

[12] Pobedrya B E 2006, Theory of thermomechanical processes *In Collection of scientific papers: Elasticity and inelasticity* (MSU Publishing House), 70–85.

[13] Novacky W 1975, *Theory of elasticity* (Moscow: Mir) (Russian translation).

[14] Nikabadze M U, Kantor M M and Ulukhanian A R 2011, *To mathematical modeling of elastic thin bodies and numerical realization of several problems on the band* (Manuscript in VINITI Ross. Akad. Nauk) 204 – B2011 207 p (in Russian).

[15] Nikabadze M U 2016, *Eigenvalue problems of a tensor and a tensor-block matrix (TMB) of any even rank with some applications in mechanics* H. Altenbach and S. Forest (eds.), Generalized continua as models for classical and advanced materials, advanced structured materials 42 279-317 DOI 10.1007/978-3-319-31721-2_14

[16] Nikabadze M U 2017, An eigenvalue problem for tensors used in mechanics and the number of independent saint-venant strain compatibility conditions *Moscow university mechanics bulletin* **72**:3 66–69. DOI: 10.3103/S0027133017030037

[17] Nikabadze M U 2017, Eigenvalue problem for tensors of even rank and its applications in mechanics *J. Math. Sci.* **221**:2 174–204. DOI 10.1007/s10958-017-3226-6

[18] Kreinin G V and Misyurin S Yu 2008, Selection of the Scheme for Incorporating a Drive into the Structure of a Mechanism in Solving Problems of Kinematic Synthesis *Journal of Machinery Manufacture and Reliability* **37**:1 1–5.

[19] Kreinin G V and Misyurin S Yu 2012, Choice of the Law for a Position Control System *Journal of Machinery Manufacture and Reliability* **41**:4 331–336.

[20] Misyurin S Yu and Kreinin G V 2015, Power Optimization Criteria of a Mechanical Unit of an Automated Actuator *Doklady Physics* **60**:1 15–18.

[21] Misyurin S Yu and Kreinin G V 2016, Dynamics and design of a power unit with a hydraulic piston actuator. *Doklady Physics* **61**:7 354–359.

IOP Conf. Series: Journal of Physics: Conf. Series **1205** (2019) 012041 doi:10.1088/1742-6596/1205/1/012041

Model of image sensor signal correction in the computer microscopy system

V G Nikitaev, O V Nagornov, A N Pronichev, E V Polyakov and V V Dmitrieva

National Research Nuclear University MEPhI (Moscow Engineering Physics Institute),), 31 Kashirskoe shosse, 115409 Moscow, Russia

vgnikitayev@mephi.ru

Abstract. The article is devoted to the analysis of the possibilities of improving the accuracy of the analysis of low-contrast images in the computer microscopy systems. The image distortion factors are considered in the system of computer microscopy with a camera based on a light-sensitive electronic matrix. The level of distortion of the image sensor signal in the computer microscopy system was estimated on the results of the experiment. The model of correction of distortions of the image sensor signal is offered. The application of the proposed model will provide an increase in the accuracy of automated analysis of microscopic images of low-contrast objects in the computer microscopy systems

1. Introduction

Light microscopy is widely used in various fields of science, industry and medicine [1, 2]. The basis of the analysis of microscopic images is the identification of the structural elements of the objects of study on the basis of their color or brightness, contrasting with the surrounding background. In a number of tasks this contrast is visually unobtrusive [3, 4, 5]. Electronic means of image recording based on optical radiation sensors can be used to increase the contrast of objects in the image and their subsequent analysis in computer microscopy systems. The formation of a digital image allows to automate image analysis, highlight objects of interest and measure their characteristics. Automatic detection of low-contrast objects is limited by the capabilities of image sensors.

2. Materials and methods

The formation of a digital image in the system of computer microscopy is as follows (in the example, the mode of transmitted light is considered): the light flux from the light source is directed by the microscope condenser to the sample under study, part of the light is absorbed and scattered by the structural elements of the sample, the transmitted light is projected onto the photosensitive matrix of the electronic camera using a microscope lens. The light energy absorbed in the cell of the photosensitive matrix is converted into an electric charge, which is converted into voltage at the input of the analog-to-digital converter, the output of which is formed by a digital code recorded in the computer memory.

The formation of the image in the computer microscopy system is influenced by a number of factors along with the light flux on the light-sensitive matrix. A set of factors determining the distortion of the signal generated by the optical radiation sensor for each cell can be represented by the model $< Er, Kr, T, Tr, P, Pr, Ur>$.

Here Er is a random component of the light energy captured in the cell (due to the fluctuation of the light flux density at a constant brightness of the light source). Kr is a characteristic of the deviation of the quantum yield of one light-sensitive cell from the average value of the quantum yield for all cells of the matrix (different cells have different quantum yield). T, Tr - constant and random components of the dark current (dark current fluctuation) due to the generation of a parasitic charge in each cell, even in the absence of lighting. P, Pr - constant and random components of other factors contributing to the formation of the charge package, regardless of the light flux. For example, in the case of a camera in which the image sensor is a CCD, such a factor is a constant and random component of the charge transfer efficiency in the CCD (part of the charge packet is lost during the transfer of charge packets on the CCD). Ur - noise caused by the process of converting the charge package into voltage and its amplification to the level required for the analog-to-digital converter, which provides the formation of a digital signal.

If the useful recorded signal is separated from the side factors, it is possible to present the image signal I (x,y,t) by the following generalized model:

$$I(x, y, t) = K*G(x,y)+S(x,y,t) \tag{1}$$

Here $G(x, y)$ is a function that determines the distribution of the light flux passing through the sample through the site, the projection of which on the photosensitive matrix corresponds to the area of the photosensitive cell of the image sensor with coordinates (x, y); (x,y) – integer spatial coordinates of the photosensitive cell of the matrix reduced to the plane of the sample (the numbers x and y correspond to the row and column of the cell of the photosensitive matrix); K - light-signal conversion factor (in the ideal system K does not depend on the coordinates (x,y), but may depend on the value of the light flux), t – the time at which the image was recorded. $S(x,y,t)$ is a function of distortions associated with the imperfection of the image recording system. Under constant lighting conditions and a fixed sample, the function $G(x,y)$ is time-independent. The dependence of the function S (x,y,t) on time reflects the random nature of the distortions on time.

Deterministic and random components can be identified in these side factors affecting the formation of the image. Generalizing influence of the specified factors, it is possible to present function of distortions in the form of:

$$S(x,y,t)=C(x,y, G(x,y))+V(x,y, t, G(x,y)) \tag{2}$$

The first term in formula (2) defines a time-independent constant distortion component for each cell with coordinates (x,y). The second term corresponds to the time-random changes in the signal under constant lighting conditions of a stationary sample.

Since the mathematical expectation of the second term in formula (2) is equal to zero, it is possible to estimate the first summand in the formula (2) with the required accuracy by experiment with statistical averaging of measurement results. The resulting estimate can be used to reduce the first term in formula (2) by subtracting the resulting estimate from the image signal. The influence of the second term on the image analysis results of low-contrast objects can be reduced by forming a calculated image, which is obtained as a result of statistical averaging of a series of images, shooting of which is performed under constant conditions.

An experiment was conducted to evaluate the components of the proposed model in a computer microscopy system using the Imperx IPX-4M15T-GCFB digital camera. The camera mode was used with a signal encoding of 24bit per pixel (8 bits per each RGB component) in the non-cooling mode. The frame size of the image is 2048x2048 pixels. The image was recorded in a file in the BMP24 format. A series of shots was taken under constant conditions to estimate the second term in formula (2). A uniformly illuminated background at different lighting levels was chosen as the subject to estimate the first term in formula (2).

3. Discussion

The results of the experiment (for example, G component with 256 brightness gradations) showed that the standard deviation of the values of the second term in the formula (2) is approximately in the range from 1.5 to 3 brightness gradations, while the values of the first term reach 5 brightness gradations. Thus, it is enough to shoot a series of 81 frames and compute the component average in each pixel for a series of frames to reduce the influence of the second term in the formula (3) to the level of less than one gradation (taking into account the spread of random values in the interval +/- 3CO). The random component of the distortion of the image sensor signal in the averaged image will not exceed two-thirds of the brightness gradation if you shoot an image in a series of 183 frames, followed by averaging the frames. The image signal can be corrected by subtracting the values of the obtained estimates from the image sensor signal based on the experimentally obtained estimates of the first term in the formula (2). The accuracy of the correction is determined by the number of frames in a series of images taken to estimate the systematic component of the signal distortion. So the error of correction of a systematic signal distortions of the image sensor did not exceed one third of component brightness gradation in shooting of images by a series from 729 frames (for a confidence interval of an assessment in +/- 3 standard deviation). This operation is performed once during system calibration. Thus, it is possible to reduce the influence of signal distortion of the image sensor in the computer microscopy system, using a camera Imperx IPX-4M15T-GCFB, to one brightness gradation if having the correction of the systematic signal distortion of the image sensor as it is showed above(it reduces the influence of the systematic signal distortion to a third of a brightness gradation), and then averaging the images in a series of 183 frames(it reduces the influence of the random signal distortion to two thirds of brightness gradation).

4. Conclusion

A model of image signal distortion in a computer microscopy system is proposed for subsequent experimental evaluation of systematic and random components of signal distortion. Experimental evaluation of the components of the image signal distortion is performed. It is revealed that the values of the random and systematic component of the signal distortion in the computer microscopy system using the Imperx IPX-4M15T-GCFB digital camera are several units of the brightness gradations in 256 gradations signal range (the random component is up to three gradations, the systematic component is up to five gradations). The distortion correction model was proposed, it ensures the correction accuracy to one brightness gradation. The proposed method of distortions correction in microscopic images automated analysis systems will improve the accuracy of measurements of the characteristics of low-contrast objects.

Acknowledgement

The reported study was funded by RFBR according to the research project № 18-07-01456

References

[1] Nikitaev VG, Nagornov OV, Pronichev AN ,Polyakov EV, Sel'chuk VY, ChistovKS, Blyndar VN, Dmitrieva VV,Zaytsev SM, Gordeev VV 2015 MeasurementTechniques 57(10) 1203

[2] Nikitaev VG, Nagornov OV, Pronichev AN,Polyakov EV, Sel'chuk VY, Chistov KS, Blyndar VN, Dmitrieva VV, Gordeev VV 2015 MeasurementTechniques 57(5) 560

[3] Nikitaev VG, NagornovOV, PronichevAN,PolyakovEV, DmitrievaVV2015 WSEAS TRANSACTIONS on BIOLOGY and BIOMEDICINE 12 16

[4] Nikitaev V G, Pronichev A N, Polyakov E V, Dmitrieva V V, Tupitsyn N N, Frenkel M A, Mozhenkova AV Journal of Physics: Conference Series vol 784, no1, p012042, 2017.

[5] Nikitaev V G, Nagornov O V, Pronichev A N, Dmitrieva V V, Polyakov E V Journal of Physics: Conference SeriesVolume 788 Issue1,2 February 2017 Article number 012056

Determination of the parameter of the differential equation of fractional order with the Caputo derivative in Hilbert space

D G Orlovsky

National Research Nuclear University MEPhI (Moscow Engineering Physics Institute), Moscow

E-mail: odg@bk.ru

Abstract. We consider an abstract differential equation of the first order with unbounded linear operator $D^\alpha u(t) = Au(t) + f(t)$ in a Hilbert space H, where D^α is the Caputo fractional derivative. For this equation the Cauchy problem is studied with initial data $u(0) = u_0$ at $t = 0$. We assume that the operator A is self-adjoint and non-positive. The inverse problem to determine the nonhomogeneous member is considered under the assumption that this term has the following representation $f(t) = \varphi(t)p$, where the scalar function $\varphi(t)$ is given and the element p is an unknown element of the space H. This problem belongs to the class of inverse problems. Inverse problems for abstract differential equations was discussed initially with this inhomogeneous structure member but for equations with an operator generating a C_0-semigroup and usual derivative. An additional condition $u(T) = u_1$ is specified for the determination of the unknown p, where u_1 is a given element of the space H. Thus we get the two-point problem which is only beginning to be studied for the case of fractional derivatives. The question of existence and uniqueness of the classical solution of the inverse problem is studied. Sufficient conditions of correct solvability of the inverse problem are obtained. Explicit formula to determine the unknown element in the differential equation is given.

1. Introduction

Consider the following Cauchy problem

$$\begin{cases} D^\alpha u(t) = Au(t) + f(t), \ 0 \leqslant t \leqslant T, \\ u(0) = u_0, \end{cases} \tag{1}$$

where A is self-adjoint and non-positive operator in a Hilbert space H, $f(t) \in C([0,T]; H)$, the element $u_0 \in H$, $0 < \alpha < 1$ and D^α is the Caputo fractional derivative

$$(D^\alpha u)(t) = \frac{1}{\Gamma(1-\alpha)} \frac{d}{dt} \left(\int_0^t \frac{u(s)}{(t-s)^\alpha} ds - u(0) \right),$$

($\Gamma(x)$ is the Gamma function).

Well-posedness of the problem (1) is connected with well-posedness of homogeneous problem

$$\begin{cases} D^\alpha u(t) = Au(t), \ 0 \leqslant t \leqslant T, \\ u(0) = u_0. \end{cases} \tag{2}$$

We will consider a strong solution for problems (1) and (2). It means that

$$(i) \quad u \in C([0,T], D(A)) \cup C([0,T], H),$$
$$(ii) \quad \int_0^t \frac{u(s)}{(t-s)^\alpha} ds - u(0) \in C^1([0,T], H).$$

Consider the following function of Wright type

$$\Phi_\alpha(z) = \sum_{n=0}^{\infty} \frac{(-1)^n z^n}{n!\Gamma(-\alpha n + 1 - \alpha)}$$

and two functions

$$\varphi_{t,\alpha}(s) = t^{-\alpha}\Phi_\alpha(st^{-\alpha}),$$
$$\psi_{t,\alpha}(s) = \alpha s t^{-\alpha-1}\Phi_\alpha(st^{-\alpha}).$$

These functions satisfy the following relations

$$\int_0^{+\infty} \varphi_{t,\alpha}(s) e^{\lambda s} ds = E_{\alpha,1}(t^\alpha \lambda),$$

$$\int_0^{+\infty} \psi_{t,\alpha}(s) e^{\lambda s} ds = t^{\alpha-1} E_{\alpha,\alpha}(t^\alpha \lambda),$$

where

$$E_{\alpha,\beta}(z) = \sum_{n=0}^{\infty} \frac{z^n}{\Gamma(\alpha n + \beta)}$$

Mittag-Leffler function.

The self-adjoint and non-positive operator A is a generator of some C_0–semigroup $V(t)$. In this case the strong solution of problem (2) exists, it is unique for every $u_0 \in D(A)$ and one has the form

$$u(t) = S_\alpha(t) u_0,$$

and the strong solution of problem (1) exists, it is unique for every $u_0 \in D(A)$, $f \in C^\beta([0,T]), H)$ $(0 < \beta \leq 1)$ and one has the form

$$u(t) = S_\alpha(t) u_0 + \int_0^t P_\alpha(t-s) f(s) ds,$$

where two strongly continuous operator functions $S_\alpha(t)$ and $P_\alpha(t)$ are related to this semigroup $V(t)$ as

$$S_\alpha(t) = \int_0^{+\infty} \varphi_{t,\alpha}(s) V(s) ds \tag{3}$$

and

$$P_\alpha(t) = \int_0^{+\infty} \psi_{t,\alpha}(s)V(s)ds. \tag{4}$$

It is known that for the bounded operator A the equality two relation

$$S_\alpha(t) = E_{\alpha,1}(t^\alpha A), \quad P_\alpha(t) = t^{\alpha-1}E_{\alpha,\alpha}(t^\alpha A),$$

are true.

Applying operator calculus of self-adjoint operators, it is easy to prove that these equations are also valid for any self-adjoint and non-positive operator. In fact, let E_λ is a spectral resolution of the operator A, then

$$S_\alpha(t) = \int_0^{+\infty} \varphi_{t,\alpha}(s)\left(\int_{-\infty}^{0} e^{\lambda s}dE_\lambda\right)ds = \int_{-\infty}^{0}\left(\int_0^{+\infty}\varphi_{t,\alpha}(s)e^{\lambda s}ds\right)dE_\lambda =$$

$$= \int_{-\infty}^{0} E_{\alpha,1}(t^\alpha\lambda)dE_\lambda = E_{\alpha,1}(t^\alpha A),$$

$$P_\alpha(t) = \int_0^{+\infty} \psi_{t,\alpha}(s)\left(\int_{-\infty}^{0} e^{\lambda s}dE_\lambda\right)ds = \int_{-\infty}^{0}\left(\int_0^{+\infty}\psi_{t,\alpha}(s)e^{\lambda s}ds\right)dE_\lambda =$$

$$= \int_{-\infty}^{0} t^{\alpha-1}E_{\alpha,\alpha}(t^\alpha\lambda)dE_\lambda = t^{\alpha-1}E_{\alpha,\alpha}(t^\alpha A).$$

Therefore, the formula for solving the direct problem (1) can be written as follows

$$u(t) = E_{\alpha,1}(t^\alpha A)u_0 + \int_0^t (t-s)^{\alpha-1}E_{\alpha,\alpha}((t-s)^\alpha A)f(s)ds. \tag{5}$$

2. Inverse problem

Consider the problem of determining the function $u(t)$ with values of H and the element $p \in H$ from the system of equations

$$\begin{cases} D^\alpha u(t) = Au(t) + \varphi(t)p, \ 0 \leqslant t \leqslant T, \\ u(0) = u_0, \quad u(T) = u_1. \end{cases} \tag{6}$$

If we choose in (5) $f(t) = \varphi(t)p$ and then put $t = T$, we get the following equation for the unknown element p

$$Bp = h,$$

where

$$B = \int_0^T (T-s)^{\alpha-1}E_{\alpha,\alpha}((T-s)^\alpha A)\varphi(s)ds,$$

$$h = u_1 - E_{\alpha,1}(T^\alpha A)u_0.$$

We introduce the concept of the characteristic function of the inverse problem (6). This function is determined by the equality

$$F(z) = \int_0^T (T-s)^{\alpha-1} E_{\alpha,\alpha}(z(T-s)^\alpha)\varphi(s)ds. \tag{7}$$

Let us transform the expression that represents the operator B. For this we use the spectral decomposition of the operator A

$$A = \int_{-\infty}^0 \lambda dE_\lambda$$

According to the operator calculus of self-adjoint operators, we have

$$E_{\alpha,\alpha}((T-s)^\alpha A) = \int_{-\infty}^0 E_{\alpha,\alpha}(\lambda(T-s)^\alpha)dE_\lambda.$$

Therefore

$$B = \int_0^T \left(\int_{-\infty}^0 (T-s)^{\alpha-1} E_{\alpha,\alpha}(\lambda(T-s)^\alpha)\varphi(s)dE_\lambda \right) ds.$$

By Fubini's theorem we have

$$B = \int_{-\infty}^0 \left(\int_0^T (T-s)^{\alpha-1} E_{\alpha,\alpha}(\lambda(T-s)^\alpha)\varphi(s)ds \right) dE_\lambda = F(A).$$

Therefore, the equation for the unknown element p takes the form

$$F(A)p = h. \tag{8}$$

Equation (8) was studied in [1]. Direct application of the results of this work leads to the following statement.

Theorem 1. *Let us assume that H is a Hilbert space, operator A is self-adjoint and non-positive, $0 < \alpha < 1$, function $\varphi \in C^1[0, T]$, it is not identically equal to zero, the elements $u_0, u_1 \in H$. The solution of problem (6) is unique if and only if point spectrum of the operator A does not contain zeroes of the characteristic function (7). The solution of problem (6) exists if and only if*

$$\int_{-\infty}^0 |F(\lambda)|^{-2} d(E_\lambda h, h) < \infty,$$

where E_λ is a spectral resolution of the operator A and the element

$$h = u_1 - E_{\alpha,1}(T^\alpha A)u_0.$$

We note that the same inverse problem for the Riemann-Liouville fractional derivative was considered in [2]. In this paper, the inverse problem is also reduced to the equation (8). Moreover, the characteristic functions of the inverse problems for the Riemann-Liouville and

Caputo derivatives are the same. The only difference is in the formulas for the element h. For the equation with the Riemann-Liouville derivative, the element h is determined by the equality

$$h = u_1 - T^{\alpha-1} E_{\alpha,\alpha}(T^\alpha A) u_0.$$

This circumstance allows to transfer the results of article [2] to the inverse problem (6) with the derivative of Caputo. In particular, we obtain the following statement.

Theorem 2. *Let us assume that H is a Hilbert space, operator A is self-adjoint and non-positive, $0 < \alpha < 1$, function $\varphi \in C^1[0,T]$, everywhere either $\varphi \geqslant 0$ or $\varphi \leqslant 0$ and $\varphi(T) \neq 0$, $u_0, u_1 \in D(A)$. Then solution of problem (6) exists and it is unique.*

Similarly, we can obtain explicit formulas for the solution of the inverse problem for the model case.

Corollary 1. *Let us assume that H is a Hilbert space, operator A is self-adjoint and non-positive, $0 < \alpha < 1$, $\beta \geqslant 0$, $u_0, u_1 \in D(A)$. Then the solution of the problem*

$$D^\alpha u(t) = Au(t) + t^\beta p, \ t > 0,$$

$$u(0) = u_0,$$

$$u(T) = u_1.$$

exists and it is unique and is given by formulas

$$p = \frac{1}{\Gamma(\beta+1) T^{\alpha+\beta}} E^{-1}_{\alpha,\alpha+\beta+1}(T^\alpha A) \left(u_1 - E_{\alpha,1}(T^\alpha A) u_0\right),$$

$$u(t) = E_{\alpha,1}(t^\alpha A) u_0 + \int\limits_0^t (t-s)^{\alpha-1} E_{\alpha,\alpha}((t-s)^\alpha A) s^\beta p ds.$$

Corollary 2. *Let us assume that H is a Hilbert space, operator A is self-adjoint and non-positive, $0 < \alpha < 1$, $u_0, u_1 \in D(A)$. Then the solution of the problem*

$$D^\alpha u(t) = Au(t) + p, \ t > 0,$$

$$u(0) = u_0,$$

$$u(T) = u_1.$$

exists and it is unique and is given by formulas

$$p = \frac{1}{T^\alpha} E^{-1}_{\alpha,\alpha+1}(T^\alpha A) \left(u_1 - E_{\alpha,1}(T^\alpha A) u_0\right),$$

$$u(t) = E_{\alpha,1}(t^\alpha A) u_0 + t^\alpha E_{\alpha,\alpha+1}(t^\alpha A) p.$$

References

[1] Orlovsky D.G. 1990 To the problem of determining the parameter of the evolution equaion. *Differential'nye uravneniya* **26** no 9, 1614 (in Russian)

[2] Orlovsky D.G. 2015 Parameter determination in a Differential Equation of Fractional Order with Riemann-Liouvile Fractional Derivative ih a Hilbert Space. *Journal of Siberian Federal University. Mathematics and Physics.* **8** no 1, 55

IOP Conf. Series: Journal of Physics: Conf. Series **1205** (2019) 012043 doi:10.1088/1742-6596/1205/1/012043

A novel approach of the grain structure modelling in the framework of polycrystalline specimen and EBSD experiment simulation

A O Ovchinnikova and T I Savyolova

Department of Applied Mathematics, National Research Nuclear University MEPhI, Kashirskoe sh. 31, Moscow, 115409 Russia

aoantonova@mail.ru

Abstract. In this work the influence of the grain structure model in polycrystalline specimen on the results of the simulated EBSD experiment is studied. In the framework of EBSD experiment simulation the improved model of the polycrystalline with the grains in form of ellipses is used. Such a model is proposed to describe structural features of the real specimen more elaborately than one with grains as circles. Matching the improved polycrystalline model with the initial one is performed by comparison of the texture entities, restored from the EBSD measurements of specimens, with reference ones, calculated for modelled specimens. The EBSD experiment simulation is conducted for both specimens models under varying of the experiment parameters: a scanning step and a tolerance angle.

1. Introduction

The EBSD (Electron Back Scattering Diffraction) experiment parameters influence study, a scanning step and a tolerance angle, on the reliability of the obtained results of the EBSD experiment is proved to be of crucial importance [1, 2]. Experimental nature of material investigations by means of EBSD does not allow to compare the experiment results with the original specimen characteristics to tune the parameters of measurements [3]. The specimen and experiment simulation makes it possible to compare texture entities calculated directly from specimen with corresponding ones obtained from simulated EBSD measurements.

In this work the attempt to improve the model of EBSD experiment proposed earlier in [4, 5] was performed by utilizing the novel approach for grain structure modelling in the framework of polycrystalline model. In the advanced grain structure model grains were considered in the form of rotated ellipses instead of circles used in [4, 5]. The proposed model of the grain structure is based on specifity of the EBSD measurements processing, as in the commercial software [6] the plane cross sections of grains are approximated by ellipses (figure 1). Matching the improved polycrystalline model with the initial one was performed by comparison of the texture characteristics, calculated from the EBSD measurements, conducted for both models under varying of the experiment parameters: scanning step and tolerance angle. As texture characteristics orientation distribution function (ODF) [7], disorientation density distribution function (DDDF) and texture index (TI) were considered [8].The dependence of TI on the experiment parameters, calculated for simulated measurements of specimens with both grain structure models, was compared with one, calculated for ferritic-martensitic steel EP823 [9].

2. Methods

2.1. Description of the polycrystalline specimen and EBSD experiment models

A model specimen was considered as a statistical object. Two distinct model specimens $\vec{S}_{cir}$ and $\vec{S}_{ell}$ consisting of $N = 1000$ grains were created:

$$\vec{S}_{cir} = \{\vec{U}_{cir\,i}, i = 1, \dots, N\}, \vec{U}_{cir} = [x, \omega, g(\varphi, \theta, \psi), \delta], \quad (1)$$

$$\vec{S}_{ell} = \{\vec{U}_{ell\,i}, i = 1, \dots, N\}, \vec{U}_{ell} = [x, c, \omega, g(\varphi, \theta, \psi), \delta]. \quad (2)$$

The vectors $\vec{U}_{cir}$ and $\vec{U}_{ell}$ so defined were proposed to be the grain models. In $\vec{U}_{cir}$ the grains' cross sections were modelled as circles sequentially located and x was proposed to be the size of the grain determined as a diameter of a corresponding circle (figure 2 (a)). In the improved specimen model the grains' cross section was considered as ellipse with grain size x, also calculated as a diameter of circle equal to the grain's cross section with respect to its area, but the following EBSD measurements were performed along chords c of ellipse approximating the plane cross section of the grain (figure 2 (b,c)). Grain sizes x were modelled by Neumann's method as random variables distributed satisfying a gamma distribution $\Gamma(2,1)$ so that the mean grain size $\bar{x}$ was equal to 2.0 μm [4, 5].

A grain boundary between the given grain and the following one δ was a variable uniformly distributed in $[0, 2\bar{x}/10)$, so that the total size of the boundaries is 10% of the grains' total size.

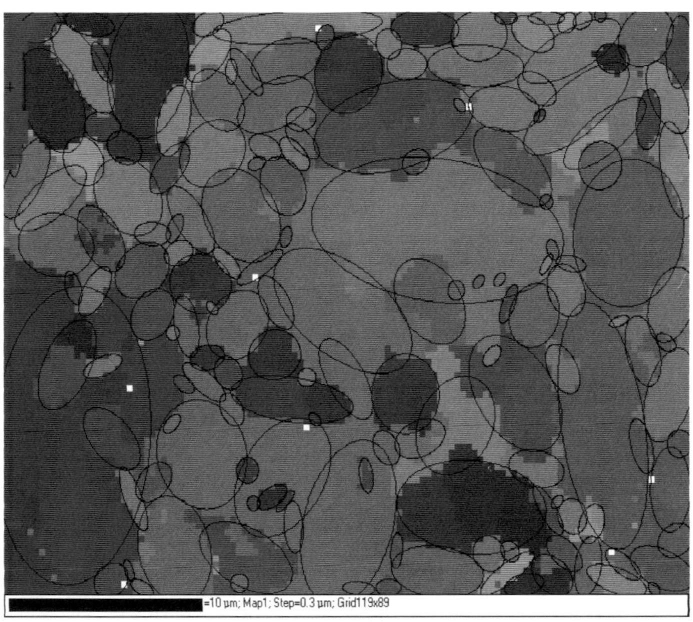

Figure 1. The EBSD map of the steel specimen, where grains' cross sections are approximated by ellipses

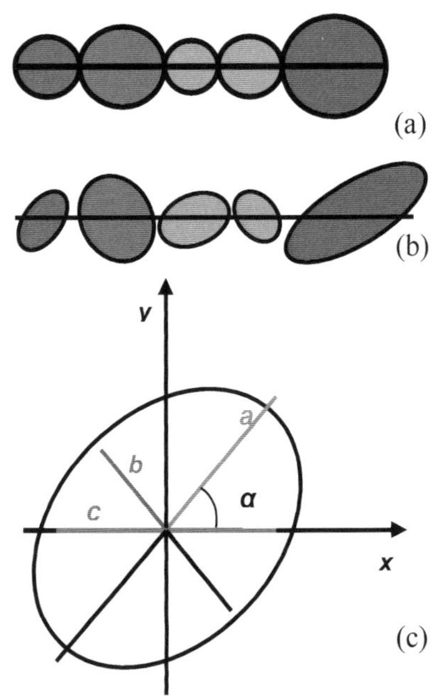

Figure 2. Schemes of the model specimens (a,b) and the individual grain in form of ellipse (c)

For both specimens the same orientations sample was utilized (equal grain orientations are shown on figure 2 (a,b) by the same color of the circle and the corresponding ellipse), orientations $g(\varphi, \theta, \psi)$ were modelled by Monte-Carlo method with ODF in the form of central normal distribution (CND) [7, 10]:

$$f(t)dt = \sum_{l=0}^{\infty}(2l+1)\exp\{-l(l+1)\varepsilon^2\}\chi_l \frac{1}{\pi}\sin^2\frac{t}{2}dt\,, \tag{3}$$

where $t \in [-\pi;\pi]$, $\cos\frac{t}{2} = \cos\frac{\varphi+\psi}{2}\cos\frac{\theta}{2}$; $\chi_l = \sin\left(l+\frac{1}{2}\right)t/\sin\frac{t}{2}$ are the characters of the rotation group SO(3) representation; $\frac{1}{\pi}\sin^2\frac{t}{2}dt$ – invariant measure, $\varepsilon = 1/8$ – texture sharpness.

ω in equations (1), (2) denotes a disorientation angle between two neighboring grains [7].

Thus, the specimens $\vec{S}_{cir}$ and $\vec{S}_{ell}$ had equal texture characteristics and equal mean grain size 2.0 μm, but different grain structure (rotated ellipses and circles, as shown on figure 2 (a,b)).

Let us define an operator $\Phi(\vec{S}, h, \omega_0)$, modelling the result of EBSD study of the specimen $\vec{S}$ under the parameters (h, ω_0), where $h \in (0, H), H > 0$ – scanning step over the modelled surface of specimen, $\omega_0 \in \Omega_0 \subseteq (0,\pi)$ – tolerance angle. We assumed, that for $\vec{S}_{cir}$ the measurements are made along circles diameters, whereas for specimen $\vec{S}_{ell}$ orientation scanning was performed along the ellipses chords. A chord of ellipse can be calculated as

$$c = \frac{x}{\sqrt{(\cos\alpha)^2/k + k(\sin\alpha)^2}}, \tag{4}$$

where x is the diameter of the circle with equal area, α – an angle of the ellipse rotation, $k = a/b$ (see figure 2 (c)). The proposed ellipse approach seemed to be more reliable according to a procedure of the real EBSD experiment (figure 1).

The results of simulated EBSD measurements of both specimens $\vec{S}_{cir}$ and $\vec{S}_{ell}$ can be presented as

$$\vec{\tilde{S}}_{cir} = \left\{\vec{\tilde{U}}_{cir\,i}, i = 1, \dots, \tilde{N}\right\}, \vec{\tilde{U}}_{cir} = [\tilde{x}_{cir}, \tilde{\omega}, \tilde{g}(\tilde{\varphi}, \tilde{\theta}, \tilde{\psi})], \tag{5}$$

$$\vec{\hat{S}}_{ell} = \left\{\vec{\hat{U}}_{ell\,i}, i = 1, \dots, \hat{N}\right\}, \vec{\hat{U}}_{ell} = [\hat{x}_{ell}, \hat{\omega}, \hat{g}(\hat{\varphi}, \hat{\theta}, \hat{\psi})]. \tag{6}$$

As it can be seen from equations (5), (6) different approaches of specimen modelling result in the difference of simulated measurements procedure, causing the change of all grain properties, restored from measurements.

2.2. Texture characteristics for verification of the novel grain structure model

To compare considered grain structure models the behaviour of texture characteristics: ODF, DDDF, TI, – under varying the parameters (h, ω_0) was investigated. For calculation of ODF and TI from the set of individual orientations the kernel estimation method was used:

$$f_N(t) = \frac{1}{\alpha}\sum_{i=1}^{N} C_i q\left(\frac{t-t_i}{\alpha}\right), C_i = \frac{x_i}{\sum_{i=1}^{N} x_i}, t \in [-\pi;\pi],\ \cos\frac{t}{2} = \cos\frac{\varphi+\psi}{2}\cos\frac{\theta}{2}. \tag{7}$$

where $q(t)$ is a Gaussian kernel. α – smoothing parameter.

For calculation of TI for ODF in form of CND (equation 3) the following estimate was used [7]:

$$\hat{J}_{mod} = \sum_{l=0}^{L}\left|\hat{C}^l\right|^2, \hat{C}^l = \frac{1}{N}\sum_{i=1}^{N}\frac{\sin\left(l+\frac{1}{2}\right)t_i}{\sin\frac{t_i}{2}}, t \in [-\pi;\pi],\ \cos\frac{t}{2} = \cos\frac{\varphi+\psi}{2}\cos\frac{\theta}{2}. \tag{8}$$

IOP Conf. Series: Journal of Physics: Conf. Series **1205** (2019) 012043 doi:10.1088/1742-6596/1205/1/012043

To compare the reference ODF and DDDF of the model specimens with corresponding ones, restored from EBSD measurements, chi-square homogeneity test was used [11].

3. Results

3.1. Comparison the circle and ellipse grains structure models with respect to the texture characteristic restored from the corresponding simulated EBSD measurements

For model specimens described by equations (1), (2) the simulation of EBSD experiment results was performed for the given set of parameters meanings [4, 5]: h = 0.5 μm; 1.0 μm; 2.0 μm; ω_0 = 5°, 10°, 20°. The application of two distinct approach for specimen modelling led to quite similar results of all texture characteristics restored from the simulated EBSD data of corresponding specimens.

DDDF and ODF were appeared to be effected by tolerance angle stronger than scanning step for both modelled specimens. This was in agreement with the chi-square homogeneity test, showing that the hypothesis on the coincidence of the DDDF of each specimen and corresponding measurements was accepted at 95% confidence level only for small ω_0 = 5° and for the steps h = 2.0, 1.0 μm. Figure 3 represents DDDF histograms for two parameters sets when tested hypothesis was accepted (figure 3 (a)) and declined (figure 3 (b)).

For both grain structure models ODF showed close behaviour under parameters varying. The hypothesis on the coincidence of the one-parameter ODFs (equation (7)) was rejected at ω_0= 20°. In this case, the ODFs, restored from modelled measurements, oscillate on the "tails" and the central minimum decrease. In the other cases of parameters sets, it was accepted at 95% confidence level.

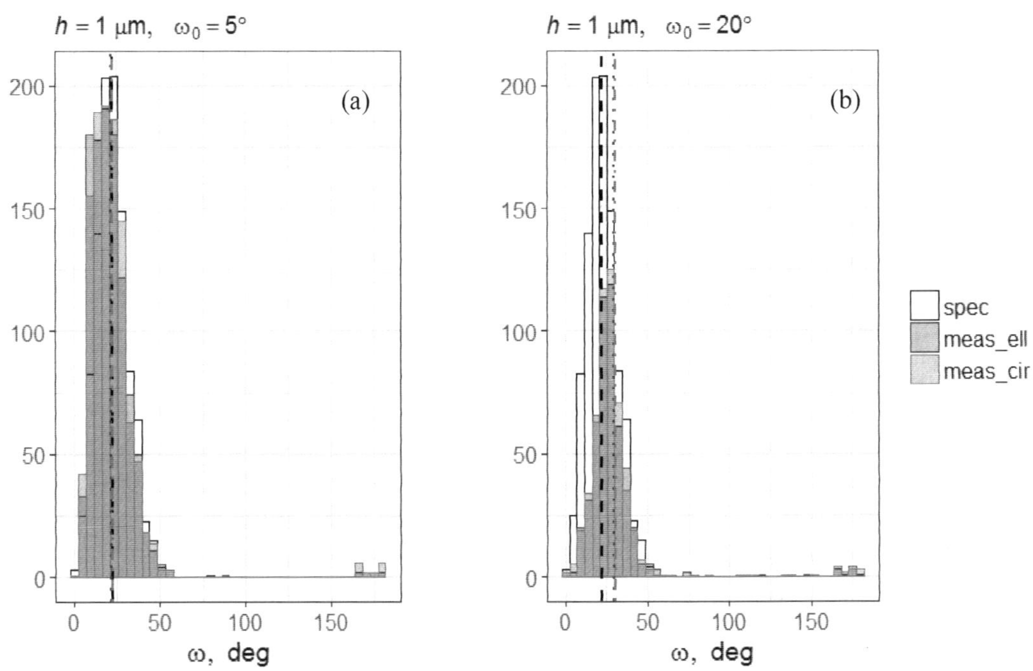

Figure 3. DDDF histograms for specimen (spec), EBSD measurements of specimen with grains as circles (meas_cir) and for corresponding ones of specimen with grains as ellipses (meas_ell), (a) the hypothesis on the coincidence of the DDDF of each specimen and corresponding measurements is accepted, (b) – rejected. The vertical lines: dotted, dashed, dash-dotted, – correspond to mean values of the specimen, meas_ell, meas_cir disorientation angles, respectively.

IOP Conf. Series: Journal of Physics: Conf. Series **1205** (2019) 012043 doi:10.1088/1742-6596/1205/1/012043

3.2. Results of the TI calculation from the modelled EBSD data of specimens with two distinct grain structure. Comparison TI of the simulated EBSD measurements with one of the steel EP823

The TI is a measure of texture severity: tending of TI to unit points out weak texture of specimen, for the case of strong texture TI shows infinite increase[1, 8, 12].

The TI calculation from the EBSD measurements demonstrated slightly different results for both modelled specimens. The dependence of the TI relative error on the experiment parameters (h, ω_0) is presented on figure 4 for both specimens.

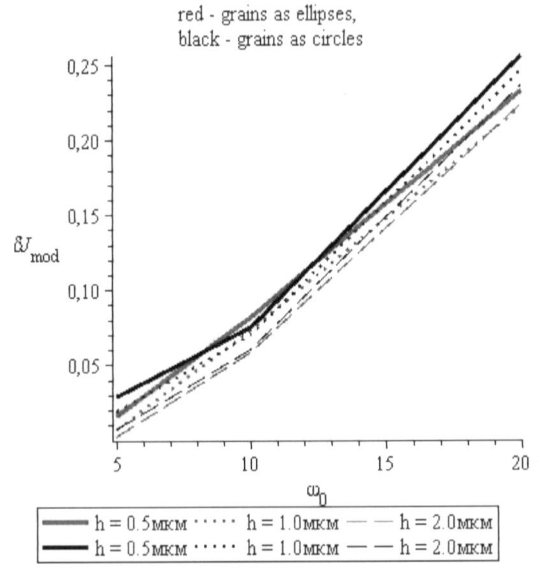

Figure 4. Dependence of the TI relative error on the experiment parameters (h, ω_0)

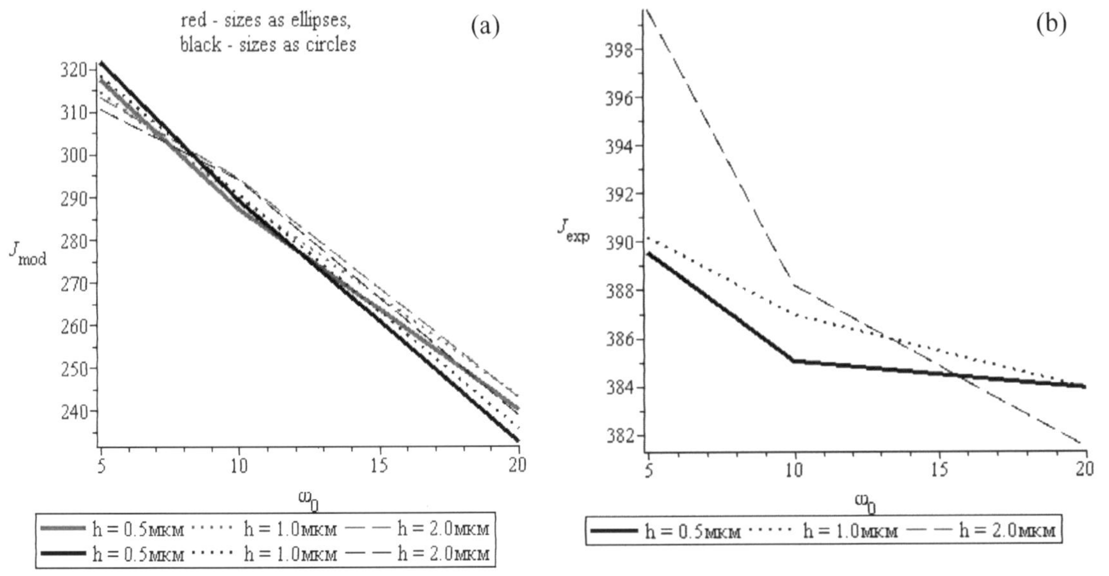

Figure 5. Dependence of the TI for simulated EBSD measurements (a) and for steel EP823 (b) on the experiment parameters (h, ω_0)

It can be noticed that the TI behaviour on figure 5 (a) under parameters varying was saved when the new approach for grain structure modelling was utilized.

The dependence of TI on experiment parameters, obtained by modelling, was compared with one, calculated for ferritic-martensitic steel EP823 (12Cr-Mo-W-Si-V-Nb-B) [9]. The EBSD measurements for steel EP823 were performed with the same parameters meanings as in modelling. The dependence of IT on measurement parameters for model experiments, presented in figure 5 (a), was confirmed by the dependence calculated for steel EP823 (figure 5 (b)). The reasonable numerical difference between model and real experiments results of TI estimation can be caused by distinct ODFs and its sharpness of models specimens and steel EP823 (the true ODF of steel is unknown).

4. Conclusions

The improved polycrystalline model with a novel grain structure simulation approach was verified by means of the texture characteristics comparison, restored from the simulated EBSD measurements, for specimens with both grains structure models: novel one with grains in form of ellipses and initial one with grains as circles.

The results of usage of the novel grain structure model allowed to achieved reliable results of texture entities behaviour under parameters varying. Both models considered led to close enough results of ODF, DDDF and TI calculation. The correlation between behaviour of texture entities, restored from simulated EBSD data with both, novel and initial, grain structure models, and really performed EBSD experiments for steel EP823 give us the ability to use novel model for further study of the EBSD parameters choice. Besides the application of the specimen model outlined makes it possible to study the influence of grains elongation factor on the scanning step choice and the reliability of the obtained experiment results. It is also supposed, that the application of the proposed specimen model in the further study of the EBSD experiment parameters choice allows to achieve results of polycrystalline properties calculation, closer to ones, obtained for steel and other real materials.

Acknowledgments

The reported study was funded by RFBR according to the research project № 18-31-00036.

References

[1] Wright S I and Field D P 2005 *Materials Science Forum* **495–497** 207–12
[2] Wright S I 2010 *Pract. Metallogr.* **47** 16–33
[3] Randle V 2010 *Introduction to texture analysis: macrotexture, microtexture and orientation mapping*: 2nd ed (Boca Raton: CRC Press) p 488
[4] Antonova A O and Savyolova T I 2015 *Comput. Math. Math. Phys.* **55** 317–29
[5] Antonova A O and Savyolova T I 2016 *Cristallogr. Rep.* **61** 523–31
[6] Software channel 5 http://hkl-technology.software.informer.com/
[7] Savyolova T I, Ivanova T M and Sypchenko M V 2012 *Methods for Solving Ill-Posed Problems in Texture Analysis and Their Applications* (Moscow: NRNU MEPhI) [in Russian] p 268
[8] Bunge H J 1982 *Texture Analysis in Material Sciences. Mathematica Methods* (London: Butterworths Publ.) p 593
[9] Antonova A O, Dzhumaev P S and Savyolova T I 2016 *MATEC Web of Conferences (ICMIE)* **75** 09007
[10] Borovkov M and SavelovaT 2007 *Mag. Appl. Cristallogr.* **40** 449–55
[11] Aivazyan S A and Mkhitaryan B S 2001 Applied *Statistics and Basics of Econometrics* (Moscow: YuNITI_DATA) [in Russian] p 656
[12] Hielscher R, Schaeben H and Chateigner D 2007 *J. of Appl. Cryst.* **40** 371–5

Nonclassical mathematical model for the mechanical analysis of additive manufacturing viscoelastic materials on rotating cylindrical substrates

D A Parshin[1,2]

[1] Laboratory of Modelling in Mechanics of Solids, Institute for Problems in Mechanics of the Russian Academy of Sciences, Prospekt Vernadskogo 101-1, Moscow 119526, Russia

[2] Department of Applied Mathematics, Bauman Moscow State Technical University, 2-ya Baumanskaya ulitza 5, Moscow 105005, Russia

E-mail: parshin@ipmnet.ru

Abstract. Technological processes of additive manufacturing viscoelastic aging axisymmetric material layers on the inner surface of a rotating rigid cylindrical substrate by means of deposition of supplementary material plies are modelled. The added plies may be preliminary tensed or compressed in the circumferential direction. A mathematical model based on general approaches of mathematical theory of accreted solids is proposed for describing the deformation process of the being formed material layer under the action of centrifugal forces in assumption of small plain strain and in quasistatic approximation. The corresponding closed-form expressions containing quadratures are constructed for describing the technological stresses evolution in the manufactured layer during and after the process of its accretion.

1. Introduction

A deal of technological processes are accompanied by an increase in the size and, possibly, by a change in the shape of solids due to attaching additional material to them, that is by accretion, or growth of these solids. Processes of such a kind are used to be called additive.

It is obvious that when studying additive processes one should take into account the kinematic and force features of the new material gradual inflow to the surface of the accreted solid simultaneously with loadings acting on this solid in the considered technological process. However, such accounting is impossible in principle to implement correctly in the framework of the classical mechanics of solids, even if we consider the traditional equations and boundary conditions in a domain variable in time. This is due to the fact that the classical formulation of the mechanical problems always implies the existence for the entire solid of such a configuration in which there are no any stresses, i.e. a natural configuration. Meanwhile, an accreted solid does not have in general case any natural configuration (unlike, for example, a solid exposed to removal of material), because while some of material elements have already deformed with the solid, others have not yet been incorporated into it. As a consequence of this fundamental fact, the problems on mechanical behaviour of accreted solids should possess a number of specific features and constitute a special class of problems in solid mechanics.

Development of common approaches to mathematical formulation of such nonclassical problems and constructing general methods of these problems solution, as well as studying various growth processes on the basis of the mechanical problems for different accreted solids solved due to the elaborated techniques present works [1–20]. Cited studies demonstrate a great number of specific examples and discuss a variety of new mechanical effects discovered thanks to the obtained solutions of the nonclassical mechanical problems.

2. The mechanical problem under study

This study is devoted to modelling the processes of gradual deposing material layers of uniform thickness on the inner surface of an axisymmetric cylindrical substrate rotating around its axis with the angular velocity $\omega(t)$ arbitrarily changing with time t. The current inner radius of the being formed material layer will be denoted by $a(t)$. Another radius of the being formed material layer, i.e. the radius of the used substrate working face will be denoted by a_0.

It is assumed that when deposing the material the rate of its inflow in the circumferential direction is incomparably higher than the rate of its inflow in the radial direction. This assumption makes it possible to simulate the considered deposing process as an axisymmetric process in which the material layer is being accreted simultaneously throughout its whole inner cylindrical surface. We will term this surface the accretion surface. We will consider the material accretion process as continuous, that means, an infinitely thin additional material ply joins to the accreted solid each infinitely small period of time. In this case the dependence of the accretion surface radius a on the time t is a continuous function.

In this paper we will consider the case of small strain arisen in the accreted material layer. So it makes no sense to take into account the strain constituent in the time-dependance $a(t)$, and we get the variable radius a of the being formed layer decreasing in the course of the time t purely due to inflow of the additional material to the accretion surface. The dependence $a(t)$ can be considered as a prescribed program of attaching the new material to the layer, which is implemented in the simulated process. Thus, in the present study the dependence $a(t)$ is a given continuous function, strictly monotonic in those time intervals in which the material is deposed, and constant in those time intervals in which the inflow of additional material to the formed layer is temporarily or finally stopped.

In the here proposed model the possible compliance of the substrate on which the material layer is being formed is not taken into consideration, and the substrate is considered to be absolutely rigid. So the working radius of the substrate a_0 is supposed to be a constant in the modelled technological process.

A sufficiently slow time variation of the substrate rotational speed is assumed: $|\dot{\omega}(t)| \ll \omega^2(t)$ (we will denote the derivative of any function of the only time-variable t by a dot on the top). Thereby the tangential inertia forces of rotation are negligible if comparing to centrifugal forces.

It is also supposed that the potentially possible dynamic effects from the attaching the additional material to the accretion surface of the being formed layer are insignificant, and therefore the forces of inertia of its deformation can also be neglected in comparison with the centrifugal forces of inertia and the problem can be considered in a quasi-static statement.

We assume that the processes and effects which accompany the continuous inclusion of the additional substance in the composition of the being formed cylindrical solid lead to the appearance of some circumferential initial stresses in this solid near the accretion surface, and these stresses can arbitrary change with the radius ρ of the currently added infinitesimal circular material ply, but not with the angular position φ on this ply. That is the newly added plies can undergo preliminary uniform tension or compression in the hoop direction.

It is stated the task to trace the evolution of the stress-strain state of the material layer being additively formed in the above described manner, under the influence of the centrifugal inertia forces caused by the layer rotational motion together with the rigid substrate, in the approximation of the plane strain state.

3. The material governing equations

In the model proposed in this paper, we consider a linear viscoelastic uniform and uniformly aging isotropic material with the same constant (independent of either the material age or the time elapsed since the application of loads to it) Poisson's ratio $\nu = \mathrm{const}$ for instantaneous elastic strain and developing over time creep strain. For the given material the relationship between the stress tensor $\mathbf{T}$ and the small strain tensor $\mathbf{E}$ at each point $\mathbf{r}$ of the solid at any time moment t, calculated from the instant of the material origination, has the form [21, 22]:

$$\mathbf{T}^{\circ}(\mathbf{r},t) = 2\mathbf{E}(\mathbf{r},t) + (\varkappa - 1)\mathbf{1}\,\mathrm{tr}\,\mathbf{E}(\mathbf{r},t), \quad g^{\circ}(\mathbf{r},t) = \frac{g(\mathbf{r},t)}{G(t)} - \int_{\tau_0(\mathbf{r})}^{t} \frac{g(\mathbf{r},\tau)}{G(\tau)} K(t,\tau)\,d\tau. \qquad (1)$$

Here $\mathbf{1}$ is the unit tensor of the second rank, $\varkappa = 1/(1-2\nu)$ is the material constant, $G(t)$ is the elastic shear modulus of the material at its age t, and $K(t,\tau)$ is the creep kernel. The latter can be expressed through various characteristics of the material, describing its behaviour in any elementary stress state. For example, using the characteristics for the pure shear state it will be $K(t,\tau) = G(\tau)\,\partial\Delta(t,\tau)/\partial\tau$, $\Delta(t,\tau) = 1/G(\tau) + \Omega(t,\tau)$, where $\Omega(t,\tau)$ is the creep measure for the pure shear, $\Omega(\tau,\tau) \equiv 0$. The function $\Delta(t,\tau)$ describes the evolution over time t of specific (per unit of the acting shear stress) shear strain caused by the constant stress state of pure shear created at the time moment $\tau > 0$. We will call the tensor $\mathbf{T}^{\circ}$ standing on the left hand side of constitutive relation (1) the operator stress tensor.

The lower limit of integration in (1) has the meaning of the time moment of occurrence of the stress state in the neighborhood of the considered point of the solid. Since the accreted solid is increased with new material elements already during the process of its deformation, the moment of occurrence of stresses at the points $\mathbf{r}$ of such a solid will change from point to point and be set by a certain function $\tau_0(\mathbf{r})$, which is taken into account in constitutive law (1). In the modelled axisymmetric additive processes of the material deposition on a rigid cylindrical substrate we have $\tau_0(\mathbf{r}) \equiv \tau_0(\rho)$, where the value $\tau_0(\rho)$ should be considered as the moment of joining the material ply of a radius ρ to the accreted solid.

The constitutive relations similar to (1) are commonly applied to describe the mechanical behaviour of such wide used engineering materials as, for example, concrete and polymers. Note that state equation (1) contains as a special case the state equation of the isotropic linearly elastic material. This case is obtained by taking $\Omega(t,\tau) \equiv 0$ and $G(t) \equiv \mathrm{const}$.

4. Nonclassical boundary value problem statement

Let the deposition of the material on the inner surface of a cylindrical substrate rotating around its axis begins at some time moment $t = t_1 > 0$. The further deposition process may be interrupted at arbitrary time moments with arbitrarily long pauses when the inflow of additional material to the material layer having already been formed on the substrate is temporarily stopped. Such a piecewise continuous process of accreting the material layer finally comes to the end after a certain number N of stages of continuous deposing the additional material. The ready-formed layer may still continue to rotate rigidly coupled with the substrate.

Let us associate with the rotating around its axis axisymmetric cylindrical rigid substrate a polar cylindrical coordinate system (ρ, φ, z) with the orthonormal frame $\{\mathbf{e}_\rho, \mathbf{e}_\varphi, \mathbf{e}_z\}$, where: $\mathbf{e}_z = \mathrm{const}$ is the direction along the axis of rotation, z is the longitudinal coordinate measured in this direction; $\mathbf{e}_\rho = \mathbf{e}_\rho(\varphi)$ and $\mathbf{e}_\varphi = \mathbf{e}_\varphi(\varphi)$ are the radial and circumferential directions in the cross-section of a rotating cylindrical solid formed on the substrate, respectively, ρ and φ are the polar radius and the polar angle in the cross-section. In this movable noninertial base the standard equilibrium equation is valid for the considered accreted solid:

$$\nabla \cdot \mathbf{T}(\mathbf{r},t) + \mathbf{f}(\mathbf{r},t) = \mathbf{0} \qquad (2)$$

where the centrifugal forces of inertia with the volume density

$$\mathbf{f}(\mathbf{r},t) = \mathbf{e}_\rho(\varphi)\,\mu\rho\,\omega^2(t) \tag{3}$$

play a part of body forces, $\mu = \text{const}$ is the mass density of the material used.

It is obvious that for the entire cylindrical solid $a(t) < \rho < a_0$ having been formed by the instant t a smooth velocity vector field $\mathbf{v}(\mathbf{r},t)$ of this solid deformation motion is determined. Due to axial and mirror symmetry of the considered process and the plane strain state of the solid, the velocity field has the following representation in the above introduced rotating coordinate system: $\mathbf{v}(\mathbf{r},t) = \mathbf{e}_\rho(\varphi)\,v(\rho,t)$ where v is the rate of the growing layer points displacement in the radial direction owing to the layer deformation. On the being formed layer boundary fastened to the rigid substrate the condition of immobility of this boundary in the rotating coordinate system must be set:

$$\mathbf{v}(\mathbf{r},t) = \mathbf{0} \quad \text{if} \quad \rho = a_0. \tag{4}$$

For formulating a mechanical problem in terms of velocities it is necessary to enter into consideration the strain velocity tensor

$$\mathbf{D} = (\nabla\mathbf{v}^{\mathrm{T}} + \nabla\mathbf{v})/2. \tag{5}$$

We can formulate the analogue of defining relation (1) for the velocity characteristics of the deformation process, i.e. for tensor $\mathbf{D}$ and the operator stress velocity tensor $\mathbf{S} = \partial\mathbf{T}^\circ/\partial t$ [1]:

$$\mathbf{S}(\mathbf{r},t) = 2\mathbf{D}(\mathbf{r},t) + (\varkappa - 1)\,\mathbf{1}\,\mathrm{tr}\,\mathbf{D}(\mathbf{r},t). \tag{6}$$

The differential equation for tensor $\mathbf{S}$ in the occupied with the solid variable domain $a(t) < \rho < a_0$ can be obtained by acting on equilibrium equation (2) with the linear integral operator of viscoelasticity $(\ )^\circ$ from (1) and by differentiating the result with respect to time t. Here it is important to note that in general case the specified integral operator and the divergence operator $(\nabla\cdot\,)$ are not permutable because of the dependence of the lower integration limit in (1) on the considered solid point $\mathbf{r}$. However it can be shown following Manzhirov [1] that for the processes in question these operators commute in relation to the stress tensor $\mathbf{T}$, that is the permutation $(\nabla\cdot\mathbf{T})^\circ = \nabla\cdot\mathbf{T}^\circ$ will be fair in the entire region occupied at the given time moment by the whole accreted solid. That means the opportunity to write out the analogue of equilibrium equation (2) for the operator stress tensor: $\nabla\cdot\mathbf{T}^\circ(\mathbf{r},t) + \mathbf{f}^\circ(\mathbf{r},t) = \mathbf{0}$, and therefore, for the operator stress velocity tensor:

$$\nabla\cdot\mathbf{S}(\mathbf{r},t) + \partial\mathbf{f}^\circ(\mathbf{r},t)/\partial t = \mathbf{0}. \tag{7}$$

In view of the transition in the mathematical problem formulation from the operator stresses $\mathbf{T}^\circ$ to their time derivatives $\mathbf{S}$ it is necessary for the closure of the problem statement to set some initial conditions for the operator stresses. This can be done taking into account the initial stresses in each newly attached material ply described by the stress tensor

$$\mathbf{T}\big(\mathbf{r},\tau_0(\mathbf{r})\big) = \mathbf{T}_0(\mathbf{r}) \tag{8}$$

inside the growing solid under consideration close to its accretion surface, where $\mathbf{T}_0(\mathbf{r})$ is a known tensor of initial stresses in the being formed solid at its point $\mathbf{r}$ at the time instant $t = \tau_0(\mathbf{r})$ of joining this point to the solid. The tensor-function $\mathbf{T}_0(\mathbf{r})$ sets the stress state of the newly added material immediately after its attachment to the growing solid. This state emerges by uniform circumferential tension–compression of the plies attached to the layer being formed and is determined by the following:

$$\mathbf{T}_0(\mathbf{r}) = \mathbf{e}_\varphi(\varphi)\,\mathbf{e}_\varphi(\varphi)\,s(\rho), \tag{9}$$

where the given hoop stress s is taken in the present study as a known function of the joined ply radius ρ. Note that tensor (9) matches with the stress-free current accretion surface during the accretion process: $\mathbf{e}_\rho(\varphi)\cdot\mathbf{T}_0(\mathbf{r})=\mathbf{0}$.

Initial condition (8) together with representation (9) implies the following condition for the operator stress velocity tensor $\mathbf{S}$ on the moving due to the inflow of additional material boundary $\rho=a(t)$ of the growing layer in any time interval of its continuous growth:

$$\mathbf{e}_\rho(\varphi)\cdot\mathbf{S}(\mathbf{r},t)=-\mathbf{e}_\rho(\varphi)\,q(t),\quad q(t)=\left[\frac{s(a(t))}{a(t)}-\mu\,\omega^2(t)\,a(t)\right]\frac{\dot a(t)}{G(t)},\quad \text{if}\quad \rho=a(t). \tag{10}$$

We omit here the proof of this fact because of the limited content of the paper.

In those time intervals when the inflow of additional material to the additively formed rotating layer is ad interim or definitively halted, the condition of loading absence on the fixed accretion surface $\rho=a(t)\equiv\text{const}$ is set by the equality $\mathbf{e}_\rho(\varphi)\cdot\mathbf{T}(\mathbf{r},t)=\mathbf{0}$ which is valid at each point $\mathbf{r}$ of this surface, starting from the moment $t=\tau_0(\mathbf{r})$ of this point inclusion into the solid composition. From this equality it obviously follows that condition (10) formally remains in force in the absence of an inflow of new material to the formed solid, that is, in any time interval of constancy of the function $a(t)$, i.e. in intervals of naught derivative $\dot a(t)$.

Thus, the entire process of piecewise continuous accreting an aging viscoelastic solid under consideration, including an arbitrary prolonged period of time after the end of accreting, will be described by the boundary value problem consisting of equation (7), (3), valid in the time-varying region $a(t)<\rho<a_0$, supplemented with relations (6), (5) and boundary conditions (4), (10). The time variable $t\in(t_1,+\infty)$ is a parameter in this boundary problem.

5. Stress evolution in the manufactured material

We can construct the exact analytical solution of the above formulated boundary value problem. The corresponding evolution with time t of the operator stress velocities, at each point $\mathbf{r}$ of the considered piecewise continuously accreted aging viscoelastic material layer on the time beam $t>\tau_0(\mathbf{r})$ covering the entire deformation history of the neighborhood of a given point $\mathbf{r}$ since it is in the composition of this layer, may be represented in the following closed explicit form:

$$\mathbf{S}(\mathbf{r},t)=\mathbf{e}_\rho(\varphi)\mathbf{e}_\rho(\varphi)\,S_\rho(\rho,t)+\mathbf{e}_\varphi(\varphi)\mathbf{e}_\varphi(\varphi)\,S_\varphi(\rho,t)+\mathbf{e}_z\mathbf{e}_z\,\nu\big[S_\rho(\rho,t)+S_\varphi(\rho,t)\big],$$

$$S_{\substack{\rho\\ \varphi}}(\rho,t)=-\Psi^{\pm}_{\varkappa}(\rho,t)+\frac{1\mp a^2(t)/\rho^2}{1/\varkappa+a^2(t)/a_0^2}\,\Psi^-_1(a_0,t)-\frac{\varkappa\pm a_0^2/\rho^2}{\varkappa+a_0^2/a^2(t)}\,q(t),$$

$$\Psi^{\pm}_{\alpha}(\rho,t)=$$
$$\frac{2\mu}{\varkappa+1}\int_{a(t)}^{\rho}\left[\frac{\omega^2(\tau_0(\xi))}{2}\,\frac{\partial\Omega(t,\tau_0(\xi))}{\partial t}+\frac{\omega(t)\dot\omega(t)}{G(t)}+\int_{\tau_0(\xi)}^{t}\omega(\tau)\dot\omega(\tau)\,\frac{\partial\Omega(t,\tau)}{\partial t}\,d\tau\right]\left(\alpha\pm\frac{\xi^2}{\rho^2}\right)\xi\,d\xi.$$

From initial condition (8) for the stress tensor we obtain according to the form of the integral operator $(\ )^\circ$ the following initial condition for the operator stress tensor at the time moments $t=\tau_0(\mathbf{r})$: $\mathbf{T}^\circ\big(\mathbf{r},\tau_0(\mathbf{r})\big)=\mathbf{T}_0(\mathbf{r})/G\big(\tau_0(\mathbf{r})\big)$. After that we can reconstruct the entire time evolution of the operator stress tensor by the known operator stress velocity tensor evolution using the integration procedure: $\mathbf{T}^\circ(\mathbf{r},t)=\mathbf{T}_0(\mathbf{r})/G\big(\tau_0(\mathbf{r})\big)+\int_{\tau_0(\mathbf{r})}^{t}\mathbf{S}(\mathbf{r},\tau)\,d\tau$.

As soon as at each point $\mathbf{r}$ of the ready-made material layer we know the evolution of the tensor $\mathbf{T}^\circ(\mathbf{r},t)$ since the moment $t=\tau_0(\mathbf{r})$ of stresses appearance at the given point, we can find the entire time-evolution of the stress tensor $\mathbf{T}(\mathbf{r},t)$ at the point $\mathbf{r}$ by the integral transformation inverse to the transformation $(\ )^\circ$: $\mathbf{T}(\mathbf{r},t)=\big[\mathbf{T}^\circ(\mathbf{r},t)+\int_{\tau_0(\mathbf{r})}^{t}\mathbf{T}^\circ(\mathbf{r},\tau)\,R(t,\tau)\,d\tau\big]\,G(t)$ where the function $R(t,\tau)$, called the relaxation kernel, is the resolvent of the kernel $K(t,\tau)$.

Acknowledgments
This study was financially supported by the Ministry of Science and Higher Education of the Russian Federation (research theme state registration number AAAA-A17-117021310381-8) and, in part, by the Russian Foundation for Basic Research (under grants Nos. 17-51-45054-IND_a, 18-01-00920-a).

References

[1] Manzhirov A V 1995 The general non-inertial initial-boundaryvalue problem for a viscoelastic ageing solid with piecewise-continuous accretion *J. Appl. Math. Mech.* **59** (5) 805–816
[2] Manzhirov A V and Parshin D A 2006 Accretion of a viscoelastic ball in a centrally symmetric force field *Mech. Solids* **41** (1) 51–64
[3] Manzhirov A V 2013 Mechanics of growing solids and phase transitions *Key Engineering Materials* **535–536** 89–93
[4] Manzhirov A V 2015 Mechanical design of viscoelastic parts fabricated using additive manufacturing technologies *Lecture Notes in Engineering and Computer Science* **2218** (1) 710–714
[5] Manzhirov A V and Parshin D A 2015 Arch structure erection by an additive manufacturing technology under the action of the gravity force *Mech. Solids* **50** (5) 559–570
[6] Manzhirov A V and Parshin D A 2015 Influence of the erection regime on the stress state of a viscoelastic arched structure erected by an additive technology under the force of gravity *Mech. Solids* **50** (6) 657–675
[7] Manzhirov A V and Parshin D A 2016 Accretion of spherical viscoelastic objects under self-gravity *Lecture Notes in Engineering and Computer Science* **2224** (1) 1131–1135
[8] Manzhirov A V and Parshin D A 2016 Application of prestressed structural elements in the erection of heavy viscoelastic arched structures with the use of an additive technology *Mech. Solids* **51** (6) 692–700
[9] Manzhirov A V 2017 Advances in the theory of surface growth with applications to additive manufacturing technologies *Procedia Engineering* **173** 11–16
[10] Manzhirov A V 2017 Fundamentals of mechanical design and analysis for AM fabricated parts *Procedia Manufacturing* **7** 59–65
[11] Manzhirov A V and Gupta N K 2017 Fundamentals of continuous growth processes in technology and nature *Procedia IUTAM* **23** 1–12
[12] Parshin D A 2017 The technological stresses in a vaulted structure built up on a falsework *Advanced Structured Materials* **46** 377–386
[13] Parshin D A 2017 Analytic solution of the problem of additive formation of an inhomogeneous elastic spherical body in an arbitrary nonstationary central force field *Mech. Solids* **52** (5) 530–540
[14] Manzhirov A V, Parshin D A and Romanov A A 2017 Additive manufacturing of conical viscoelastic parts under axial tension-compression *Lecture Notes in Engineering and Computer Science: Proc. World Congr. on Engineering 2017 (WCE 2017), 5–7 July, London* pp 934–939
[15] Parshin D A and Hakobyan V N 2017 Additive manufacturing of a cylindrical arch of viscoelastic aging material under gravity action at various modes of the process *Procedia IUTAM* **23** 66–77.
[16] Parshin D A and Manzhirov A V 2017 Quasistatic problems for piecewise-continuously growing solids with integral force conditions on surfaces expanding due to additional material influx *J. Phys.: Conf. Series* **937** 012038
[17] Manzhirov A V and Parshin D A 2018 Modeling spherical accretion of aging viscoelastic objects under self-gravitation *IAENG Transactions on Engineering Sciences: Spec. Issue for Int. Assoc. Engineers Conferences 2016* vol II (World Scientific Publishing) pp 14–27
[18] Manzhirov A V and Parshin D A 2018 Analytical solution of the mechanical problem on additive thickening of aging viscoelastic tapers under nonstationary longitudinal end forces *Eng. Letters* **26** (2) 267–275
[19] Parshin D A 2018 Impact of the rate of the additive process of forming a heavy structure deforming in creep on the development of its technological stresses *AIP Conference Proceedings* **1959** 070029
[20] Parshin D A and Manzhirov A V 2018 The mechanical problems on additive manufacturing of viscoelastic solids with integral conditions on a surface increasing in the growth process *J. Phys.: Conf. Series* **991** 012063
[21] Arutyunyan N Kh 1966 *Some Problems in the Theory of Creep* (Oxford: Pergamon Press)
[22] Arutyunyan N Kh and Manzhirov A V 1999 *Contact Problems of the Theory of Creep* (Erevan: Izdat. Inst. Mekhaniki NAN RA) [in Russian]

IOP Conf. Series: Journal of Physics: Conf. Series **1205** (2019) 012045 doi:10.1088/1742-6596/1205/1/012045

Stress concentration analysis in technical structures with boundary integral equations

M N Perelmuter

Ishlinsky Institute for Problems in Mechanics RAS, prospect Vernadskogo 101-1, Moscow, 119526 Russia

E-mail: perelm@ipmnet.ru

Abstract. The direct boundary integral equation (DBIE) method in the multi-domain formulation is used for three-dimensional stress concentration problems solution. The computational results of several new stress concentration problems are presented and analyzed. The following problems have been studied: flat plates with oblique, conical and round-edges holes under in-plane tensile loading for different angles of holes inclination. The stress concentration problem for an aviation engine turbine disk with cooling hole has also been considered.

1. Introduction

Holes of different shapes and orientations are widely used in the modern technologies. Through-thickness holes with axes inclined to surface (oblique holes) are used in cylindrical and spherical shells, turbine blades and disks [1, 2]. In the neighborhood of holes the stress state is triaxial in nature and the problem should be considered as the three-dimensional one. General analytical solutions for stress distributions around holes exist only for a few two-dimensional problems. For essentially three-dimensional stresses concentration problems (for example, oblique holes) analytical solutions do not exist and only experimental analysis and numerical modelling can be used. Stress concentration near one oblique hole or arrays of such holes have been analyzed by photoelasticity method [3, 4, 5]. The problems with one oblique hole in flat plates and in thick-walled cylinders were considered by finite and boundary element methods [6, 7, 8].

The purpose of this paper is the expansion of 3D computational modelling to the stress concentration analysis of some technical structures with one and two oblique holes, with conical holes for different angles of holes inclination.

2. Boundary integral equations formulation

Within the multi-domain formulation of the DBIE, the direct boundary integral equations for elasticity problems are used for each homogeneous subregion of the structure. The supplementary boundary conditions at interfacial boundaries of subregions are introduced and used to eliminate additional variables on joint boundaries of subregions. For elasticity problems without volume forces the DBIE for any homogeneous subregion of the 3D structure is given by [9]:

$$c_{ij}(p)\,u_i(p) = \int_{\Gamma} [G_{ij}(p,q)\,t_i(q) - T_{ij}(p,q)\,u_i(q)]d\Gamma(q)\,, \quad i,j = 1,2,3 \qquad (1)$$

here $c_{ij}(p)$ depends on local boundary geometry of the surface Γ (for a smooth boundary $c_{ij}(p) = 0.5\delta_{ij}$), $G_{ij}(p,q)$ and $T_{ij}(p,q)$ are Kelvin's fundamental solutions for displacements and tractions, respectively, $u_i(q)$ and $t_i(q)$ are displacements and traction on the subregion surface; the location of source and field points belonging to the subregion surface Γ are defined by the coordinates of points q and p.

The displacement continuity and the traction equilibrium supplementary relations are used at the interfacial boundaries of subregions with conditions of ideal contact in the following form

$$u_i^k(p) = u_i^n(p), \quad t_i^k(p) = -t_i^n(p), \quad i = 1,2 \tag{2}$$

where k and n are numbers of join subregions, $u_i(p)$ and $t_i(p)$ are displacements and traction components at the boundary point p.

For numerical solution of the BIE (1) in three-dimensional case surfaces of all subregions are subdivided into eight-node quadrilateral quadratic isoparametric elements. We use the collocation scheme to discretize BIE and collocation points coincided with the nodal points of boundary elements. The kernels in equations (1) are singular: the kernel $G_{ij}(p,q)$ is weakly singular and the kernel $T_{ij}(p,q)$ is strongly singular. In the used collocation scheme these singularity are transferred to discrete equations. In these equations the integrals with a weak singularity, associated with the kernel $G_{ij}(p,q)$, are calculated using the transformation to the polar coordinate system [10]. The sum of the integrals with a strong singularity corresponding to the kernels $T_{ij}(p,q)$ and free term $c_{ij}(p)$ in equations (1) are calculated using the rigid body motion approach, [9]. Integrating over all subregions boundaries we obtain the final discrete BIE, corresponding to equations (1). The numerical algorithm of equations system (1)-(2) solving has been implemented into the author's computer code, [10, 11, 12]. Under computer implementation this algorithm has demonstrated high accuracy and low time consuming versus the published results.

3. Results of numerical modelling

In this part of the paper we present some new results of the stress concentration analysis for plates with one and two oblique holes, plates with rounded edge of holes and plates with conical holes. The problem of stress concentration for the cooling oblique hole in the rim of the turbine disk is also considered. All computations were performed for Poisson's ratio $\nu = 0.3$ under uniaxial tension loading σ_0.

The plate with oblique holes is the model of a turbine blade wall with cooling holes. At first, we consider the computation results for in-plane uniaxial tension of the plate with one oblique hole. Note, that cuts set of this plate, parallel to its surface are ellipses with semi-axes $a = R$ and $b = Rsin\alpha$, where R is the radius of hole cylindrical surface. The plane XOZ is the plane of symmetry for uniaxial plate tension in the direction OX and OY. The boundary element model with accounting of the problem symmetry is presented in Fig. 1, axis OY is the normal to the plane XOZ. The computational analysis has been shown that under uniaxial tension in the direction OY (orthogonal to the hole axis) for hole obliquity angles more then 75° the position of the normal stress maximum (σ_y) occurs at the plate middle section (see Fig. 2). If the hole axis turns in the plane normal to the direction of tensile loading, then the position of the normal stress maximum is shifted to the hole acute-angled edge (for angles less then 75°). For tension in the orthogonal direction OX the largest stress σ_y is compressive and occurs also at the hole acute-angled edge.

For engineering applications it will be convenient to present the results of oblique holes stress analysis as a simple relation for normal stress σ_y versus a plate thickness. The maximal normal stress at the edge of an elliptical hole with semi-axes a and b in $2D$ infinite plate is

$$\sigma_y = \sigma_0(1 + 2a/b) \tag{3}$$

IOP Conf. Series: Journal of Physics: Conf. Series **1205** (2019) 012045 doi:10.1088/1742-6596/1205/1/012045

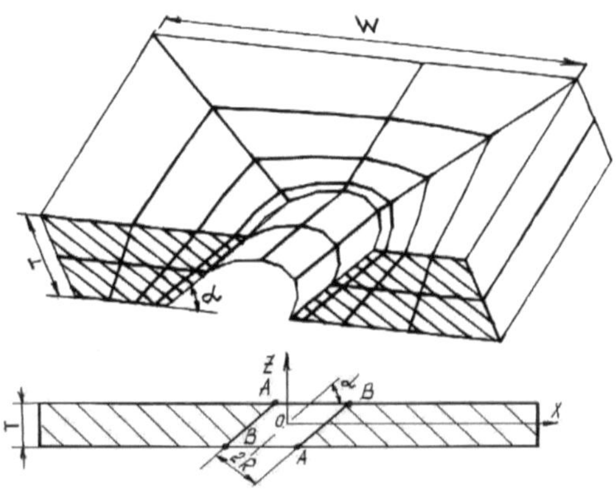

Figure 1. Plate with the oblique hole, boundary element model, number of nodes - 194, $\alpha = 30°$, $W/T = 10$, $T/R = 2$.

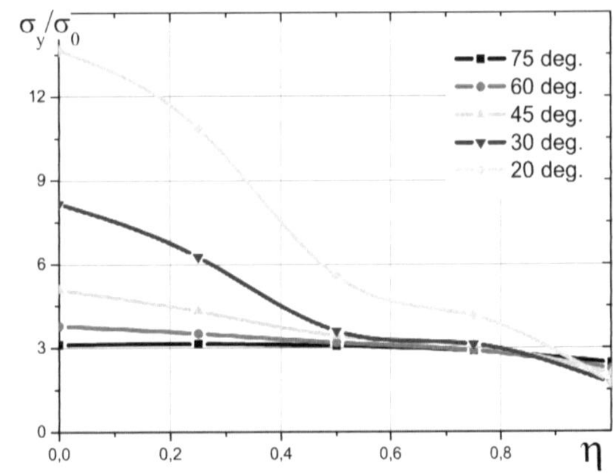

Figure 2. Normal stress distribution over thickness of the plate with the oblique hole (along line AB, Fig. 1), $\eta = (Z/T) + 0.5$, uniaxial tension.

where $b/a = sin\alpha$ for cuts of $3D$ plates with oblique holes, α is the angle of hole obliquity.

The second term in (3) corresponds to amplification of the stress σ_y (above the nominal stress) at the hole edge in $2D$ problem. Let's assume that in $3D$ problems the stress σ_y amplification (above the $2D$ stress concentration) at the hole acute-angled edge is inversely proportional to $sin\alpha$. Hence, the stress at the hole acute-angle edge (point A, Fig. 1) defines as

$$\sigma_y(-T/2) = \sigma_0 \left(1 + \frac{2}{\sin^2\alpha}\right) \tag{4}$$

The normal stress near the hole blunt angle (point B, Fig. 1) is decreased in comparison by the second term in (3) and it is assumed that this stress decreasing is proportional to $sin^3\alpha$. Consequently, stress at the hole blunt edge is

$$\sigma_y(+T/2) = \sigma_0 \left(1 + 2sin^2\alpha\right) \tag{5}$$

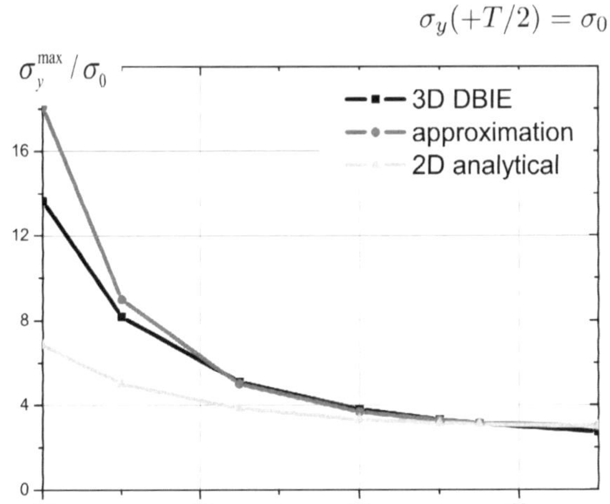

Figure 3. Stress concentration factor vs obliquity angle, comparison of the numerical and analytical results.

Note, that value (3) can be attributed as the stress at the middle section of the plate, $z = 0$. We can obtain the parabolic approximation of the normal stress at the oblique hole edge by using introduced above three values of the normal stress along the plate thickness (3), (4) and (5)

$$\sigma_y(\xi) = \sigma_0 \left(C_1\xi^2 + C_2\xi + C_3\right) f(W/T, T/R) \tag{6}$$

where $\xi = Z/T$, $|\xi| \leq 0.5$, $C_{1,2,3}$ depend on the angle obliquity α and $f(W/T, T/R)$ is the function for accounting of plates sizes, which can be obtained experimentally or numerically.

Dependencies of the normal stress at the hole acute-angle versus of hole axis obliquity, which were obtained numerically by DBIE, were calculated on the base of relation (4) (with $\xi = -0.5, f(W/T, T/R) = 1$ this relation identically to (6)) and the results (3) obtained for $2D$ problem are shown in Fig. 3. Deviation between the numerical results and the approximation by relation (6) for obliquity angles $35° \leq \alpha \leq 90°$ is less than $3 - 5\%$ that is convenient for the most technical applications.

The rounding of the hole edge is used to decrease the stress concentration at the hole acute-angle. Computations by DBIEM were performed for plates with rounded edges of oblique holes (see Fig. 4), where ϱ is the radius of rounding and values of $\varrho/T = 0.25$; 0.125 were used. The boundary conditions for this task are similar to the previous problem (see Fig. 1). The graphs of computation results are presented in Fig. 5. For rounded hole edges the normal stress at the surface of the plate becomes less than for an acute angle of hole. The position of the stress maximum shifts along the hole axis away of the hole edge. For $\alpha = 90°$ and $\alpha = 60°$ the stresses decreasing are not noticeable, but the maximum stress position is shifted from the hole edge in the distance about rounded radius. If $\alpha = 45°$ and $\alpha = 30°$ the stress decreasing are more noticeable, it is up to 20% for $\alpha = 30°$.

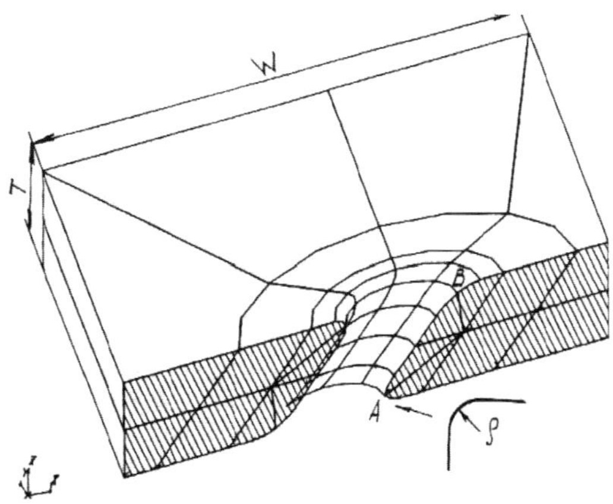

Figure 4. Plate with the hole rounded edge, $W/T = 10$, $T/R = 2$, $\alpha = 45°$, $\varrho/T = 0.25$, tension is transversally to the hole axis.

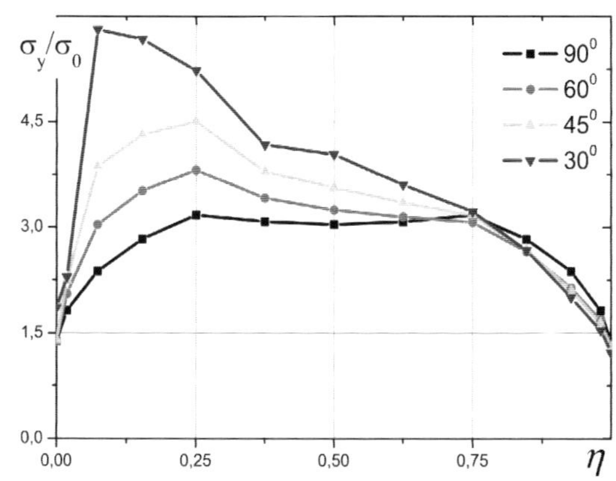

Figure 5. Normal stress distribution over thickness of the plate (along line AB, Fig. 4), rounded edge of hole, $\varrho/T = 0.25$, $\eta = Z/T$.

Effects analogous to those observed at the edges of oblique holes are also possible near the edges of conical holes. Numerical analysis of the stress state under uniaxial tension was performed for flat plates with holes in the form of a truncated cone. The radius of the small circle of the truncated cone was assigned to be equal to the size of small semi-axis of the elliptical section of oblique hole. The numerical DBIEM model of a quarter of the plate for one of the calculation variants is shown in Fig. 6. The tensile stresses distribution along the plate thickness as a function of the slope of the hole generatrix are shown in Fig. 7. In the acute edge zone of the conical hole, as in the case of oblique holes, a considerable stress concentration is observed. The maximum of stress magnitude is somewhat lower for conical holes than on the acute edge of the "normal" oblique hole.

Some products (for example, cooled blades of gas turbine engines) contain systems of closely located oblique holes. The stress state of a plate with two oblique holes was calculated to evaluate the interference of such holes. The plane XOZ is the symmetry plane in this problem, uniaxial tension is applied transversely to this plane, similarly as for a plate with one hole. A discrete model of a plate half with two oblique holes is shown in Fig. 8. The dimensions of the plate, the elastic constants and the boundary conditions are the same as those adopted for

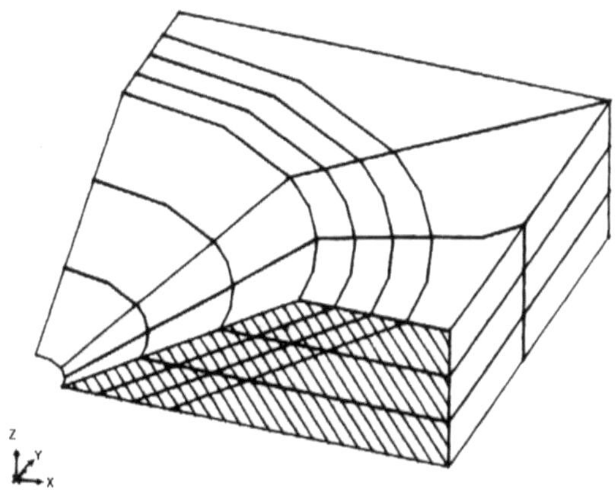

Figure 6. Plate with a conical hole, boundary element model, number of nodes - 200, the cone generatrix slope is $\alpha = 45°$, uniaxial tension.

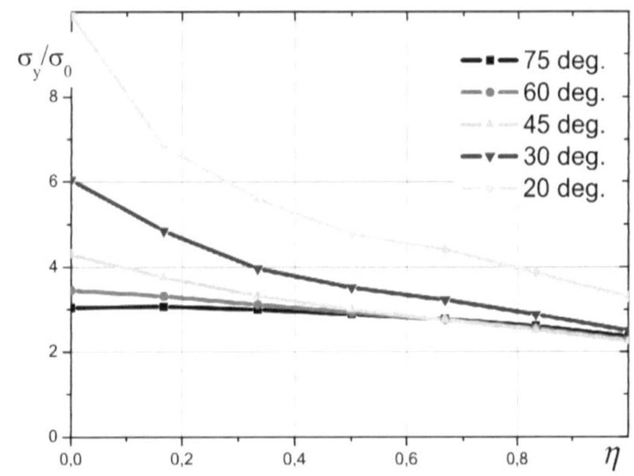

Figure 7. Distribution of normal stress over thickness of the plate at the hole edge, the cone generatrix slope variation, $\eta = Z/T$.

the problem for a plate with one hole. Calculations were performed for $\alpha = 45°$, $30°$ and for the following distances between the oblique axes of the holes $L_1 = 4R/sin\alpha$, $L_2 = 3R/sin\alpha$, $L_3 = 2.5R/sin\alpha$ (see Fig. 8). It was established that a changing in the distance between the axes of the holes from L_1 to L_3 leads to the maximum stresses increasing at the acute edge of the oblique hole, located on the bridge between the holes (point A, Fig. 8). In comparison by the plate with a single hole (see Fig. 9, where $L = \infty$ corresponds to the plate with single hole) this increasing is about $10 - 25\%$.

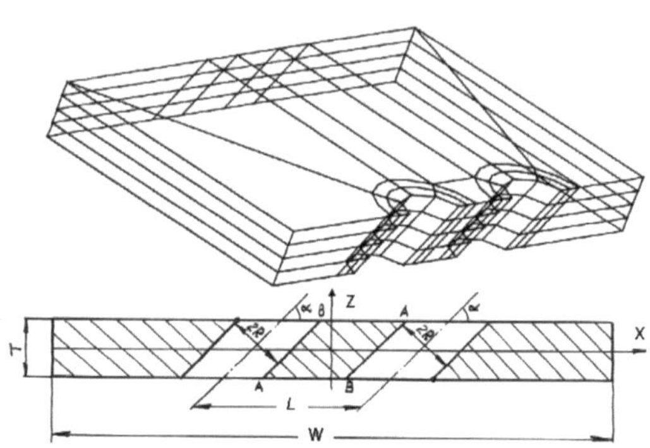

Figure 8. Plate with two oblique hole, boundary element model, number of nodes - 308, $\alpha = 30°$, uniaxial tension.

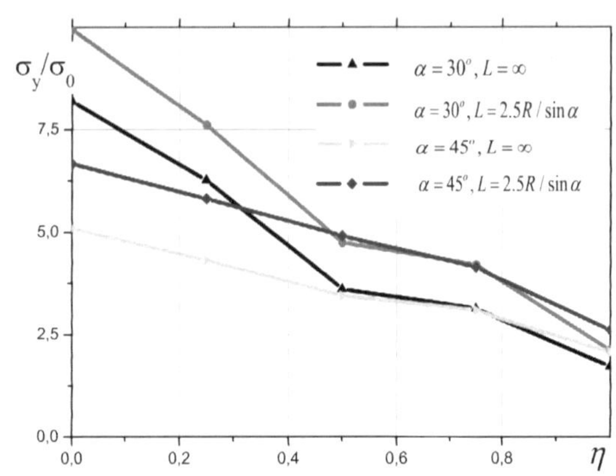

Figure 9. Distribution of normal stress over thickness of plates with the oblique holes (along lines AB, Fig. 1 and Fig. 8), $\eta = (Z/T) + 0.5$.

Oblique holes in the rim of turbine disks serve to supply of cooling air to blades. The angle between the axis of the cooling hole and the radial direction usually does not exceed $\beta = 20°$ (see Fig. 10), which corresponds to $\alpha \geq 70°$ (see Fig. 1). The effect of the hole axis inclination for such angles is relatively small (see Fig. 2), and the predominant effect is intersection of the stress concentrators (holes and wells), in the hole transition zone at the groove for attaching of the blade tail. A discrete model of this problem with the hole zone and the part of the groove for attaching the blade is shown in Fig. 11. On the rear invisible part of the model

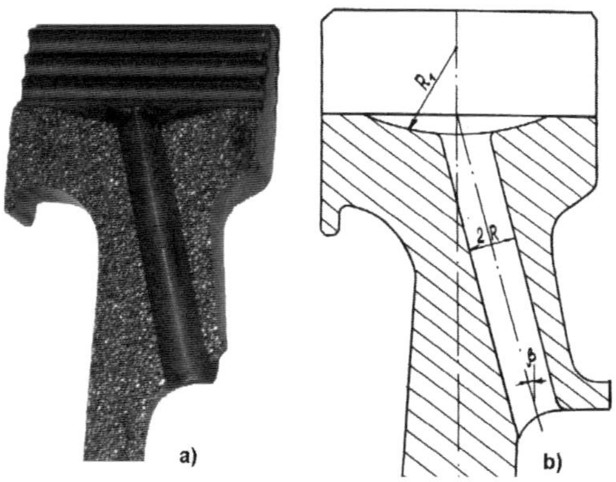

Figure 10. **a)** Photo of the section of the rim a turbine disk with a cooling hole; **b)** the simplified sketch of that section, $\beta = 20°$, $R = 3.1\,mm$, $R_1 = 3R$.

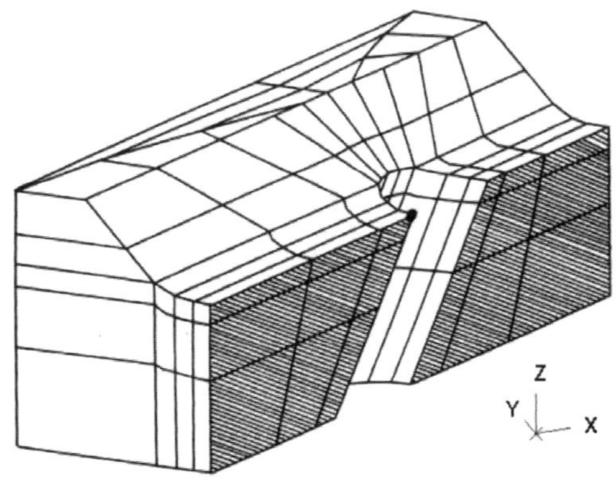

Figure 11. Discrete model of the hole zone and the groove part for the blade attaching, 2 subregions, number of nodes - 574.

(Fig. 11) uniform tensile loading simulating the action of circumferential stresses, was applied. On the lower invisible part of the model surface and in the plane passing through the axis of the hole, normal displacements and tangential tractions were assumed to be zero. The sides and top faces of the model, as well as the hole surface, were assumed as free of loads. The computation by DBIE gives the coefficient of the normal stress concentration at the hole edge in intersection between the hole and well as $K = 8.02$. This value is rather high and the hole and well intersection region must be modified to increase the structure durability.

Acknowledgments

Financial support for this research was partly provided by Russian Foundation for Basic Research (Project 17-08-01312) and also by the Research Program of RAS (Project I.2.27).

References

[1] Chung H, Park J S, Sohn H S, Rhee D H and Cho H H 2014 *International Journal of Heat and Mass Transfer* **73** 9 – 20
[2] Xu L, Bo S, Hongde Y and Lei W 2015 *Procedia Engineering* **99** 1482 – 1491 2014 Asia-Pacific International Symposium on Aerospace Technology, APISAT2014 September 24-26, 2014 Shanghai, China
[3] Stanley P and Day B V 1990 *The Journal of Strain Analysis for Engineering Design* **25** 157–175
[4] Stanley P and Day B V 1994 *Proceedings of the Institution of Mechanical Engineers, Part E: Journal of Process Mechanical Engineering* **208** 75–83
[5] McKenzie H W and White D J 1968 *Journal of Strain Analysis* **3** 98–102
[6] Abdul-Mihsein M J, Fenner R T and Tan C L 1979 *The Journal of Strain Analysis for Engineering Design* **14** 179–185
[7] Li H P and Ellyin F 1983 *Journal of Vibration, Acoustics, Stress, and Reliability in Design* **105** 206–211
[8] Nihous G C, Kinoshita C K and Masutani S M 2008 *Journal of Pressure Vessel Technology* **130** 021202–1 – 021202–6
[9] Banerjee P and Butterfield R 1981 *Boundary Element Methods in Engineering Science* (McGraw - Hill, London New-York)
[10] Perelmuter M 1989 *Proceeding of the Central Institute of Aviation Motors (CIAM)* vol 1237 (CIAM Publushing, Moscow (in Russian)) pp 74–98
[11] Genna F and Perelmuter M 2010 *Computers & Structures* **88** 845 – 858
[12] Perelmuter M 2013 *Computational Mechanics* **51** 523–534

IOP Conf. Series: Journal of Physics: Conf. Series **1205** (2019) 012046 doi:10.1088/1742-6596/1205/1/012046

Semiclassical asymptotics of the solution of the Helmholtz equation in a three-dimensional layer of variable thickness with a localized right-hand side

P N Petrov[1,2,a] **and S Yu Dobrokhotov** [1,2,b]

[1]Moscow Institute of Physics and Technology (State University), 9 Institutskiy per., Dolgoprudny, Moscow Region, 141701, Russian Federation

[2]Ishlinsky Institute for Problems in Mechanics of the Russian Academy of Sciences, Prospekt Vernadskogo 101-1, Moscow, 119526, Russia

E-mail: [a]`petr.petrov@phystech.edu`

E-mail: [b]`dobr@ipmnet.ru`

Abstract. We construct a semiclassical asymptotics of the solution of the Helmholtz equation in a three-dimensional layer of variable thickness with a localized right-hand side, assuming the absence of "trap" states and the fulfillment of radiation conditions at infinity (such as the Sommerfeld condition). Wave part of the asymptotic solution can be represented as an decomposition into a finite number of modes, each mode is connected with a pair of Lagrangian manifolds.

1. Introduction

The traditional modeling of the sound propagation in shallow water is done by solving the Helmholtz equation in a three-dimensional layer of variable thickness with δ-function on the right-hand side. In the underwater acoustics several methods for solving such a problem are developed, among these methods one can note the method of the parabolic equation[1, 2], the method of Gaussian beams[3] and methods based on separating the vertical variable and obtaining equations describing the change in modal amplitudes in the horizontal plane[4, 5]. Recently, a new approach for constructing asymptotic solutions of inhomogeneous stationary problems with localized right-hand sides was proposed in [6]. This approach is based on using the Maslov canonical operator on a pair of Lagrangian manifolds and generalizes results [7, 8, 9]. In this paper we develop the approach [6] for constructing asymptotic formulae for the solution of the Helmholtz equation in a three-dimensional layer $D(\xi_1, \xi_2) \leq z_0 \leq 0$ with slowly varying boundaries with a localized right-hand side

$$\Delta u + \frac{\omega^2}{c^2} u = F\Big(\frac{x_1 - \xi_1}{\mu}, \frac{x_2 - \xi_2}{\mu}\Big) g\Big(\frac{z - z_0}{\mu}\Big), \quad u|_{z=D(x)} = 0, \quad u|_{z=0} = 0. \qquad (1)$$

Here z-axis is directed perpendicular upwards from the horizontal plane (x_1, x_2), μ is a small parameter defining characteristic source size, values (ξ_1, ξ_2, z_0) define coordinates of the source, the source is located inside the layer $(D(x_1, x_2) \leq z_0 \leq 0)$, F and g are smooth rapidly decaying at infinity functions.

First we apply the adiabatic reduction [10] of the 3-D problem to a family of 2-D (plane) problems in $\mathbf{R}^2_x$, describing the waves corresponding to various horizontal modes. Then we apply the approach [6] to obtained 2-D problems. Note that it is possible to present the asymptotic solution in a standard form as the convolution of the right-hand side with the Green function asymptotics [7, 8, 9], but the calculation of the corresponding integral is a sufficiently complicated problem, and the realization of the approach [6] seems much more pragmatic and convenient. As we mentioned before the asymptotics is related to a pair of Lagrangian manifolds, one of them is connected with a localized ("singular") part of the solution in the neighborhood of the point ($x_1 = \xi_1$, $x_2 = \xi_2$) and the second manifold defines the oscillating ("wave") part of the solution in the entire layer (taking into account the possible appearance of caustics and focal points). For the realization of obtained formulae, it is necessary to construct rays (or trajectories in the phase space), their ends determine wave fronts, and restore the wave field on them using sufficiently effective formulae. Note that our consideration assumes the absence of so-called "trapped" (horizontal) modes.

2. Adiabatic reduction to 2-D problems

Let us pass to dimensionless variables by the following way: $x' = x/l_0$, $z' = z/d_0$, where d_0 is the characteristic thickness of the layer, and l_0 is some horizontal size. We assume that $l_0 \gg d_0$ and introduce the small parameter $h = d_0/l_0$. The equation in dimensionless variables has the form

$$\Big(h^2\Delta_{x'} + \frac{\partial^2}{\partial z'^2} + n^2\Big)u = f\Big(\frac{x'-\xi'}{\mu'}\Big)g\Big(\frac{z'-z'_0}{\mu''}\Big) \quad u|_{z'=D'(x')} = 0, \quad u|_{z'=0} = 0\,, \tag{2}$$

where $n^2 = \frac{d_0^2\omega^2}{c^2}$, $f(\frac{x'-\xi'}{\mu'})g(\frac{d_0 z - z_0}{\mu''}) = d_0^2 F(\frac{x-\xi}{\mu})g(\frac{z-z_0}{\mu})$, $D'(x') = \frac{D(x/l_0)}{d_0}$, $\mu'' = \frac{\mu}{d_0}, \mu' = \frac{\mu}{l_0}$. Further all hatches are omitted, and, for simplicity, we consider Eq. (2) with one parameter ($h = \mu' = \mu''$).

We are looking for a solution in the form:

$$u = \sum_\nu \chi_\nu\Big(\overset{2}{x}, -ih\overset{1}{\nabla}_x, z, h\Big)\psi_\nu(x,h)\,, \tag{3}$$

where $\hat{\chi}_\nu$ are pseudodifferential operators with the small parameter h, these are operator analogs of mode functions, and numbers above operators denote the order of operator action[11]. Functions $\psi_\nu(x,h)$ satisfy two-dimensional equations

$$L_\nu\Big(\overset{2}{x}, -ih\overset{1}{\nabla}_x, h\Big)\psi_\nu(x,h) = \Big(g\Big(\frac{z-z_0}{h}\Big), \chi^0_\nu(x,p,z,h)\Big)f\Big(\frac{x-\xi}{h}\Big)\,, \tag{4}$$

where $\hat{p} = -ih\nabla_x$, brackets $(v(z), w(z))$ define the scalar product $(v(z), w(z)) = -\frac{2}{D(x)}\int_{D(x)}^0 v(z)w(z)\mathrm{d}z$. $L_\nu(x,p,h)$ and $\chi_\nu(x,p,z,h)$ are symbols of operators $L_\nu(x, -ih\nabla_x, h)$ and $\chi_\nu(x, -ih\nabla_x, z, h)$. We suppose that functions χ_ν, L_ν are smooth, and these can be represented in terms of the asymptotic expansion with respect to the parameter h.

$$\begin{aligned}\chi_\nu &= \chi^0_\nu(x,p,z) + h\chi^1_\nu(x,p,z) + h^2\chi^2_\nu(x,p,z)..\\ L_\nu &= L^0_\nu(x,p) + hL^1_\nu(x,p) + h^2L^2_\nu(x,p)..\end{aligned} \tag{5}$$

With the help of perturbation theory we obtain symbols of operators $\hat{\chi}_\nu$, $\hat{L}_\nu$[10]

$$\begin{aligned}L_\nu &= -p^2 + n^2 - \frac{(\pi\nu)^2}{D^2(x)} + ih\frac{(p, \nabla_x D(x))}{D(x)} + O(h^2)\,.\\ \chi_\nu &= \sin\Big(\frac{\pi\nu z}{D(x)}\Big) + ih\frac{(p,\nabla_x D(x))z^2}{2D(\mathbf{x})}\sin\Big(\frac{\pi\nu z}{D(x)}\Big) + O(h^2)\,.\end{aligned} \tag{6}$$

The equation in the three-dimensional layer was reduced to the family of problems in $\mathbf{R}^2_x$

$$\Big(h^2\Delta - \frac{(\pi\nu)^2}{D(x)^2} + n^2 + ih\frac{(p,\nabla_x D(x))}{D(x)} + O(h^2)\Big)\psi_\nu(x,h) == f\Big(\frac{x-\xi}{h}\Big)\Big(g\Big(\frac{z-z_0}{h}\Big), \sin\Big(\frac{\pi\nu z}{D(x)}\Big)\Big). \tag{7}$$

3. Representation of rapidly decaying functions in terms of the Maslov canonical operator and elliptic modes

It is reasonable to represent right-hand sides in Eq. (7) in the form of the Maslov canonical operator, after that we can use the developed theory [6] to construct an asymptotic solution. Such functions correspond the following Lagrangian manifold(see [12])

$$\Lambda = \Big(p = \alpha, x = \xi, \alpha \in \mathbf{R}^2\Big), \quad d\mu = dp_1 \wedge dp_2\,. \tag{8}$$

Thus function $F_\nu\Big(\frac{x-\xi}{h}\Big)$ can be represented in the form:

$$F_\nu\Big(\frac{x-\xi}{h}\Big) = \frac{h}{i}K^h_{(\Lambda,d\mu)}[A_\nu(\alpha)], \quad A_\nu(\alpha) = \frac{1}{2\pi}\int_{\mathbf{R}^2} F_\nu(y)e^{-i\alpha y}dy\,, \tag{9}$$

where $K^h_{(\Lambda,d\mu)}$ is the Maslov canonical operator [13].

If the principal symbol $H_\nu(x,p) = \mathrm{L}_\nu(x,p,0)$ of the operator $\hat{L}_\nu$ does not go to zero on the vertical Lagrangian manifold Λ, i.e. $p^2 + \frac{(\pi\nu)^2}{D^2(\xi)} - n^2 \neq 0$, the operator $\hat{L}_\nu$ is invertible on functions F_ν[14]

$$\psi_\nu = \frac{h}{i}\hat{L}_\nu^{-1}K^h_{(\Lambda,d\mu)}[A_\nu(\alpha)] = \frac{h}{i}\Big(K^h_{(\Lambda,d\mu)}\Big[\frac{A_\nu(\alpha)}{H_\nu(\alpha,\xi)}\Big] + O(h)\Big)\,. \tag{10}$$

Functions ψ_ν are localized in the neighborhood of the point $(x_1 = \xi_1,\ x_2 = \xi_2)$. If $H_\nu|_\Lambda \neq 0$, the solution of the three-dimensional problem (2) can be represented in the form:

$$u = \sum_{\nu=1}^{\infty}\Big[\frac{h}{i}\Big(K^h_{(\Lambda,d\mu)}\Big[\frac{A(\alpha)}{H_\nu(\alpha,\xi)}\Big]\sin\Big(\frac{\pi\nu z}{D(x)}\Big) + O(h)\Big)\Big] =$$
$$\sum_{\nu=1}^{\infty}\Big[\frac{1}{2\pi}\sin\Big(\frac{\pi\nu z}{D(\mathbf{x})}\Big)\int_{R^2}\frac{A_\nu(p)e^{\frac{i}{h}p(x-\xi)}}{H_\nu(p,\xi)}dp + O(h^2)\Big]\,. \tag{11}$$

It is the formal asymptotic solution, it is necessary to prove that the corresponding series does converge, and the cumulative correction is the value of the order $O(h^2)$.

4. Hyperbolic modes and their asymptotics

The formula (10) is not correct in some neighborhood of the submanifold $L_0 = \Big[(x,p) \in \Lambda : H_\nu(x,p) = 0\Big]$. The operator $\hat{L}_\nu$ becomes hyperbolic on L_0, and wave part of the solution appears. The vanishing of the principal symbol $H_\nu(x,p) = L_\nu(x,p,0)$ of the operator $\hat{L}_\nu$ on some submanifold L_0 of Λ is necessary for the appearance of wave part of the solutions. *Emphasize that the identity $p^2 + \frac{(\pi\nu)^2}{D^2(\xi)} - n^2 = 0$ performed for a finite number of modes or not fulfilled at all, thus wave part of the solution contains a finite number of modes.* Wave part of the solution is connected with a new Lagrangian manifold, thus solution is connected with a pair of Lagrangian manifolds.

Under the assumption that the mode with number ν is hyperbolic, we use the construction of the asymptotic solution from [6]. Let us introduce needed objects to present the corresponding proposition:

- t_0 is a sufficiently small number, and $\alpha^* \in L_0$ is the initial point on Λ.
- $\tilde{\Lambda}$ is a small neighborhood of the submanifold L_0 in Λ. $\Big[\Lambda_t = g_H^t(\tilde{\Lambda}), \quad t \le t_0\Big]$ is the Lagrangian manifold with the measure $d\mu_t = (g_H^{t*})^{-1} d\mu$.
- $\Lambda_+ = \cup_{t\ge 0} g_H^t(L_0)$ is the two-dimensional Lagrangian manifold in $\mathbf{R}^4$ with the boundary L_0. $d\mu_+ = dt \wedge (g_H^{t*})^{-1} d\sigma$ is the measure on Λ_+, where $d\sigma$ is the measure on L_0.
- Let $\rho(\alpha) \in C^\infty(\Lambda)$ and $\theta(t) \in C_0^\infty(\mathbf{R}^1)$ be cut-off functions such that $\rho(\alpha) = 1$ in the neighborhood of the manifold L_0, and $\mathrm{supp}(\rho(\alpha)) \in \Lambda$, $\theta(t) = 1$ if $t \le t_0$, $\theta(t) = 0$ if $t \ge t_0$.

We define the function $\tilde{A}$ as a solution of the Cauchy problem for the transport equation on trajectories of the Hamiltonian system outgoing from points $(x_0, p_0) \in \Lambda$

$$\frac{d\tilde{A}_\nu}{dt} + H_{sub}(X(x_0,p_0,t),(P(x_0,p_0,t))\tilde{A}_\nu = 0\,, \quad \tilde{A}_\nu|_{t=0} = A_\nu(x_0,p_0)\,, \tag{12}$$

where A_ν is the amplitude in the formula (9), and $H_{sub}(x,p) = i\frac{\partial L_\nu}{\partial h}(x,p,0) - \frac{1}{2}\sum_{j=1}^n \frac{\partial^2 H_\nu}{\partial x_j \partial p_j}(x,p)$ is the subprincipal symbol of the operator $\hat{L}_\nu$. $A_{+\nu}$ is restrictions of functions $\tilde{A}_\nu$ on the manifold Λ_+. In our problem $H_{sub}(x,p) = -\frac{(p,\nabla_x D(\mathbf{x}))}{D(\mathbf{x})}$. The solution of the transport equation can be obtained in the explicit form:

$$\begin{aligned} A_{+\nu}(x_0,p_0,t) &= A_\nu(x_0,p_0)\exp\left(\int_0^t \frac{(P,\nabla_X D(X))}{D(X)}dt\right) = A_\nu(x_0,p_0) \\ \exp\left(-\int_0^t \frac{(\dot{X},\nabla_X D(X))}{2D(X)}dt\right) &= A_\nu(x_0,p_0)\exp\left(-\int_{D(X(x_0,p_0,0))}^{D(X(x_0,p_0,t))} \frac{dD(X)}{2D(X)}\right) \\ &= A_\nu(x_0,p_0)\frac{\sqrt{D(X(x_0,p_0,0))}}{\sqrt{D(X(x_0,p_0,t))}} = A_\nu(x_0,p_0)\frac{\sqrt{D(\xi)}}{\sqrt{D(X(x_0,p_0,t))}} \end{aligned} \tag{13}$$

Now we can apply the Theorem.1 from [6].

Proposition Eq. (7) with right-hand side(9) has an asymptotic solution $\psi_\nu = \psi_\nu(x,h)$ which satisfy the asymptotic limit absorption principle and can be represented up to lower order terms

$$\begin{aligned} \psi_\nu = \frac{h}{i}\Bigg(&K^h_{(\Lambda,d\mu)}\left[\frac{(1-\rho(\alpha)A_\nu)}{H_\nu}\right] + \frac{2\pi}{h}e^{i\pi/2}K^h_{(\Lambda_+,d\mu_+)}\Big[(1-\Theta(t))A_{+\nu}\Big] + \\ &+ \frac{i}{h}\int_0^{t_0}\exp\left(\int_{\alpha^*}^{g_H^t(\alpha^*)} pdx\right)K^h_{(\Lambda_t,d\mu_t)}\Big[\Theta(t)\rho(\alpha)A_\nu(t)\Big]dt + O(h^{1/2})\Bigg)\,. \end{aligned} \tag{14}$$

The first summand corresponds elliptic modes, the second summand is the wave "hyperbolic" part of the solution, and the third summand is the solution in the small neighborhood of the Lagrangian manifold L_0.

Remark: Let the right-hand side(9) be finite and localized in the neighborhood of the point $x = \xi$ function. Then asymptotic solution of the Eq. (7) in the region $|x - \xi| > Ch^\beta$ for any constants $C > 0$, $\beta \in \left(\frac{1}{2}, 1\right)$ for a sufficiently big $L > 0$ can be represented in the form

$$\psi_\nu(x,h) = 2\pi\Big[K^h_{(\Lambda_+,d\mu_+)}(1-\theta(Lt/h^\beta))A_{+\nu}\Big](x,h) + O(h^{2\beta-1/2})\,. \tag{15}$$

The solution of two-dimensional problem can be simplified, if we apply the representation of the canonical operator developed in the work[15]. The Maupertuis-Jacobi principle allows us to pass from one Hamiltonian system to another, if equations $H(x,p) = E$ and $F(H(x,p),x,p) = E$ define the same surface in the phase space. We pass to the Hamiltonian

$$\tilde{H} = \frac{|p|}{\sqrt{-\frac{\pi^2\nu^2}{D^2(x)} + n^2}} - 1\,. \tag{16}$$

Main features of a such passage are following

- The Jacobian depends on differentiation with respect to only the one variable $\tilde{J} = \frac{1}{\sqrt{-\pi^2\nu^2/D^2(\tilde{X})+n^2}}|\tilde{X}_\psi|$, where ψ is the coordinate on the circle L_0.
- The phase is the "proper" time of the new Hamiltonian system $s(\tau,\psi) = \tau$ and *wave fronts are ends of trajectories of the Hamiltonian system.*

Consider now Eq. (7) with the right-hand side with a small parameter μ. If $\mu \geq h$, the right-hand can be represented in the form of the canonical operator with the small parameter h

$$F_\nu\Big(\frac{x-\xi}{\mu}\Big) = \frac{h}{i}K^h_{(\Lambda,d\mu)}[A_\nu(\alpha)], \quad A_\nu(\alpha) = \frac{1}{2\pi}\int_{\mathbf{R}^2} F_\nu\Big(y\frac{h}{\mu}\Big)e^{-i\alpha y}dy\,. \tag{17}$$

Previous formulae are applicable in this case. The case $\mu < h$ is not considered in this paper.

5. Example

Consider the problem (7) with the right-hand side as the Gaussian function $F = \exp\left(\frac{-(x_1+3)^2-x_2^2}{2h^2}\right)$ and with $D(x) = -1 - \exp(-x_1^2 - x_2^2)$. This problem describes the change in modal amplitudes of sound signal generated by the (time-) monochromatic source located near the point $(\xi_1,\xi_2) = (-3,0)$ in shallow water with bathymetry function $D(x)$. The Hamiltonian systems have been solved for different (wave) modes, and corresponding trajectories and wave fronts are presented at Fig. (1,2). Rays are projections of characteristics from the 4-D phase space to the configuration space (2-D plane). The ambiguity of the preimages of points x on the Lagrangian manifold leads to intersections of wave fronts. The solution is the sum of corresponding functions connected with all wave fronts. We use the representation of the Maslov canonical operator in terms of Airy and Pearcy functions[16, 17] near caustics and focal points.

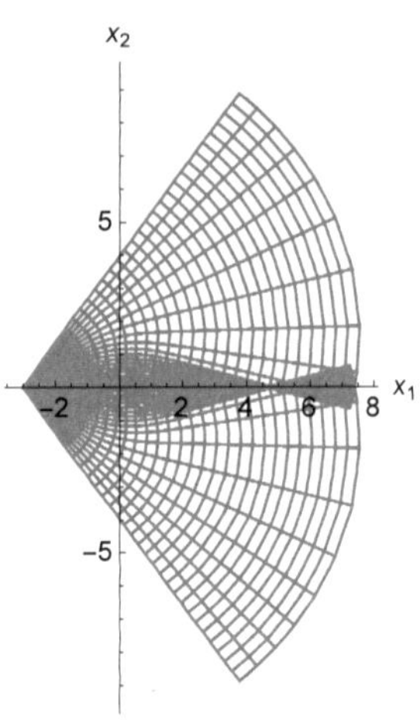

Figure 1. Rays and wave fronts for $\nu = 3$, $n = 13$, $h = 0.01$, $\psi \in [-\frac{\pi}{3}; \frac{\pi}{3}]$.

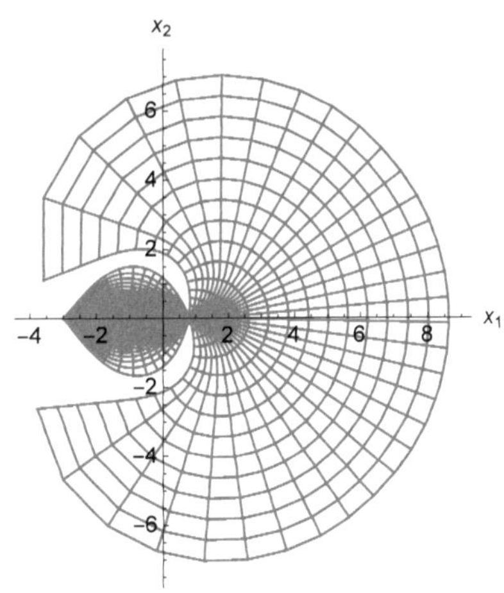

Figure 2. Rays and wave fronts for $\nu = 4$, $n = 13$, $h = 0.01$, $\psi \in [-\frac{\pi}{3}; \frac{\pi}{3}]$.

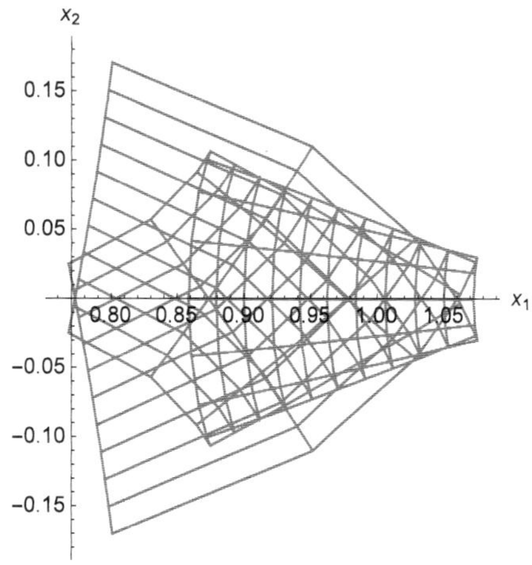

Figure 3. Rays and wave fronts for $\nu = 4$, $n = 13$, $h = 0.01$ in the neighborhood of the caustics.

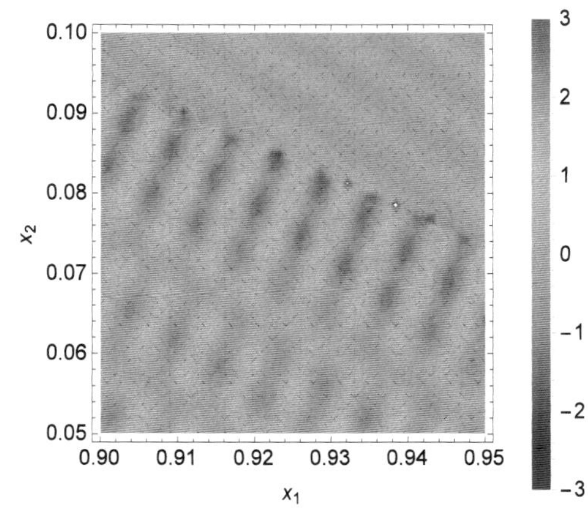

Figure 4. The solution of Eq. (7) for $\nu = 4$ in the neighborhood of the caustic.

Acknowledgments
The research was carried out within the state assignment (theme No. AAAA-A17-117021310377-1).

References
[1] Petrov P S and Sturm F 2016 *The Journal of the Acoustical Society of America* **139** 1343–1352
[2] Lin Y T, Collis J M and Duda T F 2012 *The Journal of the Acoustical Society of America* **132** EL364–EL370
[3] Jensen F B, Kuperman W A, Porter M B and Schmidt H 2011 *Computational ocean acoustics* (Springer Science & Business Media)
[4] Weinberg H and Burridge R 1974 *The Journal of the Acoustical Society of America* **55** 63–79
[5] Collins M D 1993 *The Journal of the Acoustical Society of America* **94** 2269–2278
[6] Anikin A Y, Dobrokhotov S Y, Nazaikinskii V E and Rouleux M 2017 *Doklady Mathematics* **96** 406–410
[7] Keller J B 1962 *JOSA* **52** 116–130
[8] Babich V and Buldyrev V 1972 *Asymptotic methods in problems of diffraction of short waves*
[9] Kucherenko V V 1968 *USSR Computational Mathematics and Mathematical Physics* **8** 294–302
[10] Belov V V, Dobrokhotov S Y and Tudorovskii T Y 2004 *Theoretical and mathematical physics* **141** 1562–1592
[11] Karasev M V, Maslov V P *et al.* 2012 *Nonlinear Poisson brackets: geometry and quantization* vol 119 (American Mathematical Soc.)
[12] Dobrokhotov S Y, Tirozzi B and Shafarevich A 2007 *Mathematical Notes* **82** 713–717
[13] Maslov V P and Fedoriuk M V 2001 *Semi-classical approximation in quantum mechanics* vol 7 (Springer Science & Business Media)
[14] Maslov V P 1976 *Operational methods* (Mir)
[15] Dobrokhotov S, Minenkov D and Rouleux M 2014 *arXiv preprint arXiv:1409.3170*
[16] Dobrokhotov S Y, Makrakis G, Nazaikinskii V E and Tudorovskii T Y 2013 *Theoretical and Mathematical Physics* **177** 1579–1605
[17] Reijnders K, Minenkov D, Katsnelson M and Dobrokhotov S Y 2018 *Annals of Physics* **397** 65–135

The method of nonlocal transformations: Applications to singularly perturbed boundary-value problems

Andrei D Polyanin[1,2], **Inna K Shingareva**[3]

[1] Ishlinsky Institute for Problems in Mechanics, RAS, 101 Vernadsky Avenue, bldg 1, 119526 Moscow, Russia

[2] National Research Nuclear University MEPhI, 31 Kashirskoe Shosse, 115409 Moscow, Russia

[3] University of Sonora, Blvd. Luis Encinas y Rosales S/N, Hermosillo C.P. 83000, Sonora, México

E-mail: polyanin@ipmnet.ru, inna@mat.uson.mx

Abstract. We present a method of numerical integration of singularly perturbed boundary-value problems with a small parameter, based on introducing a new nonlocal independent variable. As a result, we obtain more suitable problems that allow the application of standard fixed-step numerical methods. Two test boundary-value problems for second-order ODEs that have monotonic and non-monotonic exact or asymptotic solutions, expressed in elementary functions, are considered. Comparison of numerical, exact, and asymptotic solutions showed the high efficiency of the method based on generalized nonlocal transformations for solving singular boundary-value problems with a small parameter.

Keywords: singularly perturbed boundary-value problems, nonlocal transformations, boundary layers, nonlinear differential equations, numerical and exact solutions

1. Introduction

Singularly perturbed boundary-value problems with a small parameter, multiplying the highest derivative, are often encountered in applications (see, for example, [1–6]). An important qualitative feature of singularly perturbed boundary-value problems is that for the zero value of a small parameter the order of the differential equation under consideration decreases and some parts of the boundary conditions cannot be satisfied.

Solutions of singularly perturbed boundary-value problems with a small parameter have large gradients in the region of boundary layers, which leads to a loss of convergence of classical finite-difference schemes and makes them of little use or unsuitable for solving problems of this type. Various problems and special methods of the numerical integration for linear and nonlinear differential equations with a small parameter at the highest derivative are presented, for example, in [7–22]. For numerical solution of singularly perturbed boundary-value problems many authors use methods with a piecewise-uniform grid (two-grid methods) that is characterized by a small stepsize in the boundary layer and a large stepsize outside it (see, for example, [11, 17–20, 22]). It is important to note that in the methods based on the use of a piecewise-uniform grid, a priori information on the structure and rate of damping of asymptotic solutions in the boundary layer is explicitly or implicitly taken into account.

In this paper, for numerical integration of singularly perturbed boundary-value problems, we propose to apply (at the initial stage) nonlocal transformations that allow further integrate the reduced problem by standard numerical methods with uniform grid.

Remark 1. Nonlocal transformations were used in [23–27] for numerical integration of Cauchy problems with blow-up solutions having very large gradients in a neighborhood of a previously unknown point. Comparison of exact and numerical solutions of a number of test problems for differential equations of the first, second, third and fourth orders showed the high efficiency of this method for numerical integration of blow-up problems.

Remark 2. Nonlocal transformations of special type were used in [28–30] to obtain exact solutions, first integrals, and linearize some ordinary differential equations of the second order.

2. Qualitative features of boundary-layer type problems

2.1. An illustrating example of a linear boundary-value problem

We consider the boundary-value problem for the second-order linear differential equation with constant coefficients

$$\varepsilon y''_{xx} + y'_x + y = 0 \quad (0 < x < 1); \quad y(0) = a, \quad y(1) = b, \tag{1}$$

in which, if $\varepsilon \to 0$, a boundary layer is formed near the point $x = 0$.

The exact solution of problem (1) is determined by the formulas

$$y = \frac{ae^{\lambda_2} - b}{e^{\lambda_2} - e^{\lambda_1}} e^{\lambda_1 x} + \frac{b - ae^{\lambda_1}}{e^{\lambda_2} - e^{\lambda_1}} e^{\lambda_2 x}, \tag{2}$$

where $\lambda_1 = \frac{1}{2}\varepsilon^{-1}\big(-1 - \sqrt{1 - 4\varepsilon}\big)$ and $\lambda_2 = \frac{1}{2}\varepsilon^{-1}\big(-1 + \sqrt{1 - 4\varepsilon}\big)$. For small ε, we have $\lambda_1 \simeq -\varepsilon^{-1}$, $\lambda_2 \simeq -1$, and $y'_x(0) \simeq \varepsilon^{-1}(eb - a)$, and the corresponding asymptotic solution of the problem (1) is written as follows:

$$y_a = (a - eb)e^{-x/\varepsilon} + be^{1-x}. \tag{3}$$

For concreteness, we further assume that $a \geq 0$ and $b \geq 0$. If $a > eb$, the function (3) decreases monotonically. If $a < eb$, the function (3) increases monotonically (and very quickly) in a narrow region $0 \leq x < x_*$, where $x_* = \varepsilon \ln\Big[\frac{1}{\varepsilon}\Big(1 - \frac{a}{eb}\Big)\Big]$, $y_* = eb$, and in the remaining region, $x_* \leq x \leq 1$, the solution decreases monotonically and changes slowly enough.

In Fig. 1, the exact solutions (2) of the problem (1) are shown by the solid lines for two sets of numerical values of the defining parameters: a) $a = 1$, $b = 0$, $\varepsilon = 0.005$ and b) $a = 0$, $b = 1$, $\varepsilon = 0.005$. For the second set of parameters, the solution in the region $0 \leq x \leq 0.02649$ increases rapidly, and for $0.02649 \leq x \leq 1$ decreases slowly; in this case the maximum difference between the asymptotic solution (3) and the exact solution (2) on the whole interval, $0 \leq x \leq 1$, is 0.0127.

2.2. A relation between the derivatives

We note that the derivatives of the solution (3) on the left boundary are very large:

$$(y_a)'_x|_{x=0} \simeq -\varepsilon^{-1}(a - eb), \quad (y_a)''_{xx}|_{x=0} \simeq \varepsilon^{-2}(a - eb). \tag{4}$$

For example, for $a = 0$, $b = 1$, $\varepsilon = 0.005$, we have $(y_a)'_x|_{x=0} \simeq 543.656$. Therefore, using direct numerical methods for solving similar problems with boundary layers, the shooting procedure should begin with large values of the derivative (of order ε^{-1}), which is a complicating factor.

IOP Conf. Series: Journal of Physics: Conf. Series **1205** (2019) 012047 doi:10.1088/1742-6596/1205/1/012047

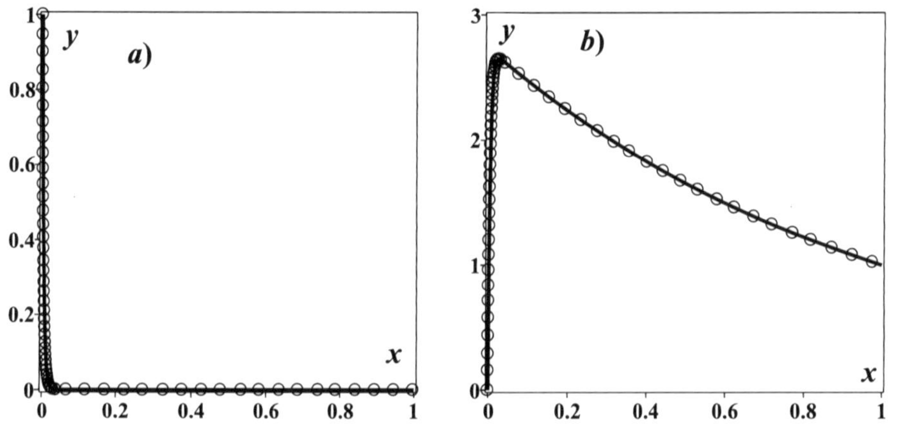

Figure 1. Exact solutions (2) of the problem (1) (solid lines) and numerical solutions of the transformed problem (9) (points) with $g = (1 + |z| + |f|)^{1/2}$ for two sets of numerical values of the determining parameters: a) $a = 1$, $b = 0$, $\varepsilon = 0.005$ and b) $a = 0$, $b = 1$, $\varepsilon = 0.005$.

Let $|a - eb| = O(1)$. Eliminating ε from (4), we obtain the order relation

$$|y''_{xx}| = O(|y'_x|^2). \qquad (5)$$

The relation between the derivatives (5) is sufficiently general and is valid if, inside the boundary layer, the leading term of the asymptotic expansion of the solution as $\varepsilon \to 0$ has the form $y_{\rm i} = \varphi(x/\Delta)$, where $\varphi = \varphi(z)$ is a smooth function having bounded and non-vanishing derivatives in some neighborhood of the point $z = 0$, and $\Delta = \Delta(\varepsilon)$ is a function having the property $\Delta \to 0$ as $\varepsilon \to 0$.

3. Numerical solution of boundary-value problems based on nonlocal transformations

3.1. General description of the method

We consider two-point problems for second-order differential equations with boundary conditions of the first kind, which in dimensionless variables have the form

$$y''_{xx} = f(x, y, y'_x) \quad (0 < x < 1); \qquad y(0) = a, \qquad y(1) = b, \qquad (6)$$

where the function f can also depend on the small parameter $\varepsilon > 0$.

We introduce a new nonlocal independent variable ξ by means of the first-order differential equation and the initial condition [23–27]:

$$\xi'_x = g(x, y, y'_x, \xi), \qquad \xi(0) = 0. \qquad (7)$$

Here $g = g(x, y, y'_x, \xi)$ is a regularizing function that can vary.

We represent the second-order equation (6) in the form of an equivalent system of two equations of the first order

$$y'_x = z, \qquad z'_x = f(x, y, z). \qquad (8)$$

Using (7), we pass from x to the new independent variable ξ in (8) and (6). As a result, the boundary-value problem (6) is transformed to the following problem for the system of three equations:

$$x'_\xi = \frac{1}{g(x, y, z, \xi)}, \quad y'_\xi = \frac{z}{g(x, y, z, \xi)}, \quad z'_\xi = \frac{f(x, y, z)}{g(x, y, z, \xi)} \quad (0 < \xi < \xi_1); \qquad (9)$$
$$x(0) = 0, \qquad y(0) = a, \qquad y(\xi_1) = b,$$

where the value ξ_1 is determined in the process of calculations according to the condition $x(\xi_1) = 1$.

If $g \equiv 1$, the first equation of the system (9), taking into account the initial condition $x(0) = 0$, gives $\xi = x$ and numerical integration of the remaining two equations is equivalent to the integration of the original problem (6). Successfully selected the regularizing function $g = g(x, y, z, \xi)$ itself will determine the position and the density of points of integration with respect to the original variables x and y and allow more accurately (for a given number of grid points) to solve the problem (9) applying the shooting method and using standard fixed-step numerical methods with respect to ξ [31–34].

3.2. Conditions to be satisfied by regularizing functions. Examples of regularizing functions
For numerical solution of boundary-value problems, as well as for solving Cauchy problems, it is reasonable to use regularizing functions of the form [23–27]:

$$g = G(|z|, |f|) \equiv G(|y'_x|, |y''_{xx}|), \tag{10}$$

where $f = f(x, y, z)$ is the right-hand side of the equation (6) and $z = y'_x$. We impose the following conditions on the function $G = G(u, v)$:

$$G > 0; \quad G_u \geq 0, \quad G_v \geq 0; \quad G \to \infty \text{ as } u + v \to \infty; \quad G(0, 0) = 1, \tag{11}$$

where $u \geq 0$, $v \geq 0$. The last relation in (11) is the normalization condition and is not mandatory.

For singularly perturbed boundary-value problems (6) with a small parameter at the highest derivative, for which the right-hand side of the equation (6) has the form $f = \varepsilon^{-1}F(x, y, z)$, when choosing regularizing functions, in addition to conditions (10)–(11), some other considerations should be taken into account. For $g = 1$ (in this particular case, nonlocal transformations are not applied) and $\varepsilon \to 0$, the right-hand sides of the last two equations of the system (9) will tend to infinity (since $|z| \to \infty$ and $|f| \to \infty$) in the boundary-layer region; in addition, the order relation $|f| = O(z^2)$ is valid, which follows from the formula (5). This circumstance considerably complicates numerical integration of the problem under consideration and leads to the need to proportionally refine the grid spacing as ε decreases.

It is possible to avoid refining the grid as $\varepsilon \to 0$ and to work with a fixed stepsize with respect to the nonlocal variable ξ by using regularizing functions satisfying the condition

$$|z|/g = O(1) \quad \text{as} \quad \varepsilon \to 0 \tag{12}$$

(in this case, the right-hand side of the second equation of the system (9) will not have singularities for small ε, and the third equation of this system in the boundary-layer region will have a substantially smaller singularity than for $g = 1$). In particular, we can choose regularizing functions having the asymptotics $g = O(|z|)$ as $|z| \to \infty$ or $g \to O(|f|^{1/2})$ as $|f| \to \infty$.

It is advisable to use, for example, regularizing functions of the form

$$g = (k_1 + k_2|z| + k_3|f|)^{1/2} \quad \text{or} \quad g = (k_1 + k_2 z^2 + k_3|f|)^{1/2}, \tag{13}$$

which satisfy the condition (12). The formulas (13) include the three constants $k_1 \geq 0$, $k_2 \geq 0$, and $k_3 \geq 0$ ($k_2 + k_3 \neq 0$), which can vary.

The use of the regularizing functions (13) allows us to suppress the unbounded growth of the right-hand side of the second equation of the system (9) as $\varepsilon \to 0$ and to reduce (in comparison with $g = 1$) the right-hand side of the third equation.

The maximum absolute error of the numerical solutions of problem (1) for $a = 1$, $b = 0$				
No.	Regularizing function	Stepsize 0.1	Stepsize 0.05	Stepsize 0.01
1	$g{=}1 + \|z\|$	0.017119347	0.006702741	0.000137030
2	$g{=}(1 + \|f\|)^{1/2}$	0.000707586	0.000160259	0.000001602
3	$g = (1 + \|z\| + \|f\|)^{1/2}$	0.000611528	0.000146118	0.000001741
4	$g = (1 + z^2 + \|f\|)^{1/2}$	0.000900004	0.000204128	0.000001775
5	$g{=}1$	process diverges	process diverges	0.193331173
The maximum absolute error of the numerical solutions of problem (1) for $a = 0$, $b = 1$				
No.	Regularizing function	Stepsize 0.1	Stepsize 0.05	Stepsize 0.01
1	$g{=}1 + \|z\|$	0.047029578	0.013710597	0.000713696
2	$g{=}(1 + \|f\|)^{1/2}$	0.000824707	0.000249922	0.000001663
3	$g = (1 + \|z\| + \|f\|)^{1/2}$	0.000570299	0.000115649	0.000000554
4	$g = (1 + z^2 + \|f\|)^{1/2}$	0.000559160	0.000109360	0.000000180
5	$g{=}1$	process diverges	process diverges	0.528189578

Table 1. Comparison of the efficiency of various regularizing functions for the transformed problem (9) used for numerical solution of the original problem (1) by nonlocal transformations with $\varepsilon = 0.005$.

4. Singularly perturbed boundary-value problems. Comparison of numerical and exact solutions

4.1. Linear problems. Numerical solutions obtained using various regularization functions

In Fig. 1, the results of numerical solutions of the transformed problem (9), used for solving the linear problem (1), for $f = -\varepsilon^{-1}(z + y)$ with the regularizing function $g = (1 + |z| + |f|)^{1/2}$, which are obtained by the shooting method (from the point $x = 0$) with the fixed stepsize $h = 0.01$ by using Maple [34], are shown by points for two sets of numerical values of the defining parameters: $a = 1$, $b = 0$, $\varepsilon = 0.005$ (monotonic solution) and $a = 0$, $b = 1$, $\varepsilon = 0.005$ (non-monotonic solution). It can be seen that there is a good coincidence between the numerical solutions and the corresponding exact solutions, which are determined by the formula (2) and are represented by solid lines.

Table 1 shows the maximum absolute errors of numerical solutions of the transformed problem (9) used for numerical integration of the original problem (1) with $\varepsilon = 0.005$ by the method based on nonlocal transformations for $a = 1$, $b = 0$ and $a = 0$, $b = 1$ for three stepsizes h and five different regularizing functions g. For comparison, similar data are also indicated for the case $g = 1$, which corresponds to the direct numerical solution (without using transformations) with the same stepsize with respect to x. It can be seen that the three regularizing functions (Nos. 2–4) make it possible to obtain numerical solutions in the entire region with high accuracy even with a sufficiently large stepsize $h = 0.1$ (with respect to ξ).

4.2. A nonlinear boundary-value problem. Exact, asymptotic, and numerical solutions

We consider the boundary-value problem with quadratic nonlinearity

$$\varepsilon y''_{xx} + (x + y)y'_x + x + y = 0 \quad (0 < x < 1); \qquad y(0) = a, \quad y(1) = b. \tag{14}$$

The general solution of this equation has the form

$$y = c\,\frac{1 - Ae^{-cx/\varepsilon}}{1 + Ae^{-cx/\varepsilon}} - x. \tag{15}$$

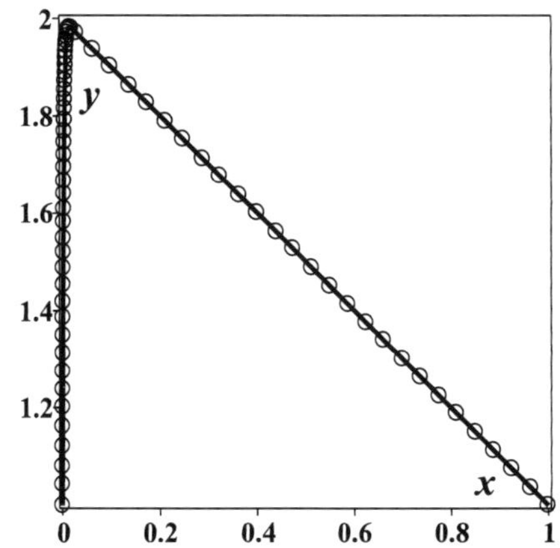

Figure 2. Exact solution (15) of the original problem (14) (solid line) and numerical solution of the transformed problem (9) (circles) obtained by nonlocal transformations with the regularizing function $g = (1 + z^2 + |f|)^{1/2}$ for $a = b = 1$, $\varepsilon = 0.005$.

	The maximum absolute error of the numerical solutions of problem (14)			
No.	Regularizing function	Stepsize 0.1	Stepsize 0.05	Stepsize 0.01
1	$g=1+\|z\|$	0.137389203	0.053399823	0.000857913
2	$g=(1+\|f\|)^{1/2}$	0.000937303	0.000228167	0.000005030
3	$g=(1+\|z\|+\|f\|)^{1/2}$	0.000786873	0.000196698	0.000003249
4	$g=(1+z^2+\|f\|)^{1/2}$	0.000607467	0.000154641	0.000003261
5	$g=1$	process diverges	process diverges	process diverges

Table 2. Comparison of the efficiency of various regularizing functions for the transformed problem (9) used for numerical solution of the original problem (14) by nonlocal transformations with $a = b = 1$, $\varepsilon = 0.005$ and three stepsizes h.

The constants of integration A and c are determined from the transcendental system of equations

$$c\,\frac{1-A}{1+A} = a, \qquad c\,\frac{1-Ae^{-c/\varepsilon}}{1+Ae^{-c/\varepsilon}} - 1 = b, \tag{16}$$

obtained by substituting the expression (15) into the boundary conditions (14). As $\varepsilon \to 0$ and $b > -1$, the asymptotic solution of the system (16) is given by the formulas

$$A = \frac{b-a+1}{b+a+1}, \qquad c = b+1. \tag{17}$$

We note that the asymptotic solution (17) exactly satisfies the first the equation of the system (16), and the discrepancy of the second equation of this system has the order $e^{-(b+1)/\varepsilon}$ as $\varepsilon \to 0$.

In Fig. 2, the exact solution of the problem (14) is shown by solid line for $a = b = 1$, $\varepsilon = 0.005$, which is determined by the formula (15) for $A = 3$, $c = -2$ (in this case the difference between the asymptotic and exact solutions is far beyond the limits of accuracy of our calculations). The circles represent the results of numerical solution of the corresponding transformed problem (9) with the regularizing function $g = (1 + z^2 + |f|)^{1/2}$, which are obtained by the shooting method (from the point $x = 0$) with the stepsize $h = 0.01$ by using Maple. The maximum modulus of the difference between the exact and numerical solution is equal to 0.000003261.

Table 2 shows the maximum absolute errors of numerical solutions of the transformed problem (9) used for numerical integration of the original problem (14) for $a = b = 1$, $\varepsilon = 0.005$ with three stepsizes h and four different regularizing functions g. It can be seen that the functions Nos. 2–4 allow one to obtain numerical solutions in the entire region with high accuracy even with a sufficiently large stepsize (with respect to ξ) equal to $h = 0.1$.

5. Brief conclusions

We offer a new method of numerical integration of singularly perturbed boundary-value problems for second-order ODEs of the form $\varepsilon y''_{xx} = F(x, y, y'_x)$ based on introducing a nonlocal independent variable ξ, which is related to original variables x and y by the auxiliary differential equation $\xi'_x = g(x, y, y'_x, \xi)$. With a suitable choice of the regularizing function g, the proposed method leads to more convenient problems that allow the application of standard numerical methods with the fixed stepsize with respect to ξ. Comparison of numerical and exact solutions of several singularly perturbed boundary-value problems showed the high efficiency of this method.

Acknowledgments

The work was supported by the Federal Agency for Scientific Organizations (State Registration Number AAAA-A17-117021310385-6).

References

[1] Lagerstrom P A 1988 *Matched Asymptotic Expansions. Ideas and Techniques* (New York: Springer)

[2] Il'in A M 1992 *Matching of Asymptotic Expansions of Solutions of Boundary Value Problems* (Providence: American Mathematical Society)

[3] Nayfeh A H 2000 *Perturbation Methods* (New York: Wiley)

[4] Polyanin A D, Kutepov A M, Vyazmin A V, Kazenin D A 2002 *Hydrodynamics, Mass and Heat Transfer in Chemical Engineering* (London: Taylor & Francis)

[5] Polyanin A D, Zaitsev V F 2003 *Handbook of Exact Solutions for Ordinary Differential Equations, 2nd Edition* (Boca Raton–London: Chapman & Hall/CRC Press)

[6] Verhulst V 2005 *Methods and Applications of Singular Perturbations, Boundary Layers and Multiple Timescale Dynamics* (New York: Springer)

[7] Bakhvalov N S 1969 On the optimization methods for solving boundary value problems with boundary layers *Zh. Vychisl. Math. Fiz.* **24** 841–859 (in Russian).

[8] Il'in A M 1969 A difference scheme for a differential equation with a small parameter affecting the highest derivative *Mat. Zametki* **6** 237–248 (in Russian).

[9] Jain M K, Iyengar S R, Subramanyam G S 1984 *Comp. Methods in Appl. Mech. Eng.* **42** (3) 273–286.

[10] Beckett G, Mackenzie J A 2000 *Appl. Numer. Math.* **35** (2) 87–109.

[11] Farrell P, Hegarty A, Miller J M, O'Riordan E, Shishkin G I 2000 *Robust Computational Techniques for Boundary Layers* (Boca Raton – London: Chapman & Hall/CRC Press)

[12] Qiu Y, Sloan D M, Tang T 2000 *J. Comput. Appl. Math.* **116** (1) 121–143.

[13] Frohner A, Roos H-G 2001 *Appl. Numerical Math.* **37** 79–94.

[14] Miranker W L 2001 *Numerical Methods for Stiff Equations and Singular Perturbation Problems* (Dordrecht: Reidel Publ)

[15] Vigo-Aguiar J, Natesan S 2006 An efficient numerical method for singular perturbation problems *J. Comput. Appl. Math.* **192** (1) 132–141.

[16] Rao S C, Kumar M 2008 *Appl. Numerical Math.* **58** 1572–1581.

[17] Shishkin G I, Shishkina L P 2009 *Difference Methods for Singular Perturbation Problems* (Boca Raton: Chapman & Hall/CRC Press)

[18] Kopteva N, O'Riordan E 2010 *Int. J. Numer. Analysis and Modeling* **7** (3) 393–415.

[19] Vulkov L G, Zadorin A I 2010 *Int. J. Numer. Analysis and Modeling* **7** (3) 580–592.

[20] Das P 2015 *J. Comput. Appl. Math.* **290** 16–25.

[21] Brdar M, Zarin H 2016 *J. Comput. Appl. Math.* **292** 307–319.

[22] Ahmadinia M, Safari Z 2018 *J. Comput. Appl. Math.* **331** 156–165.

[23] Polyanin A D, Shingareva I K 2017 *Int. J. Non-Linear Mechanics* **94** 178–184.

[24] Polyanin A D, Shingareva I K 2017 *Journal of Physics: IOP Conf. Series* **937** 012042.

[25] Polyanin A D, Shingareva I K 2018 *Appl. Math. Letters* **76** 123–129.

[26] Polyanin A D, Shingareva I K 2018 *Int. J. Non-Linear Mechanics* **99** 258–272.

[27] Polyanin A D, Shingareva I K 2018 *Appl. Math. Comput.* **336** 107–137.

[28] Kudryashov N A, Sinelshchikov D I 2016 *Appl. Math. Letters* **57** 114–120.

[29] Muriel C, Romero J L 2010 *J. Physics A, Math. Theor.* **43** (43) 434025.

[30] Meleshko S V et al. 2013 *J. Eng. Math.* **82** 17–30.

[31] Keller H B 1974 *Numerical Solution of Two Point Boundary Value Problems* (Philadelphia: SIAM)

[32] Butcher J C 1987 *The Numerical Analysis of Ordinary Differential Equations, Runge–Kutta and General Linear Methods* (New York: Wiley)

[33] Ascher U M, Petzold L R 1998 *Computer Methods for Ordinary Differential Equations and Differential-Algebraic Equations* (Philadelphia: SIAM)

[34] Shingareva I K, Lizárraga-Celaya C 2009 *Maple and Mathematica. A Problem Solving Approach for Mathematics, 2nd Edition* (Wien–New York: Springer)

Influence of compressibility on the foam fracture modeling

A V Porubov[1,2] **and I D Antonov**[1,2]

[1] Peter the Great St.Petersburg Polytechnic University (SPbPU), Polytechnicheskaya st., 29, Saint Petersburg 195251, Russia
[2] Institute of Problems in Mechanical Engineering, Bolshoy 61, V.O., Saint Petersburg 199178, Russia

E-mail: alexey.porubov@gmail.com

Abstract. The two-dimensional nonlinear model equations for hydraulic fracturing with foams are obtained. Both multi-phased compressibility and non-Newtonian rheological viscosity of the foam are taken into consideration. A local relationship between the effective pressure on the lateral surface of the fracture is obtained using a dynamic statement of the problem.

1. Introduction

Using foam for the hydraulic fracturing minimizes the volume of liquid phase injected to the formation, is very effective for proppant transport in the fracture due to its high viscosity and has other advantages. The modeling of the fracture propagation during hydraulic fracturing with foam is significantly different from the classical one. It is important that the foam contains gas part, so the compressibility of the fracturing fluid should be taken into account. Also it is necessary to account for non-Newtonian character of the foam visscosity in order to obtain the proper model of hydraulic fracturing with foam.

The modeling of the process of fracture propagation is a complicated task, and various simplifications are employed. In particular, it concerns compressibility of the fracturing fluid and its viscosity [1–5]. Also some factors affecting dynamica behavior are phenomenologically employed from the static modeling, e.g., the relationship between the fluid pressure in the fracture and the opening of the fracture [6, 7]. In this work we devlop the model equations for the hydraulic fracturing by the foam basing on the dynamic statement of the problem and taking into account both compressibility and viscosity of the foam.

2. Statement of the problem

Consider a vertical infinite in the vertical direction y fracture which is developing in the horizontal direction x. The varying width of the fracture is $w(x,t)$. The foam inside the fracture is considered as a compressible and non-Newtonian fluid. The density, pressure, and velocities along x and y axes are $\rho(x,t)$, $p(x,t)$, $u(x,y,t)$, $v(x,y,t)$ respectively.

The Navier-Stokes equations are used to describe the foam motion. The equations consist of the continuity equation,

$$\rho_t + (\rho\, u)_x + (\rho\, v)_y \;=\; 0, \tag{1}$$

and the equations of motion,

$$\rho\,(u_t + uu_x + vu_y) \;=\; -\,p_x + 2(\nu u_x)_x + (\nu v_x)_y + (\nu u_y)_y, \tag{2}$$

$$\rho\,(v_t + uv_x + vv_y) \;=\; -\,p_y + 2(\nu v_y)_y + (\nu u_y)_x + (\nu v_x)_x, \tag{3}$$

where ν is viscosity of the foam.

2.1. Compressibility

The compressibility of the foam is caused by the presence of two phases in it, gas and liquid. To characterise the multiphase nature of the foam, the quality of the foam Γ is introduced [8, 9],

$$\Gamma = \frac{V_g}{V_g + V_l}, \tag{4}$$

where V_g, V_l are the volumes of the gas and liquid phases of the foam respectively. The isothermic Boyle's law is,

$$p_0 V_{0g} \;= p\; V_g, \tag{5}$$

where p_0 is the injection pressure, allows us to get from Eq. (4)

$$\frac{p}{p_0} \;=\; \frac{\Gamma_0(1-\Gamma)V_{0l}}{(1-\Gamma_0)\Gamma V_l} \tag{6}$$

where Γ_0 is the foam injection quality. The density of the foam is

$$\rho \;=\; \frac{V_l\rho_l + V_g\rho_g}{V + g + V_l} \;=\; (1-\Gamma)\rho_l + \Gamma\rho_g, \tag{7}$$

where ρ_l and ρ_g is density of liquid phase and gas phase respectively. Equation (6) allows us to express the quality through the pressure p. Substitution of the resulting expression into Eq. (7) results in

$$\rho = \frac{(1-\Gamma_0)p\rho_l}{\Gamma_0 p_0 + (1-\Gamma_0)p} + \frac{\Gamma_0 p_0 \rho_g}{\Gamma_0 p_0 + (1-\Gamma_0)p}. \tag{8}$$

The last relationship is substituted further in the Navier-Stokes equations. It should be noticed that the second term in equation (7) can be neglected due to $\rho_g/\rho_l \ll 1$ and ρ_l is assumed to be constant due to high incompressibility of liquid phase.

2.2. Rheology

The experimental data reveal non-constant viscosity of the foam. It depends on the shear rate tensor $\dot\gamma = (1/2)(\nabla\mathbf{v} + \nabla\mathbf{v}^T)$. In our natations it is

$$\dot\gamma \;=\; \begin{pmatrix} u_x & (1/2)(u_y + v_x) \\ (1/2)(u_y + v_x) & v_y \end{pmatrix}$$

However, the scalar viscosity coefficient cannot depend directly on the tensor. Instead, it can be the function of the scalar characteristic of the shear rate tensor, or on its the invariants,

$$I \;=\; \mathrm{Tr}\;\dot\gamma \;=\; u_x + v_y, \tag{9}$$

$$II \;=\; \mathrm{Tr}\;\dot\gamma^2 \;=\; u_x^2 + v_y^2 + u_y v_x + \frac{1}{2}u_y^2 + \frac{1}{2}v_x^2, \tag{10}$$

$$III = \mathrm{Tr}\,\dot{\gamma}^3 = \frac{3}{2}u_y(u_x v_x + v_x v_y) + v_y^3 + u_x^3 + \frac{3}{4}u_y^2(u_x + v_y + v_x^2 v_y + u_x v_x^2) \quad (11)$$

For an incompressible fluid $I = 0$. Moreover, for shear flows III is also zero [10], and the dynamic viscosity coefficient can be the function of only on the second invariant. We assume such dependence for the compressible foams.

The conventional power law relationship is

$$\nu = K|\sqrt{II/2}|^{n-1}, \quad (12)$$

where K is a constant coefficient, n depends on the quality of foam and the temperature, usually its values lies between 0.5 and 1 [11]. In order to evaluate the dynamic coefficient of the foam, the known relationships are used for $K(\Gamma)$ and $n(\Gamma)$ [12].

3. Boundary conditions

The boundary conditions consist of the condition of impermeability of a solid surface $y = w(x,t)/2$,

$$w_t + u\, w_x = 2v, \quad (13)$$

the no- slip boundary condition at $y = w(x,t)/2$,

$$4u + 2v\, w_x = w_x w_t, \quad (14)$$

Also the conditions for the normal and tangential stresses hold,

$$p - \frac{2\nu}{4 + w_x^2}\left(4v_y + u_x w_x^2 - 2w_x(u_y + v_x)\right) = p^*, \quad (15)$$

$$(4 - w_x^2)(u_y + v_x) + 4w_x(v_y - u_x) = f^*, \quad (16)$$

where p^*, f^* are the boundary stresses in the solid surrounded fracture at $y = w(x,t)/2$. Moreover, one has to assume $v = 0$ at $y = 0$ from the point of view of the symmetry of the problem.

The remaining boundary conditions are zero fracture opening at $x = L$,

$$w = 0,$$

and the condition for the input pressure at $x = 0$,

$$p = p_0.$$

The solid medium surrounding the fracture mainly affects it via the normal stresses p^*. Usually the relationship for it is obtained in the framework of the static problem. Recently we developed a model of nonlinear strain waves propagation in an elastic rod surrounded by an another elastic medium [13]. In this paper the method of finding the dynamic response of the medium has been described in the framework of a dynamic problem. Now we extend it to the problem of an influence of the surrounded medium on the fracture surface.

Consider a problem of a linearly elastic isotropic medium (which models the soil surrounding the fracture) whose elastic constants are λ and μ and the constant density is ρ. We approximate the shape of the fracture as a round one, thus the cylindrical geometry can be used where axis x is directed along the fracture while the medium occupies the area $0 < r < \infty$. No variations along the polar angle are assumed. The displacements along x and r are $u(x,r,t)$ and $v(x,r,t)$ respectively. The equations of motion for the medium are

$$\rho u_{tt} - (\lambda + 2\mu)\, u_{xx} - (\lambda + \mu)\left(v_{rx} + \frac{v_x}{r}\right) - \lambda\left(u_{rr} + \frac{u_r}{r} + v_{rx} + \frac{v_x}{r}\right) = 0 \quad (17)$$

$$\rho v_{tt} - (\lambda + 2\mu)\left(v_{rr} + \frac{v_r}{r} - \frac{v}{r^2}\right) - \mu\, w_{xx} - (\lambda + \mu)\, u_{rx} \;=\; 0 \tag{18}$$

The conditions corresponding to the fracture are imposed at $r = R$.

$$v \;=\; w(x,t), \qquad \text{at } r \;=\; R, \tag{19}$$

the fracture is modeled by a continuity condition for the normal stresses and zero tangential stresses,

$$\sigma_{rr} = p^*, \qquad \text{at } r \;=\; R, \tag{20}$$

$$\sigma_{rx} \;=\; 0, \qquad \text{at } r \;=\; R, \tag{21}$$

$$u \;\to\; 0,\; w \;\to\; 0 \qquad \text{at } r \;\to\; \infty. \tag{22}$$

where $\sigma_{rr} = (\lambda + 2\mu)\, v_r + \lambda\, \frac{v}{r} + \lambda\, u_x$, $\sigma_{rx} = \mu\,(u_r + w_x)$.

Assume that all the variables depend upon the phase variable $\theta = x - c\, t$, and the radius r where c is the phase velocity of the wave of fracture. The unknown functions u, v are sought as

$$u \;=\; \Phi_\theta + \Psi_r + \frac{\Psi}{r}, \qquad v \;=\; \Phi_r - \Psi_\theta, \tag{23}$$

then Φ and Ψ satisfy the equations:

$$\Phi_{rr} + \frac{1}{r}\Phi_r + (1 - \frac{c^2}{c_l^2})\Phi_{\theta\theta} \;=\; 0, \tag{24}$$

$$\Psi_{rr} + \frac{1}{r}\Psi_r - \frac{1}{r^2}\Psi + (1 - \frac{c^2}{c_\tau^2})\;\Psi_{\theta\theta} = \;0, \tag{25}$$

where c_l and c_τ are the velocities of the bulk longitudinal and shear linear waves in the medium, respectively. They depend on the density and the Lamé coefficients, $c_l^2 \;=\; (\lambda + 2\mu)/\rho$, and $c_\tau^2 = \mu/\rho$.

To solve equations (24), (25) we introduce the Fourier transforms of Φ and Ψ:

$$\tilde{\Phi} = \int_{-\infty}^{\infty} \Phi \exp(-k\,\theta)\, d\theta, \;\; \tilde{\Psi} = \int_{-\infty}^{\infty} \Psi \exp(-k\,\theta)\, d\theta$$

that reduces Eqs. (24), (25) to the Bessel equations :

$$\tilde{\Phi}_{rr} + \frac{1}{r}\,\tilde{\Phi}_r - k^2\alpha\,\tilde{\Phi} = 0, \tag{26}$$

$$\tilde{\Psi}_{rr} + \frac{1}{r}\,\tilde{\Psi}_r - \frac{1}{r^2}\,\tilde{\Psi} - k^2\beta\,\tilde{\Psi} = 0, \tag{27}$$

with $\alpha = 1 - c^2/c_l^2$, and $\beta = 1 - c^2/c_\tau^2$. The ratios between c, c_l and c_τ define the signs of α and β, hence three possible sets of solutions to the equations (26), (27) appear, vanishing at infinity due to b.c. (22). Using the boundary conditions (19), (21), we obtain the following relationships for the Fourier images of normal stresses at the lateral surface $r = R$:

I) when $0 \;<\; c \;<\; c_\tau$,

$$\tilde{\sigma}_{rr} = \frac{\mu\,\tilde{w}}{1-\beta}\left(\frac{2(\beta-1)}{R} + \frac{k(1+\beta)^2 K_0(\sqrt{\alpha}kR)}{\sqrt{\alpha}\,K_1(\sqrt{\alpha}kR)} - \frac{4k\sqrt{\beta}K_0(\sqrt{\beta}kR)}{K_1(\sqrt{\beta}kR)}\right), \tag{28}$$

II) when $c_\tau < c < c_l$,

$$\tilde{\sigma}_{rr} = \frac{\mu\,\tilde{w}}{1-\beta}\left(\frac{2(\beta-1)}{R} + \frac{k(1+\beta)^2\,K_0(\sqrt{\alpha}kR)}{\sqrt{\alpha}\,K_1(\sqrt{\alpha}kR)} - \frac{4k\sqrt{\beta}J_0(\sqrt{-\beta}kR)}{J_1(\sqrt{-\beta}kR)}\right), \tag{29}$$

III) when $c > c_l$,

$$\tilde{\sigma}_{rr} = \frac{\mu\,\tilde{w}}{1-\beta}\left(\frac{2(\beta-1)}{R} + \frac{k(1+\beta)^2\,J_0(\sqrt{-\alpha}kR)}{\sqrt{-\alpha}\,J_1(\sqrt{-\alpha}kR)} - \frac{4k\sqrt{\beta}J_0(\sqrt{-\beta}kR)}{J_1(\sqrt{-\beta}kR)}\right), \tag{30}$$

where J_i and K_i $(i = 0, 1)$ denote the corresponding Bessel functions.

The size of the fracture is small compared with the size of the medium, then all the Bessel functions can be expanded around $R = 0$. The inverse Fourier transform results in the approximate condition for the pressure of the local form

$$p^* = k_1\,w(x,t) \tag{31}$$

where k_1 is defined for $0 < c < c_\tau$:

$$k_1 = -2\mu, \tag{32}$$

while for $c_\tau < c < c_l$:

$$k_1 = \frac{2\mu(4c_\tau^2 - c^2)}{c^2}, \tag{33}$$

and for $c > c_l$:

$$k_1 = \frac{2\mu[c^2(c_\tau^2 - c_l^2) + 3c_l^2c_\tau^2 - 4c_\tau^4]}{c_\tau^2(c_l^2 - c^2)} \tag{34}$$

One can see that the local dependence coefficient depends on the velocity of the strain wave propagating in the surrounding medium due to the development of the fracture.

4. Conclusion and future work

The model for the hydraulic fracturing by the foam is developed. It is suggested to describe the compressibility of the foam by the density-pressure relation (8), while its apparent viscosity is evaluated using equation (12). A local relation between the foam pressure in the fracture and the opening of the fracture (31) is obtained basing on the dynamic statement of the problem. An importance of considering the dynamic aspect of the hydraulic fracturing with foam should be verified in the future work using numerical simulations.

Acknowledgements

This work was supported by Ministry of Education and Science of the Russian Federation within the framework of the Federal Program Research and development in priority areas for the development of the scientific and technological complex of Russia for 2014–2020 (activity 1.2), grant No. 14.575.21.0146 of September 26, 2017, unique identifier: RFMEFI57517X0146. Industrial partner is Gazpromneft Science & Technology Centre.

IOP Conf. Series: Journal of Physics: Conf. Series **1205** (2019) 012048 doi:10.1088/1742-6596/1205/1/012048

References

[1] Economides M J *et al.* 2000 *Reservoir Stimulation, 3rd Edition* (Wiley)

[2] Nordgren R P 1972 *Society of Petroleum Engineers Journal* **12** 306–314

[3] Esipov D V *et al.* 2014 *Comput. Technologies* **19** 33–61

[4] Yew C H 2014 *Mechanics of Hydraulic Fracturing* (Gulf Professional Publishing)

[5] Spence D A and Sharp P 1985 *Proceedings of the Royal Society A: Mathematical, Physical and Engineering Sciences* **400** 289–313

[6] Geertsma J and Klerk F D 1969 *Journal of Petroleum Technology* **21** 1571–1581

[7] Protasov I I and Dontsov E V 2017 A comparison of non-local elasticity models for a bladelike hydraulic fracture 51st U.S. Rock Mechanics/Geomechanics Symposium, 25-28 June (San Francisco, California, USA: American Rock Mechanics Association)

[8] Ikoku C U 1978 *Transient flow of non-Newtonian power-law fluids in porous media* Ph.D. thesis Stanford University

[9] Escobar F H and Civan F 1996 *Journal of Petroleum Science and Engineering* **15** 379–387

[10] Bird R B 1987 *Dynamics of Polymeric Liquids, Volume 1: Fluid Mechanics* (Wiley-Interscience)

[11] Kong X, McAndrew J and Cisternas P 2016 CFD study of using foam fracturing fluid for proppant transport in hydraulic fractures *Abu Dhabi International Petroleum Exhibition & Conference* (Society of Petroleum Engineers)

[12] Khade S D and Shah S N 2004 *SPE Production & Facilities* **19** 77–85

[13] Porubov A V, Samsonov A M, Velarde M G and Bukhanovsky A V 1998 *Physical Review E* **58** 3854–3864

IOP Conf. Series: Journal of Physics: Conf. Series **1205** (2019) 012049 doi:10.1088/1742-6596/1205/1/012049

To the question of data-driven identification of author's age for Russian texts with age deceptions using machine learning

A Sboev[1,2], R Rybka[1], I Moloshnikov[1], D Gudovskikh[1] and T Litvinova[3]

[1] MEPhI National Research Nuclear University, Moscow, Russian Federation
[2] National research centre "Kurchatov institute", Moscow, Russian Federation
[3] Voronezskij Gosudarstvennyj Pedagogiceskij Universitet, Voronezh, Russian Federation

E-mail: `sag111@mail.ru`

Abstract. In this work we compare data-driven approaches to the task of author's age identification for Russian texts with age deception. The data corpus has been specially gathered with crowdsourcing for this task. Two ways to determine age deception in author texts are considered and compared: the first is a traditional task of identification of age group of a text author, the second is identification of the occurrence of age imitation in the text with its type (imitating higher age or imitating lower age). The best results obtained by LinearSVC model with vector of TF-IDF features of character n-grams as input data demonstrate the F1-score of about 80% for the second task, and for the first one it is about 44%.

1. Introduction

Nowadays a great amount of information circulates in the Internet, but only part of it is truthful. Distribution of deceitful information on the internet may mislead people, in some cases leading to serious consequences [1, 2]. Thus for many practical fields, such as forensic, security, and others, it is important to have truthful information. In our previous works we investigated the task of gender identification of Russian text author in texts with gender deception. Here we consider the task of age identification in texts with age deception, i. e. when the authors tried to distort their ages. While the case of author age identification in texts without the age deception is widely represented in scientific literature for different languages [3, 4, 5, 6, 7, 8], sources for the case with deceptions are absent. The result overview of the age identification tasks at the PAN 2016 competition [4] demonstrates the state-of-the-arts for the case of author age identification in texts without age deception. There a Twitter corpus was used for training, and different corpora from social media, blogs, essays, and reviews were used for evaluation. The following classes for age were considered: 18–24; 25–34; 35–49; 50–64; and 65+. The test samples contained two parts: texts from social media (at least 100 words per author) and blogs (25 posts of each author). Approaches based on the use of a wide range of stylistic features such as: various n-grams of symbols, words, word combinations and part of speech (PoS), measure term frequency–inverse document frequency (TF-IDF), punctuation, Out of Dictionary Words, Vocabulary Richness, Emoticons were presented in the works of winners [7]. Besides that, Second-Order Attributes were used, that represent terms and documents by vectors in a space of profiles (age, gender) to evaluate the relationship of a term with different profiles. The authors [9, 5] drew attention to the number of grammatical errors in the text as the feature clearly reflecting the literacy of the author and indicating a possible age group. In most cases the track participants used the simplest classifiers: SVM [9] and logistic regression [5]. Results of age group determination gave the accuracy levels of 0.38 for English on Social Media and 0.58 on Blogs, 0.35 for Spanish on Social Media and 0.51 on Blogs. We too investigated the task of author age identification without deception in our works [8, 10] on base of the Gender-Imitation-Crowdsource ('GI cs') corpus, collected using the crowdsourcing platform, and a corpus [6] of blogs from LiveJournal. The best result for 'GI cs' is F1 = 0.48, it was obtained using Gradient Boosting Classifier with LDR [11], LIWC features [12]. In this case 3 age groups to classify texts: 18–23, 24–29, 30+ have been specified. Regarding blogs, the best result is similar, F1 = 0.49, it was obtained using compositions of selected features: lemma, unigrams, bigrams; POS unigrams and bigrams; QUITA features [13], tenses, aspects. The formulation, closest to the task observed in this paper, was issued in [14], where the task of presence of any deception in men and women texts was investigated, but it was also far from ours.

2. Corpus Description

The corpus for this task, Age-Imitation-Crowdsource corpus (AI_cs), has been collected with a crowdsourcing platform. It expands the Gender-Imitation-Crowdsource corpus (GI_cs [10]) we used in our previous works, and so keeps the same layout and style. This time, the job the crowdsourcing respondents have been given is the following: Each respondent is asked to fill a questionnaire including their gender, age, ethnicity, native city, native language, profession, educational level, and are they left-handed or right-handed. The respondents are to choose a topic out of a few suggested, and write three texts on it.

- Text A in their natural manner.
- Text B imitating the style of someone younger.
- Text C imitating the style of someone older.

The topics to choose from are the following.

- An attempt to persuade some arbitrary listener to meet the respondent at their place;
- A story about some memorable event/acquisition/rumour or whatever else the imaginary listener is supposed to enjoy;
- A story about oneself or about someone else, aiming to please the listener and win their favour.

The task does not pass checking and is considered improper work if it contains:

- Irrelevant answers to the questionnaire,
- Incoherent jumble of words,
- Chunks of text borrowed from somewhere else,
- Texts not conforming to the above list of topics.

Texts checking is performed firstly by automated search for borrowings (by an anti-plagiarism website), and then by manual review of compliance to the task. The corpus statistic is presented in Table 1. The distribution of examples by gender feature is shown in Figure 1.

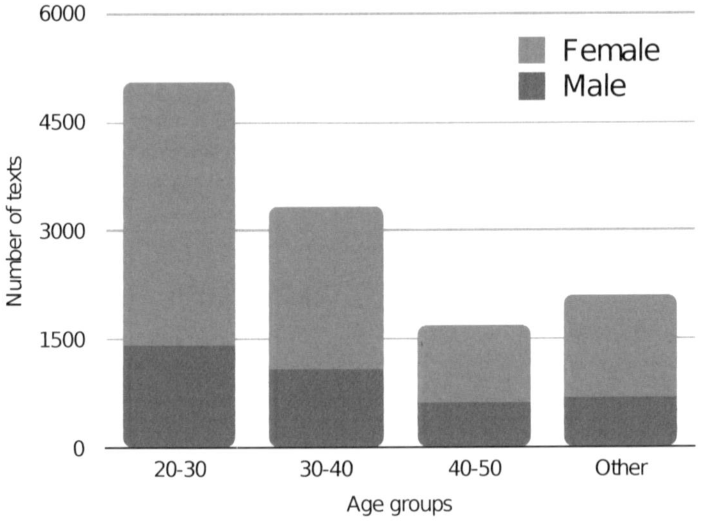

Figure 1. The distribution of texts by men/women in the age groups considered.

Table 1. The main properties of the AI–cs corpus employed.

Unique authors	Number of texts					Text length (characters)		
	Total	By age group				Min	Max	Median
		20–30	30–40	40–50	Other			
3302	12180	5049	3339	1680	2112	400	4678	492

3. Model

To obtain a base estimation we use a popular classification method based on support vector machine with linear kernel (Linear Support Vector Classifier – SVC). Due to the fact that SVC solves binary classification tasks, we use the one-vs-rest strategy: a separate classifier is created for each class of the task and is trained to determine its own class from all the remaining data. The SVC configuration is $C = 1$, with hinge squared loss function. For coding text data we use text representation based on TF–IDF vector of symbol n-grams, where n ranges from 3 to 8. The frequency vector characterizing document d is calculated with:

$$\begin{aligned} \mathrm{TF}(t,d) &= \frac{n_t}{\sum\limits_k (n_k)} \\ \mathrm{IDF}(t,D) &= \log\left(\frac{|D|}{|\{d_i \in D, t \in d_i\}|}\right) \\ \text{TF-IDF}(t,d,D) &= \mathrm{TF}(t,d) * \mathrm{IDF}(t,D), \end{aligned} \tag{1}$$

where t is an n-gram of 3-to-8-character-long sequence, $\mathrm{TF}(t,d)$ is the number of occurrences of t in the document d. $|D|$ is the total number of documents in the collection. $\sum\limits_k (n_k)$ is the number of characters in document d. $|\{d_i \in D, t \in d_i\}|$ is the number of documents in D that contain t.

Table 2. F1-score (mean ± std) of age determination after training on texts **with** age imitation, detailed by age groups and the presence of age imitation in the testing texts.

Age group	Classifying all texts	Classifying age-deceptive texts		Classifying non-deceptive texts
		Imitating higher age	Imitating lower age	
20–30	0.48 ± 0.02	0.45 ± 0.03	0.48 ± 0.04	0.50 ± 0.03
30–40	0.35 ± 0.03	0.34 ± 0.04	0.35 ± 0.04	0.36 ± 0.03
40–50	0.49 ± 0.03	0.50 ± 0.03	0.49 ± 0.04	0.48 ± 0.04
Mean	0.44	0.43	0.44	0.45

4. Experiments

4.1. Determination of age group

In the first step, the original body (AI_cs) is filtered keeping only texts of authors whose age ranges from 20 to 50 years. The resulting set is randomly split into 10 folds. In order to avoid unevenly represented classes affecting classification results, all the age groups – 20–30, 30–40, and 40–50 – are balanced by excluding samples from over-represented groups to make all the groups having similar number of samples. Then, each balanced age group is divided into training and testing sets, so that no author has their texts in both sets. Finally, we create a classifier for each balanced group and perform its training and testing on the corresponding sets. For quality assessment we use F1-score of classifying texts with and without age deception, detailed by age groups, in Table 2 – after training on age-deceptive texts, and in Table 3 – after training on non-age-deceptive texts.

To obtain the estimated values for Table 2, the training stage is accomplished using the training part of the balanced dataset AI_cs (which is 60% of it, 2982 texts), and for testing we use about 345 texts for each age group (about the 1035 texts, i.e. 20% of the balanced AI_cs).

For Table 3, the training set contained a part of the texts without imitation selected from the original balanced dataset AI_cs (about the 1725 texts), the test samples also as in the previous case contained examples of both with imitation and without it (about the 345 texts for each age group and about the 115 for each type of imitation (without, older, younger) in each group).

Table 3. F1-score (mean ± std) of age determination after training on texts **without** age imitation, detailed by age groups and the presence of age imitation in the testing texts.

Age group	Classifying all texts	Classifying age-deceptive texts		Classifying non-deceptive texts
		Imitating higher age	Imitating lower age	
20–30	0.43 ± 0.01	0.33 ± 0.04	0.47 ± 0.03	0.47 ± 0.03
30–40	0.33 ± 0.03	0.31 ± 0.05	0.30 ± 0.04	0.37 ± 0.04
40–50	0.49 ± 0.02	0.51 ± 0.03	0.47 ± 0.03	0.50 ± 0.05
Mean	0.42	0.38	0.41	0.45

Table 4. Age imitation type determination F1-scores, detailed by age groups.

Imitation type	Whole set, mean ± std	Age Groups		
		20–30	30–40	40–50
No imitation	0.74 ± 0.01	0.77 ± 0.02	0.76 ± 0.01	0.73 ± 0.05
Older	0.81 ± 0.01	0.83 ± 0.01	0.82 ± 0.01	0.81 ± 0.02
Younger	0.86 ± 0.01	0.86 ± 0.01	0.88 ± 0.01	0.86 ± 0.02
Mean	0.80	0.82	0.82	0.80

4.2. Determination of the imitation type

This task is to classify texts into 3 classes:

- Texts that do not imitate other age.
- Texts that imitate age higher than the author's.
- Texts that imitate the style of a lower age.

Like the previous experiment, texts from authors 20 to 50 year old have been used. The general experiment outline is the same. The main difference from the age group determination task is that the sets here do not need to be balanced by the number of samples from different classes. That is ensured by the crowdsourcing platform's job setting itself: each respondent will have written equal number of texts for each of the three imitation types. The classification quality in this experiment is assessed by F1-scores of classifying the three types of imitation (no imitation, pretending to be older, pretending to be younger) averaged over the whole testing set ("Whole set" column in Table 4), and specifically for certain age groups ("Age groups" column in Table 4).

For Table 4, we used 60% (about the 7340 documents) of the original AI_cs dataset for training and 20% (about the 2420 documents) for testing phases.

5. Discussion

An analysis of results of Tables 2 and 3 shows that the two approaches used to determine age deception in author texts demonstrate essentially different accuracies. The accuracy of determining the deception occurrence with identification of the type of imitation is far higher and reaches the F1-score of 0.80. The accuracy of determining the age group is only about 0.44 by F1-score, which is just about 11% higher than the baseline. Notably, the latter result practically coincidences with the result of the case without imitation (F1-score of 0.42); in other words, our age determination model is independent on the age imitation occurrence in texts.

6. Conclusion

As a result of this work, for the first time a Russian-language Age-Imitation-Crowdsource corpus (AI_cs) has been created, to solve the task of identifying the author age when the author tries to deceive it. Two approaches to the classifying solution have been presented: the traditional one, to identify the age group of a text author, and the second one, to identify the occurrence of the deception in the text with its type (overstating or lowering). The accuracies obtained demonstrate the second approach to be more efficient, achieving the F1-score of 80%. The future experiments will be focused on analyzing cross-genre text corpora and their effect on deceived age determination accuracy.

7. Acknowledgments
This research is supported by the Russian Science Foundation, project No 16-18-10050. This work has been carried out using computing resources of the federal collective usage center Complex for Simulation and Data Processing for Mega-science Facilities at NRC "Kurchatov Institute", http://ckp.nrcki.ru/

References
[1] Kocheshev P A 2016 *Problemy nauki* **6** 24–25
[2] Morozova A P 2017 *Nauchnye trudy Severo-Zapadnogo instituta upravleniya* **8** 154–163
[3] van de Loo J, De Pauw G and Daelemans W 2016 *Comput. Linguistics Netherlands* **5** 46–60
[4] Rangel F, Rosso P, Verhoeven B, Daelemans W, Potthast M and Stein B 2016 Overview of the 4th author profiling task at pan 2016: cross-genre evaluations *Working Notes Papers of the CLEF 2016 Evaluation Labs. CEUR Workshop Proceedings/Balog, Krisztian [edit.]; et al.* pp 750–784
[5] Modaresi P, Liebeck M and Conrad S 2016 Exploring the effects of cross-genre machine learning for author profiling in pan 2016. *CLEF (Working Notes)* pp 970–977
[6] Litvinova T Sboev A P P 2018 *Artificial Intelligence and Natural Language. 7th International Conference, AINL 2018, St. Petersburg, Russia, October 1719, 2018, Proceedings* 167–177
[7] Bougiatiotis K and Krithara A 2016 Author profiling using complementary second order attributes and stylometric features. *CLEF (Working Notes)* pp 836–845
[8] Sboev A, Moloshnikov I, Gudovskikh D and Rybka R 2018 *DEStech Transactions on Computer Science and Engineering*
[9] Alvarez-Carmona M A, López-Monroy A P, Montes-y Gómez M, Villasenor-Pineda L and Escalante H J 2015 *Working Notes Papers of the CLEF*
[10] Sboev A, Moloshnikov I, Gudovskikh D, Selivanov A, Rybka R and Litvinova T 2018 *Procedia Computer Science* **123** 417–423
[11] Rangel F, Franco-Salvador M and Rosso P 2016 A low dimensionality representation for language variety identification *International Conference on Intelligent Text Processing and Computational Linguistics* (Springer) pp 156–169
[12] Tausczik Y R and Pennebaker J W 2010 *Journal of language and social psychology* **29** 24–54
[13] Kubát M, Matlach V and Čech R 2014 *Studies in Quantitative Linguistics 18: QUITA-Quantitative Index Text Analyzer* (RAM-Verlag)
[14] Hong J, Mattmann C A and Ramirez P 2017 Ensemble maximum entropy classification and linear regression for author age prediction *2017 IEEE International Conference on Information Reuse and Integration (IRI)* (IEEE) pp 509–516

IOP Conf. Series: Journal of Physics: Conf. Series **1205** (2019) 012050 doi:10.1088/1742-6596/1205/1/012050

Simulation of gas release from pipelines using a new numerical method based on the Godunov approach

V A Shargatov[1], S I Sumskoi[1] and A S Pecherkin[2]

[1]National Research Nuclear University MEPhI (Moscow Engineering Physics Institute), Kashirskoye shosse 31, Moscow 115409, Russia
[2]CJSC Scientific technical center of industrial safety problems research, Block 14, Perevedenovsky Pereulok 13, Moscow 105082, Russia

E-mail: shargatov@mail.ru

Abstract. We propose a numerical method for calculating the flows arising in a pipeline during its rupture, including calculation of the release rate. Test calculations are performed and a comparison is made with the available experimental data. A satisfactory agreement between the experimentally determined flow rate and flow rate calculated by the proposed method is shown. It is also shown that the proposed method allows one to calculate with much higher accuracy than the currently used parametric formulas.

1. Introduction

Modern pipeline systems for the transportation of natural gas are characterized by high pressures, high flow rates and huge quantities of pumped hydrocarbons, including those that are simultaneously in the equipment. In some systems, the gas pressure in the pipeline may exceed 200 atmospheres. After the emergency destruction of the gas pipeline, intensive natural gas release can occur at a velocity of several hundred meters per second. In case of fire, such releases are fraught with serious consequences: the length of fire flares can be tens or even hundreds of meters. At the same time, for the gas pipelines the destruction almost always occurs at the full cross section (the guillotine rupture), which leads to the release with the maximum flow rate.

Another feature of the release following pipeline rupture is the long extension of the pipeline, the length of individual sections can reach several hundred kilometers. This means that the involvement of the gas into motion will occur sequentially due to the propagation of the unloading waves from the place of failure, with a gradual decrease in pressure in the pipe. Accordingly, the gas discharge rate in the release will gradually decrease with time. Correct calculations of such a weakening allow us not only to estimate the time variation of the outflow intensity, but also can serve as initial data for calculating processes such as dispersion, including the stage of an intensive initial release, when the starting jet is formed, which, when ignited, will lead to the formation of a fireball.

Thus, the assessment of the consequences of accidental releases from a ruptured gas pipeline is an important and urgent task, and the simulation of the outflow of gas from a destroyed pipeline is basic in modeling the entire chain of emergency consequences.

At the same time, for all the importance of obtaining information on the intensity of the emergency flow rate from the gas pipeline, a simplified approach is practically universally applied in the calculation of this process today - Bell's parametric formula [1], designed to calculate the release rate from a pipeline with a compressed single-component gas without additional gas injection. In this approach, it is assumed that the flow rate decrease is determined by two time scales:

- "fast" – at the initial stage of release; the flow rate depends on the initial pressure at the pipe;
- "slow" – when the gas is released after the pressure is reduced at the point of the rupture.

Each of these decreases is described by an exponent. In accordance with this approach, Bell proposed the first formula for the gas flow rate from a long pipeline [1, 2]. In subsequent years, several variants of the two-exponential formula were proposed for calculating the gas flow rate [3–8].

The use of two-exponential formulas is extremely attractive because of their simplicity. However, when using them, an important problem is the a priori specification of a number of parameters, for example, such parameter as the flow rate at the initial instant of time.

In addition to the parametric formulas for calculating the discharge rate from the gas pipeline during the accident (and at the same time, the entire process of gas flow in the pipeline) in recent decades the methods of computational fluid dynamics (CFD methods) based on the solution of the complete system of equations of motion expressing the laws of conservation of mass, momentum and energy.

Using the CFD methods, one can fully take into account the influence of such factors as non-isothermal flow, friction loss, heat transfer through the walls, and also changes in the specific heat of the gas and phase transitions in the gas (primarily due to the condensation of heavier natural gas fractions).

Initially, for such a simulation, a method of characteristics was used, based on the so-called Hartee's hybrid algorithm [8, 9]. The results of calculations using this approach are given in [9–12].

In addition to the method of characteristics, other numerical methods, such as the finite difference method [14–16] and the finite element method [17–19], are also used to calculate the flow rate from the destroyed gas pipeline.

One of the most effective and physically justified methods for calculating gas-dynamic flows is Godunov's method [20]. Previously, it was used to calculate the flows in oil pipelines [21–23]. In this paper Godunov's approach is proposed to be applied to the calculation of emergency flow rate from gas pipelines.

2. Basic equations

To describe the gas flow in a pipeline (including a case of its rupture and depressurization), it is proposed to use a system of gas dynamics equations describing non-stationary one-dimensional motion in a pipe of constant cross section.

$$\frac{\partial \rho}{\partial t}+\frac{\partial(\rho u)}{\partial x}=0, \tag{1}$$

$$\frac{\partial(\rho u)}{\partial t}+\frac{\partial\left(\rho u^2+p\right)}{\partial x}=-\lambda\frac{1}{D_0}\frac{\rho u|u|}{2}-\rho g\frac{dz}{dx}, \tag{2}$$

$$\frac{\partial\left(\rho\left(\varepsilon+\alpha_k\frac{u^2}{2}\right)\right)}{\partial t}+\frac{\partial\left(\rho u\varepsilon+\rho u\alpha_k\frac{u^2}{2}+pu\right)}{\partial x}=\pi D_0 q_s-\rho u g\frac{dz}{dx}. \tag{3}$$

here x – coordinate along the axis of the pipeline, t – time, ρ – density, u – velocity, p – pressure, z – height of the pipeline line, g – acceleration of gravity, D_0 – diameter of the pipeline, λ – friction factor, taking into account the effect of friction on the walls of the pipe (in this work, it is assumed equal to 0.0086), α_k – is a coefficient that takes into account the presence of the velocity pulsation component that is complementary to the mainstream flow velocity (in this work, it is assumed equal to

1.036 [24]), ε – is the specific internal energy, q_s – is the heat flux through the pipeline walls from the environment.

The energy conservation equation, in addition to the standard terms, includes a special term for describing heat exchange with the environment (q_s). The following relation is used to calculate q_s:

$$q_s = -K_T\left(T - T_s\right), \tag{4}$$

here T – the gas temperature in the pipeline, T_s – the temperature outside the pipeline, K_T – the heat transfer coefficient (in this work, it is assumed equal to 1.2 kcal /m^2/h/K).

To calculate the pressure in the pipeline, we propose to use the equation of state of an ideal gas.

As the boundary conditions, the condition of constant pressure at the inlet and outlet of the pipeline is used. The initial condition for the pipeline is the stationary pumping mode.

Only the case of the total rupture is considered in this paper when modeling the gas release. In this case, to simulate the flow, it is proposed to set at the end of each branch of the broken pipeline a velocity equal to the speed of sound in the critical flow regime, or a pressure equal to the surrounding (1 atm.) at subsonic release mode.

3. Numerical method of solution

The system of equations (1) – (4) was solved within the framework of the Gudunov approach [20]. To organize the computational procedure, the pipeline section was divided into cells of size Δx, the state in which at any time $n\Delta t$ was described using linear profiles of the corresponding parameters (pressure $p_i^n(x)$, density $\rho_i^n(x)$ and fluid velocity $u_i^n(x)$) in each i-th cell ($x_{i-1/2} \le x \le x_{i+1/2}$):

$$p_i^n(x) = \overline{p}_i^n + \Delta\overline{p}_i^n(x - x_i),\ \rho_i^n(x) = \overline{\rho}_i^n + \Delta\overline{\rho}_i^n(x - x_i),\ u_i^n(x) = \overline{u}_i^n + \Delta\overline{u}_i^n(x - x_i), \tag{5}$$

here x_i – the coordinate of the center of the i-th cell, $x_{i-1/2}$ – the coordinate of the boundary between the i-th and i-1 cells, $\overline{p}_i^n$, $\overline{\rho}_i^n$, $\overline{u}_i^n$ – pressure, density and velocity in the center of the i-th cell at the n-th time step.

The values $\Delta\overline{p}_i^n$, $\Delta\overline{\rho}_i^n$ and $\Delta\overline{u}_i^n$ are calculated according to the TVD-principle proposed in [25].

Simulation consists of two stages: the time step is split into physical processes in the same way as it was realized in [21–23].

First, at stage I, the gas flow is modeled without taking friction and gravity into account, and stage II takes into account the effect of frictional forces and gravity.

Stage I. At this stage, the mass, momentum and energy fluxes are calculated in each i-th cell, both due to convective motion, and due to the action of pressure. The fluxes are calculated by solving the problem of the discontinuity decay both on the left (index l) and right (index r) cell boundaries.

To solve the problem of a discontinuity disintegration at the boundaries of two neighboring cells, the values $p_i^n(x)$, $\rho_i^n(x)$ and $u_i^n(x)$ are calculated from formulas (5) on the right and left boundaries of each i-th difference cell. Accordingly, for i-th cell - to the right of its boundary i-$1/2$, i.e. for $x=x_{i-1/2}$, these quantities are denoted as $p_{i-1/2r}^n$, $\rho_{i-1/2r}^n$ and $u_{i-1/2r}^n$, and to the left of its boundary $i + 1/2$, $x=x_{i+1/2}$ they are denoted by $p_{i+1/2l}^n$, $\rho_{i+1/2l}^n$ and $u_{i+1/2l}^n$.

For the boundary between the i-1 and i-th cells, the Riemann problem is solved with the initial data in the form $p_{i-1/2r}^n$, $\rho_{i-1/2r}^n$ $u_{i-1/2r}^n$ and $p_{i-1/2l}^n$, $\rho_{i-1/2l}^n$, $u_{i-1/2l}^n$.

In this paper, for each i-th cell, the standard procedure for solving the Riemann problem [20] is used on each of its boundaries. The only difference between the Riemann problem in this paper and the solution algorithm in [20] is the use of quantities $\alpha_k \frac{u^2}{2}$ in the expression for the kinetic energy instead of the standard kinetic energy $\frac{u^2}{2}$.

As a result of solving Riemann problems, the pressure p^n_{ir}, density ρ^n_{ir} and velocity u^n_{ir} are determined on the right boundary of the *i*-th cell ($x=x_{i+1/2}$), and the pressure p^n_{i1}, density ρ^n_{il} and velocity u^n_{il} on the left-hand boundary of the *i*-th cell ($x=x_{i-1/2}$).

Using the obtained pressure values p^n_{ir}, p^n_{i1} and density values ρ^n_{ir}, ρ^n_{il} , the values of the specific internal energy ε^n_{ir}, ε^n_{il} are calculated at the boundaries of the *i*-th cell.

Knowing p^n_{ir}, ρ^n_{ir}, u^n_{ir}, ε^n_{ir} and p^n_{il}, ρ^n_{il}, u^n_{il}, ε^n_{il} one can calculate the values of the mass Φ^n_{mi} , momentum Φ^n_{pi} and energy Φ^n_{Ei} fluxes:

$$\Phi^n_{mi} = u^n_{il} \cdot \rho^n_{il} - u^n_{ir} \cdot \rho^n_{ir}, \tag{6}$$

$$\Phi^m_{pi} = u^{n\,2}_{il} \cdot \rho^n_{il} + p^n_{il} - u^{n\,2}_{ir} \cdot \rho^n_{ir} - p^n_{ir}, \tag{7}$$

$$\Phi^m_{Ei} = \rho^n_{il} \cdot u^n_{il} \cdot \left(\varepsilon^n_{il} + \alpha_k \frac{u^{n\,2}_{il}}{2} \right) + u^n_{il} \cdot p^n_{il} - \rho^n_{ir} \cdot u^n_{ir} \cdot \left(\varepsilon^n_{ir} + \alpha_k \frac{u^{n\,2}_{ir}}{2} \right) - u^n_{ir} \cdot p^n_{ir}, \tag{8}$$

The new values of density $\tilde{\rho}^{n+1}_i$, velocity $\tilde{u}^{n+1}_i$ and internal energy $\tilde{\varepsilon}^{n+1}_i$ in the *i*-th cell at *n + 1* time step at the end of the first stage are calculated as follows:

$$\tilde{\rho}^{n+1}_i = \frac{\tilde{m}^{n+1}_i}{\Delta x} = \frac{m^n_i + \Phi^n_{mi} \cdot \Delta t}{\Delta x}, \tag{9}$$

$$\tilde{u}^{n+1}_i = \frac{\tilde{R}^{n+1}_i}{\tilde{m}^{n+1}_i} = \frac{R^n_i + \Phi^n_{pi} \cdot \Delta t}{\tilde{m}^{n+1}_i}, \tag{10}$$

$$\tilde{\varepsilon}^{n+1}_i = \frac{\tilde{E}^{n+1}_i}{\tilde{m}^{n+1}_i} - \alpha_k \frac{\tilde{u}^{n+1\,2}_i}{2} = \frac{E^n_i + \Phi^n_{Ei} \cdot \Delta t}{\tilde{m}^{n+1}_i} - \alpha_k \frac{\tilde{u}^{n+1\,2}_i}{2}, \tag{11}$$

here $\tilde{m}^{n+1}_i$ and m^n_i is the masses, $\tilde{R}^{n+1}_i$ and R^n_i is the pulses and $\tilde{E}^{n+1}_i$ and E^n_i is the total energies in the *i*-th cell at *n + 1* (after stage I) and *n* time layers.

Stage II. The effect of friction forces on the walls, gravity and heat exchange through the walls are taken into account. Velocity and energy in the *i*-th cell are recalculated in accordance with formulas:

$$u^{n+1}_i = \tilde{u}^{n+1}_i - \lambda\left(\mathrm{Re}^{n+1}_i\right) \frac{\left|\tilde{u}^{n+1}_i\right| \tilde{u}^{n+1}_i}{2D^{n+1}_i} \Delta t - g\beta\Delta t \tag{12}$$

$$\varepsilon^{n+1}_i = \tilde{\varepsilon}^{n+1}_i + \alpha_k \frac{\tilde{u}^{n+1\,2}_i}{2} - \alpha_k \frac{u^{n+1\,2}_i}{2} + \frac{\pi D_0 \tilde{q}^{n+1}_{si}}{\tilde{\rho}^{n+1}_i} \Delta t - ug \frac{dz}{dx} \Delta t \tag{13}$$

Since the gravity and friction affects only velocity and internal energy on the new time layer, the pressure is also recalculated, and the other variables in stage II remain unchanged.

IOP Conf. Series: Journal of Physics: Conf. Series **1205** (2019) 012050 doi:10.1088/1742-6596/1205/1/012050

Thus, the required values are found on the new time step u_i^{n+1}, p_i^{n+1}, ρ_i^{n+1}, ε_i^{n+1}, after which the calculation procedure for stages I-II is repeated again to find the values on the subsequent time layers.

4. Results and discussion

The experimental data, described in [26], were taken as initial data on release gas modeling. A 160-km-long pipe with a diameter of 0.768 km is filled with methane at a pressure of 3.2 MPa. The gas is at rest. A full rupture occurs in the middle. Since the initial temperature in the pipe in [26] is not reported, in the calculations for a value of 300 K was chosen.

The results of the experimental measurements and the calculated values are shown in Fig. 1. The values obtained by calculation are given:

- according to the original Bell formula [1] without taking into account the inertia of the gas and taking into account the inertia (with different initial flow rates);
- using the numerical method proposed in this paper.

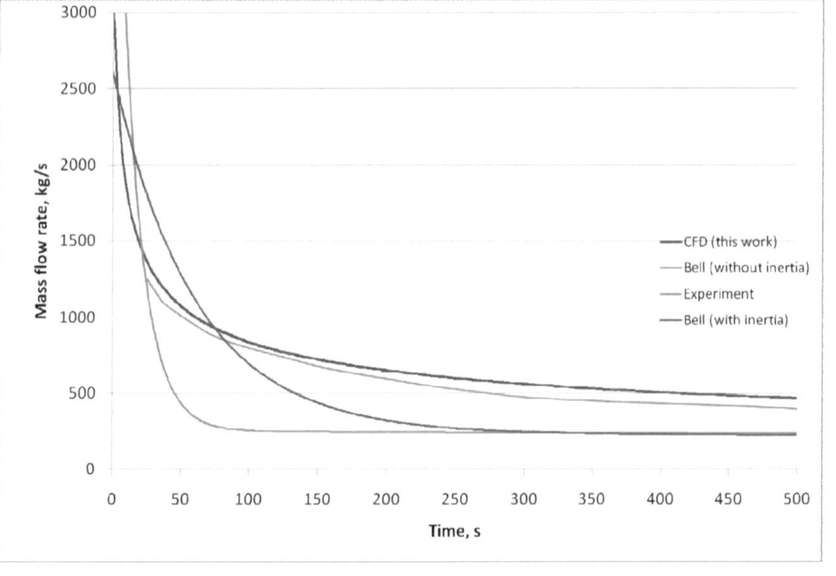

Figure 1. Experimental and calculated time dependence of the flow rate for the experiment [26]

As the boundary conditions at the release point, the atmospheric pressure for the subsonic flow regime and the sound velocity of the outflow for the supersonic regime were set. At the closed end of the pipe, the condition of a wall was specified.

As one can see from Fig. 1, the best agreement with the experimental data is given by the calculation of the complete system of gasdynamic equations. In this case, the flow rate decrease profile is reproduced throughout the interval of the times considered. In this case, the calculated values slightly exceed the observed values, which can be explained by some inaccuracies in the model (inaccuracies in calculating the friction factor, the heat transfer coefficient, neglecting the temperature dependence of the specific heat, etc.). This discrepancy in results can be decreased in subsequent works and the cause of it can be eliminated.

Calculations using the Bell formula [1] give a significantly overestimated flow at the initial instant of time and, with a rapid decay in time, reach released levels below the realistic values observed in the experiment.

Bell's formula, taking into account the inertia of the gas, gives a more realistic initial flow, but in general, a faster decline in release rate is also obtained, with an appropriate overestimation of the intensity at the initial stage and an underestimation at later stages.

IOP Conf. Series: Journal of Physics: Conf. Series **1205** (2019) 012050 doi:10.1088/1742-6596/1205/1/012050

5. Conclusions

In this paper, we propose a new numerical method for calculating flows in gas pipelines following pipeline rupture. The method was developed within the framework of the Godunov approach. The proposed method makes it possible to obtain solutions that agree well with the available experimental data. Simulation is carried out from the "first principles" without additional assumptions, as in the case of widely used parametric formulas.

Acknowledgments
This work is supported by the Russian Science Foundation under grant 16-19-00188

References

[1] Bell R P 1978 *Energy Processing Canada* **71** 36
[2] Lees F P 1996 *Loss Prevention in the Process Industries (2nd edition)* (Oxford: Butterworth-Heinemann)
[3] Wilson D J 1979 *The Release and Dispersion of Gas from Pipeline Ruptures* (Edmonton: Univ. of Alberta)
[4] Wilson D J 1981 *Expansion and Plume Rise of Gas Jets from High Pressure Pipeline Ruptures* (Edmonton: Univ. of Alberta)
[5] Wilson D J 2008 *Dispersion, Toxic Load and Outdour Calculations for Transient Releases from Pipeline Ruptures and Steady Releases from Gas Well Blowouts* (Edmonton: Alberta Energy and Utilities Board)
[6] Milne B, Kenny J A 2000 *Method for Determining Atmospheric Emissions from Ruptured Gas Gathering Systems* (Calgary: Canadian International Petroleum Conference, PETSOC-2000-070)
[7] António R, Tavares M 2011 *Accidental releases of hazardous gases modeling* (Universidade de Aveiro)
[8] Ames W F 1977 *Numerical Methods for Partial Differential Equations (2nd edition)* (New York: John Wiley & Sons)
[9] Chen J R 1993 *Modelling of Transient Flow in Pipeline Blowdown Problems* (Department of Chemical Engineering and Chemical Technology Imperial College of Science, Technology and MedicineUniversity of London Ph. D. Thesys)
[10] Cronje J S, Bishnoi P B, Svrcek W Y 1980 *Can. J. Chem. Eng.* **58** 289
[11] Picard D J, Bishnoi P R 1988 *Can. J. Chem. Eng.* **66** 3
[12] Picard D J, Bishnoi P R 1989 *Can. J. Chem. Eng.* **67** 3
[13] Zucrow M J, Hoffman J D 1976 *Gas Dynamics (vol.2)* (New York: John Wiley & Sons)
[14] Chen J R, Richardson S M, Saville G 1993 *Transactions of Institute of Chem. Eng.* **71A** 304
[15] Chen J R, Richardson S M, Saville G 1995 *Chem. Eng. Sci.* **50** 2173
[16] Bendiksen K H, Malnes D, Moe R, Nuland S 1991 *SPE Production Engineering* **6** 171
[17] Lang E 1991 *Journal of Applied Mathematics and Physics* **42** 183
[18] Bisgaard C, Sorensen H H, Spangenberg S 1987 *Int. J. Numer. Methods Fluids* **7** 291
[19] Gato L M C, Henriques J C C 2005 *Int. J. Heat Fluid Flow* **26** 817
[20] Godunov S K 1959 *Matematichesky sbornic* **47** 271
[21] Sumskoi S I, Sverchkov A M 2016 *Physics Procedia* **72** 347
[22] Sumskoi S I, Sverchkov A M, Lisanov M V, Egorov A F 2016 *J. Phys. Conf. Ser.* **751** 012024
[23] Sumskoi S I, Sverchkov A M, Lisanov M V, Egorov A F 2016 *J. Phys. Conf. Ser.* **751** 012022
[24] Lurie M V 2012 *Mathematical Modeling of Gas, Oil and Oil Product Transport* (Moscow: Gubkin University Press)
[25] Kolgan V P 1972 *Uchenye zapiski TSAGI* **3** 68 (in Russian)
[26] Safonof V S, Odisharia G E, Shvyriaev A A 1996 *Theory and practic of risk analysys in gas industry* (Moscow: NUMC)

Cluster-based energy consumption forecasting in smart grids

Eugene Yu Shchetinin

Financial University under the Government of the Russian Federation, Leningradsky pr.49, 111123 Moscow, Russia

E-mail: riviera-molto@mail.ru

Abstract. Accurate energy saving estimates are important for development of energy efficient projects and for demonstrating their cost-effectiveness. Increasingly, commercial buildings have an advanced measurement infrastructure, which has led to the availability of high-frequency interval data. This data can be used in a number of energy efficiency tasks, including anomaly detection, control and optimization of heating, ventilation, and air cooling systems. These data sets enable the application of advanced statistical training models and therefore lead to accurate estimates of energy savings. The K-means algorithm is a powerful tool that has been used in areas with intensive data mining applications, including ecology, computer vision, biology. In this paper, we propose a method for modeling the basic energy profile based on the clustering of consumers. A recently published test procedure was used to assess the accuracy of this method, and the estimate was based on analysis of data from 288 commercial buildings. The training periods of the model varied, and some predictive accuracy indicators were used to estimate the accuracy of the model. The results showed that the use of our model improved the determination and cross-validation parameters in more than 80 percent of cases in comparison with the best model used in the industry, which is based on regression, and in comparison with the random forest, decision trees.

1. Introduction

The development of intelligent networks in industry, economic, medicine and science creates new opportunities for the development and application of effective methods of machine learning and intellectual data analysis. The introduction of smart meters benefits end-users, energy suppliers and network operators by providing consumers with real-time consumption information that will help them to manage energy consumption, save money and reduce greenhouse gas emissions [1, 2, 3]. One of the most important issues in this area is the most accurate forecasting of energy consumption. Consumption tends to have a fairly complex stochastic structure that is difficult to model and predict for individual consumers. Therefore, consumption data are aggregated into energy pools called energy load profile. Engineering and statistical methods citebib4, bib5, bib6 are presented to analyze and extract the necessary information from customer load profiles. In addition, methods of statistical time series and artificial intelligence (AI) are used to estimate and predict the load in energy networks. However, these methods can be costly and difficult to implement and test when large volumes of consumption measurements are available. So, when applying different aggregation methods to a group of consumers with similar statistical characteristics of time series of electricity consumption, we can hope on significant progress in solving the tasks. One effective approach to extracting the necessary information from smart meter measurements is to use data mining techniques. Cluster analysis is one of the types of these methods [5, 6, 7]. Clustering is the grouping of load profiles into multiple clusters so that the profiles in one cluster are similar to each other.

In our paper we used such well-known methods of data representation as median, smooth averaging, splines and forecasting methods such as Holt-Winters, ARIMA model, support vector regression and some others. We produced advanced computer experiments with artificial data on modeling individual energy consumption profile and then, after clustering procedure of them, apply forecasting methods for a day and week ahead. The paper is organized as follows: section 2 describes some models for energy presentation of time series that we used for clustering and forecasting. Section 3 describes the clustering algorithm for classification and prediction. Section 4 provides an overview of our computer experiments and section 5 concludes.

2. Modeling of the energy consumption time series

The problems of applying clustering methods to energy consumption time series mainly lie in the high dimension and high level of data noise, which can be solved, as mentioned above, using machine learning methods. The papers deal with consumer aggregation typically use clustering to construct time series models [6, 7]. However, in these works little attention is paid to the use of methods of forecasting and data mining time series. For example, [6] investigated the problems of using cluster analysis to extract features from time series and its impact on the accuracy of the forecast of energy consumption. The authors used three different representations of time series: regression modeling, the extraction of average values of electricity consumption and the whole time series. The best results were obtained by clustering with regression coefficients, which showed a significant improvement in prediction accuracy by clustering. Wijaya T.K. [9] introduced the hierarchical clustering method with the optimization criterion of the forecast errors of an ARIMA model. This method was compared with a simple aggregation consumption forecast. Experiments have shown that the positive effect of consumer aggregation on the prediction accuracy of certain methods (linear regression, multilayer perceptron regression, and support vector regression) depends not only on the number of clusters, but also on the size of the customer base. To estimate this hypothesis, the Monte Carlo aggregation of consumers was used, as well as methods for predicting random walk and exponential Holt-Winters smoothing [11]. F. McLoughlin et al. [4] introduced dynamic consumer clustering. With the cluster approach, a large number of daily average profiles were created, which were then deeply analyzed. To link internal load profiles to household characteristics, SOM clustering and multi-domain logistic regression were used to analyze the characteristics of housing, comsumers and appliances [6, 7].

In this paper, we focus on the study of the impact of the consumer clustering procedure on the accuracy of various forecasting methods used for seasonal time series of electricity consumption. The paper presents a comparative analysis of two approaches to the classification of consumers: with the use of clustering and without it (aggregation). Our algorithm is as follows

(1) Normalize the data and calculate the energy consumption model for each consumer, using models based on time series representations;

(2) Calculating the optimal number of clusters for a given time series;

(3) Clustering and aggregation of consumption within clusters. For each cluster the forecasting model is learned and the forecast for the next period is run;

(4) The forecasts are aggregated and compared with the real consumption data. Next, we construct a forecast for day-ahead for the received representations of the clusters using the above-described prediction methods.

The steps (1)-(4) is iteratively repeated as measurements for a next day become available. The training window of data is changed so that the new day is added to the training window and the oldest one is removed.

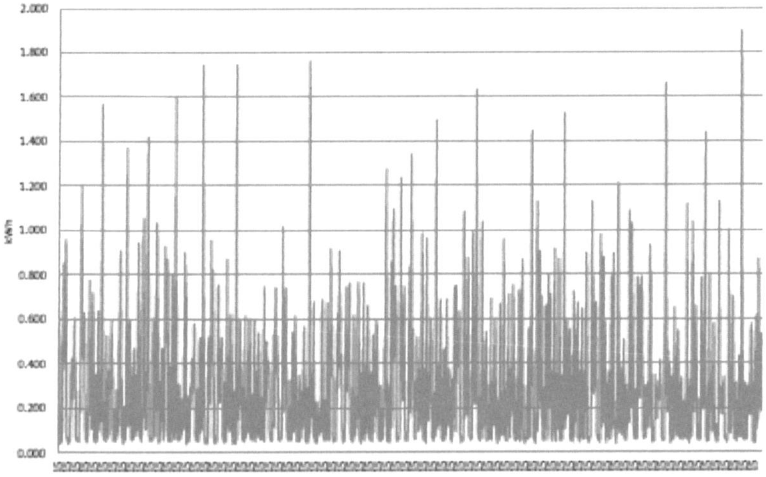

Figure 1. Energy consumption time series of typical client profile.

2.1. Energy consumption time series modeling

The main reason for the use of time series representation models is a significant reduction in the dimension of the analyzed data, respectively, reducing the required memory and computational complexity. Four time series representation methods were chosen: (a) robust linear model (RLM), (b) generalized additive model (GAM), (c) Holt-Winters exponential smoothing (Ho-W), and (d) Kalman linear filter. Presentation RLM aims next equation

$$s_i = \beta_0 + \sum_{i=1}^{l} \beta_j u_{ij} + \varepsilon_i, \quad (1)$$

where $i = 1,..,n$, (s_i) - is energy consumption time series, $\beta_1, ..., \beta_l$ are the regression coefficients. Variables $u_{i1},..,u_{il}$ are independent binary, u_{ij}, $j = 1,2,...,l$. Variable ε_i is a white noise with the normal distribution $N(0,\sigma^2)$. Extensions for RLM are generalized additive models (GAM) [10, 15]

$$E(x_i) = \beta_0 + \sum_{j=1}^{K} \beta_j f(u_{ij}), \quad (2)$$

where functions f are B-splines [15], β_j are regression coefficients, K is rank of splines.

Holt-Winters exponential smoothing (Ho-W) is another well-known method of presentations of time series that we used. It is a method that is used mainly to forecast the seasonal time series and to smoothing time series from the noise [11].

2.2. Applications of cluster analysis for smart grids

Typically, utility services divide their customers into industrial, commercial, and residential sectors based on some fixed information such as voltage level, nominal demand, etc. Based on this approach, a set of customer class load profiles has been defined and each user has been assigned one of these classes. However, this is still a fundamental issue and the procedures for dealing with customer segmentation need to be significantly revised. First, consumption data for customers who have installed smart meters is now available. Second, the measurement period is not limited and there is information on use for the next few years. These two factors affect the dimensionality of the data, which is not comparable to previously used data sets. Finally, as data is continuously recorded, they can have possible applications for real-time operation and control of electrical systems. All these factors emphasize the necessity of using new clustering methods for power consumption characteristic. To classify consumers into groups (clusters) we used the method of clustering based on K-means algorithm with mezoids as a choice of centroids [9, 12]. For each representation of the dataset, we determined the optimal number of clusters L using the empiric rule of elbow [12] and developed temporal grid search.

2.3. Time series forecasting methods

We used mostly effective methods to improve forecasting energy consumption time series:

(1) Support Vector Regression (SVR), a method that works like simple linear regression but it tries to find a real regression function that best approximates output vector. SVR technique relies on kernel functions to construct the model. The commonly used kernel functions are: a) Linear, b) Polynomial, c) Sigmoid and d) Radial Basis.

(2) Seasonal decomposition of time series based on loess regression (*Loess*) is a method, which decomposes seasonal time series into three parts: trend, seasonal component and remainder (noise) [11]. The seasonal component is found by loess (local regression) smoothing of the seasonal series, whereby smoothing can be effectively replaced by taking the mean. The seasonal values are removed, and the remainder is smoothed to find the trend. The overall level is removed from the seasonal component and added to the trend component. For the three final time series any of the forecast methods is used separately, in our case either Holt-Winters exponential smoothing or ARIMA model.

(3) Random Forest(RF). The method constructs the large number of decision trees at training time. Its output is the class that is the mode of the classes (classification) or mean prediction (regression) of the individual trees [8].

(4) The boosting algorithm GBM was first proposed for classification problems [13]. Its basic principle is that several simple models, called weak learning models, to be merged into one iterative scheme for the selection of parameters with the aim of obtaining the so-called strong learning model, i.e. models with better prediction accuracy. Thus, the GBM algorithm iteratively adds a new decision tree (i.e. "weak learner") at each step, which best reduces the loss function. Specifically, in a regression model, the algorithm starts with model initialization, which is typically a decision tree minimizing the loss function (RMSE), and then at each step, a new decision tree is adjusted to the current residual and added to the previous model to update the residuals. The algorithm continues to run until the maximum number of iterations is reached or the specified precision is reached. It means that at each new step, the decision trees added to the model in the previous steps are fixed. Thus, the model can be improved in those parts of it where it still does not assess the residuals.

(5) Bagging (Bagg) predictors generate multiple versions of predictors and use them for determination an aggregated predictor [14]. The aggregation is an average of all predictors. The bagging method gives substantial gains in accuracy, but the vital element is the instability of the prediction method. In the case that perturbing the learning set has significant influence on the constructed predictor, the bagging can improve accuracy.

(6) Regression Trees (R-Tree) are regression methods that consist of partitioning the input parameters space into distinct and non-overlapping regions following a set of if-then rules. The splitting rules identify regions that have the most homogeneous response to the predictor, and within each region a simple model, such as a constant, is fitted. The use

of decision trees as a regression technique has several advantages, one of which is that the splitting rules represent an intuitive and very interpretable way to visualize the results. In addition, by their design, they can handle simultaneously numerical and categorical input parameters. They are robust to outliers and can efficiently deal with missing data in the input parameters space. The decision tree's hierarchical structure automatically models the interaction between the input parameters and naturally performs variable selection, e.g., if an input parameter is never used during the splitting procedure, then the prediction does not depend on this input parameter. Finally, decision trees algorithms are simple to implement and computationally efficient with a large amount of data.

2.4. Metrics evaluation the accuracy forecasting

In the literature, the widespread metrics of evaluation the accuracy of forecasting is MAPE (Mean Absolute Percentage Error). But it is true, that MAPE is not robust in sence to the division by values near zero [9]. In our data set some clients some time have near zero consumption, which makes the MAPE undefined or very large, approaching infinity. Unlike the MAPE, the MAE and RMSE metrics don't suffer from this. However, they are scale-dependent, so we should suggest to apply another metrics without the above-mentioned disadvantages. Those are Normalized Mean Absolute Error (NMAE) and Normalized Root Mean Square Error(NRMSE). The NMAE is defined as

$$NMAE\left(s_t\right) = \frac{\sum\limits_{t=1}^{n} \left|s_t - \tilde{s_t}\right|}{\sum\limits_{t=1}^{n} \left|s_t\right|} \tag{3}$$

and the NRMSE is defined as

$$NRMSE\left(s_t\right) = \sqrt{\frac{\sum\limits_{t=1}^{n} \left(s_t - \tilde{s_t}\right)^2}{\sum\limits_{t=1}^{n} s_t^2}}, \tag{4}$$

where s_t is actual consumption, $\tilde{s_t}$ - model forecast, n - length of time series.

3. Computer experiments for customer energy consumption

We performed the computer experiments to evaluate the profit of using clustering procedures on four time series representation methods for one day ahead forecast. Table 1 shows average daily NRMSE forecast errors (4) of 6 forecasting methods. Each forecasting method was evaluated on 5 datasets; 4 datasets are clustered with different representation methods ($Kalman, Ho{-}W, GAM, RLM$) and aggregated electric load consumption (Sum). The following conclusions can be derived from computations. Optimized clustering of consumers significantly improves accuracy of forecast with forecasting methods $SVR, Bagging, GBM$. Despite this, clustering with $Loess\ Kalman, RF, R{-}Tree$ does not really improve accuracy of forecast. Three robust representation methods of time series Median, GAM and RLM performed best among all representations, while HW was the worst in most of the cases, because robust representations are stable and less fluctuate. The best result of all cases achieved by GBM with optimized clustering using GAM representation with mean daily MAPE error under $3,24\%$.

4. Conclusions

This paper investigates the problem of improving the accuracy of forecasting electricity consumption by users of different classes, including both individual consumers and industrial enterprises. One of the most important areas of improvement and development of energy networks is the introduction of intelligent systems for the collection, processing and analysis of data of energy consumption in order further to improve the quality of customer services, reliability of electricity supplies, reduce overall costs and improve their energy efficiency. Machine learning methods, in particular cluster analysis, were used to decide these tasks. The main purpose of this work was to show that the application of the consumer clustering procedure to the representation of time series of energy consumption can improve the accuracy of their energy consumption forecasts. Robust regression model, generalized additive model, exponential smoothing and median linear filter were used as such representations.

Modified k-means algorithm with medoids for more accurate choice of centroids and the Davis-Boldin index for estimation of clustering results was proposed . Advanced computer experiments have shown that considered methods of forecasting $Loess\ Kalman, SVR, RF, Bagging$ more effective for improving the accuracy of the forecast when used together with clustering. The most accurate prediction is obtained with gradient boosting algorithm and time series GAM presentation. Among the most promising applications of cluster analysis for smart grids are advantages for tariff design, smart demand response programming, improved load forecasting, classification of new or meterless customers, and other tasks.

5. References

[1] Haben S., Singleton C., Grindrod P.: Analysis and Clustering of Residential Customers Energy Behavioral Demand Using Smart Meter Data. IEEE Transactions on Smart Grid. v.99, pp.1-9 (2015).
[2] Chicco G., Napoli R. and Piglione F.: Comparisons Among Clustering Techniques for Electricity Customer Classification. IEEE Trans. Power Sys. vol. 21, pp.933-940 (2013).
[3] Gelling C. W.: The Smart Grid: Enabling energy efficiency and demand response. The Fairmont Press Inc. (2009).
[4] McLoughlin F., Duffy A., and M. Conlon: A clustering approach to domestic electricity load profile characterisation using smart metering data. Appl. Energy, vol. 141, pp. 190–199 (2015).
[5] Aghabozorgi S., Shirkhorshidi A. et al.: Time-series clustering: A decade review. Inf. Sys. vol. 53, pp. 16-38 (2015).
[6] Shahzadeh A., Khosravi A., Nahavandi S.: Improving load forecast accuracy by clustering consumers using smart meter data. In Int. Joint Conference on Neural Networks (IJCNN), pp. 1-7 (2015).
[7] Rodrigues P., Gama J., and Pedroso J.: Hierarchical Clustering of Time-Series Data Streams. IEEE Trans. Knowl. Data Eng., vol. 20, no. 5, pp. 615–627 (2008).
[8] Breiman L., Random forests, Machine learning vol. 45(1), pp.5–32, (2001).
[9] Wijaya T.K., Vasirani M. et al.: Cluster-based aggregate forecasting for residential electricity demand using smart meter data.Iin 2015 IEEE International Conference on Big Data. IEEE Press, pp. 879–887 (2001).
[10] Wood S., Generalized Additive Models: An Introduction with R. Chapman and Hall/CRC (2006).
[11] Hyndman R.J. Koehler A.B., Snyder R.D. and Grose S.: A state space framework for automatic forecasting using exponential smoothing methods. International Journal of Forecasting vol. 18, no. 3, pp. 439–454 (2002).
[12] Arthur D. and Vassilvitskii S., K-means++: The Advantages of Careful Seeding. In SODA '07 Proceedings of the eighteenth annual ACM-SIAM symposium on discrete algorithms. pp. 1027–1035 (2007).
[13] J.H. Friedman, Stochastic gradient boosting, Comp. Stat. Data Anal. vol.38 (4), pp.367–378 (2002).
[14] Breiman L., Bagging predictors, Machine learning, vol. 24(2), pp. 123–140 (1996).
[15] Shchetinin Eu.Yu., Lyubin P.G.: Fast two-dimensional smoothing with discrete cosine transform. Springer Communications in Computer and Information Science (CCIS), vol. 678, pp. 646-656. Springer, Berlin (2016).

Global uniqueness of the compact support source identification problem

V V Soloviev[1] **and D S Tkachenko**[1]

[1] Department 30, NRNU MEPhI, Moscow, Russian Federation

E-mail: VVSoloviev@mephi.ru, DSTkachenko@mephi.ru

Abstract. In this paper, we formulate a global theorem of uniqueness of the solution of the inverse problem fro the recostruction of a special-kind source in heat equation. As a overdetermination we take comact-support trace of the solution on an internal segment inside the domain at the final moment of time. We assume the symmetry of the domain along the coordinate axis and the validity of some restrictions on the signs of a given term in the source and its derivatives. No restrictions on the size of the domain or on the absolute value of the given functions assumed. We prove corresponding facts for the solution of the direct problem. These are maximum principle and belonging of the solution to the peculiar class of functions.

1. Direct problem statement

Let $T > 0$, $0 < \alpha < 1$ – fixed numbers, on the plane with Cartesian coordinates of points $x = (x_1, x_2)$ a bounded domain Ω with a smooth boundary of class $C^{2,\alpha}$is given. In the point space (x, t) we define a cylinder $\Omega_T = \Omega \times (0, T]$ with a side boundary $\Gamma_T = \partial\Omega \times [0, T]$. In a closed cylinder $\overline{\Omega}_T = \overline{\Omega} \times [0, T]$ consider the first initial-boundary value problem for the heat equation with a source of a special form, namely, the problem of determining the function $u: \quad \overline{\Omega}_T \mapsto \mathbb{R}$ from the conditions:

$$(Lu)(x, t) = u_t(x, t) - \ Deltau(x, t) = f(x_1)h(x, t) + g(x, t), \qquad (x, t) \in \Omega_T, \qquad (1)$$

$$u(x, t) = \mu(x, t), \quad (x, t) \in \Gamma_T, \qquad u(x, 0) = \phi(x), \quad x \in \overline{\Omega}. \qquad (2)$$

In the equation (1) and conditions (2) functions $f: \mathbb{R} \mapsto \mathbb{R}$, $h, g: \overline{\Omega}_T \mapsto \mathbb{R}$, $\mu: \Gamma_T \mapsto \mathbb{R}$, $\phi: \overline{\Omega} \mapsto \mathbb{R}$ are considered to be known.

On the function f we assume that it has a compact support. We assume that there are such numbers a, b that this function differs from 0 only on the segment $[a, b]$. In addition, we assume that there is such a number c that the interval $\omega = (a, b) \times c$ lies entirely in the region Ω and moreover $\overline{\omega} = [a, b] \times c \subset \Omega$.

2. Direct problem solution properties

If the functions defined in the conditions (1) – (2) satisfy the constraint $h, g \in C^{\alpha,\frac{\alpha}{2}}(\overline{\Omega}_T)$, $\mu \in C(\Gamma_T)$, $\phi \in C(\overline{\Omega})$, $f \in C^{\alpha}(\mathbb{R})$, the conditions of matching $\mu(x, 0) = \phi(x)$, $x \in \partial\Omega$, it is known (see, e.g., [1], p. 95), task (1) – (2) has a unique solution $u \in C^{2+\alpha,1+\frac{\alpha}{2}}(\Omega_T) \cap C(\overline{\Omega}_T)$. In the future, when studying the inverse problem for the problem (1) – (2), we assume that f satisfies only the condition $f \in C^{\alpha}[a, b]$, that is, generally speaking $f \notin C^{\alpha}(\mathbb{R})$. In this case, as it is known ([2]), the function $u(x,\,t)$ – solution of the problem (1) – (2) may not have second derivatives of x_1. In this case, the solution of problem 1 should be understood in a generalized sense. For the convenience of further reasoning, we define the sets:

$$P = \Big\{(x_1,\,x_2):\ \ a < x_1 < b,\ \ x_2 \in \mathbb{R}\Big\},$$
$$\overline{P} = \Big\{(x_1,\,x_2):\ \ a \le x_1 \le b,\ \ x_2 \in \mathbb{R}\Big\},$$
$$P_t = P \times (0,\,T], \qquad \overline{P}_T = \overline{P} \times [0,\,T].$$

We define the space of functions $U(\Omega_T)$ by the rule:

$$U(\Omega_T) = C(\overline{\Omega}_T) \cap C^{2+\alpha,\,1+\frac{\alpha}{2}}(\overline{P}_T \cap \Omega_T) \cap C^{2+\ alpha,\,1+\frac{\alpha}{2}}(\Omega_T \setminus \overline{P}_T) \cap C^{1,\,0}_{x,t}(\Omega_T).$$

We consider the solution of the problem (1) – (2) in the class of functions $U(\Omega_T)$.

Remark 1. *Everywhere further when writing the equation 1 as it is usually accepted we will write $(x,\,t) \in \Omega_T$ though as it follows from definition of class $U(\Omega_T)$ at $x_1 = a$, $x_1 = b$ the equation (1) can be not executed.*

Remark 2. *the solution of equation (1), generally speaking, is not a generalized solution of the problem (1) – (2) in the space $W_2^{1,\,0}(\Omega_T)$, since the functions ϕ, μ can only be continuous. But at $\phi = 0$, $\mu = 0$ this solution will be, of course.*

Remark 3. *an Important property of the class of functions $U(\Omega_T)$, which will then be used principally in the formulation of the inverse problem for equation(1) is that $u(x_1,\,c,\,T) \in C^{2,\,\alpha}[a,\,b]$ at $u \in U(\Omega_T)$.*

To consider the inverse problem for equation (1), we impose additional restrictions on the domain Ω. We say that the domain Ω satisfies the condition (A) when

(i) there is a function $x_2 = \nu(x_1)$, $0 \le x_1 \le l$, such that $\Omega = \{(x_1,\,x_2):\ x_2 = \nu(x_1),\ \ 0 < x_1 < l\}$,
(ii) $\partial\Omega \in C^{2,\,\alpha}$.

Clearly, the class of such areas is not empty. All conditions are satisfied, for example, a circle of radius $\frac{l}{2}$ centered at $\left(\frac{l}{2}\right)$.

We assume that the interval ω lies on the axis Ox_1, that is $c = 0$, $0 < a < b < l$.

In the space $U(\Omega_T)$ for solutions of the equation (1), as well as for classical solutions, the maximum principle is valid.

Lemma. *Let the domain Ω satisfy the condition (A), function $u \in U(\Omega_T)$. Then if $(Lu)(x,\,t) \le 0$ ($(Lu)(x,\,t) \ge 0$), $(x,\,t) \in \Omega_T$, the function u takes the value of the positive maximum (negative minimum) (if any of them, of course) on the parabolic boundary, i.e. either at $t = 0$, or if $x \in \partial\Omega$.*

The formulated Lemma implies the uniqueness of the solution of the problem (1) – (2) in the class $U(\Omega_T)$ and the positivity of the solution of the problem (1) – (2) in Ω_T at $(Lu)(x,\,t) \ge 0$, $(Lu)(x,\,t) \ne 0$ in $\ Omega_T$.

Using thermal potentials (volume and double layer) (see [2]), we can prove the following theorem of existence and uniqueness of the solution of the problem (1) – (2) in the class Ω_T.

Theorem 1. *Let $\mu \in C(\Gamma_T)$, $\phi \in C(\overline{\Omega})$, $f \in C^{\alpha}[a, b]$, $h\,g \in C^{\alpha, \frac{\alpha}{2}}(\overline{\Omega}_T)$, the matching conditions $\mu(x, 0) = \phi(x)$, $x \in \partial\Omega$. Then there is a unique solution to the problem (1) – (2) in the class of functions $U(\Omega_T)$.*

3. Uniqueness for the inverse problem solution

For the formulation of the inverse problem for equation (1) we assume that not only the function $u(x, t)$ is unknown in equations(1), but also the function $f(x_1)$. In this case, we assume that for the solution of equation (1) the information about the function $u(x, t)$ is given not only in the form of condition (2), but also in the form of additional conditions (so-called overrides) in the following form:

$$u(x_1, 0, T) = \chi(x_1), \qquad x_1 \in [a, b]. \tag{3}$$

Thus, we obtain the following inverse problem:
Define a pair of functions $(u, f) \in U(\Omega_T) \times C^{\alpha}[a, b]$ from conditions (1) – (3).

The inverse problem of determining the source with two compact carriers and a slightly different redefinition for the heat equation was studied in [3]. In [4] the inverse problem (1) – (3) for the case of a source with two compact supports was considered, and for it in the class $U(\Omega_T)$ a local (a little $b - a$) theorem of existence and uniqueness was prooven. In this paper, we formulate a global theorem of uniqueness of the solution of the inverse problem (1) – (3). A little $b - a$ is not assumed. Instead, we assume the symmetry of the Ω domain along the Ox_1 axis and the validity of some restrictions on the signs of the function $h(x, t)$ and its derivatives. Other formulations of inverse problems of determining the source of the history of the study of these problems with him in [5] – [10].

For the problem (1) – (3) the following global uniqueness theorem holds.

Theorem 2. *let $h, h_t, h_{x_2} \in C^{\alpha, \frac{\alpha}{2}}(\overline{\Omega}_T)$, $\left|h(x_1, 0, T)\right| \geq h_T > 0$, $x_1 \in [a, b]$., function*

$$\tilde{h}(x, t) = \frac{1}{2}\left(h(x_1, x_2, t) + h(x_1, -x_2, t)\right), \qquad (x_1, x_2) \in \Omega$$

satisfies the following conditions: $\tilde{h}(x, t) \geq 0$, $\tilde{h}_t(x, t) \geq 0$, $(x, t) \in \Omega$, $x_2 \leq 0$. Then problem (1) – (3) cannot have 2 different solutions.

References

[1] Friedman A 1967 *Partial differential Equations of parabolic type (in Russian)* (Moscow: Nauka)
[2] Ladyzhenskaya O, Solonnikov V and Uraltseva N 1967 *Linear and quasi-linear equations of parabolic type (in Russian)* (Moscow: Nauka)
[3] Denisov A 2016 *J. of Comp. Math. and Math Phys.* **56 No. 10** 1754-59
[4] Soloviev V 2018 *J. of Comp. Math. and Math Phys.* **58 No. 5** 778-89
[5] Denisov A 1994 *Introduction to the theory of inverse problems (in Russian)* (Moscow: MSU publishing)
[6] Lavrent'ev M 1973 *Conditionally well-posed problems for differential equations (in Russian)* (Novosibirsk: NSU publishing)
[7] Prilepko A., Orlovsky D. and Vasin I 2010 *Methods for solving problems in mathematical physics* (New-York, Basel: Marcel Dekker, Inc.)
[8] Isakov V 2006 *Inverse problems for differential equations* (New-York: Springer)
[9] Kostin A 2013 *Matematicheskii Sbornik (in Russian)* **204 No. 10** 3-46
[10] Soloviev V 1989 *Differential equations (in Russian)* **29 No. 9** 1577-83.

Nonlinear partial differential equations with delay: linear stability/instability of solutions, numerical integration

V G Sorokin[1,2] **and A D Polyanin**[1,2,3]

[1] Ishlinsky Institute for Problems in Mechanics, Russian Academy of Sciences, Moscow, Russia
[2] Bauman Moscow State Technical University, Moscow, Russia
[3] National Research Nuclear University MEPhI, Moscow, Russia

E-mail: vsesor@gmail.com, polyanin@ipmnet.ru

Abstract. The paper deals with partial differential equations of parabolic and hyperbolic types that contain a nonlinear kinetic function with delay. Conditions for the linear stability and instability of stationary solutions of reaction-diffusion equations and conditions for the instability of solutions of more complicated equations with delay are formulated. A nonstationary solution of the model initial-boundary value problem with delay and quadratic nonlinearity is investigated for stability/instability. A numerical solution of the test problem in the domain of stability is obtained by the method of lines.

Keywords: partial differential equations with delay, delay reaction-diffusion equations, exact solutions, stability and instability of solutions, numerical integration

1. Introduction

Differential equations with delay arise in biology, biophysics, biochemistry, chemistry, medicine, control, climate model theory, fluid mechanics, heat conduction and other areas [1–5]. It is noteworthy that these equations occur in the mathematical theory of artificial neural networks, whose results are used for signal and image processing and in image recognition problems [6, 7].

Differential equations with delay are used to describe phenomena and processes, the state of which depends not only on the current time t, but also on some moment $t - \tau$ in the past. The initial conditions of the corresponding initial-boundary value problems must be specified on the interval $-\tau \leq t \leq 0$ (or $0 \leq t \leq \tau$); the boundary conditions are given in the same way as for problems without delay.

Many methods of numerical integration of partial differential equations without delay after certain modification can be used to solve initial-boundary value problems of the corresponding equations with delay (see article [9] and references therein). The modification, development and use of numerical methods must be carried out taking into account the qualitative features of the equations with delay, associated primarily with the possible instability of solutions.

The most obvious and very effective way to evaluate the accuracy of numerical methods is a direct comparison of numerical and exact solutions of test problems. Many exact solutions of nonlinear partial differential equations with delay (as well as systems of such equations) are given in [4, 5, 10–19, 21]. These equations and their exact solutions contain a number of free parameters (which can be varied) and can be used to formulate test problems that allow constructive investigation of the region of applicability of various numerical methods.

In this paper we discuss questions related to the linear instability of solutions of a certain class of nonlinear partial differential equations with delay; we give an example of numerical integration of the equation of this class by the method of lines.

2. Linear stability (instability) of stationary solutions of problems with delay

Let us consider a fairly general class of nonlinear partial differential equations with delay:

$$\varepsilon u_{tt} + \sigma(u,w)u_t = au_{xx} + f(u,w), \qquad w = u(x,t-\tau), \tag{1}$$

where $a > 0$, $\varepsilon \geq 0$, $\sigma = \sigma(u,w) \geq 0$ ($\varepsilon + \sigma > 0$), $\tau > 0$. The values $\varepsilon = 0$, $\sigma = 1$ define a subclass of reaction-diffusion equations with delay; the values $\varepsilon = 1$, $\sigma = 0$ define a subclass of nonlinear equations of Klein-Gordon type.

Let the constant $u = u_0$ be a solution of Eq. (1), i.e. we have the equality

$$f(u_0, u_0) = 0. \tag{2}$$

The constant u_0 is also a solution of the problem for Eq. (1) on the interval $0 \leq x \leq h$ with the following initial data and Dirichlet boundary conditions:

$$u(x,t) = u_0, \quad u_t(x,t) = 0 \quad \text{if} \quad -\tau \leq t \leq 0, \tag{3}$$

$$u(0,t) = u(h,t) = u_0 \quad \text{if} \quad t > 0. \tag{4}$$

(For $\varepsilon = 0$, the second initial condition for the derivative u_t in (3) should be omitted.)

We seek perturbed solutions of the form

$$u = u_0 + \delta e^{-\lambda t}\sin(\pi n x/h), \qquad n = 1,\, 2,\, \dots, \tag{5}$$

where δ is a small parameter, λ is the spectral parameter (note that solution (5) satisfies the boundary conditions (4)). Substituting (5) into the considered equation (1) and neglecting terms of the order of smallness δ^2 and higher, we obtain the dispersion (characteristic) equation for determining the spectral parameter λ:

$$Q(\lambda) \equiv -\varepsilon\lambda^2 + \sigma(u_0,u_0)\lambda - a(\pi n/h)^2 + f_u(u_0,u_0) + f_w(u_0,u_0)e^{\lambda\tau} = 0, \tag{6}$$

where $n = 1,\, 2,\, \dots$ We note that the function $Q(\lambda)$ for $f_w(u_0,u_0) \neq 0$ is a quasi-polynomial of the special form [20] and has an infinite set of zeros.

If the real part of at least one root of the transcendental equation (6) is negative, then the solution $u_0 =$ const of the initial-boundary value problem for Eq. (1) is unstable (since $\lim_{t\to\infty} |\bar{u}| = \infty$ at any point x_* satisfying condition $\sin(\pi n x_*) \neq 0$). For the linear stability of this problem, the real parts of all the roots of the transcendental equation (6) must be positive.

We specify a simple (but rather coarse) criterion for the instability of a stationary solution $u_0 =$ const. To do this, consider the function $Q(\lambda)$, which defines the dispersion equation (6). We have $Q(\lambda) \to -\infty$ for $\lambda \to -\infty$. If $Q(0) > 0$, then there exists a real value $\lambda = \lambda_* < 0$, for which $Q(\lambda_*) = 0$. This value yields the unstable perturbed solution (5). The described sufficient criterion for the linear instability of a stationary solution can be represented in the form of inequality

$$f_u(u_0,u_0) + f_w(u_0,u_0) > a\pi^2/h^2, \tag{7}$$

where the minimum value $n = 1$ was taken.

Let us consider in more detail the reaction-diffusion equation

$$u_t = au_{xx} + f(u, w), \qquad w = u(x, t - \tau), \tag{8}$$

which is obtained by substituting the values $\varepsilon = 0$, $\sigma = 1$ into Eq. (1). In this case the dispersion equation (6) can be represented in the form

$$p - \zeta + qe^{-\zeta} = 0, \tag{9}$$

where we use the notation:

$$p = \tau f_u(u_0, u_0) - a\tau(\pi n/h)^2, \quad q = \tau f_w(u_0, u_0), \quad \zeta = -\lambda\tau. \tag{10}$$

For what follows we need the results of the theorem formulated below.

Hayes theorem [20]. All the roots of Eq. (9) with real coefficients p and q have negative real parts ($\operatorname{Re}\zeta < 0$) if and only if the following three inequalities are simultaneously satisfied:

$$\begin{aligned} &\text{(i)} \quad p < 1, \\ &\text{(ii)} \quad p + q < 0, \\ &\text{(iii)} \quad q + \sqrt{p^2 + \mu^2} > 0, \end{aligned} \tag{11}$$

where μ is the root of the transcendental equation $\mu = p\tan\mu$ that satisfies the condition $0 < \mu < \pi$. For $p = 0$, we must take $\mu = \pi/2$.

If inequalities (11) with p and q defined in (10) are fulfilled, then the stationary solution $u = u_0$ of problem (1), (3), (4) for $\varepsilon = 0$, $\sigma = 1$ under condition (2) will be stable in the linear approximation.

Substituting (10) into (11), we can formulate the stability conditions of the stationary solution in terms of partial derivatives of the kinetic function. To analyze the arising inequalities, it is convenient to introduce the notation:

$$U = \tau f_u(u_0, u_0), \qquad W = \tau f_w(u_0, u_0), \qquad s = \pi^2 a h^{-2}\tau, \tag{12}$$

where s is the characteristic dimensionless parameter of the problem, $\tau > 0$. As a result, conditions (11) are transformed to the form

$$\begin{aligned} &\text{(i)} \quad U < sn^2 + 1, \\ &\text{(ii)} \quad U + W < sn^2, \\ &\text{(iii)} \quad W + \sqrt{(U - sn^2)^2 + \mu^2} > 0, \end{aligned} \tag{13}$$

where μ is the root of the transcendental equation $\mu = (U - sn^2)\tan\mu$ that satisfies the condition $0 < \mu < \pi$.

The stationary solution $u_0 = \text{const}$ of the problem under consideration will be stable if three inequalities (13) are satisfied simultaneously for all integer values $n = 1, 2, \ldots$

The equation of the curve

$$W + \sqrt{(U - sn^2)^2 + \mu^2} = 0, \tag{14}$$

which describes part of the boundary of the stability region in the plane (U, W) and is obtained by replacing the sign $>$ by the sign $=$ in the last condition (13), is conveniently represented in the parametric form

$$U = sn^2 + \mu\cot\mu, \quad W = -\mu/\sin\mu \quad (0 < \mu < \pi). \tag{15}$$

IOP Conf. Series: Journal of Physics: Conf. Series **1205** (2019) 012053 doi:10.1088/1742-6596/1205/1/012053

For a given n, the curve (15) passes through the point $(sn^2, -\pi/2)$ of the plane (U, W). In addition, we have $U \to sn^2 + 1$, $W \to -1$ as $\mu \to 0$, and $U \to -\infty$, $W \simeq U - sn^2 \to -\infty$ as $\mu \to \pi$.

The analysis shows that the condition of simultaneous fulfillment of three inequalities (13) for all integer values $n = 1, 2, \ldots$ reduces to a simpler condition of fulfillment of three inequalities (13) for $n = 1$.

Example. Let us consider the reaction-diffusion equation with delay

$$u_t = au_{xx} + bu - b[ru + (1-r)w]^\beta, \qquad w = u(x, t-\tau), \tag{16}$$

which is a special case of Eqs. (1) and (8).

Eq. (16) admits two stationary solutions $u_0 = 0$ and $u_0 = 1$ for any admissible values of the defining parameters b, r, $\beta > 0$, $\tau > 0$. Using the results of Section 2, we analyze the stability/instability of the solution $u_0 = 1$ of the initial boundary value problem, which is described by Eq. (16), the first initial condition (3) and the boundary conditions (4). For concreteness, we now set $a = h = 1$. Then the stability conditions (13) for Eq. (16) will have the form:

$$\begin{aligned}
&\text{(i)} \quad b(1-\beta r) < \pi^2 + \tau^{-1},\\
&\text{(ii)} \quad b(1-\beta) < \pi^2,\\
&\text{(iii)} \quad -b\beta(1-r) + \sqrt{[b(1-\beta r) - \pi^2]^2 + \mu^2\tau^{-2}} > 0,
\end{aligned}$$

where μ is the root of the transcendental equation $\mu = [b(1-\beta r) - \pi^2]\tau \tan\mu$ that satisfies the condition $0 < \mu < \pi$.

Remark 1. All mentioned above in Section 2 is also applicable to the analysis of the linear stability/instability of nonstationary solutions of a broader class of nonlinear initial boundary value problems that are described by Eq. (1) with initial and boundary conditions of the general form:

$$u(x,t) = \varphi_1(x,t), \quad u_t(x,t) = \varphi_2(x,t) \quad \text{if} \quad -\tau \le t \le 0, \tag{17}$$

$$u(0,t) = \psi_1(t), \quad u(h,t) = \psi_2(t) \quad \text{if} \quad t > 0 \tag{18}$$

(for $\varepsilon = 0$, the second initial condition for the derivative u_t in (17) should be omitted), if the functions $\psi_1(t)$ and $\psi_2(t)$, appearing in the boundary conditions (18), have asymptotic properties

$$\lim_{t\to\infty} \psi_1(t) = \lim_{t\to\infty} \psi_2(t) = u_0, \tag{19}$$

where the constant u_0 satisfies condition (2).

3. Stability/instability analysis of the nonstationary solution of a nonlinear test problem with delay

Consider a nonlinear reaction-diffusion equation with delay

$$u_t = u_{xx} + bu - b\left(\frac{u - kw}{1-k}\right)^2, \qquad w = u(x, t-\tau), \quad b = \frac{\ln k}{\tau} - 1, \quad k > 0, \quad k \ne 1, \tag{20}$$

which is a particular case of Eq. (16) for $r = (1-k)^{-1}$ and $\beta = 2$ and in degenerate cases $k = 0$ or $\tau = 0$ goes into the non-normalized Fisher's equation [22].

Eq. (20) admits an exact solution

$$u(x,t) = U_1(x,t) \equiv 1 + \frac{e^{ct+1}}{e^2-1}(e^x - e^{-x}), \quad c = \frac{\ln k}{\tau}. \tag{21}$$

Remark 2. The parameters b and $c = b+1$ in Eq. (20) and solution (21) depend singularly on the delay time τ, since $\lim\limits_{\tau\to 0} |b| = \infty$.

It is not difficult to show that the function $u = U_1(x,t)$ is also a solution of the problem for Eq. (20) in the domain $0 \le x \le 1$, $t > 0$ with the following initial data and boundary conditions:

$$\begin{aligned} &u(x,t) = U_1(x,t), \quad \text{if} \quad -\tau \le t \le 0; \\ &u(0,t) = 1, \quad \text{if} \quad u(1,t) = 1 + e^{ct}, \quad t > 0. \end{aligned} \tag{22}$$

For $0 < k < 1$, solution (21) and functions from (22) defined on the boundaries tend asymptotically to the stationary solution $u_0 = 1$, which makes it possible to analyze the linear stability/instability of the nonstationary solution (21) (see Remark 1).

The stability conditions (13) for the stationary solution $u_0 = 1$ of Eq. (20) with the Dirichlet boundary conditions (4) for $a = h = n = 1$ ($k > 0$, $k \ne 1$) take the form

$$\begin{aligned} &\text{(i)} \quad \tau > \zeta(k) \ \text{ if } \ k < \frac{\pi^2-1}{\pi^2+1} \ \text{ and } \ k > 1; \quad \tau < \zeta(k) \ \text{ if } \ \frac{\pi^2-1}{\pi^2+1} < k < 1, \\ &\text{(ii)} \quad \tau > \frac{\ln k}{1-\pi^2}, \\ &\text{(iii)} \quad \theta(k,\tau) > 0, \end{aligned} \tag{23}$$

where

$$\begin{aligned} \zeta(k) &= \Big(\frac{k+1}{k-1}\ln k - 1\Big)\Big(\pi^2 + \frac{k+1}{k-1}\Big)^{-1}, \\ \theta(k,\tau) &= \frac{2k}{1-k}(\ln k - \tau) + \sqrt{\Big(\frac{k+1}{k-1}(\ln k - \tau) - \pi^2\tau\Big)^2 + \mu^2}, \end{aligned}$$

and μ is the root of the transcendental equation $\mu \cot \mu = \frac{k+1}{k-1}(\ln k - \tau) - \pi^2\tau$ that satisfies the condition $0 < \mu < \pi$.

In particular, the following stability conditions follow from inequalities (23): $\tau > 0.455059$ for $k = 0.5$, $\tau > 0.205634$ for $k = 2$.

In Fig. 1 the regions of the plane (k,τ) highlighted in gray correspond to the case when the stationary solution $u_0 = 1$ of Eq. (20) is stable (in the linear approximation), and the regions of instability of this solution are not painted; the stability/instability region of the nonstationary solution (21) of the problem under consideration is determined by the values $0 < k < 1$. It is clear that for $k = 0.5$, the small delay times $\tau < 0.455$ lie in the instability region of the nonstationary solution (21) (see point $B(0.5, 0.455)$).

4. Numerical integration of a nonlinear test problem with delay by the method of lines

The essence of the method of lines is in reducing the partial differential equation to a system of ordinary differential equations by approximating the spatial derivatives by finite differences [23, 24].

Let us set the spatial grid $x_m = m\Delta x$, where $m = 0, 1, \ldots, M$, $\Delta x = 1/M$ is the space step. Let us carry out a discretization of the initial-boundary value problem and obtain a system,

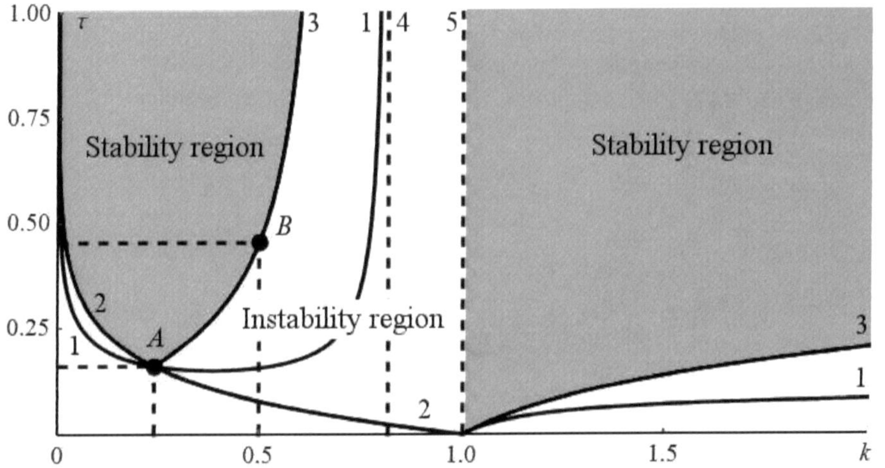

Figure 1. Regions of stability (shaded) and instability of the stationary solution $u_0 = 1$ of problem (20)–(22). Point $A(0.239, 0.161)$ denotes the intersection point of the three curves. Point $B(0.5, 0.455)$ denotes the boundary of the stable region for the problem under consideration for a fixed $k = 0.5$. The curves are indicated by numbers: 1 — $\tau = \zeta(k)$, 2 — $\tau = \frac{\ln k}{1-\pi^2}$, 3 — $\theta(k,\tau) = 0$, 4 — $k = \frac{\pi^2-1}{\pi^2+1}$, 5 — $k = 1$

consisting of $M - 1$ ordinary differential equations and two boundary conditions. The resulting problem, which also contains the initial conditions, can be solved numerically, for example, by implicit Runge-Kutta methods.

The resulting discretized system for test problem (20)–(22) has the form

$$\begin{aligned} &(u_m)'_t = a\delta_{xx}u_m + bu_m - b(1-k)^{-2}(u_m - kw_m)^2, \quad m = 1, \dots, M-1, \quad 0 \le t \le T; \\ &u_0(t) = 1, \quad u_M(t) = 1 + e^{ct}, \quad 0 \le t \le T; \\ &u_m(t) = 1 + \frac{e^{ct+1}}{e^2 - 1}(e^x - e^{-x}), \quad m = 0, 1, \dots, M, \quad -\tau \le t \le 0, \end{aligned} \tag{24}$$

where δ_{xx} is the difference operator, $\delta_{xx}u_m = (u_{m+1} - 2u_m + u_{m-1})/(\Delta x)^2$, T is the time period of calculations.

A numerical experiment was conducted in the computing system Mathematica (for a brief description, see [25]) by means of a combination of the method of lines with the Runge — Kutta method and the BDF method (Gear's method).

At moderate and large delay times (the values $\tau = 0.5, 1, 5$ were tested), both methods demonstrate a good approximation of the exact solution of problem (20)–(22) for $a = 1$, $k = 0.5$ on the whole interval $0 \le t \le T = 50\tau$. The relative errors are of the order 10^{-5} for $M = 10$ and 10^{-7} for $M = 100$. In Fig. 2 the exact solutions of the problem for $a = 1$, $k = \tau = 0.5$ are represented by solid lines, and the numerical solutions for $M = 100$, obtained by a combination of the method of lines and the Runge — Kutta method, are represented by circles.

At sufficiently small delay times $\tau = 0.05$ and $\tau = 0.1$ the numerical solution of problem (20)–(22) for $a = 1$, $k = 0.5$ begins to deviate strongly from the exact one after reaching the stationary mode. Then execution of the program is interrupted with an error. This is due to the instability for small τ of the stationary solution $u = 1$, to which the solution of the considered test problem (20)–(22) tends as $t \to \infty$ (see Section 3).

We note that for $\tau = 0.1$ the Runge — Kutta method and the BDF method provide a sufficiently accurate approximation of the desired solution over a fairly large time interval

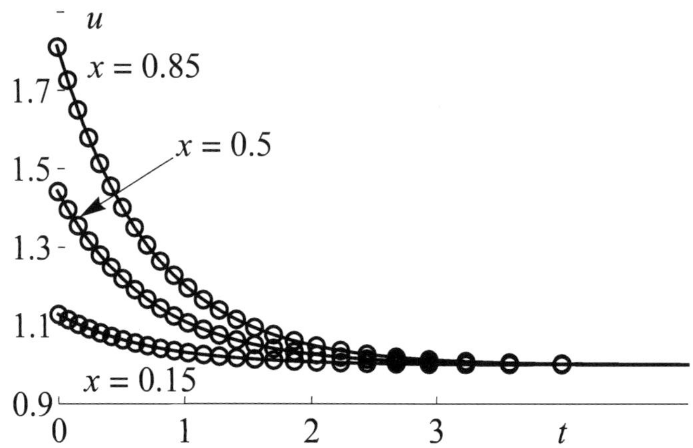

Figure 2. Exact solutions (solid lines) and numerical solutions obtained by a combination of the method of lines and the Runge — Kutta method (circles) of the test problem (20)–(22) for $a = 1$, $k = \tau = 0.5$, $M = 100$

$0 \leq t \leq 1.7$ (practically reaching the stationary mode $u_0 = 1$), and the BDF method has a slightly larger range of applicability for t.

Acknowledgments

The work was supported by the Federal Agency for Scientific Organizations (State Registration Number AAAA-A17-117021310385-6) and was partially supported by the Russian Foundation for Basic Research (project No. 18-29-03228 mk).

References

[1] Wu J 1996 *Theory and Applications of Partial Functional Differential Equations* (New York: Springer)
[2] Huang J and Zou X 2002 *J. Math. Anal. Appl.* **271** 455–466
[3] Wang X and Li Z 2007 *Nonlinear Anal. Theory Methods Appl.* **67** 2699–2711
[4] Polyanin A D and Zhurov A I 2013 *Int. J. Non-Linear Mech.* **54** 115–126
[5] Polyanin A D and Zhurov A I 2013 *Int. J. Non-Linear Mech.* **57** 116–122
[6] Wang L and Gao Y 2006 *Phys. Lett.* A **350** 342–348
[7] Lu J G 2008 *Chaos Solitons Fractals* **35** 116–125
[8] Polyanin A D and Nazaikinskii V E 2016 *Handbook of Linear Partial Differential Equations for Engineers and Scientists, Second Edition* (Boca Raton: CRC Press)
[9] Polyanin A D and Sorokin V G 2017 *Vestnik NIYaU MIFI* **6** (2) 126–142
[10] Polyanin A D and Zhurov A I 2014 *Int. J. Non. Linear. Mech.* **59** 16–22
[11] Polyanin A D and Zhurov A I 2014 *Commun. Nonlinear. Sci. Numer. Simul.* **19** 409–416
[12] Polyanin A D and Zhurov A I 2014 *Appl. Math. Lett.* **37** 43–48
[13] Polyanin A D 2015 *Theor. Found. Chem. Eng.* **49** (1) 107–114
[14] Polyanin A D 2015 *Theor. Found. Chem. Eng.* **49** (2) 169–175
[15] Polyanin A D, Sorokin V G and Vyazmin A V 2015 *Theor. Found. Chem. Eng.* **49** (5) 622–635
[16] Polyanin A D and Sorokin V G 2015 *Appl. Math. Lett.* **46** 38–43
[17] Polyanin A D and Zhurov A I 2014 *Commun. Nonlinear. Sci. Numer. Simul.* **19** 417–430
[18] Polyanin A D and Zhurov A I 2014 *Int. J. Non-Linear Mech.* **67** 267–277
[19] Polyanin A D and Zhurov A I 2015 *Int. J. Non-Linear Mech.* **71** 104–115
[20] Bellman R and Cooke K L *Differential-Difference Equations* 1963 (New York: Academic Press)
[21] Polyanin A D and Zhurov A I 2014 *Int. J. Non-Linear Mech.* **62** 33–40
[22] Polyanin A D and Zaitsev V F 2012 *Handbook of Nonlinear Partial Differential Equations* (Boca Raton: CRC Press)
[23] Van Der Houwen P J, Sommeijer B P and Baker C T H 1986 *IMA J. Num. Anal.* **6** 1–23
[24] Rihan F A 2010 *Numer. Methods Partial Differ. Equ.* **26** 1556–1571
[25] Polyanin A D and Sorokin V G 2017 *J. Phys.: Conf. Ser.* **937** 012041

IOP Conf. Series: Journal of Physics: Conf. Series **1205** (2019) 012054 doi:10.1088/1742-6596/1205/1/012054

Regression analysis of the results of planned computer experiments in machine mechanics

I N Statnikov and G I Firsov

Blagonravov Mechanical Engineering Research Institute of RAS, Moscow, 101990, Russian

E-mail: firsovgi@mail.ru

Abstract. The possibility is considered on the basis of the planned computational experiment at the stage of mathematical modeling to efficiently construct convolutions of the received information. This allows us to build a direct functional relationship between the parameters and the quality criteria of the dynamic system. formulas for calculating the coefficients for some types of regression dependencies are obtained, pre-calculated estimates of the variances from below and from above are given both for the coefficients and for the values of the regression function (i.e., the error forecast of the quality criterion). The results of research on the mathematical model of acoustic emission of loom elements are presented; a mathematical model for the propagation of acoustic energy by the design of the machine is constructed; Such assessments made it possible to develop recommendations for reducing machine noise.

1. Introduction

Conducting mathematical experiments should be purposeful and optimal. The specific nature of the various fields of scientific research led to the development of a variety of methods and their modifications to the planning of extreme experiments [1].

In the study of machines and mechanisms, the specificity of this field of scientific knowledge is associated with the construction on the basis of classical mechanics of mathematical models in the form of a set of differential equations (ordinary, linear and nonlinear, in partial derivatives), and analysis of the solutions of these equations, depending on the parameters of the system and the specified quality criteria cars [2,3,4]. Taking into account the specific nature of the investigation of the problems of machine mechanics, the authors proposed the PLP search method (planned LP search) [5], which, on the one hand, makes it possible, on the basis of computational experiments, to perform a quasi-uniform survey of the region of the investigated parameters $G(\overline{\alpha})$, on the other hand, as a result of special planning of these experiments, quantitative estimates of the influence of the variable parameters α_j on the analyzed properties of the machine (estimates developed in mathematical statistics) are applied.

With the use of simulation in the problems of analysis and synthesis of dynamic systems, they face the problem of obtaining a large amount of information, and hence the problems of processing and interpreting the results obtained. One possible way to solve these problems is to reduce the information obtained on the basis of statistical processing of the results of simulation modeling. In particular, widely used dependencies in the form of generalized. of the power Kolmogorov-Gabor polynomial

$$\widehat{\Phi}_k(\overline{\alpha}) = \beta_0 + \sum_1^r \beta_j \alpha_j + \sum_1^r c_j \alpha_j^s + \sum_{i \neq j}^r d_{ij} \alpha_j^s \alpha_i^t, \tag{1}$$

where $\widehat{\Phi}_k(\overline{\alpha})$ is the value of the k-th criterion of system quality ($k = \overline{1,m}$); α - a vector of design parameters of the system ($\overline{\alpha} = \alpha_1, ..., \alpha_r$); s and t are the numbers of the natural series, greater than or equal to two; $j = \overline{1,r}$; β_0, β_j, c_j and d_{ij} are the coefficients to be estimated from the results of the experiments. We introduce the vector of coefficients $\overline{b} = (b_1, ..., b_p)$, estimated in expression (1), where $p = \overline{1, n-1}$, and n in the general case is defined by the formula

$$n = (r^2 + 3r + 2)/2. \tag{2}$$

It follows from (2) that even at r >> 3 n >> 10. Obviously, such a power-law dependence, useful in some cases, is most often inefficient due to the cumbersomeness of expression (1) and, therefore, does not lend itself to operational interpretation. There is a problem of reducing the number of terms in expression (1), which is directly related to the decrease in the dimension r of the space of variable parameters α_j. In the PLP-search method, this problem is solved on the basis of the constructed procedure for carrying out the experiments and using the variance analysis. So we assume that the real number of parameters $r' < r$.

2. Estimation of regression model parameters

The standard method for determining the unknown coefficients of expression (1) is the method of least squares (LSM). Using the LSM procedure (1), we obtain a b_p system of linear equations of the form

$$dQ/db_p = 0, \tag{3}$$

where $Q = \sum_{l=1}^N [\widehat{\Phi}_{k,l}(\overline{\alpha}) - \Phi_{k,l}^Э(\overline{\alpha})]^2$; N is the number of experiments on a computer, $\Phi_{k,l}^Э(\overline{\alpha})$ the value of the quality criterion obtained on a computer as a result of calculations on the mathematical model. To represent the form of the system of equations (3), we write the first equation of this system:

$$\begin{aligned} &Nb_0 + b_1 \sum_1^N \alpha_{1,l} + ... + b_r \sum_1^N \alpha_{r,l} + b_{r+1} \sum_1^N \alpha_{1,l}^s + ... + b_{2r} \sum_1^N \alpha_{r,l}^s + \\ &+ b_{2r+1} \sum_1^N \alpha_{1,l}^s \alpha_{2,l}^t + ... + b_{n-1} \sum_1^N \alpha_{r-1,l}^s \alpha_{r,l}^t = \sum_1^N \Phi_{k,l}^Э. \end{aligned} \tag{4}$$

Already from equation (4) it is clear that to find the coefficients b_p it is necessary to calculate various functions of the parameters $\varphi(\alpha_j)$, for example, $\sum_1^N \alpha_{r,l}^s$, $\sum_1^N \alpha_{r-1}^s \alpha_{r,l}^t$. In addition, it is necessary to calculate the right-hand sides of equations (3). For an arbitrary arrangement of vectors $\overline{\alpha}$ in the space of variable parameters, all the indicated functions φ (αj) and right-hand sides are usually counted by a computer and then use some standard procedure for solving a system of linear equations.

In the case of a study of a dynamical system on a computer with the help of a PLP-search, it is possible to calculate every function of $\varphi(\alpha_j)$ in advance using the formulas proposed below. This is possible, since in PLP-search all the number of N experiments is carried out on a computer in series by M_j experiments in each, i.e. $N = M_j T$, where T is the number of series of experiments, and pseudo-random numbers q_{il} of the LP_τ-sequence are used.

Let's consider in detail the derivation of the formula $\varphi(\alpha_j) = \sum_1^{M_j} \alpha_{j,l}$. Let us write down the expression for calculating $\alpha_{j,l}$:

$$\alpha_{j,l} = \alpha_{j*} + q_{j,l}\,(\alpha_{j**} - \alpha_{j*}), \tag{5}$$

where $0 < q_{j,l} < 1$, $\alpha_{j,l} \in (\alpha_{j,*}, \alpha_{j,**})$ and $\Delta\alpha_j = \alpha_{j**} - \alpha_{j*}$. In the PLP-search $M_j = 2^{2v}$ (ν=2,3,...,), and the index l ranges from 2^v to 2^{v+1}-1. In view of the above, the expression for $\varphi(\alpha_j)$ takes the form

$$\varphi(\alpha_j) = \sum_{2^v}^{2^{v+1}-1} \alpha_{j,l} = \sum_{2^v}^{2^{v+1}-1} (\alpha_{j0} + q_{j,l}\Delta\alpha_j) = 2^v \alpha_{j*} + \Delta\alpha_j \sum_{2^v}^{2^{v+1}-1} q_{j,l}. \tag{6}$$

On the interval $[2^v; 2^{v+1}]$, the number $q_{j,l}$ is represented as a fraction

$$q_{j,l} = (2(l - 2^v) + 1) / 2^{v+1} \tag{7}$$

Taking into account (7), expression (6) takes the form

$$\varphi(\alpha_j) = 2^v \alpha_{j,*} + \Delta\alpha_j \sum_{l=2^v}^{2^{v+1}-1} \frac{2(l-2^v)+1}{2^{v+1}} = 2^v \alpha_{j,*} + \frac{\Delta\alpha_j}{2^{v+1}}[1+3+...+(2^{v+1}-1)] = \frac{M_j}{2}(\alpha_{j*} + \alpha_{j,**}). \tag{8}$$

Analogously to (8), using the formulas (5) and (7), we can find expressions for the sum $\sum_1^N \alpha_{j,l}^s$ for any $s \geq 1$. In [6] the formulas for these sums are given for the most commonly used values of s. In order to obtain the values of $\varphi(\alpha_j)$ for all N experiments on a computer, we need to multiply the values calculated by tabular formulas by the number of series T. Otherwise it is the sums of the form $\sum_1^N \alpha_{j,l}^s \alpha_{i,l}^t$. Exact expressions for calculating such sums can not be found. However, in the implementation of the matrix of planned experiments in series with a large number of experiments in each (M_j = 16, 32,...), it turned out that these sums can be calculated from approximate formulas with a sufficient degree of accuracy:

$$\sum_{\substack{j\neq i \\ l=M_j}}^{2M_j-1} \alpha_{j,l}^s \alpha_{i,l}^t \approx M_j E(\alpha_{j,l}^s \alpha_{i,l}^t), \tag{9}$$

where $E(\alpha_{j,l}^s \alpha_{i,l}^t)$ is the mathematical expectation of the quantity $y_l = \alpha_{j,l}^s \alpha_{i,l}^t$. Taking into account that $\alpha_{j,l}$ and $\alpha_{i,l}$ are independent pseudo-random uniformly distributed numbers computed by formula (5), we obtain $E(\alpha_{j,l}^s \alpha_{i,l}^t)$ the following expression:

$$E(\alpha_{j,l}^s \alpha_{i,l}^t) = E(\alpha_{j,l}^s)E(\alpha_{i,l}^t) = \frac{(\alpha_{j**}^{s+1} - \alpha_{j*}^{s+1})(\alpha_{i**}^{t+1} - \alpha_{i*}^{t+1})}{(s+1)(t+1)\Delta\alpha_j \Delta\alpha_i}. \tag{10}$$

When calculating the sums Σy_i, for all N experiments formula (9) takes the following form:

$$\sum_1^T \Big(\sum_{M_j}^{2M_j-1} \alpha_{j,l}^s \alpha_{i,l}^t\Big) \approx M_j T E(\alpha_{j,l}^s \alpha_{i,l}^t). \tag{11}$$

The error in the computation by formula (11) can be estimated as follows:

$$\delta = \left| 1 - \Big[\sum_1^T \Big(\sum_{M_j}^{2M_j-1} (\alpha_{j,l}^s \alpha_{i,l}^t)\Big)\Big] / M_j T E(\alpha_{j,l}^s \alpha_{i,l}^t) \right|. \tag{12}$$

Obviously, $E(\delta) = 0$. Let us calculate the variance $\sigma^2(\delta)$:

$$\sigma^2(\delta) = \sigma^2\Big[\sum_1^T \Big[\sum_{M_j}^{2M_j-1} \alpha_{j,l}^s \alpha_{i,l}^t\Big]\Big] / [M_j^2 T^2 E^2(\alpha_{j,l}^s \alpha_{i,l}^t)]. \tag{13}$$

For the internal sum of the numerator in formula (13), we can write

$$\sigma^2\Big(\sum_{M_j}^{2M_j-1} \alpha_{j,l}^2 \alpha_{i,l}^2\Big) = \frac{(\alpha_{j**}^{2s+1} - \alpha_{j*}^{2s+1})(\alpha_{i**}^{2t+1} - \alpha_{i*}^{2t+1})}{(2s+1)(2t+1)\Delta\alpha_j \Delta\alpha_i} - \frac{(\alpha_{j**}^{s+1} - \alpha_{j*}^{s+1})^2 (\alpha_{i**}^{t+1} - \alpha_{i*}^{t+1})^2}{(s+1)^2 (t+1)^2 (\Delta\alpha_j \Delta\alpha_i)^2}. \tag{14}$$

The substitution of (14) in (13) gives an expression for the variance $\sigma^2(\delta)$:

$$\sigma^2(\delta)=[\frac{(s+1)^2(t+1)^2(\alpha_{j**}^{2s+1}-\alpha_{j*}^{2s+1})(\alpha_{i**}^{2t+1}-\alpha_{i*}^{2t+1})\Delta\alpha_j\Delta\alpha_i}{(2s+1)(2t+1)(\alpha_{j**}^{s+1}-\alpha_{j*}^{s+1})^2(\alpha_{i**}^{t+1}-\alpha_{i*}^{t+1})^2}-1]\frac{1}{M_jT}. \quad (15)$$

As can be seen from formula (15), the quantity $\sigma^2(\delta)$ does not depend on the dimension r of the space of variable parameters α_j, is symmetric with respect to the exponents s and t. It also follows from (15) that the variance of the error δ depends on the values of $\Delta\alpha_j$ and $\Delta\alpha_i$ and tends $M_jT\to\infty$ to zero as t increases. However, it is clear from (10) and (12) that if $|\alpha_{j*}| = |\alpha_{j**}|$ or $|\alpha_{i*}| = |\alpha_{i**}|$ and at least one of the exponents s or t is an odd number, then we can not use formula (11), since in accordance with formula (15) $\sigma^2(\delta)\to\infty$. Analysis of the functional dependences of $\sigma(\delta)$ on T for various M_j, carried out for the cases $s = t = 1$ and $s = 2$ and $t = 1$, shows that $\sigma(\delta)$ decreases significantly after 5-7 series, does not change significantly. To significantly reduce the variance of the error δ, it is necessary to increase the number of experiments M_j in the series.

3. Solution of the two-criteria problem of optimal design of the weaving system of the loom

Let us consider a concrete example of the application of the above procedure for investigating the sensitivity of criteria to changes in parameters in the problem of multicriteria design of a loom according to acoustic criteria. Any task of designing a modern technical device is multicriteria, even if a hierarchy of criteria is established. However, I would like to emphasize that when using the decision-making methodology in design problems of technical devices, two groups of quality criteria should be distinguished. The first group includes all those criteria that characterize (assess) directly the quality of the required work and the corresponding technical characteristics (accuracy, speed, vibration, size, etc.). The second group includes all those criteria that characterize the technical and economic properties of the projected device (cost, labor, reliability, environmental adaptability to the environment, etc.). At the same time, analysis of the first group of criteria provides quite a lot of opportunities in formalizing the very process of selecting a set of permissible variants, and hence complete automation of the entire process with the help of a computer. Vulnerable in this process is in many cases the lack of functional dependence of the criteria of the first group on the parameters of the device being designed. However, on the basis of simulation experiments on computers, this gap can be eliminated in each specific case by constructing approximation models of criteria.

In [7], based on the use of the method of the planned PLP search experiment in the space of variable parameters η_j $(1,...,6)$, the region $G(\overline{\eta})$, where a probability of $P\geq 95\%$ can find variants of the vibration proof coating of the weaving loom system, good from a natural point of view, was identified numerically. And this point of view consists in the fact that a variant with standardized quality criteria λ_k (κ = 1,2), estimating the noise level ($0\leq\lambda_1\leq 1$) and the consumption of the vibration-isolating material ($0\leq\lambda_2\leq 1$), would be close to the ideal, albeit unrealistic, version of the coverage. As the criteria for identifying the region in the space of variable parameters, the utility function $\Phi_3(\overline{\eta})=\sum_1^2 c_k\lambda_k$ and function $\Phi_n(\overline{\eta})=|c_1\lambda_1-c_2\lambda_2|/\sqrt{c_1^2+c_2^2}$, criteria λ_1 and λ_2, satisfying the condition. In the future, to simplify the analysis and calculations, it is assumed that $c_1 = c_2 = 0,5$.

The fact that the region $G(\overline{\eta})$ does contain variants that satisfy the definition of a good variant is indicated by the results of a numerical experiment carried out in this area. 190 variants were calculated on the computer, in which the maximum and minimum values of λ_1 and $\lambda_2\lambda_1$ and λ_2 determined the following intervals of change:

$$0{,}690\leq\lambda_1\leq 0{,}952 \text{ and } 0{,}539\leq\lambda_2\leq 0{,}835. \quad (16)$$

Such a final result in [7] has the disadvantage that knowing a region's $G(\overline{\eta})$ boundaries by solving a particular design problem of a vibration protection device is of little use in itself. After all, for every new option in this area, all the calculations must be carried out again, and the question of the sensitivity of λ_1 and λ_2 to variations of the damping coefficients η_j remains unresolved.

Therefore, the problem of restoring the immediate functional connection between the criteria λ_1 and λ_2 and the parameters η_j is solved below; in the selected area $G(\overline{\eta})$. The success in constructing such a dependence is related to the depth of the physical representations about the modeled phenomenon and the successful choice of its form, the completeness of the experimental data. Before proceeding to a description of the construction of such dependences, we classify the parameters η_j whose boundaries of variation form the region $G(\overline{\eta})$:

$$\begin{aligned} &0,010 \le \eta_1 \le 0,055;\ 0,010 \le \eta_2 \le 0,060;\ 0,010 \le \eta_3 \le 0,156;\\ &0,010 \le \eta_4 \le 0,105;\ 0,015 \le \eta_5 \le 0,156;\ 0,024 \le \eta_6 \le 0,156. \end{aligned} \quad (17)$$

The normalized values of the parameters η_{nj} are calculated from formula $\eta_{nj} = (\eta_j - \eta_j^*)/(\eta_j^{**} - \eta_j^*)$, and the region $G(\overline{\eta})$ is transformed into a single six-dimensional hypercube K^6 with an edge [0; 1].
The construction of polynomial dependencies $\lambda_1(\overline{\eta}_n)$ and $\lambda_2(\overline{\eta}_n)$ is associated with the use of multiple regression methods. However, such dependencies are often of little value at least because of the large number of parameters and their combinations that are included in these dependencies.
In this case, a different path was used. It is known that the product $V = \Pi\eta_{nj}$ indirectly characterizes the consumption of a vibration-insulating material. It is also known that an increase in V leads to an increase in λ_1 and a decrease in λ_2. Therefore, the following dependencies were chosen for the construction:

$$\tilde{\lambda}_{1i} = a\lg(V_i + b) + c,\quad \tilde{\lambda}_{2i} = d - e\lg(V_i + f). \quad (18)$$

By minimizing the expressions $(\lambda_{ei,2} - \tilde{\lambda}_{2i})^2$ and $(\lambda_{ei,1} - \tilde{\lambda}_{1i})^2$ with a given accuracy δ, for each one they $i = \overline{1,80}$, find their own set of parameters (a; b; c) and (d; e; f), and then the values a_i, b_i, c_i, d_i, e_i, f_i are averaged over 80 realizations. Here $\lambda_{ei,1}$ and $\lambda_{ei,2}$ - the values of the criteria λ_1 and λ_2, obtained in experiments on a computer. As a result, the following expressions are obtained:

$$\tilde{\lambda}_1 = 0,0525\lg(V + 2\bullet10^{-6}) + 0,9852,\quad \tilde{\lambda}_2 = 0,4555 - 0,0955\lg(V + 2\bullet10^{-6}). \quad (19)$$

The approximate dependences (19) obtained are checked for the significance of predictive properties (adequacy) in the region of K^6. Calculating the mean values $\overline{\lambda}_1, \overline{\lambda}_2$ and variances σ_{1e}^2 and σ_{2e}^2 using numerical experiments, we obtain: $\overline{\lambda}_1 = 0,8514$, $\overline{\lambda}_2 = 0,7106$, $\sigma_{1e} = 0,0521$, $\sigma_{2e} = 0,0796$. Using these same data, determining the confidence intervals in ±3σ for λ_1 and λ_2, we obtain:

$$0,695 \le \lambda_1 \le 1 \text{ and } 0,472 \le \lambda_2 \le 0. \quad (20)$$

Inequalities (16) and (20) are in good agreement with each other. Values of the variances of the predicted values $\tilde{\lambda}_1$ and $\tilde{\lambda}_2$ are calculated by the formulas $\sigma_{1T}^2 = \frac{\sum_{i=1}^{80}(\lambda_{1i} - \tilde{\lambda}_{1i})^2}{N-n-1}$, $\sigma_{2T}^2 = \frac{\sum_{i=1}^{80}(\lambda_{2i} - \tilde{\lambda}_{2i})^2}{N-n-1}$, where n is the number of coefficients defined in each of formulas (18). By calculations, we obtained $\sigma_{1T} = 0,0262$ and $\sigma_{2T} = 0,0675$. Using the Fisher criterion, we check the significance of equations (19) $F_1 = \sigma_{1e}^2/\sigma_{1T}^2 = 3,95$ and $F_2 = \sigma_{2e}^2/\sigma_{2T}^2 = 1,39$. These values of the Fisher test for degrees of freedom $f_1 = N - 1 = 79$ and $f_2 = N - n - 1 = 76$ and significance level $p = 0.05$ are larger than the corresponding theoretical value $F_T = 1,34$. This result ($F_1 > F_T$ and $F_2 > F_T$) indicates that with a probability of $P \ge 95\%$, the predictive properties of the equations for $\tilde{\lambda}_1$ and $\tilde{\lambda}_2$ are sufficiently reliable. What useful information does the dependency (19) give directly to the designer? First of all, if the coverage option is estimated using some utility function, for example $\Phi_3(\overline{\eta}_n)$, an analogous function $\Phi_3(\overline{\eta})$, it is possible to predict the value $\Phi_3(\overline{\eta}_n)$ at specific values η_{nj}, knowing in advance about the absolute value of the discrepancy in the values of $\tilde{\lambda}_1$ and $\tilde{\lambda}_2$. If $\Phi_4(\overline{\eta}_n)$ = ε, then $\Phi_3(\overline{\eta}_n)$ depends on ε as follows:

IOP Conf. Series: Journal of Physics: Conf. Series **1205** (2019) 012054 doi:10.1088/1742-6596/1205/1/012054

$$\Phi_3(\overline{\eta}_n) = 0,7204 + 0,2054\varepsilon, \tag{21}$$

where $|\varepsilon| \leq m$. It follows from the dependence (21) that for $\varepsilon = \pm 20\%$ the difference between the maximum and minimum values $\Phi_3(\overline{\eta}_n)$ is only 0.08. In this case, to each pair of values of $\Phi_3(\overline{\eta}_n)$ and $\Phi_4(\overline{\eta}_n)$ there corresponds a specific value of V. It is important that this value of V can be achieved by any combination of values η_{nj}, and hence also η_j. It is also important that the values $\tilde{\lambda}_1$ and $\tilde{\lambda}_2$ are insensitive to the variations of V. Indeed, on the basis of (19), the sensitivity functions S_1 and S_2 with respect to V, are equal to $S_1 = \frac{d\tilde{\lambda}_1}{dV} = \frac{0,0525V}{V + 2 \cdot 10^{-6}}$; $S_2 = \frac{d\tilde{\lambda}_2}{dV} = -\frac{0,0955V}{V + 2 \cdot 10^{-6}}$.
Analysis of the functions S_1 and S_2 shows that, starting from a certain value $V \geq 0,0001$, both criteria become insensitive to changes in the values of this parameter η_j up to its maximum value $V_{\max} = 1$
Significant changes in the sensitivities of S_1 and S_2 in the zone of small values of V are easily interpreted physically. The point is that the minimum value of V will correspond to the minimum value of at least one of η_j, that is, at least one of the parameters will assume its lower boundary value from the region (17), and there will be a tendency to decrease $\tilde{\lambda}_1$ and increase $\tilde{\lambda}_2$.

4. Conclusion

Based on the results of this paper we draw the following conclusions. First, on the basis of the application of the PLP-search method, when simulations were carried out on computers, it was possible to isolate the region $G(\overline{\eta})$, that is the generator of such variants of the vibration protection coating of the loom, in which the standardized quality criteria λ_1 and λ_2, with a high degree of probability exceed 0.5. Secondly, on the basis of identifying this area and understanding the physical processes, it was possible to construct mathematical models that reliably relate the values of the criteria to the values of the variable parameters. Third, on the basis of the constructed mathematical models of criteria λ_1 and λ_2, the functions of the sensitivity of the criteria with respect to variable parameters are constructed; thereby creating objective opportunities for the preliminary selection of a compromise cover option according to the criteria of the first λ_1 and second λ_2 groups. It should be noted that the approximated dependences obtained were sufficiently well coordinated with the approximated mathematical models of the systems. At the same time, these dependencies turned out to be compact, easily interpreted and informative.

References

[1] Montgomery D C 2001 *Design and Analysis of Experiments* (Hoboken, NJ, John Wiley and Sons)
[2] Kreinin G V and Misyurin S Yu 2008 *Journal of Machinery Manufacture and Reliability* **37**, 1
[3] Kreinin G V and Misyurin S Yu 2012 *Journal of Machinery Manufacture and Reliability* **41**, 331
[4] Misyurin S Yu and Kreinin G V 2015 *Doklady Physics* **60**, 15
[5] Statnikov I N and Firsov G I 2017 *Journal of Physics: Conference Series* **937,** 012050. 1
[6] Statnikov I N and Firsov G I 1983 Automation of scientific research in the field of engineering (Moscow, Nauka)) p 19 (In Russian)
[7] Pobol O N, Statnikov I N, Firsov G I and Chernyavsy I T 1979 Methods for solving problems of machine science on computers (Moscow, Nauka)) p 62 (In Russian)

Steady-state transalfvenic MHD flows in coaxial channels with longitudinal magnetic field

E V Stepin

Keldysh Institute of Applied Mathematics, Russian Academy of Sciences, Miusskaya sq. 4, 125047 Moscow, Russia

eugene.v.stepin@gmail.com

Abstract. Transalfvenic magnetohydrodynamic (MHD) flows in coaxial channels of plasma accelerators in the presence of a longitudinal magnetic field are considered. Transalfvenic flows are characterized by the transition of a plasma flow through the alfvenic velocity corresponding to the longitudinal field. The main results are obtained for narrow nozzle-type flow tubes of curvilinear geometry in the quasi-one-dimensional approach. The analytical expressions for plasma parameters in the transition point and for its geometric place are presented. The numerical researches of flow regime changing from superalfvenic to subalfvenic in a narrow flow tube of two configuration variants are provided.

1. Introduction

1.1. Classification of MHD flows in channels of plasma accelerators with a longitudinal magnetic field

Plasma acceleration in crossed electric and magnetic fields [1] is one of the most promising effects in terms of its research and applied importance. This made it possible to develop a new type of devices – plasma accelerators [2]. Various acceleration schemes and a wide range of its parameters allow to use them across several actual and perspective technological directions.

In the present work, plasma is considered as a continuous medium, which behavior is described in terms of magnetic hydrodynamics (MHD). Plasma motion in an accelerator channel is due to the interaction between an electric current flowing in it and a transverse magnetic self-field generated in the plant. The acceleration mechanism under consideration can be enhanced by creation of a longitudinal magnetic field using external electric current conductors.

Depending on the relation between the plasma velocity υ and the alfvenic one C_{A} (corresponding to the longitudinal magnetic field), steady-state MHD flows in an accelerator channel belong to one of the two classes [3]: superalfvenic ($\upsilon > C_{\mathrm{A}}$) or subalfvenic ($\upsilon < C_{\mathrm{A}}$). A general classification of steady-state flow types in a coaxial channel of a plasma accelerator is provided on figure 1 [4].

The notion of the transalfvenic flows in which the transition through the alfvenic velocity C_{A} takes place is introduced in the present paper. In this case, combined flow regimes simultaneously containing areas of superalfvenic and subalfvenic classes are realized in an accelerator channel.

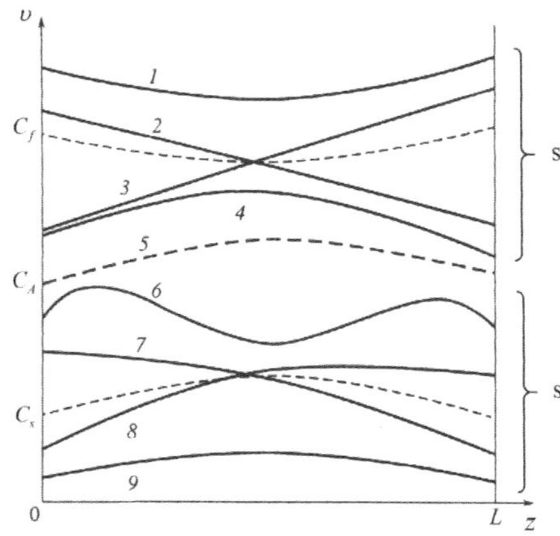

Figure 1. Classification of steady-state flows in plasma accelerator with longitudinal field

1.2. Objective of the work

The main attention of the previous researches was focused on the study of the superalfvenic flow regime with acceleration (type 3 on figure 1). The analyses of how electrode geometry and a longitudinal magnetic field influence on properties of this flow type are presented in [5, 6].

In the paper [7], features of the relaxation process and characteristics of the alfvenic regime (type 5) in which the plasma velocity coincides with the alfvenic one along the channel had been determined. This flow type can exist only in a narrow tube with a horizontal boundary.

The objective of the present work is the research of the previously unexplored transalfvenic flows with the transition through the alfvenic velocity in the accelerator channel space. Coaxial narrow nozzle-type tubes formed by close flow trajectories are considered, therefore the formulation of the corresponding problem is provided in the quasi-one-dimensional approach (see, e.g., [8]).

2. Mathematical model

2.1. Object of modelling

Preshaped electrodes connected to an external electric current source create the coaxial nozzle-type channel of the plasma accelerator (figure 2). Working medium in the plasma state enters the channel. Plasma as an electrically conductive medium completes the electric circuit formed, and the discharge current with the density $\mathbf{j}$ arises in it. The electric current flowing through the anode creates the transverse magnetic self-field $\mathbf{H}_\perp$. Thus, the Ampere force $\mathbf{F}_\parallel = (1/c)\,\mathbf{j}\times\mathbf{H}_\perp$ appears in the system and causes plasma acceleration along the channel axis. External electric current conductors induce the longitudinal magnetic field $\mathbf{H}_\parallel$. It causes rotation of the plasma flux in the transverse direction of the channel with the force $\mathbf{F}_\perp = (1/c)\,\mathbf{j}\times\mathbf{H}_\parallel$.

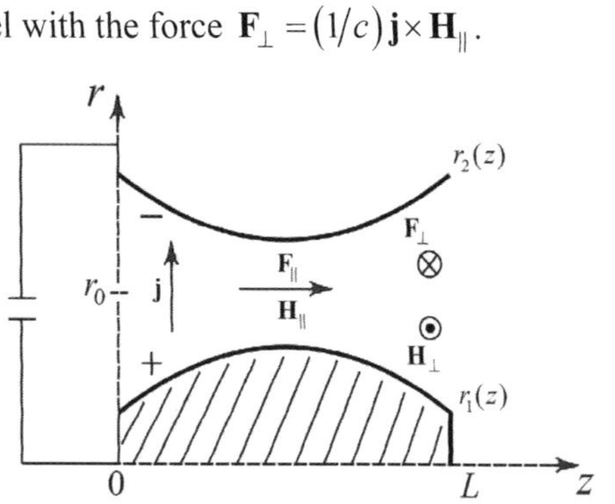

Figure 2. Coaxial plasma accelerator scheme

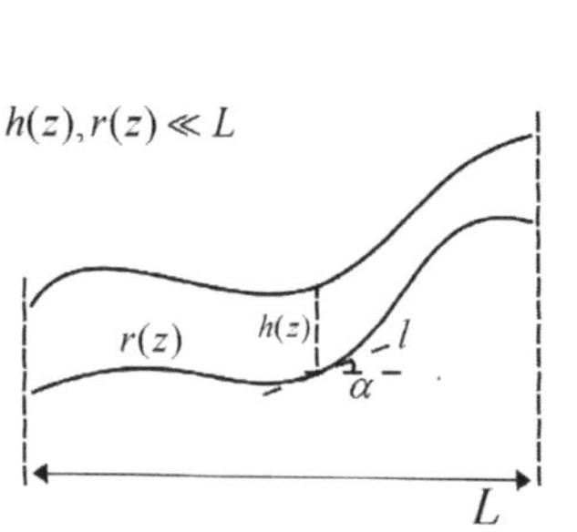

Figure 3. Coaxial narrow flow tube scheme

2.2. Statement of the steady-state problem

The axisymmetric $(\partial/\partial\varphi \equiv 0)$ MHD flows in narrow tubes with a ring cross-section $S(z)$ formed by close trajectories of the flux motion (figure 3) in the plasma accelerator channel are considered.

Plasma enters the channel of the length L with the density ρ_0, the temperature T_0 and the pressure p_0 in the absence of rotation $\upsilon_{\varphi 0} = 0$. The azimuthal magnetic field $H_{\varphi 0}$ is specified in the accelerator input section. The longitudinal field can be defined by a some constant value at the channel entrance.

Steady-state flows of a continuous infinitely conducting medium consisting of ions and electrons with common macro-parameters in the quasi-one-dimensional approach are described by the following system of algebraic equations – conservation laws (in the dimensionless form):

$$\rho\upsilon_z S = C_1,\ H_z S = C_2,\ \frac{p}{\rho^\gamma} = C_3,\ \left(C_1\upsilon_\varphi - C_2 H_\varphi\right) r = C_4,$$
$$\left(\frac{C_1 H_\varphi}{\rho} - C_2\upsilon_\varphi\right)\frac{1}{r} = C_5,\ \frac{\mathbf{v}^2}{2} + \mathrm{W} + \frac{C_5}{C_1} H_\varphi r = C_6, \tag{1}$$

where $S = r(z)h(z)$, $\mathrm{W} = \gamma p/(\gamma-1)\rho$ – entalphy.

The system of algebraic equations (1) completely describes steady-state plasma flows in coaxial narrow tubes. The values of all constants, except C_1, are determined by the input boundary conditions:

$$\rho = 1,\ T = 1,\ H_\varphi = r_0/r(0),\ \upsilon_\varphi = 0,\ H_z = H_{z0}, \tag{2}$$

where r_0 – the average accelerator channel radius.

According to the classification (figure 1), the flows of the superalfvenic subsonic (type 4) and subalfvenic supersonic (type 6) regimes are adjacent to the alfvenic velocity. Therefore, the transalfvenic flows located between these types are considered in the present paper.

These flow types are required an additional boundary condition at the channel exit to be defined. Since the main interest is focused on the process of the transition through the alfvenic velocity, it makes sense to specify the following relation at the tube exit:

$$\upsilon_l = C_{Al} + \delta, \tag{3}$$

where $\upsilon_l = \upsilon_z \cos\alpha = \upsilon_z\sqrt{1 + r'(z)^2}$, $C_{Al} = H_z\sqrt{1 + r'(z)^2}/\rho$ – the plasma velocity and the alfvenic one along the flow tube.

The value δ acts as a parameter in the problem under consideration. It allows to vary the difference between the plasma velocity and the alfvenic one at the tube output.

It is known [4, 7] that in a coaxial narrow nozzle-type tube symmetrical relative to its center, distributions of the basic plasma parameters of the flow types 4 and 6 are also symmetrical. Therefore, the constant C_1 (that equals to the flow input velocity to within a factor) is defined by the condition (3) on the plasma output velocity.

Excluding all the variables, except ρ, the system (1) is reduced to the single algebraic equation:

$$\Phi(\rho, z, C_1) \equiv F_0(\rho, z, C_1) + F_1(\rho, z, C_1) - \frac{C_1^2\left(1 + r'(z)^2\right)}{2S^2(z)} = 0, \text{ where}$$
$$F_0 = \rho^2\left(C_6 - \frac{\gamma}{\gamma-1} C_3\rho^{\gamma-1} - \rho\left(\frac{C_5}{C_1}\right)^2 r(z)^2\right) \tag{4}$$
$$F_1 = \frac{\rho^2 k^2}{\left(1 - k^2\rho\right)^2}\left(-\frac{C_4}{C_2 r(z)} - \rho\frac{r(z)C_5}{C_1}\right)\left(\frac{C_4}{2C_2 r(z)} + \rho\frac{r(z)C_5}{C_1}\left(\frac{3}{2} - \rho k^2\right)\right)$$
$$k = C_2/C_1$$

After solving the equation (4), all other flux parameters are determined through $\rho(z)$ by using the equations (1).

3. Results

The results of the calculations are represented by two series, differing from each other by a value of the longitudinal field and by geometry of the flow tubes identified in the channels of two different configurations: convex outward and inward relative to the channel axis (figure 4). The dimensionless parameter, the ratio of the characteristic gas and magnetic pressures, $\beta = 0.05$ corresponds to a strong azimuthal field and electric current.

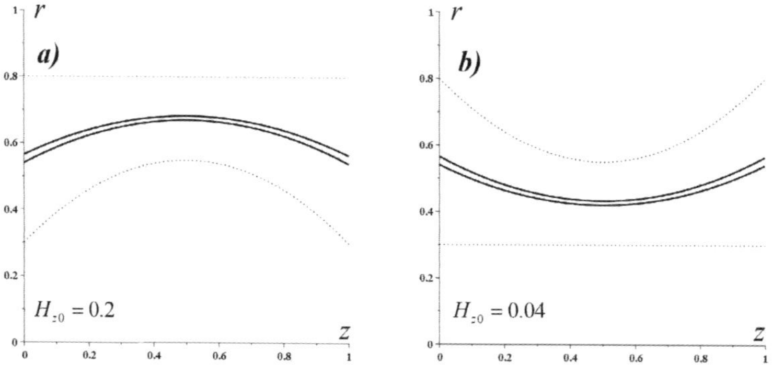

Figure 4. Narrow nozzle-type flow tubes convex outward (a) and inward (b) relative to $z-$axis

3.1. *Plasma parameters in the transition point through the alfvenic velocity and its place*

The function $\Phi(\rho, z, C_1)$ has a pole of order 2 in the transition place through the alfvenic velocity. A numerical solution of the equation (4) therefore doesn't exist in this point.

Operating with the system (1), the analytical expressions for the desired plasma flux parameters in the transition point had been obtained:

$$\rho^{\text{alf}} = \frac{1}{k^2},\ \upsilon_z^{\text{alf}} = k\frac{C_2}{S(z^{\text{alf}})},\ \upsilon_\varphi^{\text{alf}} = k\left(\frac{C_4}{C_2 r(z^{\text{alf}})} + H_\varphi^{\text{alf}}\right)$$
$$H_\varphi^{\text{alf}} = \frac{1}{k}\sqrt{2C_6 - k^2\left[\left(\frac{C_2}{S(z^{\text{alf}})}\right)^2\left(1 + r'(z^{\text{alf}})^2\right) + \left(\frac{C_4}{C_2 r(z^{\text{alf}})}\right)^2 + 2C_3\frac{\gamma}{(\gamma-1)k^{2\gamma}}\right]} \quad (5)$$

The geometric place of the transition is defined as follows:

$$r(z^{\text{alf}}) = \sqrt{-k\frac{C_4}{C_5}} \Rightarrow z^{\text{alf}} = f^{-1}\left(\sqrt{-k\frac{C_4}{C_5}}\right),\ \text{where } r = f(z) \quad (6)$$

3.2. *Narrow flow tube convex outward*

In this section, the results of the calculations of the transalfvenic flows in the narrow flow tube convex outward (figure 4a) are presented.

Positive values of δ correspond to the superalfvenic subsonic flow regime (type 4) with distributions of plasma parameters symmetrical relative to the tube center. Decrease of δ results in the plasma velocity decreasing along the channel.

When $\delta < 0$, the flow transition through the alfvenic velocity takes place. The areas of the subalfvenic regime begin to form in the input and output sections of the tube. With increasing δ towards negative values, the subalfvenic areas expand to the channel center until the flow type will not be completely changed to the subalfvenic supersonic one (type 6).

It is important to note that the opportunity of the transition through the alfvenic velocity is entirely based on curvilinear geometry of the tubes under consideration. In a narrow tube with a horizontal boundary, this transition is impossible. In this case, the plasma velocity can only coincide with the alfvenic one along the entire channel length, and so-called alfvenic regime (type 5) is realized.

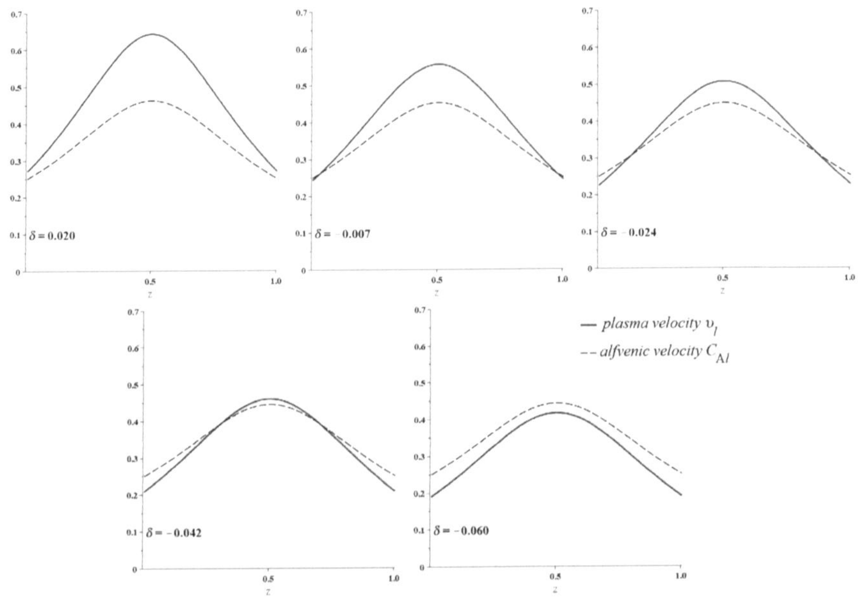

Figure 5. Process of transition through alfvenic velocity in narrow tube convex outward

The plasma density and the azimuthal magnetic field, meanwhile, increase in the channel center, but the rotational velocity decreases (figure 6).

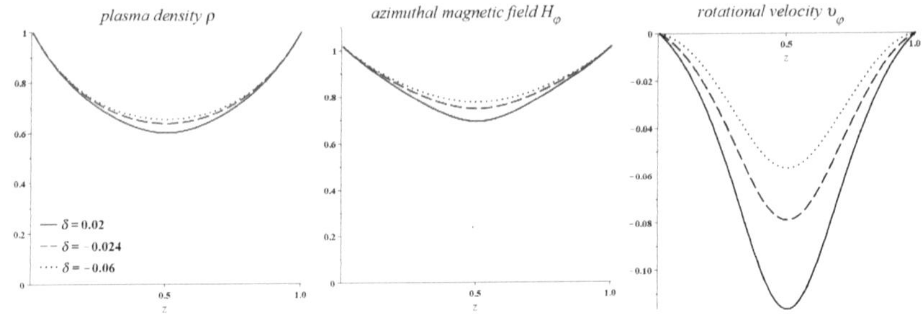

Figure 6. Distributions of basic plasma parameters in process of transition through alfvenic velocity in narrow flow tube convex outward

3.3. Narrow flow tube convex inward

In this section, the reverse geometry of the tube is considered: now, it is convex inward relative to the channel axis (figure 4b).

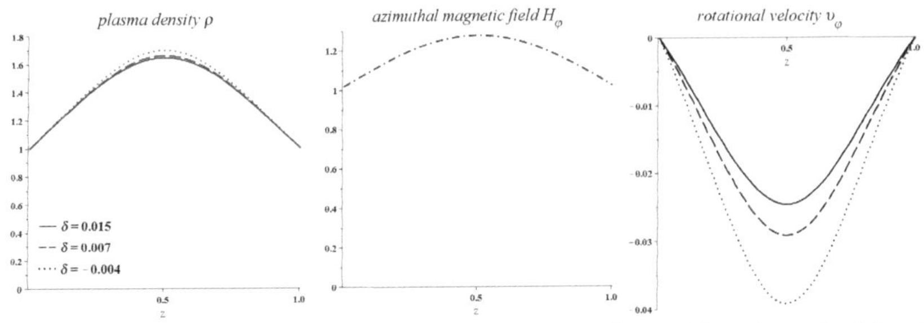

Figure 7. Distributions of basic plasma parameters in process of transition through alfvenic velocity in narrow flow tube convex inward

IOP Conf. Series: Journal of Physics: Conf. Series **1205** (2019) 012055 doi:10.1088/1742-6596/1205/1/012055

The plasma density and rotational velocity increase in the central part of the channel (figure 7), while the azimuthal field is remained almost unchanged (i.e. insignificantly decreases).

Contrary to the previous case, while reducing δ, i.e. reducing the output velocity, the transition through the alfvenic velocity initially takes place in the tube center. The flow regime changing arises here and then symmetrically moves to the tube boundaries (figure 8).

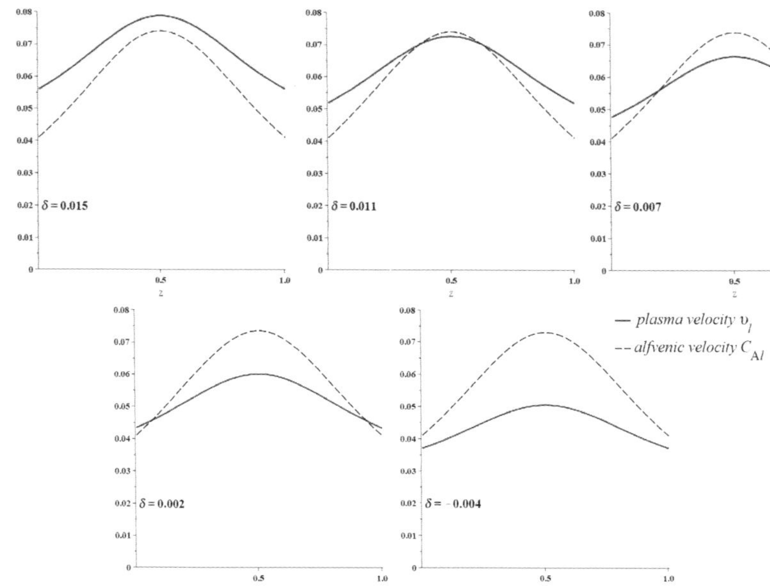

Figure 8. Process of transition through alfvenic velocity in narrow tube convex inward

4. Conclusions

In the present study, the results of the theoretical and numerical researches of transalfvenic flows in coaxial channels of plasma accelerators with a longitudinal magnetic field are presented. The steady-state quasi-one-dimensional MHD model in narrow flux tubes bounded by close flow trajectories is considered.

The following results are highlighted as the most significant:

- The analytical expressions for the plasma parameters in the transition point through the alfvenic velocity and for its geometric place are obtained;
- The characteristics of the transalfvenic flows in the coaxial narrow nozzle-type tubes of the two configurations are numerically studied.

Acknowledgments

The reported study was funded by RFBR according to the research project № 18-31-00351.

References

[1] Morozov A I 1957 *JETP* **5** (2) 215–20
[2] Artsimovich L A, Luk'ianov S Iu, Podgornyi I M, Chuvatin S A 1958 *JETP* **6** (1) 1–5
[3] Morozov A I and Solov'ev L S 1980 vol 8 ed by M A Leontovich *Reviews of Plasma Physics* (NY, London: Consultants Bureau) pp 1–103
[4] Brushlinskii K V 2009 *Mathematical and Computational Problems in Magnetohydrodynamics* (Moscow: Binom) [in Russian]
[5] Brushlinskii K V and Styopin E V 2017 *J. Phys.: Conf. Ser.* **937** 012007
[6] Brushlinskii K V, Zhdanova N S, Stepin E V 2018 *Comp. Math. and Math. Phys.* **58** (4) 593–603
[7] Styopin E V 2015 J. Plasma Phys. **81** 905810309
[8] Brushlinskii K V and Styopin E V 2017 *J. Phys.: Conf. Ser.* **788** (1) 012009

Lyapunov functions and asymptotic analysis of a complex analogue of the second Painlevé equation

Oskar Sultanov

Institute of Mathematics, Ufa Federal Research Centre, Russian Academy of Sciences, 112, Chernyshevsky str., Ufa, Russia

E-mail: oasultanov@gmail.com

Abstract. We consider the ordinary differential equation on the real axis that is a complex analogue of the second Painlevé equation. The solutions with a growing amplitude at positive infinity and the solutions that tend to zero at negative infinity are investigated. By applying Lyapunov function method we analyze the stability of such solutions and construct the long-term asymptotics for general solutions.

1. Introduction

Consider the ordinary differential equation:

$$w''_{zz} + (2|w|^2 - z)w = 0, \quad z \in \mathbb{R}, \quad w \in \mathbb{C}, \tag{1}$$

that is a complex analogue of the second Painlevé equation. Such an equation appears in the description of the transition layer in the problems of self-steepening [1, 2] and self-focusing [3]. In this paper, we study the solutions with unbounded growing amplitude as $z \to +\infty$ and the solutions that tend to zero as $z \to -\infty$. It follows from [4] that the formula $f = (\log w)'_z$ connects such functions with complex singular solutions of the second Painlevé equation $f''_{zz} = 2f^3 - 2zf + \kappa$, where $\kappa = 2(w_z w^* - w^*_z w) + 1$. Our goal is the construction of asymptotics for general solutions to equation (1) as $z \to \pm\infty$.

Note that the substitution $w(z) = R(z)\exp(i\Phi(z))$ in equation (1) leads to the following system: $R''_{zz} - R(\Phi'_z)^2 + (2R^2 - z)R = 0$, $R\Phi''_{zz} + 2R'_z\Phi'_z = 0$. It follows from the second equation that

$$\Phi(z) = c_2 + c_1 \int_{z_0}^{z} \big(R(\zeta)\big)^{-2} d\zeta, \quad z_0 = \text{const}, \tag{2}$$

with arbitrary parameters c_1, c_2. The substitution of the expression for $\Phi(z)$ in the first equation leads to the ordinary differential equation for $R(z)$:

$$R'' + (2R^2 - z)R - c_1^2 R^{-3} = 0. \tag{3}$$

Asymptotic solutions of (3) can be constructed in the form:

$$R_+ = \sqrt{\frac{z}{2}} + \sum_{k=2}^{\infty} a_k z^{-3k/2+1/2}, \quad z \to +\infty; \quad R_- = \sum_{k=0}^{\infty} b_k(-z)^{-3k/2-1/4}, \quad z \to -\infty. \tag{4}$$

IOP Conf. Series: Journal of Physics: Conf. Series **1205** (2019) 012056 doi:10.1088/1742-6596/1205/1/012056

Substituting these series in (3) and grouping the expressions of the same powers of z and $-z$ give the recurrence relations for determining the constant coefficients a_k and b_k. In particular, $a_2 = (1+16c_1^2)/(8\sqrt{2})$, $b_0 = |c_1|^{1/2}$, $b_1 = -|c_1|^{3/2}/2$. The existence of solutions $R_+(z)$ as $z > z_*$, $z_* = \text{const} > 0$ and $R_-(z)$ as $z < -z_*$ with such asymptotics follows from [5].

In the first step, we investigate the stability of the particular solutions $R_+(z)$ and $R_-(z)$ as $z \to +\infty$ and $z \to -\infty$, respectively. The stability will ensure the existence of a family of solutions to equation (3) with a similar long-term behaviour. In the last step, asymptotics for such solutions will be constructed by averaging method [6, 7] with some modifications.

2. Stability of solutions $R_-(z)$ and $R_+(z)$

Theorem 1 *The solution $R_+(z)$ to equation (3) is asymptotically stable as $z \to +\infty$. Moreover, for all $\epsilon \in (0,1)$ there exist $\delta_0 > 0$ and $z_0 > 0$ such that for all (R_0, R_1): $(R_+(z_0) - R_0)^2 + (R_+'(z_0) - R_1)^2 < \delta_0^2$ the solution to equation (3) with initial data $R(z_0) = R_0$, $R'(z_0) = R_1$ has the asymptotics:*

$$R(z) = \sqrt{\frac{z}{2}} + \mathcal{O}\Big(z^{\frac{1}{2}-\frac{3(1-\epsilon)}{4(1+\epsilon)}}\Big), \quad R'(z) = \frac{1}{2}\sqrt{\frac{1}{2z}} + \mathcal{O}\Big(z^{1-\frac{3(1-\epsilon)}{4(1+\epsilon)}}\Big), \quad z \to +\infty. \tag{5}$$

Proof. In equation (3) we make the change of variables:

$$R(z) = R_+(z) + \frac{z^{1/2} r(t)}{\sqrt{2}}, \quad t = \frac{2z^{3/2}}{3}, \tag{6}$$

and for new functions $r(t)$, $v(t) = r'(t)/\sqrt{2}$ we study the stability of the fixed point $(0,0)$ in the following system:

$$\frac{dr}{dt} = \partial_v H(r,v,t), \quad \frac{dv}{dt} = -\partial_r H(r,v,t) - t^{-1} v, \tag{7}$$

where

$$H = \frac{v^2}{\sqrt{2}} + \frac{r^2(3p_*^2 - 1)}{2\sqrt{2}} + \frac{r^3 p_*}{\sqrt{2}} + \frac{r^4}{4\sqrt{2}} + t^{-2} r^2 \Big(\frac{4\sqrt{2}c_1^2(2r + 3p_*)}{9p_*^3(r+p_*)^2} - \frac{1}{18\sqrt{2}}\Big),$$

$p_*(t) = \sqrt{2}R_+(z)z^{-1/2} = 1 + \mathcal{O}(t^{-2})$ as $t \to \infty$. It is easy to calculate that $H(r,v,t) = \big(r^2 + v^2 + r^3 + r^4/4\big)/\sqrt{2} + \mathcal{O}(d^2)\mathcal{O}(t^{-2})$, uniformly as $t \to \infty$ and $d = \sqrt{r^2+v^2} \to 0$. We see that the fixed point $(0,0)$ is a center in the asymptotic limit ($t \to \infty$), that, however, does not guaranty the stability in the complete system [8]. A Lyapunov function candidate is constructed of the form [9, 10]:

$$V(r,v,t) = \sqrt{2}\Big[H(r,v,t) + t^{-1}\frac{rv}{2}\Big]. \tag{8}$$

The function V is positive definite in the vicinity of the fixed point: $V = d^2 + \mathcal{O}(d^3) + \mathcal{O}(d^2)\mathcal{O}(t^{-2})$ as $t \to \infty$ and $d \to 0$. The total derivative of this function with respect to t along the trajectories of system (7) has the following asymptotics:

$$\frac{dV}{dt}\Big|_{(7)} \stackrel{def}{=} \partial_t V + \partial_r V \partial_v H + \partial_v V[-\partial_r H - t^{-1}v] = -t^{-1}[d^2 + \mathcal{O}(d^3)][1 + \mathcal{O}(t^{-1})].$$

Therefore for all $\epsilon > 0$ there exist $d_0 > 0$ and $t_0 > 0$ such that the function V satisfies the inequalities: $(1-\epsilon)d^2 \le V(r,v,t) \le (1+\epsilon)d^2$, $dV/dt|_{(7)} \le -t^{-1}\sigma V$, $\sigma = (1-\epsilon)/(1+\epsilon)$, as $d \leq d_0$ and $t \geq t_0$. Moreover, for all $\varepsilon \in (0, d_0)$ there exists $d_\varepsilon = \varepsilon\sqrt{\sigma/2}$ such that $\sup_{d\leq d_\varepsilon} V(r,v,t) \leq (1+\epsilon)d_\varepsilon^2 \leq (1-\epsilon)\varepsilon^2 \leq \sup_{d=\varepsilon} V(r,v,t)$ as $t > t_0$. The last estimates and the negativity of the total derivative of the function V ensure that any solution of system (7) with initial data from the ball $d \leq d_\varepsilon$ cannot leave ε-neighbourhood of the equilibrium as $t > t_0$. Consider the solution $r(t)$, $v(t)$ to system (7) with initial data $r^2(t_0) + v^2(t_0) \leq d_0^2$, then the function $\ell(t) \equiv V(r(t), v(t), t)$ satisfies the inequality $d\ell/dt \leq -\sigma t^{-1}\ell$ as $t > t_0$. Integrating the last expression with respect to t, we obtain $0 \leq \ell(t) \leq \ell_0 t^{-\sigma}$ with the parameter ℓ_0, depending on d_0 and t_0. This implies the asymptotic estimate: $r^2(t)+v^2(t) = \mathcal{O}(t^{-\sigma})$ as $t \to \infty$. Returning to the original variables we derive asymptotic estimates (5) as $z \to +\infty$.

Theorem 2 *The solution $R_-(z)$ to equation* (3) *is stable as $z \to -\infty$. Moreover, there exist $\delta_0 > 0$ and $z_0 > 0$ such that for all (R_0, R_1): $(R_-(-z_0) - R_0)^2 + (R_-'(-z_0) - R_1)^2 < \delta_0^2$ the solution to equation* (3) *with initial data $R(-z_0) = R_0$, $R'(-z_0) = R_1$ has the asymptotics:*

$$R(z) = \mathcal{O}\Big((-z)^{-1/4}\Big), \quad R'(z) = \mathcal{O}\Big((-z)^{1/4}\Big), \quad z \to -\infty. \tag{9}$$

Proof. The exchange of variables $z = -y$, $R(z) = R_-(z) + y^{-1/4}r(t)/2$, $t = 2y^{3/2}/3$ leads to the non-autonomous Hamiltonian system:

$$\frac{dr}{dt} = \partial_v H(r,v,t), \quad \frac{dv}{dt} = -\partial_r H(r,v,t), \tag{10}$$

where

$$H(r,v,t) = v^2 + \frac{r^2}{4} + \frac{c_1^2 r^2(r+3q_*)}{q_*^3(2q_*+r)^2} + t^{-1}\Big(r^2q_*^2 + \frac{r^3 q_*}{3} + \frac{r^4}{24}\Big) + t^{-2}\frac{5r^2}{144},$$

$q_*(t) = R_* y^{1/4} = b_0 + 2b_1 t^{-1}/3 + \mathcal{O}(t^{-2})$ as $t \to \infty$. For the new variables $r(t)$ and $v(t) = r'(t)/2$ we study the stability of the fixed point $(0,0)$ in the Hamiltonian system (10). Note that the function $H(r,v,t)$ has the following asymptotics as $t \to \infty$ and $d \to 0$: $H(r,v,t) = r^2+v^2+\mathcal{O}(d^3)+t^{-1}[2|c_1|r^2+\mathcal{O}(d^3)]+\mathcal{O}(t^{-2})\mathcal{O}(d^2)$. As above, the fixed point $(0,0)$ is a center in the asymptotic limit. A Lyapunov function candidate for system (10) is constructed in the form: $V(r,v,t) = H(r,v,t) - t^{-2}|c_1|rv/2$. The function $V(r,v,t)$ is positive in a punctured neighborhood of the fixed point: $V = d^2[1+\mathcal{O}(d)+\mathcal{O}(t^{-1})]$ as $t \to \infty$ and $d \to 0$. The total derivative of $V(r,v,t)$ with respect to t along the the trajectories of system (10) has the following asymptotics: $dV/dt|_{(10)} = -t^{-2}|c_1|[r^2+v^2+\mathcal{O}(d^3)]+\mathcal{O}(d^2)\mathcal{O}(t^{-3})$. Moreover, for all $\epsilon > 0$ there exist $d_0 > 0$ and $t_0 > 0$ such that $(1-\epsilon)d^2 \leq V(r,v,t) \leq (1+\epsilon)d^2$, $dV/dt|_{(10)} \leq -|c_1|\sigma t^{-2}V$ as $d \leq d_0$ and $t \geq t_0$. Consider the solution to system (7) with initial data from the ball $d \leq d_0$, then the function $\ell(t) \equiv V(r(t),v(t),t)$ satisfies the inequality: $d\ell/dt \leq -t^{-2}\sigma|c_1|\ell$. Hence $0 \leq \ell(t) \leq \ell_0$ as $t > t_0$ and $r^2(t) + v^2(t) = \mathcal{O}(1)$ as $t \to \infty$. Returning to the original variables we derive asymptotic estimates (9) as $z \to -\infty$.

3. Asymptotics for general solutions

The stability of the particular solutions $R_+(z)$, $R_-(z)$ ensures the existence of a family of solutions to (3) with a similar behaviour as $z \to +\infty$ and $z \to -\infty$. At this section we construct the asymptotics for such solutions by averaging method with a use of the Lyapunov functions found in the previous section.

Theorem 3 *For all $c_1 \in \mathbb{R}$ equation* (3) *has two-parameter family of solutions $R_+(z; c_3, c_4)$ with the asymptotics:*

$$R(z) = \sqrt{\frac{z}{2}} + z^{-1/4}\sum_{k=0}^{\infty}\rho_k(S(z);c_3)z^{-3k/4}, \quad R'(z) = z^{1/4}\sum_{k=0}^{\infty}\vartheta_k(S(z);c_3)z^{-3k/4} \tag{11}$$

as $z \to +\infty$*, where the functions* $\rho_k(S; c_3)$*,* $\vartheta_k(S; c_3)$ *are* 2π*-periodic in* S*,* $S(z) = c_4 + 2\sqrt{2}z^{3/2}/3 + 3c_3^2(\log z)/(4\sqrt{2})$*,* $\rho_0 = c_3\cos S$*,* $\vartheta_0 = -c_3\sqrt{2}\sin S$*.*

Proof. In the first step, we make change of variables (6) in system (3) and for new functions $r(t)$, $v(t) = r'(t)/\sqrt{2}$ we consider the non-autonomous near-Hamiltonian system (7) in a neighbourhood of the stable fixed point $(0,0)$. Let us consider a Hamiltonian system: $r' = \partial_v H_0$, $v' = -\partial_r H_0$, where $H_0(r,v) = \lim\limits_{t\to\infty} H(r,v,t)$ and $H_0 = (v^2 + r^2 + r^3 + r^4/4)/\sqrt{2}$. It is easy to see, that the level lines $\{(r,v) : H_0(r,v) = h\}$ define a family of closed curves on the phase space (r,v) parameterized by the parameter $h \in (0,h_0)$, $h_0 = \text{const} > 0$. To each closed curve there corresponds a $T(h)$-periodic solution $r_0(t,h)$, $v_0(t,h)$ of the Hamiltonian system such that $T(h) = 2\pi/\omega(h)$, $\omega(h) = \sqrt{2} + 3h/8 + \mathcal{O}(h^2)$ as $h \to 0$. Consider the transformation of variables $r = \widetilde{r}_0(s,h)$, $v = \widetilde{v}_0(s,h)$, where $\widetilde{r}_0(s,h) = r_0(s/\omega(h),h)$, $\widetilde{v}_0(s,h) = v_0(s/\omega(h),h)$. Then in new variables (s,h) system (7) takes the form: $dh/dt = -\omega(h)[\partial_s\widetilde{H} + t^{-1}\partial_s\widetilde{r}_0\widetilde{v}_0]$, $ds/dt = \omega(h)[\partial_h\widetilde{H} + t^{-1}\partial_h\widetilde{r}_0\widetilde{v}_0]$, where $\widetilde{H}(s,h,t) \equiv H(\widetilde{r}_0(s,h),\widetilde{v}_0(s,h),t)$ is 2π-periodic functions with respect to s. Moreover, the last system is asymptotically Hamiltonian with decaying non-Hamiltonian terms as $t \to \infty$. Solutions of this systems can be investigated by the averaging method. To simplify asymptotic constructions, we make the additional transformation of variables. Introduce a new variable $\ell(t)$ associated with the Lyapunov function (8) such that $\ell(t) = \widetilde{V}(s(t),h(t),t)$, where $\widetilde{V}(s,h,t) \equiv V(\widetilde{r}_0(s,h),\widetilde{v}_0(s,h),t)$ is 2π-periodic function in s. It follows from the properties of the Lyapunov function that $\widetilde{V} = \sqrt{2}h[1 + \mathcal{O}(t^{-1})]$, $d\widetilde{V}/dt|_{(7)} = -t^{-1}[\sqrt{2}h + \mathcal{O}(h^{3/2})] + \mathcal{O}(h)\mathcal{O}(t^{-2})$ as $t > t_0$, $h < h_0$, uniformly with respect to $s \in \mathbb{R}$. Hence the transformation $(s,h) \mapsto (s,\ell)$ is invertible: $\partial_h\widetilde{V} \neq 0$. System (7) in new variables takes the form:

$$\frac{d\ell}{dt} = F(s,\ell,t), \quad \frac{ds}{dt} = G(s,\ell,t), \tag{12}$$

where $F(s,\widetilde{V},t) = \partial_r\widetilde{V}\partial_v\widetilde{H} + \partial_v\widetilde{V}(\partial_r\widetilde{H} - t^{-1}\widetilde{v}_0) + \partial_t\widetilde{V}$, $G(s,\widetilde{V},t) = \omega(h)\big[\partial_h\widetilde{H} + t^{-1}\partial_h\widetilde{r}_0\widetilde{v}_0\big]$. It is not difficult to deduce the asymptotics of the functions F and G as $t \to \infty$: $F = \sum_{k=1}^{\infty} f_k(s,\ell)t^{-k}$, $G = g_0(\ell) + \sum_{k=1}^{\infty} g_k(s,\ell)t^{-k}$, where $f_k(s,\ell)$ and $g_k(s,\ell)$ are 2π-periodic functions with respect to s, $f_1(s,\ell) = -\ell + \mathcal{O}(\ell^{3/2})$, $g_0(\ell) = \omega(\ell/\sqrt{2}) = \sqrt{2} + 3\ell/(8\sqrt{2}) + \mathcal{O}(\ell^2)$, $g_1(\ell,s) = -\sin s\cos s + \mathcal{O}(\ell)$ as $\ell \to 0$, $s \in [0,2\pi)$. To construct asymptotic solutions to system (12), it is convenient to single out a Hamiltonian part:

$$\frac{d\ell}{dt} = -\partial_s\mathcal{H} + \mathcal{F}, \quad \frac{ds}{dt} = \partial_\ell\mathcal{H}, \tag{13}$$

where $\mathcal{H}(s,\ell,t) = \int\limits_0^\ell G(s,l,t)\,dl$, $\mathcal{F}(s,\ell,t) = F(s,\ell,t) + \partial_s\mathcal{H}(s,\ell,t)$ are 2π-periodic with respect to s. The asymptotic solution to the first equation in (13) is sought in the form: $\ell(t) = \lambda(t) + \Lambda(s,\lambda(t),t)$, where $\lambda(t)$ is the solution of the averaged equation:

$$\frac{d\lambda}{dt} = \langle\mathcal{F}(s,\lambda + \Lambda(s,\lambda,t),t)\rangle_s, \tag{14}$$

and $\Lambda(s,\lambda,t)$ is 2π-periodic function in s with a zero average $\langle\Lambda(s,\lambda,t)\rangle_s \stackrel{def}{=} (2\pi)^{-1}\int\limits_0^{2\pi}\Lambda(s,\lambda,t)ds \equiv 0$, and satisfies the equation $\partial_s\mathcal{H}(s,\lambda+\Lambda,t) + \partial_\ell\mathcal{H}(s,\lambda+\Lambda,t)\partial_s\Lambda + \partial_t\Lambda = \mathcal{F}(s,\lambda+\Lambda,t) - [1 + \partial_\lambda\Lambda]\langle\mathcal{F}(s,\lambda+\Lambda,t)\rangle_s$. We integrate the last equation with respect to s and choose the constant of integration in such a way that the result has a zero average:

$$\mathcal{H}(s,\lambda+\Lambda,t)-\langle\mathcal{H}(s,\lambda+\Lambda,t)\rangle_s+\int\limits_{s_*}^{s}\partial_t\Lambda(\sigma,\lambda,t)\,d\sigma-\Big\langle\int\limits_{s_*}^{s}\partial_t\Lambda(\sigma,\lambda,t)\,d\sigma\Big\rangle_s=$$
$$\int\limits_{s_*}^{s}\mathcal{F}(\sigma,\lambda+\Lambda,t)-\langle\mathcal{F}(\sigma,\lambda+\Lambda,t)\rangle_s d\sigma-\Big\langle\int\limits_{s_*}^{s}\mathcal{F}(\sigma,\lambda+\Lambda,t)-\langle\mathcal{F}(\sigma,\lambda+\Lambda,t)\rangle_s d\sigma\Big\rangle_s \quad (15)$$
$$-\langle\mathcal{F}(s,\lambda+\Lambda,t)\rangle_s\Big(\int\limits_{s_*}^{s}\partial_\lambda\Lambda(\sigma,\lambda,t)d\sigma-\Big\langle\int\limits_{s_*}^{s}\partial_\lambda\Lambda(\sigma,\lambda,t)d\sigma\Big\rangle_s\Big),$$

$s_*=\text{const}$. The asymptotic solution of (15) is sought in the form: $\Lambda(s,\lambda,t)=\sum_{k=1}^{\infty}\lambda_k(s,\lambda)t^{-k}$. Substituting this series into equation (15), and equating the terms of the same power of t, we obtain the chain of equations: $g_0(\lambda)\lambda_k=L_k(s,\lambda)-\langle L_k(s,\lambda)\rangle_s$, $k\geq 1$, where each function L_{k+1} is expressed through $\lambda_1,\ \dots,\ \lambda_k$. For example, $L_1=\int\limits_{s_*}^{s}\{f_1-\langle f_1\rangle_s\}\,d\sigma$, $L_2=\int\limits_{s_*}^{s}\{f_2+\lambda_1(1+\partial_\ell f_1+\partial_s g_1)-\langle f_2+\lambda_1(\partial_\ell f_1+\partial_s g_1)\rangle_s-\langle f_1\rangle_s\partial_\lambda\lambda_1\}\,d\sigma-\lambda_1 g_1-\lambda_1^2\partial_l g_0/2$. It can easily be checked that all coefficients ϑ_k are uniquely determined in the class of 2π-periodic functions with zero average $\langle\lambda_k(s,\lambda)\rangle_s=0$.

The asymptotic solution of the equation for the angle s is sought in the form $s(t)=\psi(t)+\Psi(\psi(t),\lambda(t),t)$, where

$$\frac{d\psi}{dt}=\big\langle\partial_\ell\mathcal{H}\big(\phi+\Psi(\phi,\lambda,t),\lambda+\Lambda(\phi+\Psi(\phi,\lambda,t),\lambda,t),t\big)\big\rangle_\phi, \quad (16)$$

and the function Ψ satisfies the equation: $(\partial_\psi\Psi+1)\big\langle\partial_\ell\mathcal{H}(\phi+\Psi,\lambda+\Lambda,t)\big\rangle_\phi+\partial_t\Psi=\partial_\ell\mathcal{H}\big(\psi+\Psi,\lambda+\Lambda,t\big)-\partial_\lambda\Psi\langle\mathcal{F}(s,\lambda+\Lambda,t)\rangle_s$, with the additional condition: $\langle\Psi\rangle_\psi=0$. Asymptotics for Ψ is constructed in the form: $\Psi(\psi,\lambda,t)=\sum_{k=1}^{\infty}\psi_k(\psi,\lambda)t^{-k}$. Substituting the series into the last equation and grouping the expressions of the same power of t give the following chain of differential equations: $g_0(\lambda)\partial_\psi\psi_k=P_k(\psi,\lambda)-\langle P_k(\phi,\lambda)\rangle_\phi$, $k\geq 1$. Note that each function P_{k+1} is expressed through $\psi_1,\ \dots,\ \psi_k$. For example, $P_1=g_1+\partial_\lambda g_0\lambda_1$, $P_2=g_2+\partial_\lambda g_1\lambda_1+\partial_\lambda g_0\lambda_2+\partial_\lambda^2 g_0\lambda_1^2/2+(1+\partial_\psi g_1)\psi_1-\partial_\psi\psi_1\langle f_1+g_1+\partial_\lambda g_0\lambda_1+\partial_s\int\limits_0^\ell g_1 dl\rangle_\psi$. It follows that all coefficients ψ_k are uniquely determined in the class of 2π-periodic functions with $\langle\psi_k(\psi,\lambda)\rangle_\psi=0$.

In the last step, we integrate the averaged equations (14) and (16). Note that the stability of the trivial solution to system (7) ensures that $\ell(t)\to 0$ as $t>t_0$. Therefore, we can use the asymptotic behavior of the functions $F(s,\ell,t)$ and $G(s,\ell,t)$ as $\ell\to 0$ and $t\to\infty$ in the integration of the averaged equations. Thus we have

$$\frac{d\lambda}{dt}=-t^{-1}[\lambda+\mathcal{O}(\lambda^{3/2})][1+\mathcal{O}(t^{-1})].$$

It follows from the last equation that $\lambda(t)=\ell_0^2 t^{-1}+\sum_{k=2}^{\infty}\ell_k t^{-k}$ as $t\to\infty$, where ℓ_0 is an arbitrary parameter and the coefficients $\ell_k=\text{const}$ are uniquely determined. The averaged equation (16) for the angle is integrated trivially: $\psi(t)=\hat{\psi}_0+\int\limits_{t_0}^{t}\big\langle\partial_\ell\mathcal{H}\big(\psi+\Psi(\psi,\lambda(\tau),\tau),\lambda(\tau)+\Lambda(\psi+\Psi(\psi,\lambda(\tau),\tau),\lambda(\tau),\tau),\tau\big)\big\rangle_\psi d\tau$, where $\hat{\psi}_0=\text{const}$. The asymptotics of $\psi(t)$ is defined by the expansions for λ, Λ, and Ψ. Thus we have $\psi(t) = \psi_0 + \sqrt{2}t + \omega_{-1}\log t + \sum_{k=1}^{\infty} s_k t^{-k}$ as $t \to \infty$, where ψ_0 is an arbitrary parameter, $\omega_{-1} = 3\ell_0^2/(8\sqrt{2})$ and the coefficients $s_k = \text{const}$ are uniquely determined. Note that the justification of the constructed asymptotics follows from [11]. Thus, a two-parameter family of solutions $\ell(t;\ell_0,\psi_0)$, $s(t;\ell_0,\psi_0)$ to system (12) has the following asymptotics: $\ell(t) = \ell_0^2 t^{-1} + \sum_{k=2}^{\infty} \hat{\ell}_k(\hat{S}(t))t^{-k}$, $s(t) = \hat{S}(t) + \sum_{k=1}^{\infty} \hat{s}_k(\hat{S}(t))t^{-k}$ as $t \to \infty$, where $\hat{\ell}_k$ and $\hat{s}_k$ are 2π-periodic functions in $\hat{S}$, and the function $\hat{S}(t)$ has the following form: $\hat{S}(t) = \psi_0 + \sqrt{2}t + \omega_{-1}\log t$. The transformation formulas lead to the asymptotic estimates for solutions to system (7): $r(t) = \ell_0 t^{-1/2}\big[\cos\hat{S}(t) + \mathcal{O}(t^{-1})\big]$, $v(t) = -\ell_0 t^{-1/2}\big[\sin\hat{S}(t) + \mathcal{O}(t^{-1})\big]$. Returning to the original variables we obtain estimates (11).

Theorem 4 *For all* $c_1 \neq 0$ *equation* (3) *has two-parameter family of solutions* $R_-(z;c_3,c_4)$ *with the asymptotics:*

$$R(z) = (-z)^{-1/4}\sum_{k=0}^{\infty}\rho_k(S(z);c_3)(-z)^{-3k/2}, \quad R'(z) = (-z)^{1/4}\sum_{k=0}^{\infty}\vartheta_k(S(z);c_3)(-z)^{-3k/2},$$

as $z \to -\infty$, *where the functions* $\rho_k(S;c_3)$, $\vartheta_k(S;c_3)$ *are* 2π*-periodic in* S, $S(z) = c_4 + \omega_0(c_1,c_3)(-z)^{3/2} + \omega_{-1}(c_1,c_3)\log(-z)$, $\rho_0 = |c_1|^{1/2} + (c_3/2)\cos S$, $\vartheta_0 = -c_3\sin S$, $\omega_k(c_1,c_3) = \text{const}$.

Proof. The arguments are similar to that of Theorem 3.

The asymptotic estimates for $\Phi(z)$ are derived from the expansions for $R(z)$ as $z \to \pm\infty$. In particular, the solutions $R_+(z)$, $R_-(z)$ with asymptotics (4) lead to

$$\Phi_+ = 2c_1\log z + c_2 + c_1\sum_{k=2}^{\infty}\alpha_k z^{-3k/2}, \quad z \to +\infty,$$

$$\Phi_- = -\frac{2\text{sgn}(c_1)}{3}(-z)^{3/2} - c_1\log(-z) + c_2 + c_1\sum_{k=1}^{\infty}\beta_k(-z)^{-3k/2}, \quad z \to -\infty.$$

Here the coefficients α_k, β_k are uniquely defined, for example, $\alpha_2 = 4a_2\sqrt{2}/3$, $\beta_1 = -2(2b_2b_0 - 3b_1^2)/(3b_0^2)$.

Thus, Theorems 3 and 4 with (2) give the asymptotics for general solutions to equation (1) as $z \to \pm\infty$. The connection formulae for asymptotics of equation (1) have not been considered here. This will be discussed elsewhere.

Acknowledgments

The author is grateful to Suleimanov B. I. for discussions.

References

[1] Kudashev V R and Suleimanov B I 1995 *JETP Lett.* **62** 382–388
[2] Kudashev V R and Suleimanov B I 1999 *Theor. Math. Phys.* **118** 325–332
[3] Suleimanov B I 2017 *JETP Lett.* **106** 400–405
[4] Jimbo M and Miwa T 1981 *Physica D* **2** 407–448
[5] Kuznetsov A N 1989 *Funct. Anal. Appl.* **23** 308–317
[6] Neishtadt A I 1984 *J. Appl. Math. Mech.* **48** 133–139
[7] Brüning J, Dobrokhotov S Yu and Poteryakhin M A 2001 *Math. Notes* **70** 599–607
[8] Kalyakin L A 2015 *Math. Notes* **98** 752–764
[9] Sultanov O A 2010 *Ufimsk. Mat. Zh.* **2** 88–98
[10] Kalyakin L A and Sultanov O A 2013 *Differ. Equat.* **49** 267–281
[11] Kalyakin L 2008 *J. Nonlin. Math. Phys.* **15** suppl. 3, 210–216

IOP Conf. Series: Journal of Physics: Conf. Series **1205** (2019) 012057 doi:10.1088/1742-6596/1205/1/012057

On discrete form of the mean value inequality for subharmonic functions

D S Telyakovskii[1]

[1]National Research Nuclear University MEPhI (Moscow Engineering Physics Institute), 31 Kashirskoe shosse, 115409 Moscow, Russia

E-mail: dtelyakov@mail.ru

Abstract. We obtain a sufficient condition for the subharmonicity of a function $u(x, y) = u(z)$, $z \in G \subset \mathbb{R}^2$, in which the mean value inequality has discrete form. Namely, it is assumed that for each point $\zeta \in G$ there are a circle of arbitrarily small radius centered at ζ and a set of nodes lying on this circle for which the value $u(\zeta)$ does not exceed the arithmetic mean of the function values in the nodes of this set. A necessary and sufficient condition for the location of the nodes of the set is established when executed, the function $u(z)$ satisfying at each point of G such discrete form of mean value inequality, and, additionally, some condition of summability and continuity in the directions is subharmonic in the domain G.

1. Statement of the problem

There is obtained a sufficient condition for subharmonicity of functions $u(x, y) = u(z)$, $z \in G \subset \mathbb{R}^2$, in which the mean value inequality is considered in discrete form.

Let points $\zeta_1, \dots, \zeta_n$, $n \geq 3$ lie on a circle with the center at ζ_0. We search conditions on the location of points $\{\zeta_j\}$, under which as the mean value inequality of can be considered an inequality

$$u(\zeta_0) \leq \frac{u(\zeta_1) + \cdots + u(\zeta_n)}{n} \tag{1}$$

between the value of the function at ζ_0 and the arithmetic mean of its values in the set nodes $\{\zeta_j\}$.

In the book by I.I. Privalov [1], pp. 57–59, it was shown that continuous functions $u(z)$ are subharmonic if at each point $\zeta_0 \in G$ the inequality (1) is satisfied for all possible sets of points, lying at the vertices of regular n-gons inscribed in each circle of sufficiently small radius with center at ζ_0. In the present work we generalized this sufficient condition for subharmonicity: the assumption of continuity of the function is significantly weakened and there is obtained a necessary and sufficient condition on location of the nodes of the set $\{\zeta_j\}$, under which the inequality (1) is a discrete form of the mean value inequality for subharmonic functions.

To formulate the condition of continuity in the sense, which is considered in this paper, definitions of an asterisk are required and h-regular functions [2].

Star $S_\zeta = S(\zeta, \sigma_\zeta)$ with k rays ($k \geq 3$) and center at point ζ we will call the collection of non coinciding intervals $s^{(j)}$, $j = 1, \dots, k$ of lenght σ, starting from point ζ. We will consider only those stars S_ζ, that inside of every half plane containing point ζ at its bounding, contain at least one ray. It is clear that this condition is equivalent to the statement that point ζ is inside of a polygon $P_\zeta = P(\zeta,\sigma)$ with vertexes at the end of the rays of star S_ζ and that also there can not be less than three rays for such star.

Let $h(t)$, $t \geq 0$, be a modulus of continuity type function and and let function $u(z)$ be defined on some set B_ζ, $z \in B_\zeta$, where point ζ is the limiting point. If for some value $L_\zeta > 0$ at every point $z \in B_\zeta$ the following inequality holds $|u(z)-u(\zeta)| \leq L_\zeta h(|z-\zeta|)$, then we say that function $u(z)$ is *h-regular at point ζ with respect to set B_ζ with coefficient L_ζ*, or simply *h-regular at ζ with respect to B_ζ*, if specific value of L_ζ is not important.

Denote by $C(\zeta,r)$ a circle with center at ζ and radius r.

The following theorem is proved.

Theorem. *Let function $u(z)$ be defined in region $G \subset \mathbb{R}^2$ and for each point $\zeta_0 \in G$ there exist a circle $C(\zeta_0,\rho) \subset G$ of arbitrery small radius and a set of nodes on this circle $\zeta_j = \zeta_0 + \rho e^{i\varphi_j}$, $j = 1,\ldots,n$, $n = n(\zeta_0,\rho) \geq 3$, such that*

- 1° *for the set $\{\zeta_j\}$ the inequality (1) satisfies,*
- 2° *the point ζ_0 is the center of gravity of the system of unit masses located at points $\{\zeta_j\}$ and systems of unit masses, located at $\{\zeta'_j := \zeta_0 + \rho e^{2i\varphi_j}\}$.*

Let, further, for some function $h(t)$ of modulus of continuity type the function $u(z)$ is h-regular at each point $\zeta \in G$ with respect to some star S_ζ with center at point ζ, and the angles between the rays of the star do not exceed $\alpha\pi$, $\alpha \geq 1/2$, and the function $\left(u^+(z)\right)^{2\alpha}$ locally summable in G. Then the function $u(z)$ is subharmonic in the region G.

2. Proofs of Lemmas

To prove this theorem we need several lemmas.

Lemma 1. *Assume that function $u(z)$ has a second Peano differential at point $z_0 = (x_0,y_0)$, i.e. for some numbers p, q, r, s and t holds*

$$\begin{aligned} u(z) = u(z_0) + \Big[p(x-x_0)+q(y-y_0)\Big] + \frac{1}{2}\Big[r(x-x_0)^2+2s(x-x_0)(y-y_0)+t(y-y_0)^2\Big]+ \\ +\varepsilon(x,y)\big((x-x_0)^2+(y-y_0)^2\big) \quad \text{where} \quad eps(x,y) \to 0 \quad \text{when} \quad (x,y) \to (x_0,y_0). \end{aligned} \tag{2}$$

Let $C(z_0,\rho)$ be a circle of sufficiently small radius. Then for any collection of nodes $z_1,\ldots,z_n$, $n = n(\rho) \geq 3$, of $C(z_0,\rho)$ that satisfies the condition 2° of the theorem, the following estimate holds

$$\frac{u(z_1)+\cdots+u(z_n)}{n} - u(z_0) = \frac{r+t}{4}\rho^2 + o(\rho^2) \quad \text{when } \rho \to 0. \tag{3}$$

Proof. We represent coordinates of each point z_j in the form $x_j = x_0+\rho\cos\varphi_j$, $y_j = y_0+\rho\sin\varphi_j$, $j = 1,\ldots,n$, and substitute them in the formula (2). We have

$$\begin{aligned} u(z_j) &= u(z_0) + \rho\big[p\cos\varphi_j+q\sin\varphi_j\big]+\frac{1}{2}\rho^2\big[r\cos^2\varphi_j+2s\cos\varphi_j\sin\varphi_j+t\sin^2\varphi_j\big] + \varepsilon_j\rho^2 = \\ &= u(z_0) + \rho\big[p\cos\varphi_j+q\sin\varphi_j\big]+\frac{1}{2}\rho^2\left[r\frac{1+\cos 2\varphi_j}{2} + s\sin 2\varphi_j + t\frac{1-\cos 2\varphi_j}{2}\right] + \varepsilon_j\rho^2 = \\ &= u(z_0) + \rho\big[p\cos\varphi_j+q\sin\varphi_j\big]+\frac{1}{2}\rho^2\left[\frac{r+t}{2} + s\sin 2\varphi_j + \frac{r-t}{2}\cos 2\varphi_j\right] + \varepsilon_j\rho^2. \end{aligned}$$

We substitute resulting expressions for $u(z_j)$ in the left part of the formula (3). We obtain

$$\frac{1}{n}\big(u(z_1)+\cdots+u(z_n)\big) - u(z_0) = \rho\left[\frac{p}{n}\sum_{j=1}^{n}\cos\varphi_j + \frac{q}{n}\sum_{j=1}^{n}\sin\varphi_j\right]+$$

$$+\rho^2\left[\frac{r+t}{4}+\frac{r-t}{4n}\sum_{j=1}^{n}\cos 2\varphi_j+\frac{s}{4n}\sum_{j=1}^{n}\sin 2\varphi_j\right]+\rho^2\frac{1}{n}\sum_{j=1}^{n}\varepsilon_j.$$

Sums $\rho\sum\cos\varphi_j$ and $\rho\sum\cos 2\varphi_j$ are equal to the sums of moments of vertical forces applied to the points of systems $\{z_j\}$ and $\{z_j'\}$ ($z_j' := (x_0+\rho\cos 2\varphi_j, y_0+\rho\sin 2\varphi_j)$, $j=1,\dots,n$), respectively relative to the point z_0, and $\rho\sum\sin\varphi_j$ and $\rho\sum\sin 2\varphi_j$ — to the sums of moments of horizontal forces applied to the points of these systems, relative to z_0. The set of points $z_1,\dots,z_n$ is selected on the circle $C(z_0,\rho)$ so that the center of gravity of the systems $\{z_j\}$ and $\{z_j'\}$ is at z_0. So all these four the amounts vanish and therefore,

$$\frac{1}{n}\big(u(z_1)+\dots+u(z_n)\big)-u(z_0)=\rho^2\frac{r+t}{4}+\rho^2\frac{1}{n}\sum_{j=1}^{n}\varepsilon_j. \tag{4}$$

According to the statement of Lemma for the function $u(z)$ at point z_0 holds the estimate (2), so for sufficiently small radii ρ modules all the values ε_j do not exceed arbitrary predetermined positive ε. In this case, modulus of the sum $\left|\sum\varepsilon_j\right|$ on the right side of formula (4) is not greater than $n\varepsilon$ and hence (3) is evaluated. Lemma 1 is proved.

Note that for second-order polynomials the equality (3) is exact, that is, it does not contain the term $o(\rho^2)$.

Lemma 2. *Let function $u(z)$ be continuous in the domain $G\subset\mathbb{R}^2$ and in any nighborhood of each point $\zeta_0\in G$ there exists a set of n nodes $\{\zeta_j\}$ (number of nodes in sets for different points $\zeta_0\in G$ and in different sets for the same point can be different, but in any case of $n\geq 3$) lying on a circle with center at ζ_0, for which the conditions 1° and 2° of the Theorem hold. Then the function $u(z)$ is subharmonic in the region G.*

Proof of this lemma follows the approach similar to the theorem 2, I.I. Privalov [3]. Proof of the Privalov's theorem is also provided in the footnote in the work of I.G. Petrovskii [4], pg. 107.

Remark 1. The condition 1° of the theorem can be loosened somewhat by assuming that for set $\{\zeta_j\}$ of circle points $C(\zeta_0,\rho)$ completed not the inequality (1), but the next one for the negative part of the difference between the arithmetic mean of the values of the function $u(z)$ at nodes of the collection $\{\zeta_j\}$ and its value at ζ_0

$$\left[\frac{u(\zeta_1)+\ldots+u(\zeta_n)}{n}-u(\zeta_0)\right]^-=o(\rho^2)\quad\text{when}\quad\rho\to 0. \tag{5}$$

It is sufficient to prove the Lemma 2 under this assumption. We shall prove this version of Lemma 2.

Proof. Suppose the Lemma is not true and the function $u(z)$ is subharmonic not everywhere in the domain G. Violation of subharmonicity means that there exist a closed disk $D\subset G$ and function $v(z)$, continuous at D and harmonic in the interior of D, such that the inequality $v(z)\geq u(z)$ holds on ∂D, but in some points of the interior of D is executed the opposite inequality $v(z)<u(z)$. The function $u(z)$ is continuous, so as a function of $v(z)$ we can take the solution of the Dirichlet problem in the disk D with values $u(z)$ on ∂D.

On the disk D we set $w(z):=u(z)-v(z)$. The function $w(z)$ is continuous on the closed disk D, equals zero on its boundary ∂D and is positive at some interior points of D.

Let z_0 be an arbitrary inner point of the disk D. According to the statement of the Lemma inside D there exist circles of arbitrarily small radius with center at z_0 and sets of points $\{z_1,\dots,z_n\}$ of these circles, such that the function $u(z)$ satisfies the conditions of 1° and 2° of the Theorem. Consider the expression included in the inequality (1) for the function $w(z)$ on such sets of nodes $\{z_j\}$. We have

$$\frac{w(z_1)+\cdots+w(z_n)}{n} - w(z_0) =$$
$$= \left[\frac{u(z_1)+\cdots+u(z_n)}{n} - u(z_0)\right] + \left[\frac{v(z_1)+\cdots+v(z_n)}{n} - v(z_0)\right]. \quad (6)$$

According to the Lemma condition on the set $\{z_j\}$ the negative part of the value in the first pair of brackets on the right side of the formula (6) satisfies the estimate of $o(\rho^2)$ as $\rho \to 0$ (we take the radii of the circles $C(z_0, \rho)$ on which collections $\{z_j\}$ lie). The function $v(z)$ is harmonic in a sufficiently small neighborhood of the point z_0. So according to the Lemma 1 the value in the second pair of brackets on the right side of the formula (6) also satisfies the estimate $o(\rho^2)$ as $\rho \to 0$. Therefore, the negative part of (6) satisfies (5) on some sequence $\rho_n \to 0$.

The function $w(z)$ is continuous in the disk D, positive in some internal points of D and is identically zero on ∂D. Hence, the maximum on D of the function $w(z)$ is positive and this maximum is located strictly inside D. Let ζ_0 be the maximum point: $\max_{z\in D} w(z) = w(\zeta_0)$. We set

$$\tilde{w}(z) := w(z) + \frac{w(\zeta_0)}{2\operatorname{diam}^2 D}|z-\zeta_0|^2.$$

According to the definition of the function $\tilde{w}(z)$ the inequality $\tilde{w}(z) \geq w(z)$ holds everywhere in the disk D, so

$$\max_{z\in D} \tilde{w}(z) \geq \max_{z\in D} w(z) = w(\zeta_0).$$

There holds the identically $w(z) \equiv 0$ on the boudary ∂D so we have

$$\tilde{w}(z) := w(z) + \frac{w(\zeta_0)}{2\operatorname{diam}^2 D}|z-\zeta_0|^2 < \frac{w(\zeta_0)}{2\operatorname{diam}^2 D}\operatorname{diam}^2 D = \frac{w(\zeta_0)}{2}$$

Therefore, the maximum of the function $\tilde{w}(z)$ on the disk D is achieved strictly inside D. Let us denote z_0 a point of maximum: $\max_{z\in D} \tilde{w}(z) = \tilde{w}(z_0)$.

The expression $|z-\zeta_0|^2 = (x-\xi_0)^2+(y-\eta)^2$ is a second order polynomial. Then by Lemma 1 and a remark after it, there exist a circle $C(z_0, \rho)$ of arbitrarily small radius and a set of nodes $\{z_j\} \subset C(z_0, \rho)$ such that

$$\frac{\tilde{w}(z_1)+\ldots\tilde{w}(z_n)}{n} - \tilde{w}(z_0) = \left[\frac{w(z_1)+\cdots+w(z_n)}{n} - w(z_0)\right] + \frac{w(\zeta_0)}{4\operatorname{diam}^2 D}\rho^2, \quad (7)$$

and the negative part of the expression in brackets in (7) satisfies the estimate of $o(\rho^2)$ as $\rho \to 0$ (estimate (5). Since $w(\zeta_0) > 0$, for sufficiently small ρ the value (7) is positiv. It is clear that in this case the value of the function $\tilde{w}(z)$ at least one of the nodes of the set $\{z_j\}$ exceeds value $\tilde{w}(z_0)$. This contradicts the assumption that the function $\tilde{w}(z)$ has maximum in the disk D at z_0.

Thus, the assumption that the function $u(z)$ is subharmonic is not everywhere in the region G leads to a contradiction. Hence, the function $u(z)$ subharmonic. Lemma 2 is proved.

Further, the proof of the Theorem from the present work follows the proof of the Theorem by author of [2] on the harmonicity of functions satisfying discrete Laplace equation. So why we don't bring it.

The author was partially supported by the Program of competitiveness increase of the National Research Nuclear University MEPhI (Moscow Engineering Physics Institute); contract No 02.a03.21.0005, 27.08.2013.

IOP Conf. Series: Journal of Physics: Conf. Series **1205** (2019) 012057 doi:10.1088/1742-6596/1205/1/012057

References

[1] I.I. Privalov, Subharmonic Functions (ONTI, Moscow–Leningrad, 1937), Vol. 2 [in Russian].
[2] D.S. Telyakovskii. A sufficient condition for the harmonicity of a function of two variables satisfying the Laplace difference equation, Tr. Inst. Mat. Mekh. UrO RAN, 2016, 22:4, 269–283, [in Russian].
[3] I. Priwaloff. Sur les fonctions harmoniques. — Mat. Sb., 1925, 23:3, 464–471.
[4] I.G. Petrovskii, Perron's method for solving the Dirichlet problem, Uspekhi Mat. Nauk, 1941, 8, 107–114 [Selected Works, Gordon and Breach Publishers, Amsterdam, 1996, Vol. 2, 20–29].

Using Machine Learning to predict survival in patients with brain metastases after Gamma Knife radiosurgery

G Vazhenin[1], S Banov[3], A Golanov[2], P Ryabov[1], V Kostjuchenko[3] and A Dalechina[3]

[1]National Research Nuclear University "MEPhI", Kashirskoe highway, 31, Moscow, Russia
[2]N.N. Burdenko Neurosurgical Institute, 1-st Tverskoy-Yamskoy lane, 13/5, Moscow, Russia
[3] Moscow Gamma Knife Center, 1-st Tverskoy-Yamskoy lane, 13/5, Moscow, Russia

avdalechina@gmail.com

Abstract. In this work machine learning approach was used to predict the patients overall survival after Gamma Knife radiosurgery. We constructed the regression and multiclass classification models to predict the time interval from the onset of the oncological disease to the unfavourable outcome and the patient's survival class. The models were built on data of 916 patients with 26 different primary features. The train set included patients with known clinical outcomes (445 patients). The median deviation in determining of the time from the onset of cancer to the date of death was 1.4 months. According to the regression model the most significant feature was the largest volume of the lesion by the date of the first radiosurgery. The most important feature was identified as the time interval from the date of birth to the onset of the oncological disease. The mean accuracy, according to the confusion matrix was 0,76. The accuracy of the study can be improved by the increasing of the patients number and completeness of the patient data.

1. Introduction

The quantity of detected cases of oncology diseases is being on arise all over the world. According to the World Health Organization's data the oncology diseases are within the leading mortality factors. On a conservative estimate the brain metastases are in progress in 8 - 10 % of cancer patients, what determines principal clinical symptomatology. The key method to treat the multiple brain metastases combines the various types of radiotherapy with modern pharmacotherapy (immunotherapy, target therapy etc.).

At the moment the most efficient and secure method in case of limited brain metastases is considered to be radiosurgery. The essential feature of radiosurgery is the locality of radiation exposure which is substantially decreasing the toxicity. It enables to undertake the radiosurgery procedure multiply controlling the growth of the lesions which could be re-appeared in 50 % of the cases.

It is extremely important for cancer research to predict outcomes after a given therapy and detect the key features from complex datasets and identify the prognostic factors effecting patients survival. But traditionally prediction has been performed by clinicians using only statistically significant clinical factors such as age, performance status, diagnosis and etc. Every patient is complex and presents with a great number of clinical factors that may or may not be relevant to survival [1].

Machine Learning (ML) was proposed as an alternative method of data analysis. It is a powerful tool to overcome the limitations of classical statistical methods while analyzing and interpreting massive amounts of data [2]. ML enables computers to learn from previous data to make accurate predictions on new data.

The purpose of this study is to apply the ML to predict survival in patients with brain metastases (BS) after Gamma Knife radiosurgery (GKRS). We applied ML algorithms to the tasks of classification and regression to predict overall survival. ML methods were implemented using open-source Python libraries: scikit learn, catboost, pandas, numpy, scipy, pylab.

2. Materials and methods

2.1. Patients selection

We retrospectively evaluated the total data of 916 patients with brain metastases undergone the stereotactic Gamma Knife radiosurgery treatment in Moscow Gamma Knife Center (Burdenko neurosurgical institute) within the period from 2005 to 2017. These patients had 26 different primary features: Karnofsky Performance Scale (KPS), oncological diagnosis, primary tumor histology, number of brain tumors and etc. The data could be divided into 4 classes: real, categorical, date, text feature (Table 1).

Table 1. Primary features[a].

Categorical variables	Real variables	Date	Text
1. Sex	10. Number of GKRS	15. Date of birth	26. Histology
2. OD	11. KPS on the date of the first RS	16. Diagnosis date	
3. Activating mutations	12. Number of brain tumors on the date of the first RS	17. Date of the primary tumor surgery	
4. EM by the date of the first RS	13. Total volume of the BM on the date of the first RS	18. Date of the BM appearance	
5.Pharmacotherapy by the date of the first RS	14. Maximum volume of the brain tumor by the date of the first RS	19. WBRT date	
6. IP		20. Date of the BM neurosurgery	
7. LR treatment after the first RS		21. Date of the first GKRS	
8. LR after the first RS		22. Date of the LR appearance after the first RS	
9. DM after the first RS		23. Date of the DM appearance after the first RS	
		24. Date of the last follow-up	
		25. Date of the patient's death	

[a] Abbreviations used in the table: OD - oncological diagnosis, EM - extracranial metastases, RS - radiosurgery, IP - intracranial progression, LR - local recurrence, DM - distant metastases, WBRT - whole brain radiation therapy.

The data were divided into training and test sets. The target variable was time measured in months from the oncological diagnosis date (the onset of the oncological disease) to the date of the patient's death. There was information about the date of death of 479 patients in the data. But the train set contained only 445 patients with information about both the date of death and the oncological diagnosis date. The test set in turn did not include any information about the date of death for these patients or they had been alive by the date of the last observation. Predictions were made for the test set.

For the classification task we divided the training set into 3 classes: 0 (297 patients) – time interval from the first RS to the death less than 1 year, 1 (95 patients) – time interval from the first RS to the date of death from one to two years, 2 (87 patients) – time interval from the first RS to the date of death more than 2 years or the patient was alive by the date of the last observation. Thereby, in classification, alive patients by the date of the radiosurgery were included into analysis.

2.2. Data pre-processing

Missing data is a common problem in real ML tasks. Our dataset had 24 % of the missing values. The visualization of the missing data is presented in the Fig.1.

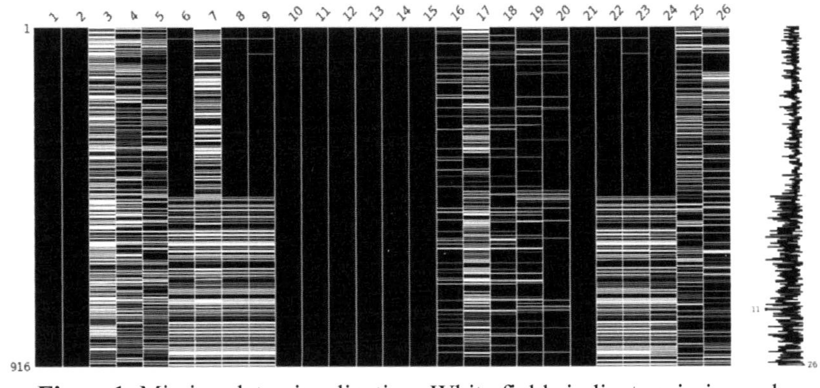

Figure1. Missing data visualization. White fields indicate missing values.

The missing values were replaced with 0 (in case of the categorical features) and with median value (in case of the missing values in the date features). Categorical features were transformed to the binary format. The "date" features were transformed into the differences between dates. Quantities of words, letters, numbers were calculated for the text feature "Histology". The appearance of the most frequently used words of the "Histology" feature was normalized to binary features. The data filled less than 10 % were excluded from the analysis.

2.3. Data pre-processing

Feature selection is the process of identifying and removing from the training data set irrelevant and redundant features. Training time increases with number of features. Besides model has risk of overfitting with increasing number of features. Using Random Forest classifier and greedy algorithm the most relevant features were selected [3].

For the regression task, we identified 19 features: 1- maximum volume of the brain tumors on the first RS date, 2 - total volume of the BM by the date of the first RS , 3 - time interval from the birth to the onset of the oncological disease, 4 - absence of the intracranial progression, 5 - KPS by the date of the first RS, 6 - IR treatment after the first RS, 7 -GKRS number, 8 - IR without treatment after the first RS, 9 - LR, 10 - absence of the EM by the date of the first RS, 11- DM, 12 - chemotherapy, 13 - male, 14 - EM by the date of the first RS, 15 - Breast cancer, 16 - Melanoma, 17 - NSCLC (non-small cell lung cancer), 18 - IR treatment – WBRT, 19 - Kidney cancer. The bar chart is presented in the Fig.2.

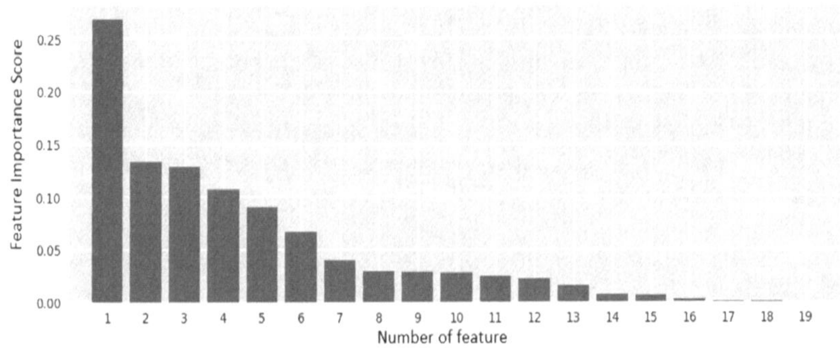

Figure 2. Feature Importance bar chart (regression model).

For the classification task the relevant features were:1 - time period from the birth to the onset of the oncological disease, 2 - total volume of the BM on the date of the first RS, 3 - maximum volume of the brain tumors on the first RS date, 4 – time interval from the disease onset to the BM appearance, 5 – time from the BM appearance to the first GKRS,6 - IR treatment - RS, 7 - number of BM treated with RS, 8 - KPS by the date of the first RS, 9 - number of words in the "histology feature", 10 - GKRS number, 11 – quantity of numbers in the "histology feature", 12 - Breast cancer, 13 - Melanoma (Fig.3).

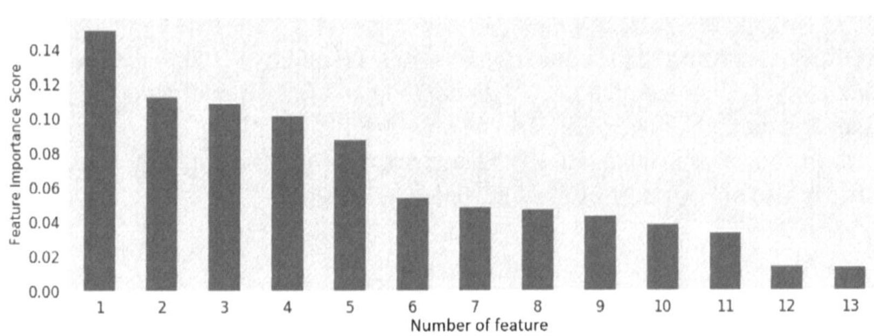

Figure 3. Feature Importance bar chart (classification task).

2.4. Tuning model hyperparameters and training

Predictive models were trained using Gradient Boosting Regressor and Gradient Boosting Classifier in cases of regression and classification tasks respectively [4]. Using variational principle, we found the optimal Gradient Boosting hyperparameters for the classification task: minimum number of samples required to split an internal node = 200, minimum number of samples required to be at a leaf node = 30, maximum depth of the tree = 7, number of the features to consider when looking for the best split = 7. For the regression task default scikit-learn hyperparameters were used.

3. Results

Predictive models were constructed to predict overall survival in patients with BM using ML approach.

It was founded that the most significant features to predict the time interval from the oncological diagnosis date to the date of the patient's death were the largest volume of the lesion by the moment of the first RS, total volume of the BM by the date of the first RS , time interval from the birth to the onset of the oncological disease, absence of the intracranial progression, KPS by the date of the first RS.

Five most significant features in classification were: time period from the date of birth to the onset of the oncological disease, total volume of the BM by the date of the first RS, maximum volume of the brain tumors by the first RS date, time interval from the disease onset to the BM appearance, time from the BM appearance to the first GKRS.

The 5-fold cross validation technique was used to estimate the accuracy of the ML models. The mean and the median errors of the regression predictive model were 1.8 and 1.4 months, respectively.

The mean accuracy of the classifier was 0.72. For the classification problem the confusion matrix was used to evaluate the quality of the output of the classifier (Fig.4.). It allowed to evaluate correctness of the model for each class (mean accuracy - 0,76).

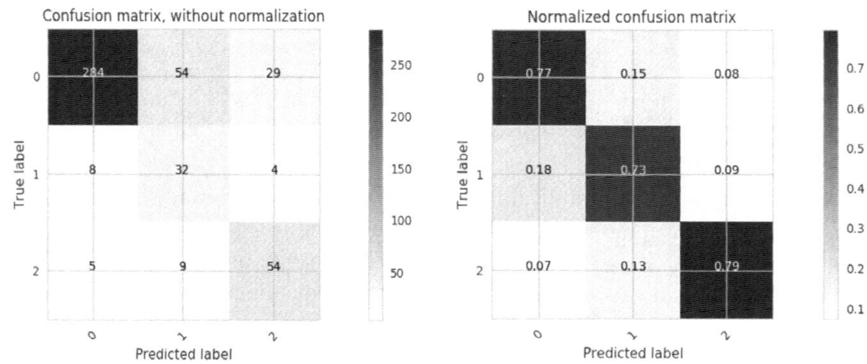

Figure 4. Confusion matrix (non-normalized and normalized)

The mean accuracy according to the confusion matrix was roughly equal to the mean accuracy of the classifier without reference to the classes. It demonstrated that we succeeded in correcting of the imbalanced classes and the imbalance of the data had no significant influence on the prediction accuracy. The presented confusion matrix has showed the approximate equality of the accuracies ref to each separate class, that is also an indicator of the well-balanced classes.

4. Conclusion

This research is just a first step to apply the ML algorithms to prediction of treatment results in case of patients with BS in the Moscow Gamma Knife Center. It has been found that the most important features are the largest volume of the lesion by the moment of the first radiosurgery (the regression task) and the time interval from the date of birth to the onset of the oncological disease (the classification task). The prediction error of the regression model was 1,4 months and the mean accuracy of the classification was 0,76. The accuracy of the study can be improved by the increasing number of patients and completeness of the patient data.

References

[1] Kondziolka D *et al* 2014 *J. Neurosurg.* **120** 24–30
[2] Celtikci E 2018 *Turk Neurosurg.* **28** 167 - 73
[3] Kam Ho T 1995 ICDAR '95 Proceedings of the Third International Conference on Document Analysis and Recognition **1** 278
[4] Friedman J 2001 *Stat.* 1189–32.

IOP Conf. Series: Journal of Physics: Conf. Series **1205** (2019) 012059 doi:10.1088/1742-6596/1205/1/012059

Simplified mathematical model and comparative analysis of possible ways of realizing the residual energy resource when decommissioning a channel-type reactor

A Zagrebayev[1], G Zaluzhnaya[1] and V Naumov[1]

[1]National Research Nuclear University MEPhI (Moscow Engineering Physics Institute), Kashirskoe shosse 31, 115409 Moscow, Russia

Zaluzhnaya_galya@mail.ru

Abstract. We consider the problem of realization of the residual energy resource of fuel during decommissioning of the High-Power Channel-Type reactor. A mathematical model of the fuel assemblies spectrum for energy production is provided. Two methods of reactor decommissioning are compared. The first possible way to preserve the criticality of the reactor during decommission is the collapse of the core, when fuel assemblies with high energy production are discharged. The second possible way of decommissioning is formation of a two-zone loading: fresh fuel assemblies are put in the centre, burnt out - on the periphery. It is shown that the second method gives a greater gain in power and unprocessed energy resources. The results are obtained on the point model of the fuel assemblies spectrum dynamics in the linear approximation of the with linear dependences of the power and the multiplication factor of FAs from energy productionn and indicate only the possibility of optimizing the fuel use in the process of removing the RBMK from service.

1. Introduction

The problem of nuclear power plant decommission is a complex task, and one of the aspects of which is optimal utilization of the fuel assemblies (FA) of decommissioned reactors. In relation to High-Power Channel-Type reactors (HPCTR), the problem of optimal use of shutdown reactors residual energy resources was considered in [1]. It has been shown that the re-use of fuel from a shutdown reactor in other nuclear power units can save a significant amount of fresh fuel, from tens to hundreds and even thousands of FAs. At the same time, the process of decommissioning of nuclear power plants from several power units was optimized. However, it is possible to use the fuel resource inside each power unit of nuclear power plants through internal permutations.

The physical and structural features of the HPCTR [2] (possibility of FA rearrangements during operation) make it possible to implement various approaches to the formation of a critical charge without fuelling, for example, removing the most burnt fuel ("collapsing" reactor) or forming a multi-zone (in particular, two-zone) charge, when the burnt fuel is placed on the periphery. That is, there is a principal possibility to prolong the operation of the decommission reactor being due to the appropriate structuring of core charge and thereby use the energy resource of unburnt fuel

IOP Conf. Series: Journal of Physics: Conf. Series **1205** (2019) 012059 doi:10.1088/1742-6596/1205/1/012059

The purpose of this article is to use a simple model to pay attention to the physical possibility and efficiency of using the residual nuclear fuel resource due to internal core restructuring with a decrease in power during decommissioning.

2. Mathematical model of the FA spectrum dynamics in energy production.

To conduct research on the use of energy resources of the decommissioned reactor, the following mathematical model of the FA spectrum dynamics for energy production is proposed:

$$\frac{\partial n(E,t)}{\partial t} = -\frac{\partial q(E,t)}{\partial E} - S(E,t) + F(E,t), \tag{1}$$

where: E - energy production of FAs, MW · day / FA, $n(E,t)dE$ - the number of FAs with energy production in the interval from E to $E + dE$ at time t; $S(E,t)dE$ is the number of FAs that have energy production in the range from E to $E + dE$ discharged from the reactor at time t (unloading rate); $F(E,t)dE$ is the number of FAs with energy production in the range E to $E + dE$, loaded into the reactor at time t (loading rate); $q(E,t)$ is the number of FAs increasing their energy production per level E per unit time at time t (burn-up flow).

Equation (1) is derived from the following balance considerations: the rate of change in the number of FAs$\left(\frac{\partial n}{\partial t}dE\right)$whose energy production lies in the interval from E to $E + dE$ at time t, is determined by the rate of decrease due to $q(E + dE, t) - q(E,t) = \frac{\partial q}{\partial E}dE$, the unloading rate $S(E,t)dE$, and the loading rate due to possible loading other reactors $F(E,t)dE$.

The connection between the fuel burnup flow $q(E,t)$ and the FA number $n(E,t)$ is determined from the following considerations: for a unit of time, the FA energy production changes by the value $W(E) \cdot 1$, where $W(E)$ is the power of the FA with energy production E. Thus, all FAs with energy production ranging $E - W(E) \cdot 1$to E will cross the energy production level E, that is, $q(E,t) = n(E,t) \cdot W(E)$.

Then we can reduce equation (1) to the first-order partial differential equation for the function $q(E,t)$:

$$\frac{1}{W(E)}\frac{\partial q(E,t)}{\partial t} = -\frac{\partial q(E,t)}{\partial E} - S(E,t) + F(E,t). \tag{2}$$

The initial condition is: $q(E,0) = n(E,0)W(E)$. (3)

The boundary condition: $q(0,t) = q_0(t)$. (4)

Equation (3) is defined on the interval $(0, E_m)$. The physical formulation of the problem assumes that when the maximum depth of energy production is reached, E_m FA is extracted from the core. In the "ideal" mode of steady overloads, when only fresh FAs are loaded and only FAs with maximum energy output $E_m$are unloaded, the mathematical model (2-4) looks like:

$$0 = -\frac{\partial q(E,t)}{\partial E}$$

$$q(0) = q_0.$$

Then the fuel distribution in the energy production is:

$$n(E) = \frac{q_0}{W(E)}. \tag{5}$$

Power of the FA has a linear dependency on energy production and neutron flux density

$$W(E,r) = \frac{\phi(r)}{\phi_{max}}(W_0 - B \cdot E)$$

in conditions of maximum flow. The distribution of FAs for energy production $n(E)$ does not depend on the neutron flux density.

From physical considerations, it is clear that if fresh fuel is stopped, then the previously loaded fresh fuel will have some energy output $E(t) = E_0$ after a time t, and the FAwith an energy output less than E_0 in the reactor will not.

Let's consider two ways of prolonging the reactor operation mentioned above due to the core structuring. In this case, the estimates given assume a linear dependence of the multiplication factor and the power of the FA on energy production. A model was constructed with parameters close to those of the HPCTR [2].

$$K(E) = K_0 - AE, \overline{K_\infty} = 1.02, K_0 = 1.2, A = 1.1535 \cdot 10^{-4}\, FA\, /\, (\text{MWt} \cdot \text{day})$$
$$W(E) = \frac{\phi(r)}{\phi_{max}}(W_0 - B \cdot E); W_0 = 3\frac{MWt}{FA}, B = 5.3571 \cdot 10^{-4}\ \text{day}^{-1}$$

The following physical values are used in calculations: the number of FAs in the core $N = 1600$; the square of the migration length $M^2 = 0.04\ m^2$; the area of the cell with FA $a^2 = 0.06\ m^2$.

3. The "collapsible" reactor

After cessation of feeding with fresh fuel as the burn-out proceeds, the left boundary of the spectrum shifts to the right and the right border to the left.

The distribution of the neutron flux density in such a reactor obeys the equation:

$$\begin{cases} \Delta\phi + \dfrac{\frac{\overline{K}_\infty}{K_{эфф}} - 1}{M^2}\phi = 0 \\ \phi'(0) = 0 \\ \phi(R) = 0 \end{cases}, \tag{6}$$

where $\overline{K}_\infty = \frac{\int_{E_0}^{E_{max}} N(E)K(E)dE}{\int_{E_0}^{E_{max}} N(E)dE}$.

The radius of the reactor core is found from the relation $\pi R^2 = a^2 \int_{E_0}^{E_{max}} n(E)dE$.

Criticality criterion for the reactor:

$$\frac{\frac{\overline{K}_\infty}{K_{эфф}} - 1}{M^2} = \left(\frac{2.405}{R}\right)^2 \tag{7}$$

It follows from this that for a given square of the migration length M^2, the effective multiplication factor K_{eff}, the cell area, and the known spectrum, $n(E) = \frac{q_0}{W(E)}$, it is possible to find the dependence of the energy output of the unloaded fuel $E_{max}(t)$ on the energy production $E_0(t)$ consequently, on the time of operation of the reactor without recharging with fresh fuel . At the same time, it has been obtained that as fuel burns out it is necessary to extract fuel with less energy production in order to maintain criticality. Finally, there comes a point where maintaining criticality becomes impossible. Thus, the concept of a "collapsing" reactor assumes the extraction of fuel assemblies with less than the regular energy production E_m.

4. Two-zone reactor

It is proposed to consider a model of a two-zone fuel loading reactor. In the central zone is fuel with energy production from $E_0(t)$ to $E_x(t)$ and in the peripheral from $E_x(t)$ to $E_m(t)$. As the fuel burns out, the left and right boundaries of the spectrum $E_0(t)$ shift to the right, and the boundary between the zones $E_x(t)$- to the left. The critical state of the reactor is described by a system of equations:

$$\begin{cases} \Delta\phi_1 + \dfrac{\frac{K_1}{K_{эфф}} - 1}{M^2}\phi_1 = 0 \\ \Delta\phi_2 + \dfrac{\frac{K_2}{K_{эфф}} - 1}{M^2}\phi_2 = 0 \end{cases}, \tag{8}$$

where for the inner zone:

$$\begin{cases} K_1 = \dfrac{\int_{E_0}^{E_x} N(E)K(E)dE}{\int_{E_0}^{E_x} N(E)dE} \\ N_1 = \displaystyle\int_{E_0}^{E_x} N(E)dE - \text{FA number in the central zone} \\ \pi r_1^2 = a^2 N_1 - \text{the size of the first zone} \end{cases},$$

And for the outer zone:

$$\begin{cases} K_2 = \dfrac{\int_{E_x}^{E_m} N(E)K(E)dE}{\int_{E_x}^{E_m} N(E)dE} - \text{medium multiplication factor} \\ N_2 = \displaystyle\int_{E_x}^{E_m} N(E)dE - \text{FA number in the zone} \\ \pi R^2 = a^2(N_1 + N_2) - \text{the size of the zone)} \end{cases}$$

For identical diffusion coefficients in the zones, the equations of uniqueness have the form:

$$\begin{cases} \phi_1'(0) = \phi_2(R) = 0 \\ \phi_1(r_1) = \phi_2(r_1) \\ \phi_1'(r_1) = \phi_2'(r_1) \end{cases}$$

The solution of the two-zone problem allows one to obtain a critical condition that will contain in addition to K_{eff}, the parametersr_1, R, K_1, K_2, which in turn depend on $E_0(t), E_x(t), E_{max}(t)$. Thus, for a given$K_{eff}$, one can find $E_0(t)$ at which it is possible to maintain criticality in the reactor. At the same time, the concept of a two-zone reactor does not involve the extraction of FAs, but only their rearrangement to the periphery. If we assume that the reactor maintains a neutron flux density such that the power W_0 is removed from the most stressed FA, then the reactor power can be calculated as follows:

$$W = \overline{W_1}N_1 + \overline{W_2}N_2,$$

Where $\overline{W_1}, \overline{W_2}$ is the average power of the FA in the center and at the periphery, respectively:

$$\overline{W_1} = \frac{W_0}{K_1^E K_1^\phi}, \overline{W_2} = \frac{W_0 \cdot \frac{\phi_2^{max}}{\phi_1^{max}}}{K_2^E K_2^\phi}$$

$K_1^E, K_1^\phi, K_2^E, K_2^\phi$ are the coefficients of unevenness in energy production and flux density in each zone.

5. Results

The simulation of the decommissioning of a HPCTR was carried out using the two strategies described above. The minimum permissible effective multiplication factor was assumed equal to 1.02, (considering the margin for axial leakage and the operational reserve of reactivity). It has been obtained that the "collapsing" reactor can, generate heat energy of about 170 GW per day until the critical loss is reached after fueling stop, however, its power is reduced to 25% of the nominal one. The residual resource (the number of FAs remaining in the core that could be used) after stopping the "collapsing" reactor was 366 FAs in terms of fresh fuel. This is 48% less than if the reactor had simply been stopped.

Two-zone charged the reactor can provide energy up to 400 GW by the time of total loss of criticality, but by the time it stops, power is reduced to 14% of the nominal. The amount of unprocessed resource decreases in comparison with the "collapsing" reactor - 311 FAs in terms of fresh fuel. This is 56% less than in the case of a simple stop.

These estimates are limiting until the complete exhaustion of the reactivity reserve and are obtained on the basis of a simplified model of a HPCTR with linear dependences of the power and the multiplication factor of FAs from energy production. In this case, the long operation of the reactor is explained by a significant (4-5 times) decrease in power. There may be reasons limiting the reactor's decommissioning - technological and economic constraints. In general, it has been shown that by organizing the appropriate core charge, it is possible to effectively use energy resources in a HPCTR being decommissioned, using such simple operations as the FA rearrangements in the core. With the imposition of technical, economic and other constraints that do not allow the full utilization of the residual energy resources to be fully utilized with a reduction in reactor power, a known method of moving unburned FAs from the shutdown power unit to the operating one can be used [1].

IOP Conf. Series: Journal of Physics: Conf. Series **1205** (2019) 012059 doi:10.1088/1742-6596/1205/1/012059

References

[1] Fedosov A Optimal fuel usage during HPCTR decommission *Atomic energy* **102** iss.5 284-290
[2] Dollezhal N, Emelyanov I Channel-type nuclear reactor *Atomizdat*

IOP Conf. Series: Journal of Physics: Conf. Series **1205** (2019) 012060 doi:10.1088/1742-6596/1205/1/012060

Analytical properties and exact solution of the Aliev-Panfilov model

A I Zemlyanukhin and A V Bochkarev
Yuri Gagarin State Technical University of Saratov,
Politekhnicheskaya, 77, Saratov, Russia

E-mail: zemlyanukhinai@sstu.ru, ab2009sar@list.ru

Abstract. Some analytical properties of the Aliyev – Panfilov model describing the propagation of impulses through the cardiac tissue are studied. It is shown that the Aliyev – Panfilov system generally does not pass the Painleve test. The relations between the parameters of the model for which there is an exact solution in the form of a kink corresponding to the process of switching the system from one state of stable equilibrium to another are established.

1. Introduction

Statistical studies of the World health organization show that the main cause of premature death worldwide is cardiovascular disease. One of the most important tasks is to build a complete and consistent model of the heart as an autowave medium controlled by a neural network. Historically, the first model that adequately describes the dynamics of a neuron in terms of its physical and chemical structure was the Hodgkin-Huxley (HH) model. The HH-model included 4 differential equations establishing the relationship between the membrane potential of the neuron and the variables of its ion channels. The numerical simulation showed a satisfactory agreement with the experimental data, but the HH-model turned out to be too complicated for the analytical study [1]. The analysis of the asymptotic simplification of the HH-model revealed the similarity of its phase portrait with the Van der Pol equation which is the 2nd order equation for the oscillator with nonlinear attenuation [2]. This allowed to reduce the number of equations to two in the FitzHugh-Nagumo (FHN) model, the dynamic modes of which have been exhaustively studied to date [3-5]. However, the features of this model, namely, the effect of repolarization and the absence of the phenomenon of restitution, in which the duration of the period of excitation of the neuron depends on the time elapsed since the previous excitation, do not allow the use of FHN-model for cardiac myocardium [6]. These shortcomings of the FHN-model were eliminated in the Aliev-Panfilov (AP) model:

$$\begin{aligned}&\frac{\partial u}{\partial t}-\frac{\partial^2 u}{\partial x^2}+ku(u-a)(u-1)+uv=0,\\&\frac{\partial v}{\partial t}+\left(\varepsilon_0+\frac{\mu_1 v}{u+\mu_2}\right)(v+ku(u-a-1))=0,\end{aligned}\tag{1}$$

which includes fast variable $u(x,t)$ and slow variable $v(x,t)$, as well as five parameters $k, a, \varepsilon_0, \mu_1, \mu_2$. The behavior of the model is correlated with experimental data to the greatest extent if the following values are selected for the parameters [6]

$$k=8, a=0.15, \varepsilon_0=0.002, \mu_1=0.2, \mu_2=0.3.\tag{2}$$

The numerical experiments have shown that a twofold change in the values of its parameters does not violate its qualitative properties [7].

2. The Painleve analysis

To perform a qualitative analysis of the system (1), let's express $v(x,t)$ from the first equation of the system and substitute it to the second one. After the transition to the running variable $z=x-C_0t$ we obtain a 3rd order nonlinear differential equation

$$\begin{gathered}-C_0u(u+\mu_2)\frac{\partial^3u}{\partial z^3}+\mu_1\left(\frac{\partial^2u}{\partial z^2}\right)^2+\left[C_0(2\mu_1+\mu_2+u)\frac{\partial u}{\partial z}-k\mu_1u^3+\left(\varepsilon_0+\mu_1k(1+a)-C_0^2\right)u^2+\right.\\ \left.+\left(\mu_2\varepsilon_0-2ak\mu_1-\mu_2C_0^2\right)u\right]\frac{\partial^2u}{\partial z^2}+(\mu_1+\mu_2+u)C_0^2\left(\frac{\partial u}{\partial z}\right)^2+\left[2ku^3-k(a+\mu_1-2\mu_2+1)u^2+\right.\\ \left.+\left[(\mu_1-\mu_2)(a+1)k+\varepsilon_0\right]u-2ak\mu_1+\varepsilon_0\mu_2\right]C_0u\frac{\partial u}{\partial z}-ak\left[-k\mu_1u^2+\left(\mu_1(a+1)k+\varepsilon_0\right)u-\right.\\ \left.-ak\mu_1+\mu_2\varepsilon_0\right]u^2=0.\end{gathered} \tag{3}$$

Substitution $u(z)=a_0z^{-p}$ reveals five leading terms [8]

$$C_0u\frac{\partial^2u}{\partial z^2}\frac{\partial u}{\partial z}-k\mu_1u^3\frac{\partial^2u}{\partial z^2}+2C_0ku^4\frac{\partial u}{\partial z}+\mu_1\left(\frac{\partial^2u}{\partial z^2}\right)^2-C_0u^2\frac{\partial^3u}{\partial z^3}, \tag{4}$$

the balance of which is achieved when $p=1$ and

$$a_0=\pm\sqrt{\frac{2}{k}};\ a_0=-\frac{\mu_1}{C_0}. \tag{5}$$

Replacing $u(z)=a_0z^{-p}+\beta z^{r-p}$ in the leading terms (4) results in the following expressions for the Fuchs indices for the first pair of constants $(a_0,p)=\left(\pm\sqrt{2/k},1\right)$:

$$r_1=-1,\ r_2=4,\ r_3=2\pm\frac{\mu_1\sqrt{2k}}{C_0}, \tag{6}$$

so that r_3 takes integer values only when the relation $\mu_1\sqrt{2k}/C_0$ is an integer. Thus, in general, the AP system does not pass the Painleve test [8] and one can only hope for the existence of particular exact solutions at certain ratios between the coefficients of the model. The second pair of constants $(a_0,p)=(-\mu_1/C_0,1)$ gives only two variants of Fuchs indices having integer values:

$$r_1=-1,\ r_2=0,\ r_3=4 \text{ and } r_1=-1,\ r_2=1,\ r_3=6. \tag{7}$$

Further passing of the Painleve test stages reveals the conditions that the model parameters $k, a, \varepsilon_0, \mu_1, \mu_2$ and the traveling wave velocity C_0 must satisfy. In all considered integer variants of the index r_3 these conditions lead to far from optimal (2) or even to zero or negative values of the parameters. The best result corresponds to the combination of indices

$$r_1=-1,\ r_2=4,\ r_3=5, \tag{8}$$

in which the fulfillment of the conditions

$$C_0=\pm\frac{1}{3}\mu_1\sqrt{2k},\ \varepsilon_0=0,\ \mu_1=3\left(1-\frac{a}{2}\right) \tag{9}$$

provides the presence of three arbitrary constants in the decomposition of the solution of equation (3) in the Laurent series, and, consequently, conditional passing of the Painleve test.

3. Exact solutions

To find the exact solution of equation (3) taking into account the equations (9), we use the hyperbolic tangent method [8]. The pole order of the solution is one, so the solution should be found in the form

$$u(z) = A\tanh(Bz) + C. \tag{10}$$

Substitution of expression (10) in (3) reveals the conditions under which this expression is an exact solution:

$$A = -\frac{3B}{2}\sqrt{\frac{2}{k}},\ C = -\frac{9}{2}\mu_1 + 5\mu_2 + \frac{27}{2},$$
$$B = \frac{\sqrt{k}}{6}\left(1022\mu_1^2 - (2270\mu_2 + 5727)\mu_1 + 1260\mu_2^2 + 6345\mu_2 + 80284\right)^{1/2}. \tag{11}$$

At the same time, there are 3 variants for the parameters of the AP-system (1), when they remain real:

$$(\mu_1 \approx 6.056,\ \mu_2 \approx 2.751),\quad (\mu_1 \approx 3.172,\ \mu_2 \approx 0.355),\quad (\mu_1 = 3,\ \mu_2 = 3/10). \tag{12}$$

It can be seen that the obtained values are far from optimal: parameter μ_1 in all variants exceeds the optimal value (2) by more than an order; in the third variant parameter a, calculated in accordance with (9), turns to zero.

Let us find out whether there is an exact solution to the AP-model (1) with less stringent requirements on its parameters. The nullclines analysis of the system (1) reveals 4 stationary points, of which only one point $u = 0, v = 0$ is stable at $u \geq 0, v \geq 0$. When the condition $\varepsilon_0 = 0$ is met, the number of stationary points is reduced to 3 and one of them, namely $u = 1, v = 0$, becomes stable. Switching between two stable stationary points can be performed by the kink propagating along Ox-axis.

Substitution of the ansatz (10) into (3), when $\varepsilon_0 = 0$ is true, reveals the exact solution:

$$u(z) = \frac{1}{2}\tanh\left(\frac{1}{4}\sqrt{2k}\,z\right) + \frac{1}{2},\quad C_0 = \sqrt{2k}\left(a - \frac{1}{2}\right), \tag{13}$$

where the constant k, a, μ_1, μ_2 may remain arbitrary.

4. Acknowledgments

This work was supported by the Russian Foundation for Basic Research (project no. 16-01-00176-a)

References

[1] Casten R G, Cohen H and Lagerstrom P A 1975 *Quarterly of Appl. Math.* **32** 365
[2] FitzHugh R 1961 *Biophysics J.* **1** 445
[3] Kudryashov N A, Rybka R B and Sboev A G 2018 *Appl. Math. Lett.* **76** 142
[4] Kudryashov N A 2018 *Regul. Chaotic Dyn.* **23** 152
[5] Demina M V and Kudryashov N A 2018 *Appl. Math. Lett.* (in press) https://doi.org/10.1016/j.aml.2018.02.012
[6] Aliev R R and Panfilov A V 1996 *Chaos Soliton Fract.* **7** 293
[7] Pavelchak I A 2012 *Vychisl. Metody Programm.* **13** 172 (in Russian)
[8] Kudryashov N A 2010 *Methods of nonlinear mathematical physics* (Dolgoprudnyj: Izd. dom "Intellekt") (in Russian)

AUTHOR INDEX

Institute of Physics
Dirac House, Temple Back
Bristol BS1 6BE UK

ISSN: 1742-6588
ISBN 978-1-5108-8721-3

CURRAN ASSOCIATES INC.
proceedings.com